Prefixes used with the Metric System

d	deci	10^{-1}			
c	centi	10^{-2}			
m	milli	10^{-3}			
μ	micro	10^{-6}			
n	nano	10^{-9}	k	kilo	10^{3}
p	pico	10^{-12}	M	mega	10^{6}
f	femto	10^{-15}	G	giga	10^{9}
a	atto	10^{-18}	T	tera	10^{12}

FREQUENTLY USED CONVERSION FACTORS

1 radian = 57.3°

1 meter = 10^2 cm = 10^3 mm = 10^6 μm = 10^9 nanometers (nm) = 10^{10} angstroms (Å)
 = 39.37 in. = 3.281 ft

1 kilogram = 6.852×10^{-2} slug

1 atomic mass unit (amu) = 1.660×10^{-27} kg

1 day = 86,400 s

1 newton = 10^5 dyn = 0.2248 lb

1 meter per second = 3.6 km/h = 3.281 ft/s = 2.237 mi/h

60 miles per hour = 88 ft/s

1 atmosphere = 1.013×10^5 N/m^2 = 76 mm Hg = 14.70 lb/in.^{2}

1 joule = 0.2389 cal = 0.7376 ft-lb = 9.481×10^{-4} Btu

1 electron volt (eV) = 1.601×10^{-19} J

1 watt = 1.341×10^{-3} hp = 0.7376 ft-lb/s

(*See Appendix B for a more complete list.*)

Fundamentals of Mechanics and Heat

McGRAW-HILL SERIES IN FUNDAMENTALS OF PHYSICS: AN UNDERGRADUATE TEXTBOOK PROGRAM

E. U. Condon, Editor, University of Colorado

Members of the Advisory Board
D. Allan Bromley, Yale University
Arthur F. Kip, University of California, Berkeley
Hugh D. Young, Carnegie-Mellon University

Introductory Texts
Beiser   Concepts of Modern Physics
Kip   Fundamentals of Electricity and Magnetism
Young   Fundamentals of Mechanics and Heat
Young   Fundamentals of Optics and Modern Physics

Upper-Division Texts
Barger and Olsson   Classical Mechanics: A Modern Perspective
Beiser   Perspectives of Modern Physics
Cohen   Concepts of Nuclear Physics
Elmore and Heald   Physics of Waves
Kraut   Fundamentals of Mathematical Physics
Longo   Fundamentals of Elementary Particle Physics
Meyerhof   Elements of Nuclear Physics
Reif   Fundamentals of Statistical and Thermal Physics
Tralli and Pomilla   Atomic Theory: An Introduction to Wave Mechanics

SECOND EDITION

FUNDAMENTALS OF MECHANICS AND HEAT

HUGH D. YOUNG
Associate Professor of Physics
Carnegie-Mellon University

McGRAW-HILL BOOK COMPANY
New York   St. Louis   San Francisco   Düsseldorf   Johannesburg
Kuala Lumpur   London   Mexico   Montreal   New Delhi
Panama   Rio de Janeiro   Singapore   Sydney   Toronto

**FUNDAMENTALS
OF MECHANICS AND HEAT**

Copyright © 1964, 1974 by McGraw-Hill, Inc.
All rights reserved.
Printed in the United States of America. No part of this publication may be reproduced,
stored in a retrieval system, or transmitted,
in any form or by any means,
electronic, mechanical, photocopying, recording, or otherwise,
without the prior written permission of the publisher.

1 2 3 4 5 6 7 8 9 0   MAMM   7 9 8 7 6 5 4 3

This book was set in Bodoni Book by York Graphic Services, Inc.
The editors were Jack L. Farnsworth and Carol First;
the designer was Nicholas Krenitsky;
and the production supervisor was Bill Greenwood.
The new drawings were done by Felix Cooper.
The printer and binder was The Maple Press Company.

Library of Congress Cataloging in Publication Data

Young, Hugh D
    Fundamentals of mechanics and heat.

   (McGraw-Hill series in fundamentals of physics: An
Undergraduate textbook program)
   1. Mechanics. 2. Heat. I. Title.
QC125.2.Y66 1974   531   73-11435
ISBN 0-07-072638-8

Contents

Preface to the Second Edition   xiii
Preface to the First Edition   xv

PROSPECTUS   1

1   What Is Physics?   3
   1-1   *Objects and Methods of Physics   3*
   1-2   *Measurements and Units   9*
   1-3   *Equations in Physics   13*
   1-4   *Models   14*
   1-5   *Macroscopic and Microscopic Views   17*
   1-6   *Mathematical Language   18*
          Problems   22

2   Vector Language   25
   2-1   *Position of a Point   25*
   2-2   *Vectors   29*
   2-3   *Addition of Displacements   31*
   2-4   *Components and Unit Vectors   34*
   2-5   *Scalar Product   39*
   2-6   *Vector Product   42*
   2-7   *Representation of Physical Quantities by Vectors   48*
          Problems   49

3 Describing Motion — 52
- 3-1 Velocity 52
- 3-2 Acceleration 56
- 3-3 Motion with Constant Acceleration 59
- 3-4 Velocity in Space 62
- 3-5 Acceleration in Space 66
- 3-6 Relative Motion 69
- Problems 72

PERSPECTIVE I 77

4 Principles of Dynamics — 79
- 4-1 Inertia 79
- 4-2 Fundamental Experiments 82
- 4-3 Inertial Mass 84
- 4-4 Force and Motion 86
- 4-5 Superposition of Forces 89
- 4-6 Action and Reaction 94
- 4-7 Frames of Reference 98
- Problems 100

5 Dynamics of a Particle I — 104
- 5-1 Mass and Weight 104
- 5-2 Free Fall 106
- 5-3 Ballistic Trajectories 110
- 5-4 Examples 114
- 5-5 Contact Force 119
- 5-6 Equilibrium 122
- 5-7 Nature of Forces 124
- Problems 126

6 Dynamics of a Particle II — 132
- 6-1 Circular Motion 132
- 6-2 Centripetal Acceleration 135
- 6-3 Centripetal Force 139
- 6-4 Vector Angular Velocity 145
- 6-5 Velocity and Acceleration in Polar Coordinates 147
- Problems 151

PERSPECTIVE II 155

7 Momentum — 157
- 7-1 Simple Collisions 157
- 7-2 Momentum 159
- 7-3 Variable Mass 162
- 7-4 Massless Particles 166
- 7-5 Collisions 167
- 7-6 Impulse 170
- 7-7 Generalizations 173
- Problems 174

8 Dynamics of a Particle III — 179
- 8-1 Gravitation 179
- 8-2 Force Fields 183
- 8-3 Planetary Motion 185
- 8-4 Central-Force Problems 189
- 8-5 Electric and Magnetic Forces 192
- 8-6 Scattering 197
- Problems 198

9 Mechanical Energy — 203
- 9-1 Work 203
- 9-2 Work and Kinetic Energy 209
- 9-3 Power 214
- 9-4 Potential Energy 215
- 9-5 Conservation of Energy 219
- 9-6 Collisions 226
- 9-7 Conservative Force Fields 228
- 9-8 Nonmechanical Energy 235
- Problems 237

10 Periodic Motion — 242
- 10-1 The Harmonic Oscillator 242
- 10-2 Characteristics of Motion 246
- 10-3 Energy Relations 250

10-4 Simple Pendulum   252
10-5 Molecular Vibrations   254
10-6 Damped Oscillations   262
10-7 Forced Oscillations   267
    Problems   270

PERSPECTIVE III   275

11 Systems of Interacting Particles   277
11-1 Center of Mass   277
11-2 Motion of the Center of Mass   285
11-3 Center-of-Mass Coordinate System   288
11-4 Kinetic Energy   291
11-5 Rigid Bodies   293
11-6 Moments of Inertia   295
11-7 Examples   302
    Problems   306

12 Angular Momentum   312
12-1 Planets and Satellites   312
12-2 Dynamics of Planetary Motion   315
12-3 Rutherford Scattering   318
12-4 Angular Momentum of a Rigid Body   323
12-5 Vector Angular Momentum   328
    Problems   331

13 Dynamics of a Rigid Body   337
13-1 Rotation about the Center of Mass   337
13-2 Gravitational Torques   341
13-3 Examples   342
13-4 The Gyroscope   348
13-5 Spin and Orbital Angular Momenta   353
13-6 Generalizations   358
13-7 Equilibrium of a Rigid Body   361
    Problems   364

PERSPECTIVE IV   373

14   Relativity   375
  14-1   Invariance of Physical Laws   375
  14-2   Galilean Transformation   378
  14-3   Relativity of Time   381
  14-4   Relativity of Length   386
  14-5   The Lorentz Transformation   389
         Problems   394

15   Relativistic Dynamics   397
  15-1   Momentum   397
  15-2   Force and Motion   401
  15-3   Work and Energy   404
  15-4   Relativity and Newtonian Mechanics   407
         Problems   410

16   Fluid Mechanics   414
  16-1   Hydrostatics   414
  16-2   Fluid Flow   420
  16-3   Bernoulli Equation   423
  16-4   Applications of Bernoulli's Equation   425
  16-5   Viscosity   428
  16-6   Turbulence   432
         Problems   433

17   Properties of Matter I   436
  17-1   Structure of Matter   436
  17-2   Molecular Mass   440
  17-3   Temperature   442
  17-4   Gas Thermometer and Absolute Temperature   445
  17-5   Quantity of Heat   447
  17-6   Heat Transfer   451
  17-7   Heat and Energy   458
         Problems   459

18 Kinetic Theory of Gases 464
- 18-1 Ideal-Gas Law 464
- 18-2 Microscopic Model of an Ideal Gas 467
- 18-3 Pressure Exerted by an Ideal Gas 469
- 18-4 Specific Heat of an Ideal Gas 474
- 18-5 Distribution of Molecular Speeds 477
- 18-6 Maxwell-Boltzmann Distribution 482
- 18-7 Molecular Collisions and Transport Phenomena 485
- 18-8 Limitations of Kinetic Theory 489
- Problems 490

19 Properties of Matter II 496
- 19-1 State of a Thermodynamic System 496
- 19-2 Real Gases and Equations of State 498
- 19-3 Phases of Matter 502
- 19-4 Elasticity 506
- 19-5 Specific Heats of Solids 513
- 19-6 Thermal Expansion 515
- 19-7 Interatomic Forces 518
- Problems 520

PERSPECTIVE V 525

20 Heat and Work 527
- 20-1 Conversion of Energy 527
- 20-2 Internal Energy 529
- 20-3 Changes of State of an Ideal Gas 533
- 20-4 Specific Heats of an Ideal Gas 536
- 20-5 Adiabatic Process for an Ideal Gas 539
- 20-6 Microscopic View of an Adiabatic Process 541
- 20-7 Thermodynamic Properties of Matter 543
- 20-8 Quasi-Static Processes 543
- 20-9 A Practical Problem 546
- Problems 549

21 Second Law of Thermodynamics — 554
- 21-1 Directionality of Natural Processes   554
- 21-2 Carnot Cycle   556
- 21-3 Refrigerators   561
- 21-4 Second Law of Thermodynamics   563
- 21-5 Entropy   566
- 21-6 Entropy and the Second Law   571
- 21-7 Thermodynamic Temperature Scale   572
- 21-8 Entropy and Probability   573
- Problems   577

22 Mechanical Waves — 582
- 22-1 The Nature of Waves   582
- 22-2 Mathematical Description of Waves   585
- 22-3 Dynamics of Waves on a String   591
- 22-4 Sinusoidal Waves   593
- 22-5 Energy in Wave Motion   596
- 22-6 Longitudinal Waves   600
- 22-7 Superposition of Waves   605
- 22-8 Standing Waves   609
- 22-9 Extensions and Generalizations   612
- Problems   617

EPILOGUE   621

Appendixes — 623
- A   Physical Constants   623
- B   Conversion Factors   624
- C   The Solar System   627
- D   The Greek Alphabet   627
- E   Common Logarithms   628
- F   Value of Trigonometric Functions   630
- G   Mathematical Formulas   631

Answers to Odd-Numbered Problems   635

Index   645

Preface to the Second Edition

The basic philosophy and outline of the book, as described in the original Preface, remain unchanged, but the author has incorporated a number of changes to enhance the usefulness of the book as a teaching and learning tool. Several changes have been suggested by actual classroom experience of the author and others with the first edition, reflecting in turn the merciless but usually constructive scrutiny which beginning students bring to bear on a textbook.

The most conspicuous changes in subject matter are the addition of a chapter on fluid mechanics and the expansion of the treatment of relativity into two chapters, including simple but complete derivations of the Lorentz transformation and the relativistic momentum and energy relations. Many portions of the book have been rewritten to improve clarity of exposition; in particular, Chapter 9 (Mechanical Energy) has been completely rewritten and the treatment of potential energy improved substantially.

Users of the first edition observed that it tended to give only lip service to calculus, in the sense that calculus was used freely in derivations but much less frequently in applications. In this new edition many examples and problems making explicit use of elementary calculus have been added; the author hopes that those instructors who include among their goals the development of facility in the applications of calculus to physics will find this new material useful. Conversely, however, those wishing to make only minimal use of calculus will find this book as adaptable to their needs as was the first edition. Other examples and problems have been added to broaden the spectrum of difficulty available and to provide a gradation in each subject area from very straightforward problems to those of sufficient subtlety to challenge the best students.

The position of the fluid mechanics and relativity material in the outline is somewhat arbitrary. Any or all of this material may of course be omitted without loss of continuity. Those who want to get into relativity early may want to use Chapter 14 following Chapter 3, and Chapter 15 following Chapter 9. Similarly, Chapter 16 can be taken up any time after Chapter 9, and Chapter 22 any time after Chapter 10.

More generally, the book is adaptable to a variety of course outlines. Used in its entirety, it would occupy a full year of a 3-hour course or a semester of a very intensive 5 or 6-hour course. For a usual 3- or 4-hour semester-long course several chapters would ordinarily be omitted. Chapters 1 to 10 and 17 to 21 are sequential in nature, but a wide variety of topic outlines is still possible.

In conclusion, the author would like again to thank his colleagues at Carnegie-Mellon University and elsewhere, and their students, for the many valuable suggestions they have contributed to this revision. Their kindness and generosity are greatly appreciated.

Hugh D. Young

Preface to the First Edition

No one needs to be persuaded nowadays that a thorough course in the fundamentals of physics is an essential part of the curriculum of every beginning student of engineering or the physical sciences. However, the addition of a new text to an already large accumulation of basic physics books might well be questioned, since a number of excellent texts already exist.

Nevertheless, new books are needed and will continue to be needed, for several reasons. The explosive growth of scientific knowledge and the increasing reliance of engineering technology on basic science and analytical methods make it more urgent than ever that the basic physics course give a thorough grounding in essential principles and associated analytical tools, stripped of gadgetry and lengthy discussions of current technology which will soon grow obsolete. Furthermore, it is essential to raise the level and goals of the introductory course whenever possible, to take fullest advantage of the improved preparation in physics which many high schools now offer as a result of the fine work of the Physical Science Study Committee and similar groups.

This book is intended as an introductory text, elementary yet thorough, which concentrates on fundamentals, uses analytical techniques such as calculus and vector analysis where they are needed, is thoroughly up to date and contemporary in viewpoint, and exhibits the spirit of scientific inquiry and the empirical basis of natural science.

Specifically, this book is intended as the beginning of a series of courses of total length one to two years, starting in the freshman or sophomore college year. For the continuation of the sequence, the other volumes in the McGraw-Hill Series in Fundamentals of Physics are particularly suitable because of their uniformity of level and viewpoint, but there are several other suitable combinations.

As the table of contents shows, the order of topics is not revolutionary. The author believes strongly that physics should be presented as an empirical science and that frequent contact with experience is essential. Thus newtonian mechanics is the most logical place to begin, inasmuch as everyday experience provides an abundance of examples of mechanical principles, more so than for any other branch of physics. Furthermore, the closely knit logical structure of mechanics and its close relations to all the other branches of physics enhance its suitability as a starting point.

The table of contents indicates the scope of the book, but we also wish to point out the following features:

1 The book assumes a concurrent course in differential and integral calculus, and elementary calculus is used wherever necessary. It is not our aim to teach calculus as such, but we recognize that for most students calculus is a new and still somewhat unfamiliar tool; for this reason, some applications of calculus are spelled out in more detail than would ordinarily be necessary, especially in the early chapters of the book. No previous exposure to vector algebra is assumed; a systematic introduction is provided in Chap. 2, and further explanations of various essential points accompany their application to physical situations. As is customary, vector quantities are denoted by boldface symbols; in addition boldface **+**, **−**, **Σ**, and **=** signs are used in vector equations to denote vector addition and equality.

2 The book attempts to exhibit the inductive and empirical nature of physics along with its deductive aspects. Care has been taken to distinguish clearly between principles which are generalizations from experience and those which are derived, and to indicate the experimental basis of the former. The relative status of each principle in the whole logical structure is thus made clear.

3 Throughout the book the importance of *models* is stressed. We rarely deal directly with physical reality, but rather with simplified models designed to retain the essential features of a physical situation and eliminate the unessential ones, to facilitate analysis. The student is constantly reminded of the process of constructing models as idealizations of reality, and of the limitations of analytical results imposed by the limitations of validity of the model.

4 In the chapters on elementary thermodynamics we abandon completely the classical approach with its confinement to a macroscopic viewpoint. Instead, the macroscopic and microscopic approaches are thoroughly interwoven; a common theme in these chapters is the relation between the macroscopic behavior of matter and its microscopic structure. Thus the kinetic-molecular model of an ideal gas is discussed in detail, and its relation to the first law of thermodynamics is exhibited. The concept of entropy is discussed as a measure of the microscopic disorder of a system as well as a macroscopic thermodynamic quantity.

5 An attempt has been made to exhibit the essential *unity* of various branches of physics; in particular, mechanical principles and the various conservation laws are applied to atomic and nuclear phenomena and to the structure of matter, as well as to problems in macroscopic mechanics. As an aid to understanding the status and relationships of various principles, short Summary and Perspective sections are inserted at intervals of three or four chapters.

6 With a few exceptions, the mks system of units is used exclusively. There seems little doubt that it will eventually be used universally in scientific work. In addition, the author feels strongly that the burden of mastering several systems of units simultaneously should not be added to that of mastering new physical concepts. If the British system of units is needed in later engineering courses, it can easily be learned once the basic principles are well in hand. Several tables for conversion of units, both metric and British, are included in Appendix B.

7 An unusually large collection of almost 600 problems is included. A few of these are simple substitution exercises, designed to illustrate definitions, but most require some thought and insight on the part of the student. A majority of the problems are literal or algebraic rather than numerical; some are too difficult for all except the best students, and there is plenty of material for "honors" sections. In several cases a sequence of problems can be used to explore additional subject matter. For example, Problems 19-4, 20-14, 20-20, 20-21, 20-25, and 21-11 together constitute a brief introduction to the thermodynamics of paramagnetism, a topic which is not discussed explicitly in the text.

This book grew out of the author's experience with the freshman physics course for engineering and science students at Carnegie Institute of Technology, which is accompanied by a concurrent calculus course. This book is also suitable for the first semester of a two- or three-semester sequence of four or five credit hours each semester; when used for a one-semester course, it may be shortened somewhat. Several sections and a few entire chapters may be omitted without interrupting the continuity. These include Chapters 13, 14, 15, 21, and 22 and Sections 6-5, 7-4, 7-6, 8-5, 8-6, 9-6, 9-7, 10-5 to 10-7, 12-5, 13-4 to 13-7, 18-5 to 18-8, 19-4 to 19-6, 20-6 and 20-9. By omitting selected combinations of these sections, courses with a variety of length and emphasis can be constructed.

Although this book is primarily intended as a high-level introductory text, it may also be used for an intermediate course following a first physics course given without calculus. A thorough and detailed exposition of principles and a plentiful supply of challenging problems make it useful for such intermediate course.

ACKNOWLEDGMENTS

This book has benefited greatly from the advice of a number of people. The following have reviewed part or all of a preliminary form of the manuscript: Prof. Stanley Ballard, Dr. Arthur Beiser, Prof. Owen Chamberlain, Prof. Edward Condon, Prof. Arthur Kip, Prof. Hans Mark, and Dr. William Wolfe. Their critical comments and suggestions have been invaluable. The author is also grateful to his colleagues at Carnegie Institute of Technology, whose suggestions have greatly increased the book's usefulness as a teaching tool, and to the two generations of freshmen at Carnegie who mercilessly but constructively pointed out errors and obscure passages in the preliminary editions. Particular thanks are due to the various members of the secretarial staff at Carnegie who aided in the preparation of several successive versions of the manuscript. Finally and most important, the author acknowledges his great debt to his wife Alice for her unending patience, confidence, and moral support throughout the writing of this text.

Hugh D. Young

PROSPECTUS

To both the intellect and the emotions the study of physics is exciting, satisfying, and even beautiful; it is the intention of this book to convey these qualities to the student.

In learning new and fundamental principles, and especially in comprehending the interplay among them, the student may share some of the pleasure of discovery felt by scientific giants such as Galileo, Newton, Einstein, and Fermi, whose work forms the foundation for our present knowledge of the behavior of the physical world. In a sense the student rediscovers the principles of physics in learning their power and usefulness as the building blocks of engineering, as sources of new insight into phenomena of everyday experience, and most of all as achievements of the human mind in its struggle to understand the physical world.

In this book, the first of a series providing an introduction to the most important principles and generalizations of physics, we begin with the study of mechanics, which is the study of the relationship of motion to its causes. This is a logical starting point, inasmuch as everyday experience provides an abundance of examples of mechanical principles, more so than for any other branch of physics.

The problems of mechanics are typically concerned with predicting the motion of a particle, a body, or a collection of particles under given conditions or with understanding the influences which must have been responsible for a given motion. Starting with principles governing the motion of one particle, we can compose logical structures enabling us to understand much more complicated systems, including many of immediate practical importance. The unity and coherence of the resulting body of theory are properly described as beautiful.

Next we proceed to topics associated with heat. It may seem that mechanics and heat are strange traveling companions, and yet they are closely allied. Starting with such familiar concepts as temperature and heat, we go quickly to more fundamental topics, including the relationship of heat to mechanical energy (embodied in the subject of thermodynamics), the relation of the motion of gas molecules to the behavior of the gas, and the understanding of various aspects of the behavior of matter on the basis of the mechanical structure of atoms and molecules. The part of the book dealing with heat is really in large measure concerned with the relationships of thermal phenomena to mechanics.

In order to start off well equipped, several important tools are provided in the first three chapters. These are not really physics, but they are part of its language and, as such, are essential for its study. They are of several kinds: rules of procedure, in which we stress the importance of observations of the physical world, a discussion of the need for simplifying apparently complicated situations to facilitate analysis, and several items of mathematical language.

The author has tried throughout to exhibit his personal conviction that physics is beautiful, exciting, and satisfying. It is his hope that the student will enjoy understanding new principles and grasping their power and usefulness and will feel the satisfaction and personal achievement that come of struggling with and solving challenging problems.

What Is Physics? | 1

As an introduction to the study of mechanics and heat, it is valuable to consider briefly the general objects and methods of all of physics. In this opening chapter we discuss several fundamental ideas about physics which will make our study of mechanics and heat more meaningful. These same ideas appear in many other areas of physics and give unity and structure to the whole science.

1-1 OBJECTS AND METHODS OF PHYSICS

It is difficult to state concisely just what physics is. Physics deals with all phenomena which occur in the material world. The fundamental assumption of physics is that all such phenomena, without exception, behave according to a set of general principles, or physical laws. The goal of physics is to discover these principles, so that various phenomena can be correlated and understood and to use them to predict the outcome of further observations not yet made.

Traditionally, the phenomena which form the subject matter of physics have been divided into the categories of mechanics, electricity and magnetism, thermodynamics, optics, and atomic, molecular, and nuclear physics. It is now known that these divisions are to a large extent arbitrary and artificial; at the most fundamental level lie principles which are applicable to *all* these areas. This essential unity of its various branches and of the principles underlying them is, in fact, one of the most exciting and intellectually stimulating aspects of physics.

The role of physics is to correlate observations of physical phenomena.

4 WHAT IS PHYSICS?

From the welter of phenomena occurring in the world around us, we try to extract common features and characteristics, so that several different events can be understood as examples of some single general principle. Suppose several objects having different weights are permitted to fall freely. By measuring the time required for each to fall a given distance, one can discover that different objects fall at nearly the same rate, almost independent of their weights.

The essence of these observations is contained in the generalization "the rate of free fall of an object is independent of its weight." But there is an important difference between this generalization and the experiments on which it is based, for the generalization predicts that when *more* experiments of this kind are performed, the results will be in agreement with the generalization. Thus it is both a summary of experimental observations and a prediction of the results of experiments not yet performed.

The process of making a general statement on the basis of limited observations is called *induction*. Once the generalization has been stated, it can be used to predict the results of further experiments by a process of *deduction*. Then these experiments can be performed, and it can be determined whether or not their results agree with the predictions. If they do, the original generalization is given further support; if not, the source of the disagreement must be sought. Perhaps errors were made in the experiments, or there was faulty logic in the derivation of the predictions from the gener-

Fig. 1-1 Two laboratories for research in physics. *Left:* The leaning tower of Pisa. According to legend, Galileo experimented with falling bodies dropped from the top of this tower, although there is no reference to it in this connection in Galileo's own writings. *Right:* The 400-GeV accelerator at the National Accelerator Laboratory at Batavia, Illinois. The world's highest-energy accelerator, this machine accelerates protons to very high speeds for research in fundamental-particle interactions. (*NAL Photo.*)

alization. If neither of these is the source of the disagreement, the generalization itself must be at fault and in need of modification or rejection.

The essential point is that the development of a new physical theory is always a two-way process that starts and ends with observations of physical phenomena. The development of a new theory nearly always takes a devious path. There is always considerable guesswork, much of which turns out to be wrong, following of blind alleys, and discarding of unsuccessful guesses in favor of more promising ones. Furthermore, no theory is ever regarded as the final or ultimate truth. The possibility that a theory may need further modification or generalization as a result of new experimental evidence is always present.

The point of view that a physical theory must stand or fall on the basis of its agreement or disagreement with observed phenomena may seem so natural and obvious as not to require comment. It has not always been so, however; the Greek philosophers of the Hellenic age relied on reason rather than observation in their efforts to discover truth. Socrates, in the fifth century B.C., said:

If we are ever to know anything absolutely, we must be free from the body and must behold actual realities with the eye of the soul alone.

6 WHAT IS PHYSICS?

In modern terms, we would describe Socrates' belief as the view that reason, not observation, is the essential tool of the physicist. This belief led one of his successors, Aristotle, to the following deduction:

A given weight falls a given distance in a given time; a greater weight moves the same distance in a lesser time, the times being in inverse proportions to the weights. For instance, if one weight is twice another, it will take half as long over a given movement.

There is no record that Aristotle actually tested this statement experimentally; even a very simple experiment demonstrates its lack of validity, but Aristotle did not regard experiments as the most important tool in understanding the physical world. The viewpoint of nearly all present-day scientists is, of course, that experimental observations are of paramount importance in developing new scientific principles.

Sometimes it is possible to arrive at two different generalizations *both* of which agree with a given set of observations. A classic example is the description of the motion of planets in the solar system. Many of the ancients regarded the earth as the center of the universe. The motions of the sun, moon, and planets around the earth were described by the Egyptian astronomer Claudius Ptolemy (second century A.D.) by means of a complicated system in which each planet was thought to move in an epicycloidal path, with the earth displaced from the center of the epicycloid. His scheme is illustrated in Fig. 1-2. This system was involved and cumbersome, but it did describe the motion of the planets and the sun quite accurately.

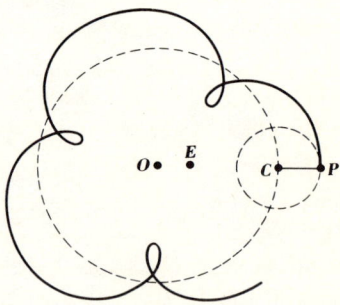

Fig. 1-2 Ptolemaic system for describing the motion of planets. The planet P is assumed to move uniformly in a circular path around the point C, which in turn moves in a circular path around a fixed point O. The earth E is displaced somewhat from O. One motion which can be described by this scheme is shown as a solid line.

During the sixteenth century Copernicus, Kepler, and others discovered that the description could be made simpler by regarding the *sun*, not the earth, as the center of the solar system. Then the path of each planet around the sun could be described as an ellipse, with the sun at one focus. The idea that the sun is the center of the solar system was not new; it had been proposed by Aristarchus, a contemporary of Aristotle. Kepler was, however, the first to show that the sun-centered picture provides a *simpler* description of the motion of the planets than the earth-centered picture.

Which of these theories is correct? Both agree with the observed phenomena. Kepler's system is simpler than Ptolemy's, and we are inclined to prefer it for its simplicity. But a much clearer advantage of the sun-centered picture is that it lends itself easily to further generalization. Sir Isaac Newton, in one of the most remarkable scientific achievements of all time, showed that if one adopts the sun-centered picture, then the motion of the planets about the sun is consistent with the principles which describe the motion of objects on the earth. Thus, Newton's laws of motion contain the discoveries of Kepler as consequences; they predict many other kinds of behavior as well. No such generalizations have been found possible for the earth-centered theory, and so it has been discarded.

The idea of *simplicity* of physical theories has played a very important role in the development of physics. Strictly an article of faith, but nevertheless very widely accepted, is the view that not only does the physical world behave in accordance with definite laws but that it is always possible to describe this behavior in a simple way. For many physicists this is one of the most appealing aspects of physics. One may find a principle, such as the principle of conservation of energy, which is stated very simply, yet which is applicable in a very wide variety of situations and is thus very powerful. This underlying simplicity and elegance in what appears at first glance to be a very complicated world yield for many physicists an aesthetic pleasure and satisfaction which may be compared to the pleasure experienced in listening to a great musical composition or viewing the masterwork of a great painter.

The statement is often made that a certain physical principle "explains" a phenomenon. This implies that the principle tells us *why* the phenomenon takes place. Strictly speaking, this is not what physics does at all. To ask "why" very quickly takes us outside the realm of physics, into metaphysics and teleology. We do not know, for example, *why* the universe exists at all. Theology may supply an answer to this question, but physics cannot. Ultimately, the role of physics is always to *describe* phenomena in terms of general principles but not to *explain* why the phenomenon takes place or why the principle is valid.

A related point is the question whether physical laws are *discovered* or *invented*. In the preceding discussion we have implicitly taken the point of view that laws have always existed and are waiting for man to discover them. But this is not by any means the only possible point of view. Consider the discovery of the electron. In the latter part of the nineteenth century various physicists made observations which could be understood if one assumed that electric current consists of motion of small particles, called *electrons*, having definite mass and electric charge. Since that time a great deal more information has been accumulated which is consistent with this picture. However, it must be admitted that the concept of an electron is not a discovery but an invention of the mind of man. Such an enormous body of information has accumulated which agrees with this view of the electron that we are inclined to say that the electron really has an objective existence and that we have discovered it. Still, an equally valid point of view is that the electron is a concept which has been invented by man and which has been retained because it happens to describe effectively a large number of experimental observations.

The necessity of a close relationship between the principles of physics and the observed behavior of the physical world constitutes an important and fundamental difference between physics and pure mathematics. Mathematics deals with logically coherent structures which must be internally consistent, but there is no necessity for any relation between such a structure and anything outside it. Examples of such structures are found in the various non-euclidean geometries developed by Riemann, Lobachevsky, and several others during the latter part of the nineteenth century. In these, one or more postulates of Euclid's geometry are replaced with different postulates. As long as the result is internally consistent, the system of geometry thus constructed is a valid mathematical system. But when we inquire which of these geometric systems, if any, actually describe the physical world, we leave the domain of mathematics and enter that of physics.

Since mathematics is constructed without regard to its relationship to the physical world, it may seem somewhat strange that it is so useful in physics. We must keep in mind, however, that not all branches of mathematics are utilized in physics. Of the many geometric systems which have been invented, for example, only a few seem to have any relevance to the physical world. In a similar way, we must extract from the entire body of mathematics those parts which seem helpful in formulating physical principles. Furthermore, developments in mathematics are often motivated by physical problems, as in the development of the calculus by Newton.

Physics provides the basic principles which, together with appropriate mathematical techniques, are the essential building blocks for all engineering work. In present usage, the terms *engineering* and *technology* usually apply to the *application* of scientific principles to the solution of practical problems.

The development of the transistor, for example, necessitated advances in the fundamental theory underlying the behavior of semiconductor materials; this was basic research in physics. But the design of television sets using transistors is not a problem of physics but one of electrical engineering. Every new step forward in space exploration is hailed by the press as a great *scientific* achievement, whereas in fact it is mainly a great *engineering* achievement. Of course, it is neither possible nor necessary to draw a very firm line between physics and engineering. Many physicists are engaged in work which is at least in part of an engineering nature. But the central goal of physics is understanding the principles of the physical world, not in applying them to practical situations.

Lest the reader receive an erroneous impression, we must state emphatically that although physics is concerned with fundamental principles, exercise in the application of these principles is essential for thorough understanding. In this text there are many worked-out examples, and at the ends of the chapters are many other problems for the student to solve. Exercise in solving a variety of problems is an essential part of learning physics. Equally important is a variety of laboratory work in which the principles discussed in this text are experienced at first hand, so that the importance of comparing theory and experience can be firmly grasped.

1-2 MEASUREMENTS AND UNITS

In observing physical phenomena, we often observe *magnitudes* of the quantities involved. Any magnitude which can be described quantitatively by means of one or more suitably chosen numbers and which is relevant to the description of a physical phenomenon is called a *physical quantity*. The term *quantity* is used throughout this book, and it always means a physical quantity. In order to define a physical quantity precisely, we must specify, at least in principle, a means of *measuring* this quantity, by which we mean the process of assigning a number or numbers to the quantity. In the case of length, it is obvious how to do this. We first decide on a standard or *unit* to use for the measurement, such as a 1-ft ruler. Then any other distance can be measured by successively laying off the ruler a number of times, subdividing it geometrically if necessary for fractions of a foot, until the required distance is found. If an object is dropped from a height of 16 ft, it falls 16 times the length of the 1-ft ruler.

A definition of a physical quantity which defines the quantity by describing how to measure it is called an *operational definition*. The above example forms the basis for an operational definition of length; an operational definition of weight can be based on the use of a chemical balance and set of standard objects, which constitute the units of weight. Physical quantities must always be defined operationally; a quantity which cannot be measured or observed in some manner cannot have any physical meaning. It is always necessary to specify how to measure a new physical quantity or how to compute it from physical quantities which can be measured.

Thus the description of physical quantities involves numbers with associated units. If different experimenters are to compare their results, it is essential that they use the same system of units. In the scientific world today, the accepted unit of length is the *meter*. This was originally defined as the distance between marks engraved on a certain bar of platinum-iridium alloy measured under certain conditions of temperature, but in 1960, the meter was redefined in terms of the wavelength of light emitted by atoms of krypton under particular conditions. This wavelength can be compared directly with the distance on the platinum-iridium bar by methods which make use of the interference of light waves.

The universally accepted unit of time is the *second*. This was originally defined as the amount of time required for a simple pendulum 1 m long to swing from one extreme to another. Later a more exact standard was defined in terms of the solar year, the time required for the earth to make one complete revolution around the sun. An even better time standard, officially adopted in 1967, defines the second in terms of the frequency of radiation associated with a certain transition in the cesium-133 atom. Units of length and time based on atomic and molecular processes have the great advantage that they are much less affected than the others by changes in external conditions such as temperature and humidity. They are much easier to reproduce precisely in various environments and so are much better approximations to *universal* definitions of fundamental units than the original definitions.

Lengths, times, and other directly observable quantities are often used to calculate other quantities. An example is *area*. We know that the area of a rectangle is the product of its length and its width. Thus, the *units* of area must be the same as the units of length squared. A square sheet of paper 0.5 m on a side has an area of 0.25 m^2; this means that its area is 0.25 that of a sheet 1 m on a side. The unit of area, the square meter, is *derived* from the unit of length already defined. Similarly, a unit of volume, the cubic meter, can be derived from the unit of length, and a unit of speed, the meter per second, from the units of length and time. These are not new independent units but are derived from the fundamental units of length and time. We often use exponents in notation for derived units, such as m^2 for square meters, m^3 for cubic meters, and m/s or m s^{-1} for meters per second, a unit of speed.

Some derived quantities are *ratios* of two numbers having the same units. A familiar example is the slope of a hill, defined as the ratio of vertical rise to horizontal distance, with both measured in the same units. This is a quotient of two quantities of the same type, namely length. If the same unit is used for both, it is immaterial what unit is used. The result is the same when we measure the lengths in feet as when we use meters.

A quantity which is independent of the system of units used is said to be a *unitless* or *dimensionless* quantity. Another familiar example is the radian measure of an angle. An angle in radians is defined as the quotient of the arc length subtended by the angle on a circle and the radius of the circle. It is assumed that both lengths are measured in the same units, and so an angle measured in radians is a unitless quantity.

In any physical measurement or calculation it is very important to use a consistent set of units. In most of this book, the system used is the meter-kilogram-second (mks) system. In this system, there are four fundamental units in all. Two of these have already been mentioned, the meter and the second. The other two are a unit of mass, the *kilogram*, which is discussed in detail in Chap. 4, and a unit of electric charge, the *coulomb*.

In commerce and in some practical engineering work in the United States, the British engineering system of units is frequently used. In this system the basic units of length, time, and mass are the foot, the second, and the slug, respectively. There is no British system of electrical units, so in any problem involving electrical quantities the mks system is usually used. One other system, the cgs system, is used to some extent in scientific work, although it is gradually being supplanted. In the cgs system the basic units are the centimeter (1 cm = 10^{-2} m), the gram (1 g = 10^{-3} kg), and the second. There seems little doubt that the United States will eventually adopt some form of the metric system for commercial as well as scientific use, as Britain is now doing, and the consensus of informed opinion is that this conversion should be initiated as soon as possible.

Physical quantities are usually represented by numbers having units. Whenever such numbers appear in an equation, it is necessary for all terms of the equation to have the same units. This consistency of units is referred to as *dimensional consistency* in equations. In calculations with equations involving numerical quantities, it is almost always advisable to carry along the units of the quantities as well as the numbers. This procedure often facilitates the detection of errors and blunders in the calculation; if at any

12 WHAT IS PHYSICS?

stage in the computation two terms are dimensionally inconsistent, an error has been made.

Here is an example. The distance y through which a freely falling object has fallen during a time t, starting from rest, is given by

$$y = \tfrac{1}{2}gt^2$$

where g is the acceleration of gravity, about 9.8 m/s^2. Suppose a student asked to find how far an object falls in 4 s remembers this formula incorrectly and uses instead $y = \tfrac{1}{2}gt$. He then finds

$$y = (\tfrac{1}{2})(9.8 \text{ m/s}^2)(4 \text{ s}) = 19.6 \text{ m/s}$$

Because y is a distance, it ought to have units of length, i.e., meters. Instead, it comes out with units of meters per second, which shows immediately that a mistake has been made somewhere. Indeed, this equation is dimensionally consistent *only* when the exponent on t is 2, as the reader can easily verify.

In many observations it is necessary to specify other conditions in

Fig. 1-3 Description of any motion depends on the state of motion of the observer who describes the motion. As the brakeman on the train lowers his lantern, it appears to him to follow the vertical path A. To a stationary observer standing beside the track while the train moves past, it appears to follow the diagonal path B.

addition to the system of units used. A familiar example is the observation of motion of any type. In describing the motion of a lantern held by a brakeman on a moving railroad car, it is important to state whether the observer is standing in the railroad station or moving with the train; the motion looks quite different to these two observers, as shown in Fig. 1-3. We must specify what we call the *frame of reference* in which the observations are made. The concept of frame of reference is discussed in considerable detail in connection with the principles of mechanics.

1-3 EQUATIONS IN PHYSICS

Precise descriptions of phenomena usually involve physical quantities whose magnitudes are expressed in terms of numbers, and it is natural that these relationships should find expression in equations. An equation relating two quantities is not itself a physical principle, but it is a shorthand, a symbolic way of expressing the essence of a physical principle.

One of the simplest forms of relationship between two physical quantities is a simple proportionality. Two quantities are *proportional* if they vary in such a way that their quotient is constant. For example, if the force F required to stretch a spring a distance x is proportional to x, then F/x is constant. To double x, it is necessary to double F, and so forth. Denoting the constant quotient by k, we express this relationship in an equation $F/x = k$, or $F = kx$. The constant k, determined by the material and design of the spring, is called a *proportionality constant*, since it appears in an equation expressing a proportionality.

A simple extension of this idea is a relationship expressed in the form $y = ax + b$, where a and b are constants and x and y are physical quantities which may vary. An example is the relationship between velocity and time for a falling object: $v = v_0 + at$, where v is the velocity at time t and v_0 and a are constants. This is called a *linear* relationship because a graph of v versus t is a straight line, and the equation expressing it is a *linear equation*.

The constants which appear in equations are of several different types. Some, such as the constant k describing the properties of a spring, are unique to a particular system. Others characterize properties of particular materials, such as elasticity, thermal conductivity, density, and so on. Still others are more general constants which apply to the entire universe; examples are the gravitation constant G, the speed of light c, and the mass of the electron m. These are called *universal* constants. In addition, there are numerical constants such as 2 and π, which have physical meaning only in combination with physical quantities. All the constants mentioned (except numerical constants) have units, and their units must be such that the equations in which they appear are dimensionally consistent.

There is always a temptation to adopt the viewpoint that physics consists of equations and that to master the subject of physics one need only remember a collection of equations. This is a naïve and dangerous point of view. The student whose approach to physics is to page through the book to find a formula which fits the problem he is trying to solve will never master the subject. Equations provide a concise means of summarizing results, but they can never substitute for thorough understanding of the concepts and quantities concerned. Thus it is never wise to concentrate on memorizing formulas. When the basic concepts have been mastered, equations are retained automatically and without conscious effort. Similarly, in solving problems it is far better to use one's fundamental understanding than merely to substitute numbers into a formula containing the proper physical quantities.

1-4 MODELS

In discussing the foundations of physical theories we have stressed the importance of making observations of physical phenomena as complete and quantitative as possible. But it is easy to see that following this admonition can lead to ridiculous extremes. For example, in observing the motion of a falling body we might record the hour of the day, the phase of the moon, the temperature of the air, the experimenter's blood pressure, and a million other quantities which may or may not be relevant. If we then try to construct a physical theory which correlates the observations and takes *all* these quantities into account, we quickly reach a hopeless situation.

This may seem somewhat paradoxical; we have expressed the belief that physical principles are simple, and yet even very elementary experiments seem to be hopelessly complicated. Is there any way out of this dilemma?

Of course there is. In this example, the essential point is that most of the quantities we have mentioned are actually irrelevant. What we must do is separate the important quantities from the unimportant ones and concentrate on the former.

The necessity for simplifying and idealizing in order to formulate general principles was seen clearly by Galileo Galilei (1564–1642), often called the father of modern physical science. Galileo frequently presented his thoughts in dialogue form, showing the influence of the writings of Aristotle. Following is an excerpt from one of these. The characters are Simplicio, representing the old aristotelian view, Salviati, representing Galileo's ideas, and Sagredo, an intelligent and inquisitive man who has not yet committed himself to either view.[1]

Salviati *I greatly doubt that Aristotle ever tested by experiment whether it be true that two stones, one weighing ten times as much as the other, if allowed to fall, at the same instant, from a height of say, 100 cubits, would so differ in speed that when the heavier had reached the ground, the other would not have fallen more than 10 cubits.*

Sagredo *I, who have made the test, can assure you that a cannon ball weighing one or two hundred pounds, or even more will not reach the ground by as much as a span ahead of a musket ball weighing only half a pound. . . .*

Salviati presents an argument to show why Sagredo's result should have been expected, and then the discussion continues.

Simplicio *Your discussion is really admirable; yet I do not find easy to believe that a bird-shot falls as swiftly as a cannon ball.*

Salviati *Why not say a grain of sand as rapidly as a grindstone? But, Simplicio, I trust you will not follow the example of many others who divert the discussion from its main intent and fasten upon some statement of mine that lacks a hairsbreadth of the truth and, under this hair, hide the fault of another that is as big as a ship's cable. Aristotle says that "an iron ball of one hundred pounds falling from a height of 100 cubits reaches the ground before a one-pound ball has fallen a single cubit." I say that they arrive at the same time. You find, on making the experiment, that the larger outstrips the smaller by two fingerbreadths . . . ; now you would not hide behind these two fingers the 99 cubits of Aristotle, nor would you mention my small error and at the same time pass over in silence his very large one.*

It is certainly true that two different objects falling freely in air *do not* reach the ground at precisely the same instant. But much more important than this, as Galileo points out, is the fact that they reach the ground at *nearly* the same time; the small difference in arrival time is an unimportant detail, perhaps attributable to air resistance, rather than an essential feature of the motion.

[1] From Galileo's "Two New Sciences," translated by Crew and de Salvio, The Macmillan Company, New York, 1914.

In analyzing a physical situation it is nearly always useful to construct mentally an *idealized* situation which is similar to the actual one but which is simplified by the omission of some of the relatively unimportant influences. In the falling-body experiment, if we think that air resistance is relatively unimportant, we may imagine an *idealized* experiment which takes place in a total vacuum. If we think that the size and shape of the object are unimportant, we may imagine it as a *point* in the idealized experiment. The reason for constructing this imaginary situation, of course, is to eliminate enough unimportant factors for it to be practical to make a detailed analysis of the simplified situation which remains.

Such a simplified and idealized version of a complicated physical problem is called a *model* of the original situation. A model is an *approximate* representation of a physical situation which is simplified to facilitate analysis and calculation. A classic example of such a model is Newton's first approximate analysis of the motion of the solar system. The spatial extensions of the sun and the planets were neglected, and each was represented as a point. The gravitational interactions of each planet with other planets and with its own satellites were also neglected. In the simplified model of the motion of the earth around the sun, what is really considered is the motion of a point mass under the action of a gravitational force directed toward a fixed point. Clearly, this system is much simpler to analyze than the real one, in which the gravitational interactions of several extended bodies of complicated shape must be taken into consideration.

The assumptions of the model must be based on the nature of the physical situation. The dimensions of both the earth and the sun are quite small compared to the distance between them; the gravitational forces on the earth due to the other planets are considerably smaller than that exerted by the sun. Thus, at least in a first attempt to understand the motion of the earth, it is reasonable to omit these relatively small effects. The result is an idealized model which is simple enough to be analyzed quite precisely.

The precision with which the results of this analysis agree with observed behavior depends in part on the accuracy of the numerical calculations, but the limitations imposed by the model are equally important. For example, a model in which the moon's gravitational force on the earth is neglected can never predict the small but measurable variations in the earth's speed in its orbit which result from the moon's motion around the earth. To take this effect into account, one would have to use a more detailed model which includes the moon's gravitational force, at least in some approximate manner.

A more contemporary example of a simplified model is the analysis of the wavelengths of light emitted by hydrogen atoms. In the simplest model consistent with contemporary quantum mechanics, we neglect the size of the nucleus and represent it as a point. We consider only one isolated atom, neglecting the effects of interactions with other nearby atoms, and we assume that the electron behaves according to a certain equation called the *Schrödinger equation*. A mathematical analysis based on this model can predict the wavelengths with considerable precision. It cannot, however, give the small corrections resulting from the finite size of the nucleus or the *pressure broadening* of the spectrum lines, the small shifts in wavelength resulting from interaction of an atom with its neighbors. Theoretical treatment of these effects requires a more detailed model.

The concept of an idealized, simplified model of a complicated situation is of the utmost importance in all of physics. We rarely deal directly with physical reality; instead we usually deal with an idealized version of physical reality which is simple enough to permit analysis. Indeed, the basic principles of physics are always stated in terms of idealized models. In this book we frequently stress the construction of models to facilitate the analysis of physical situations.

1-5 MACROSCOPIC AND MICROSCOPIC VIEWS

Many important and interesting branches of physics are concerned with the relationship between the behavior of bulk aggregations of matter and the detailed structure of the matter in terms of its atomic and nuclear constituents. In this connection we introduce the terms *macroscopic* and *microscopic*. A macroscopic phenomenon is one involving the behavior of matter in bulk, in which ordinarily a very large number of atoms participate. A microscopic phenomenon, on the other hand, is concerned with the interactions and motions of individual atoms and fundamental particles within them. Such motions usually are not directly observable.

It is often possible to gain insight into a macroscopic phenomenon by considering its microscopic basis. A classic example is the kinetic-molecular theory of gases. By observing the behavior of various gases, one can formulate general laws regarding the relationships between pressure, volume, and temperature, such as Boyle's and Charles' laws. But by considering the microscopic structure of a gas and the motions of its individual molecules, it can be shown that these relationships can be deduced from an analysis of the molecular motions, which in turn are governed by Newton's laws of motion.

Similarly, the phenomenon of *heat*, which can be observed macroscopically by a thermometer or by the sense of touch, takes on new meaning when we realize that adding heat to a body corresponds to increasing the motions of the molecules in the body. Heat is, in fact, associated directly with mechanical energy on a microscopic scale. The macroscopic electrical conductivity of solid matter can be understood on the basis of the motion of electrons. In some materials the electrons are tightly bound to particular atoms; such materials are insulators. In others, particularly the metals, electrons are relatively free to move; such materials are conductors.

In analyzing the relationship between macroscopic and microscopic phenomena we often make use of models as discussed in Sec. 1-3. For example, in the kinetic-molecular theory of a gas mentioned above, we use a model in which the gas molecules are thought of as particles of negligible size, interacting only by means of elastic billiard-ball-type collisions with each other and with the walls of the container. These and other simplifying assumptions constitute a *microscopic model* of a gas. With such models we try to represent the microscopic behavior accurately enough for the results of calculations to show a degree of agreement with observed physical reality and yet in a manner sufficiently simple to permit mathematical analysis.

Much of the most important recent research in physics has been concerned with the relationship between the macroscopic and microscopic behavior of matter. Such considerations always involve the use of a microscopic model. In this book, especially in the sections dealing with thermal phenomena, we make considerable use of microscopic models in understanding the macroscopic behavior of matter.

1-6 MATHEMATICAL LANGUAGE

As already mentioned, certain branches of mathematics are particularly appropriate for expressing quantitative relationships in physics. It is assumed that the reader of this volume has some mastery of algebra, geometry, and plane trigonometry at the level usually attained in secondary schools and that he is studying elementary differential and integral calculus concurrently with physics. Considerable use is made of calculus, wherever it is appropriate. Since many readers may have very limited acquaintance with vector language, Chap. 2 is devoted to useful techniques of vector algebra. In addition, we include here a few details of mathematical language.

First, a few remarks are in order concerning numerical magnitudes. It is usually advantageous to use one system of fundamental units consistently in physical calculations; the mks system is used almost exclusively in this volume. But while the meter may be an appropriate unit to use in measuring the length of a desk, it is not so appropriate to use in measuring the solar system, whose diameter is many billions of meters, or a hydrogen atom, whose diameter is only a few billionths of a meter.

Nevertheless, the advantages of a uniform system are so great that it is usually preferable to cope with these very large and very small numbers rather than introduce new units. Thus, it is handy to have a shorthand system for expressing very large and very small quantities. The system usually used makes use of powers of 10. For example, we may use the fact that $1{,}000{,}000 = 10^6$ to write

$$2{,}300{,}000 = 2.3 \times 10^6$$

The advantage of this notation appears when we try to multiply two large or small numbers together. For example,

$$\begin{aligned} 2{,}300{,}000 \times 4{,}970{,}000{,}000{,}000 &= 2.3 \times 10^6 \times 4.97 \times 10^{12} \\ &= 11.431 \times 10^{18} \\ &= 11{,}431{,}000{,}000{,}000{,}000{,}000 \end{aligned}$$

This technique is especially important when using a slide rule; we multiply the two numbers without the powers of 10 and then add the exponents of the two powers of 10:

$$2.3 \times 10^6 \times 4.97 \times 10^{12} = (2.3 \times 4.97) \times 10^{6+12}$$

Similarly,

$$\frac{6.6 \times 10^{-27}}{(9.1 \times 10^{-28}) \times (3.0 \times 10^{10})} = \frac{6.6}{9.1 \times 3.0} \times 10^{-27+28-10}$$

$$= 0.24 \times 10^{-9} = 2.4 \times 10^{-10}$$

where we have used the fact that $1/10^{-28} = 10^{28}$.

The last example serves to illustrate another advantage of the powers-of-10 notation. When this notation is used, the number of digits in the number multiplying the power of 10 indicates the precision with which the number is known. For example, 9.1×10^{-28} indicates that we know the value of this quantity to be between 9.05×10^{-28} and 9.15×10^{-28}. Similarly writing a number as 6.653×10^{-27} indicates that it lies within the range 6.6525×10^{-27} to 6.6535×10^{-27}.

The number of digits known to be correct is called the *number of significant figures* in the quantity. Clearly, it makes no sense to write a number using 10 digits if only the first three are known with certainty. Similarly, when two numbers are multiplied or divided, the result should be rounded off to leave at most one more significant figure than in the factor having the fewer significant figures. One may measure the diameter and circumference of a circle, obtaining the values $d = 1.50$ m and $C = 4.71$ m. From these one can calculate $\pi = 3.1400000$; only the first three digits (3.14) in this value of π are significant, and it agrees with the correct value 3.14159265 to this number of significant figures. In the first power-of-10 example above, the result is properly stated as 11.4×10^{18}, not 11.431×10^{18}.

Often we must describe *changes* in physical quantities. As an example, if the temperature of a solution at one time is represented by the symbol T_1 and at a later time by T_2, then the *difference* $T_2 - T_1$ represents the *change* in temperature which took place. Such expressions occur so often that it is very useful to have a shorthand notation to denote changes. The usual notation makes use of the Greek letter Δ (capital delta). In this example, we denote the temperature change as ΔT. One may think of Δ as an abbreviation for *difference*, in this case the difference in T between the final and initial values. That is,

$$\Delta T = T_2 - T_1$$

When we describe a change in a quantity between some time and another later time, it is customary to define the change as the final value minus the initial value, and not the reverse. In general, if a variable x initially has a value x_1 and later a different value x_2, we define the *change* in x as

$$\Delta x = x_2 - x_1$$

This notation is widely used in calculus; those who have studied elementary calculus should be familiar with it.

Another useful shorthand notation is used to represent a *sum* of several quantities. Suppose we have a set of numbers to be added. We may represent these numbers symbolically by using the letter a with a subscript which is different for each one. For example, if there are five numbers, we might represent them as

$$a_1, a_2, a_3, a_4, a_5$$

If there are N in all, they may be represented as

$$a_1, a_2, a_3, \ldots, a_{N-1}, a_N$$

where N may be any positive integer and the dots represent the intermediate numbers which have not been written explicitly.

Now suppose we want to compute the *sum* of these numbers,

$$a_1 + a_2 + \cdots + a_{N-1} + a_N$$

It is cumbersome to write out this expression each time it is needed. The shorthand usually used is

$$a_1 + a_2 + \cdots + a_{N-1} + a_N = \sum_{i=1}^{N} a_i$$

If this symbol is not familiar, it may seem complicated; it is really very simple. The most important part is the Greek letter Σ (capital sigma) which is an abbreviation for *sum*. Instead of putting a particular subscript on a, we use the letter i, which may stand for any of the subscripts of the set of numbers and which is called the *summation index*. The whole expression says that we should sum all the a's, starting with the first one (for which i has the value 1) and ending with the last one (for which i has the value N), and including all the possible subscripts in between, i.e., all values of i between 1 and N. The expressions written above and below Σ are called the *limits* of the sum; they simply tell where to start and stop in the values of i we use. For example,

$$\sum_{i=3}^{6} a_i = a_3 + a_4 + a_5 + a_6$$

Sometimes, when it is clear from the context what the limits of the sum should be, they are omitted to save writing. If we have a collection of N particles whose masses are m_1, m_2, and so on, then the total mass M, which is the sum of the individual masses, can be written simply as $M = \Sigma m_i$ instead of the complete expression

$$M = \sum_{i=1}^{N} m_i$$

The expression Σm_i simply means the sum of the masses of *all* the particles, of which a typical one is m_i. If they are numbered from 1 to N, then the understood limits of the sum are $i = 1$ to $i = N$.

If a given set of numbers is to be added, the summation notation does not in itself save any labor. Its real advantage is in denoting the *operation* of adding a set of quantities. It is also used to denote addition of mathematical quantities more complicated than ordinary numbers. We shall often use it to represent addition of vectors, in which case it always denotes a *vector* sum of the vector quantities following the summation symbol.

Problems

1-1 Find the units of acceleration, frequency, and density.

1-2 In problems of supersonic flight, speeds are sometimes given in Mach numbers. The Mach number of a speed is the quotient of the speed and the speed of sound in air under the same conditions. What units does the Mach number have?

1-3 Comment on the dimensional consistency of the equation

$$\cos \theta = \frac{3R + 2}{R}$$

where R is the radius of a certain circle and 2 and 3 are pure numbers.

1-4 Does the expression cos (3 ft) have any meaning? Explain.

1-5 Suppose that the speed v of a certain moving object varies with time t according to the equation

$$v = at + \frac{b}{t + c}$$

where a, b, and c are constants. What *units* must each of these constants have in order to make this equation dimensionally consistent?

1-6 A point moves in such a way that its distance x from a fixed point varies with time t according to the equation

$$x = at + bt^2$$

where a and b are constants. What *units* must each constant have to make the equation dimensionally consistent? What physical quantities have these units?

1-7 A student calculated the period T of a simple pendulum, which is the time required for a complete back-and-forth swing. He obtained the result

$$T = 2\pi \sqrt{\frac{g}{l}}$$

where l is the length of the pendulum and g is the acceleration of gravity.

PROBLEMS

    a Does this result seem reasonable, disregarding any consideration of units? Explain.

    b Check the equation for dimensional consistency. If it is inconsistent, make a guess at the correct form of the equation.

1-8 The slope of a certain trail was described as 500 ft/mi. Show how this can be expressed in a dimensionless (unitless) form. What is the number describing the slope in this form?

1-9 Celsius and Fahrenheit temperatures are related by a linear equation. Derive an equation relating them, using the fact that $0°C = 32°F$ and $100°C = 212°F$.

1-10 The velocity v of a freely falling object varies with time t according to the equation $v = v_0 + at$, where v_0 and a are constants. In one particular experiment the object was observed to have $v = 64$ ft/s at $t = 1$ s and $v = 96$ ft/s at $t = 2$ s. From this information, find v_0 and a; be sure to express them using the correct units.

1-11 If $m = 1.67 \times 10^{-27}$ kg and $c = 2.997 \times 10^8$ m/s, compute the quantity mc^2 using powers of 10, the correct number of significant figures, and the correct units.

1-12 If $h = 6.63 \times 10^{-34}$ kg·m²/s, $m = 9.11 \times 10^{-31}$ kg, and $c = 3.00 \times 10^8$ m/s, compute the quantity h/mc using powers of 10 with the correct units and number of significant figures.

1-13 If $a_1 = 1$, $a_2 = 2$, $a_3 = 3$, and $a_4 = 4$, evaluate $\sum_{i=1}^{4} a_i$. Is this equal to $\sum_{i=1}^{4} i$ in this case? Explain. Is this true in general?

1-14 If $b_1 = 1$, $b_2 = 4$, $b_3 = 9$, and $b_4 = 16$, evaluate $\sum_{i=1}^{4} b_i$. Is this equal to $\sum_{i=1}^{4} i$? If not, is there a similar expression to which it is equal?

1-15 Evaluate $\sum_{n=2}^{4} (3n^2 - n)$.

1-16 Evaluate $\sum_{m=0}^{3} (m^2 - 3)$.

1-17 Are the expressions $\sum_{n=1}^{5} n^2$ and $\left(\sum_{n=1}^{5} n\right)^2$ equal? Explain.

1-18 Given a set of N numbers a_i, where i may be 1, 2, 3, . . . , N (N numbers in all), write an expression for the *average* (*or mean*) of the set.

1-19 Show that $\sum_{n=1}^{N} n = N(N+1)/2$.

Vector Language | 2

Many physical quantities require both a numerical magnitude and a direction for their descriptions. Such quantities are conveniently represented by vectors. To introduce the concept of a vector, we consider various ways to describe the position of a point in space and a displacement from one point to another. Displacement vectors are used to illustrate various operations with vectors, including addition and subtraction, multiplication by a scalar, and two kinds of products of vectors. A number of applications to physical principles are discussed briefly.

2-1 POSITION OF A POINT

Many physical quantities have *direction* as well as magnitude. To describe the motion of an airplane in flight, we must give not only its speed but also the direction of its motion. To describe a force, we use a number to describe how hard the force is pushing or pulling and then describe separately the direction of the push or pull. Other examples of physical quantities having both magnitude and direction are acceleration, momentum, electric or magnetic field intensity, and displacement. Any quantity having both magnitude and direction which is combined with other like quantities according to certain definite rules to be introduced in this chapter is called a *vector* quantity.

Conversely, other quantities can be described completely by a single number without any direction; examples are temperature, mass, volume, time, and electric charge. Some of these may be positive or negative, but they have no direction in space. Any such physical quantity is called a *scalar* quantity.

26 VECTOR LANGUAGE

The principal object of this chapter is to discuss ways of describing vector quantities and the operations relating them. We begin with the simplest vector quantity, displacement. A displacement is a change in the position of a point, so the first order of business is to discuss how the position of a point can be described. The term *position* has meaning only when used with a particular reference system, or *frame of reference*. Similarly, the coordinates of a point have meaning only when the positions of the coordinate axes are known.

A simple example is a situation in which the point is known to lie along a given straight line. Suppose we want to describe the motion of a car on a straight speed-trial track. We use as a model of the car a point moving on a straight line. Its position can be described by measuring its distance from some fixed point on the track. In Fig. 2-1, we represent the position of the car by giving its distance from a starting flag. If the car starts on one side of the flag and moves past it, we must also say whether the car is to the right or to the left of the flag. A convenient way to do this is to use positive numbers for one side (say the right) and negative numbers for the other.

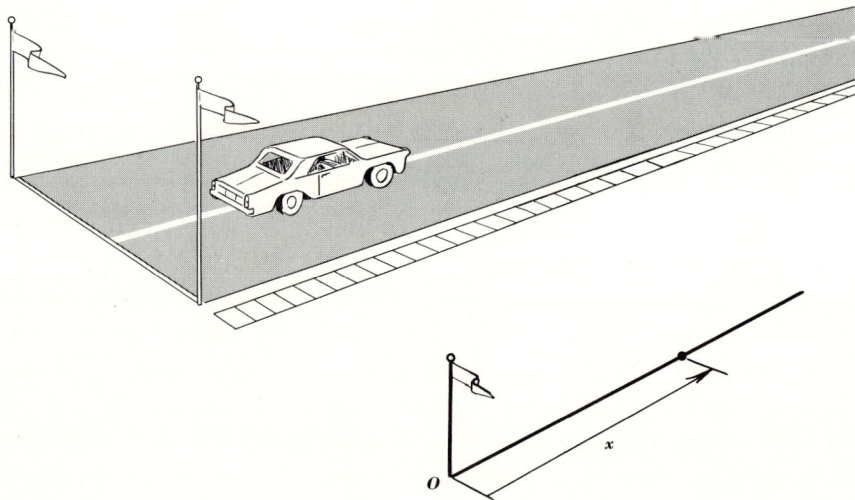

Fig. 2-1 Distance x represents the position of the car in the simplified model shown. The arrow indicates that x is positive when the point is on the right side of O and negative when it is on the left.

The above situation is the simplest possible example of a *coordinate system*. Its essential ingredients are (1) the line, on which the point is known to lie; (2) the fixed point O on the line, used as a point of reference, called the *origin* of the coordinate system; (3) a means of measuring the distance from the origin to the position of the point; (4) a designation of positive and negative regions.

If the car swerves from side to side on the track, rather than moving in a straight line, we can represent the motion with a model in which the point moves in a plane, rather than along one straight line. There are various ways of assigning numbers to a position in a plane. One familiar method is shown in Fig. 2-2; we draw two perpendicular lines, called *coordinate axes*. The point at which they cross, called the *origin of coordinates*, is labeled O. One axis is labeled x, the other y. We specify the position of a point P in the plane by measuring the distances x and y, as shown in Fig. 2-2. The coordinates of P are x and y.

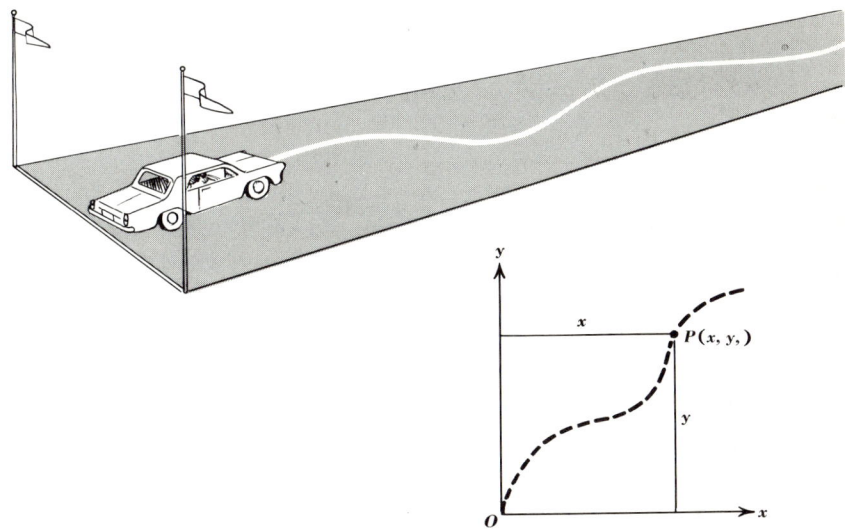

Fig. 2-2 Two numbers x and y describe the position of the car, represented by P in this simplified model.

When specifying the x and y coordinates of a point, the usual practice is to write the numbers in parentheses, the x coordinate first. The position is then specified by the *pair* of numbers (x,y). The label on the x axis is always placed on the side corresponding to positive values of x, and similarly for the y axis. Several points are shown in Fig. 2-3, illustrating the signs of the coordinates and the order in which they are written.

Since x and y are distances from the coordinate axes, they have units of length, such as feet or meters. In any physical problem, therefore, we must specify not only the numerical values of the coordinates but also the units in which the distances are measured.

28 VECTOR LANGUAGE

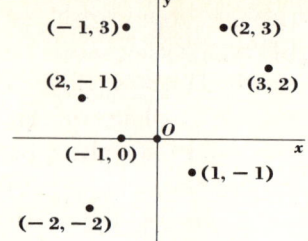

Fig. 2-3 Several points in a plane with their coordinates. One pair of coordinates is incorrect. Which is it?

The use of x and y coordinates, usually called *rectangular* or *cartesian* coordinates, is not the only way of specifying the position of a point in a plane. Another scheme uses the *distance* from the origin to the point P as one coordinate and the *angle* between the line joining O and P and some fixed line as the other. It is conventional to use the positive x axis as the fixed line. Then the location of the point P in Fig. 2-4 is specified by the distance r (always positive) and the angle θ. The arrow shows that θ is positive if measured counterclockwise from the positive x axis, negative if clockwise.

The coordinates of a point in this scheme are called *polar coordinates*. If the cartesian coordinates (x,y) of a point are known, the polar coordinates (r,θ) of the same point can be found, and conversely. Reference to Fig. 2-4 shows that the relations between the coordinates are

$$x = r \cos \theta$$
$$y = r \sin \theta \qquad (2\text{-}1)$$

or, alternatively,

$$r = \sqrt{x^2 + y^2}$$
$$\theta = \tan^{-1} \frac{y}{x} \qquad (2\text{-}2)$$

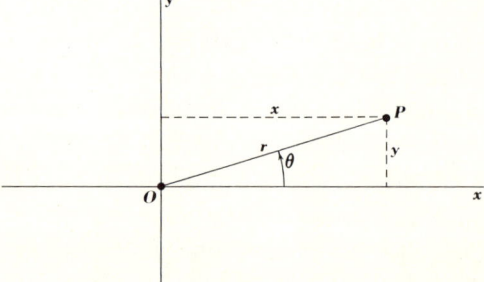

Fig. 2-4 The position of point P can be described by either the pair of numbers (x,y) or the pair of numbers (r,θ).

which is read "θ is the angle whose tangent is y/x" or "θ is the inverse tangent of y/x." An alternative notation is $\tan^{-1}(y/x) = \arctan(y/x)$.

To describe the position of a point not just in a plane but anywhere in space, a scheme very similar to the xy coordinate system uses *three* mutually perpendicular axes, as shown in Fig. 2-5. The set of three coordinates describing the point P is usually written in the form (x,y,z). For brevity we sometimes refer to this coordinate system as *system Oxyz*. By applying the theorem of Pythagoras twice, we find that the distance from the origin O to the point P, denoted in the figure as r, is given by

$$r = \sqrt{x^2 + y^2 + z^2} \tag{2-3}$$

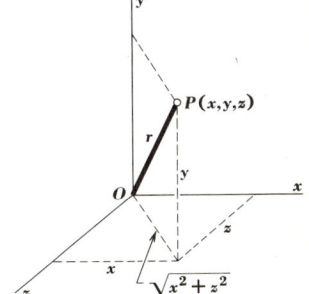

Fig. 2-5 Representation of a point P in space by means of its coordinates (x,y,z).

There are many other ways of specifying the position of a point in space. Some use three distances, like the system just discussed; others use combinations of distances and angles, analogous to the polar coordinate system in a plane. All these systems have in common the fact that three numbers are required to specify a point completely. For this reason, space is said to be *three-dimensional*. It is impossible to specify a point in space with fewer than three numbers.

2-2 VECTORS

Having discussed various ways of describing the position of a point, we are now ready to discuss *changes* in position, which we call *displacements*. Describing a change in position is the first step in describing motion, the principal subject of Chap. 3. In addition, a discussion of displacements leads naturally to the concept of addition of vectors, introduced in Sec. 2-3. This concept will be seen later to be very useful not only for displacement but for force, momentum, and several other fundamental quantities in mechanics.

Consider again the motion of a point on a line; suppose the point moves from a position $+100$ ft to another position $+150$ ft. We say that it has undergone a *displacement* of $+50$ ft. A displacement is positive if it is directed toward the positive end of the line and negative if in the opposite direction. In a displacement from $+150$ ft to $+100$ ft, the displacement is -50 ft, although the coordinates of both the starting and end points are positive. Similarly, a displacement from -15 ft to -10 ft is $+5$ ft.

In motion along a line, the displacement is described completely by the change in the coordinate x, usually denoted by Δx, which means simply a change in the value of x. If a point is displaced from x_1 to x_2, the displacement is

$$\Delta x = x_2 - x_1 \tag{2-4}$$

The sign in this expression comes out correctly automatically; if x_2 is more positive or less negative than x_1, the displacement is toward greater positive values of x, and Δx is positive, and conversely.

For displacements of a point in a plane, it is not sufficient to say *how far* the point is displaced; we must also say in *what direction* it is displaced. In describing a general displacement from a point P_1 to a point P_2, as in Fig. 2-6, it is useful to think of a *directed line segment* joining P_1 and P_2, with a direction from P_1 to P_2. Such a directed line segment is called a *vector*. One way to specify a vector is to write the labels for the beginning and end points below an arrow showing the direction of the displacement, such as

$$\overrightarrow{P_1P_2}$$

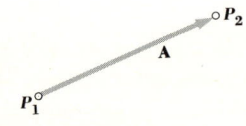

Fig. 2-6 Directed line segment used to describe a displacement from P_1 to P_2. This is a displacement vector, which can be represented by a single symbol **A**.

For our purposes, it is usually handier to denote a vector with a single symbol, such as **A**. To differentiate between a vector and an ordinary number, we always use boldface type to denote a vector. Thus we may refer to the vector in Fig. 2-6 simply as **A**.

The vector **A** describes the net result of the motion of the point from P_1 to P_2, not necessarily the actual path taken by the point, which in general need not move in a straight line at all. Several possible paths are shown in Fig. 2-7. Each path, however, leads to the same result, namely the displacement from P_1 to P_2, and so by definition the *displacement* is the same for all these paths. A displacement vector is always a directed *straight*

31 ADDITION OF DISPLACEMENTS [2-3]

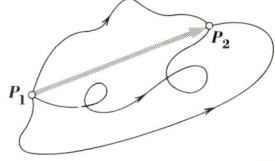

Fig. 2-7 Several paths from P_1 to P_2. The displacement is the same for all these paths, by the definition of displacement.

line, regardless of the actual path of the motion. Later, in representing the actual path of the motion, we shall find it useful to think of the path as made up of a very large number of very small displacements; this view is especially useful in defining velocity and acceleration, to which we return in Chap. 3.

Example
Consider the motion of the end of the minute hand on a clock. Find the displacements which occur between 12:00 and 12:15, between 12:00 and 12:30, and between 12:00 and 1:00.

Solution
The displacement from 12:00 to 12:15 is given by the vector **A** in Fig. 2-8. The displacement from 12:00 to 12:30 is given by the vector **B** in this

Fig. 2-8 Displacements of the end of the minute hand of a clock.

figure. In neither case does the vector lie along the path followed by the end of the hand. The vector denotes only the final result of the displacement, not the details of the motion between the beginning and end points. Between 12:00 and 1:00, there is *no net displacement* because the point at which the minute hand ends is the point at which it started.

2-3 ADDITION OF DISPLACEMENTS
When a point undergoes two displacements in succession, such as displacement **A** followed by displacement **B** in Fig. 2-9, what is the total displacement?

32　VECTOR LANGUAGE

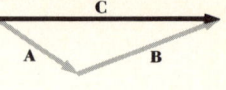

Fig. 2-9 The end result of two successive displacements **A** and **B** is equivalent to a single displacement **C**. In vector language, **C = A + B**.

By total displacement we mean the single displacement, such as **C**, which is equivalent to the final result of two other displacements, such as **A** and **B**, carried out in succession.

To specify concisely the relationship of **C** to **A** and **B**, we call **C** the *vector sum* of **A** and **B**. It is convenient to write this relationship symbolically as follows:

C = A + B　　　　　　　　　　　　　　　　　　　　　　　　　　　(2-5)

The plus sign does *not* denote the same operation as the addition of ordinary numbers. In Fig. 2-9, if **A** has a length of 3 in. and **B** a length of 2 in., it is not in general true that **C** has a length of 5 in. In fact, this is true only if **A** and **B** are in the same direction, which is not always the case. Whenever the boldface **+** appears in an equation whose terms are vectors, the vector sum indicated by Fig. 2-9 is meant. Similarly, the presence of the boldface symbol **=** between two vector quantities means that the quantities are equal in magnitude and have the same direction.

What happens if the two vectors **A** and **B** are combined in the opposite order? Figure 2-10 shows that the result of **B** followed by **A** is the same

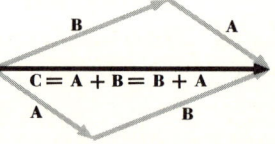

Fig. 2-10 The order in which two vectors **A** and **B** are added is immaterial. Vector addition is commutative.

as the result of **A** followed by **B**. The vector sum of *two vectors does not depend on their order.* Symbolically,

A + B = B + A　　　　　　　　　　　　　　　　　　　　　　　　　(2-6)

Addition of vectors is *commutative*.

Figure 2-10 illustrates another important property of displacement vectors: A vector can be shifted from one position to another without changing its meaning, provided its direction and length are not altered. A displacement vector does not specify where to start and where to stop but only how far to go and in what direction. Thus two cars which travel 1 mi east have the same displacement even if one starts in Pittsburgh and the other in Philadelphia (if we neglect the curvature of the earth).

Although the previous illustrations have been confined to the addition of vectors in a plane, we can easily extend the notion of vector addition to vectors which have any orientation in space. Furthermore, the rule for adding vectors can be extended to any number of vectors. Suppose we have four vectors **A, B, C,** and **D** as shown in Fig. 2-11. The vector sum can be

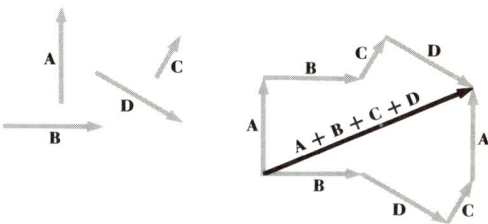

Fig. 2-11 Vector sum of four vectors. Two different orders of addition are shown; the result is the same for both. Try drawing a third!

obtained geometrically by starting with any one of the vectors, moving the next, without changing its direction, so that its tail is on the head of the first, and so on, and then drawing a line from the tail of the first vector to the head of the last, as shown in Fig. 2-11. By an extension of the argument already given, it is easy to show that the *order* of addition is immaterial. In adding vectors by this geometrical method, we can move any vector wherever desired to arrange a diagram for the vector sum, provided we do not alter the direction of any vector.

In dealing with two vectors **A** and **B,** we may need the vector sum of **A** and a vector with the same length as **B** but the opposite direction. The vector opposite to **B** is denoted by **−B.** This notation is analogous to that used for ordinary numbers; we have **B + (−B) = 0,** just as $2 + (-2) = 0$. Using this notation, we write the sum of the two vectors mentioned as **A + (−B).** This can be further abbreviated

$$\mathbf{A} + (-\mathbf{B}) = \mathbf{A} - \mathbf{B} \tag{2-7}$$

which defines the operation of subtraction of one vector from another.

One other useful operation is the multiplication of a vector by a number. This operation multiplies the length of the vector by the number without changing its direction. For example, 3**A** is a vector in the same direction as **A** but three times as long. The vector **−2B** is a vector in the direction *opposite* that of vector **B** and twice as long.

Example

In Fig. 2-12 are shown various combinations of the two vectors **A** and **B**.

It sometimes happens that only the distance involved in a displacement is of interest, not the direction. This is the length of the vector describing the displacement, called the *magnitude* of the vector or its *absolute value*. The magnitude of a vector is a single positive number; it is denoted by the same symbol as used for the vector but in lightface type. Thus A is the magnitude of **A**. Another notation sometimes used for the magnitude of **A** is $|\mathbf{A}|$.

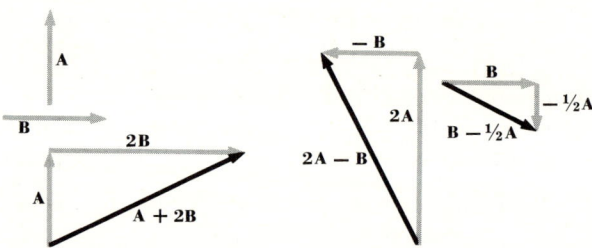

Fig. 2-12 Various sums and differences involving two vectors **A** and **B**.

2-4 COMPONENTS AND UNIT VECTORS

Although the geometrical discussion of the preceding section is sufficient to *define* the operation of vector addition, it is often somewhat clumsy to apply this definition directly in calculations. The accuracy with which a vector sum can be computed using this method is limited by the accuracy with which lines can be drawn and distances and angles measured. For three-dimensional problems, where diagrams cannot be drawn on a flat sheet of paper, the method is not at all practical. Fortunately, there is a much handier technique, making use of *components* of a vector.

The components of a vector are based on a set of coordinate axes. Let us consider first only vectors lying in a single plane; we use an xy coordinate system in this plane. Any vector **A** which lies in this plane can be represented as the vector sum of a vector $\mathbf{A}_x$ parallel to the x axis and a vector $\mathbf{A}_y$ parallel to the y axis, as shown in Fig. 2-13. The vectors $\mathbf{A}_x$ and $\mathbf{A}_y$ are called the x and y component vectors of the vector **A** in the coordinate system Oxy, and $\mathbf{A} = \mathbf{A}_x + \mathbf{A}_y$. This decomposition of vector **A** into its components along the x and y directions is *unique*; there is only one such pair of components which can be added to give the vector **A**.

Fig. 2-13 Representation of a vector **A** as the vector sum of two component vectors, one parallel to the x axis, the other parallel to the y axis.

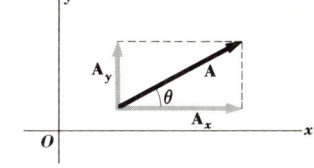

Because of the way in which the component vectors have been defined, the x-component vector is always parallel to the x axis, no matter what the direction of **A**, and similarly for the y-component vector. To facilitate indicating the directions of various vectors, we introduce special vectors whose sole function is to describe directions. We define three vectors **i**, **j**, and **k** which point in the directions of the positive x, y, and z axes, respectively, and which have the property that the magnitude of each is just unity, without any physical units. Then the vector $\mathbf{A}_x$ can be represented as some multiple of the unit vector **i**, that is, **i** multiplied by a number with the same physical units as the magnitude of **A**. Suppose we denote this number by A_x. Then, just as the vector **A** is represented in terms of its component vectors, $\mathbf{A} = \mathbf{A}_x + \mathbf{A}_y$, these in turn can be represented in terms of the unit vectors **i** and **j**, as follows:

$$\mathbf{A}_x = A_x \mathbf{i} \qquad \mathbf{A}_y = A_y \mathbf{j} \tag{2-8}$$

The relationships among the various vector quantities are shown in Fig. 2-14.

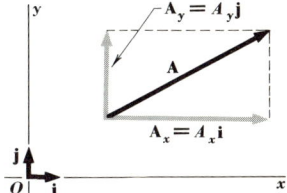

Fig. 2-14 Each component vector is a scalar multiple of the corresponding unit vector.

One is tempted to call A_x and A_y the magnitudes of the corresponding component vectors; strictly speaking, this is not quite correct. The reason is that A_x and A_y are algebraic quantities which may be either positive or negative. If the component vector $\mathbf{A}_x$ is in the negative x direction, then A_x is a negative number, since the unit vector **i** must be multiplied by a negative quantity to produce a vector whose direction is opposite that of **i**. The scalar quantities A_x and A_y are usually called the *components* of the vector **A**.

From Fig. 2-13 the following relations can be obtained:

$$|\mathbf{A}| = A = \sqrt{A_x^2 + A_y^2} \qquad A_x = A \cos \theta$$

$$\theta = \tan^{-1} \frac{A_y}{A_x} \qquad A_y = A \sin \theta \tag{2-9}$$

These have the same form as the relationships between rectangular and polar coordinates given by Eqs. (2-1) and (2-2).

Although Fig. 2-13 shows a vector for which θ is between zero and 90°, Eqs. (2-9) give the correct components for *any* angle provided it is measured counterclockwise from the positive x axis. For example, a vector whose direction is downward and to the right has a positive x component but a negative y component. Its direction is given by an angle between 270 and 360°, or between 0 and $-90°$. In either case, the sine of such an angle is negative and its cosine positive, so the signs of the components come out correctly.

The representation of a vector in terms of components in a rectangular coordinate system can be immediately extended to three dimensions. We can represent any vector in space as the sum of three vectors, one in each of the coordinate axis directions. Conversely, a vector is described completely by its components; if we know the three components of a vector in a given coordinate system, we know all there is to know about the vector.

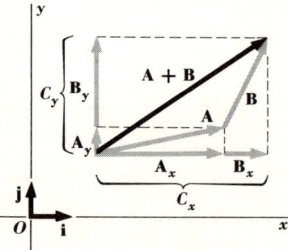

Fig. 2-15 The x component of the vector sum **C** of the two vectors **A** and **B** is equal to the sum of the x components of **A** and **B**, and similarly for the y components.

Given two vectors, **A**, with components A_x and A_y, and **B**, with components B_x and B_y, how can we *calculate* the components C_x and C_y of the vector sum $\mathbf{C} = \mathbf{A} + \mathbf{B}$? Reference to Fig. 2-15 shows that they are given simply by

$$C_x = A_x + B_x$$
$$C_y = A_y + B_y \tag{2-10}$$

More generally, if several vectors are to be added, such as $\mathbf{T} = \mathbf{A} + \mathbf{B} + \mathbf{C} + \cdots$, then it must be true that

$$T_x = A_x + B_x + C_x + \cdots$$
$$T_y = A_y + B_y + C_y + \cdots$$
(2-11)

Again we can immediately generalize to the case in which the vectors also have components in the direction of the z axis. In that case we have the additional equation

$$T_z = A_z + B_z + C_z + \cdots \qquad (2\text{-}11a)$$

Once the components T_x, T_y, T_z of the vector $\mathbf{T}$ have been found, the magnitude of this vector can be found:

$$T = \sqrt{T_x^2 + T_y^2 + T_z^2} \qquad (2\text{-}12)$$

The relationships between the components of the vectors in a sum and the components of the vector sum, given by Eqs. (2-11), were derived by geometrical considerations, but they can also be derived by purely algebraic means. For example, for the vector sum $\mathbf{T} = \mathbf{A} + \mathbf{B}$ of two vectors $\mathbf{A}$ and $\mathbf{B}$, not necessarily in the xy plane, we have

$$\mathbf{T} = (A_x \mathbf{i} + A_y \mathbf{j} + A_z \mathbf{k}) + (B_x \mathbf{i} + B_y \mathbf{j} + B_z \mathbf{k}) \qquad (2\text{-}13)$$

Because vector addition is commutative, we are at liberty to rearrange the terms in the sum as follows:

$$\mathbf{T} = (A_x + B_x)\mathbf{i} + (A_y + B_y)\mathbf{j} + (A_z + B_z)\mathbf{k} \qquad (2\text{-}14)$$

Now by definition the components of a vector are the numbers which multiply the unit vectors when the vector is represented in terms of unit vectors, so if the components of $\mathbf{T}$ are T_x, T_y, and T_z, then $\mathbf{T} = T_x \mathbf{i} + T_y \mathbf{j} + T_z \mathbf{k}$. Comparing this with Eq. (2-14), we obtain immediately the results given by Eqs. (2-10). The derivation can easily be extended to the sum of any number of vectors.

Example

Using the method of components, find the vector sum of the two vectors $\mathbf{A}$ and $\mathbf{B}$ if the vector $\mathbf{A}$ makes an angle of $45°$ with the x axis and has a length of 6 in. and vector $\mathbf{B}$ makes an angle of $135°$ with the x axis and has a length of 8 in.

Solution

Using Eqs. (2-9) we find the components:

$$A_x = (6 \text{ in.}) \cos 45° = \frac{6}{\sqrt{2}} \text{ in.} \qquad B_x = (8 \text{ in.}) \cos 135° = \frac{-8}{\sqrt{2}} \text{ in.}$$

$$A_y = (6 \text{ in.}) \sin 45° = \frac{6}{\sqrt{2}} \text{ in.} \qquad B_y = (8 \text{ in.}) \sin 135° = \frac{8}{\sqrt{2}} \text{ in.}$$

Denoting the vector sum by **C** = **A** + **B**, we find the components of **C**:

$$C_x = \frac{6}{\sqrt{2}} \text{ in.} - \frac{8}{\sqrt{2}} \text{ in.} = \frac{-2}{\sqrt{2}} \text{ in.}$$

$$C_y = \frac{6}{\sqrt{2}} \text{ in.} + \frac{8}{\sqrt{2}} \text{ in.} = \frac{14}{\sqrt{2}} \text{ in.}$$

The magnitude of **C** is

$$C = \sqrt{\left(\frac{-2}{\sqrt{2}} \text{ in.}\right)^2 + \left(\frac{14}{\sqrt{2}} \text{ in.}\right)^2} = 10 \text{ in.}$$

and its angle θ with the positive x axis is given by

$$\theta = \tan^{-1} \frac{14/\sqrt{2} \text{ in.}}{-2/\sqrt{2} \text{ in.}} = 97.9°$$

Note that the angle (97.9° + 180°) has the same tangent as 97.9° but this alternative angle is eliminated because it is in the fourth quadrant, corresponding to a positive x component and negative y component, just the opposite of the present situation.

Example

Given two vectors **A** and **B**, as follows:

$$\mathbf{A} = 3\mathbf{i} + \mathbf{j} - \mathbf{k} \qquad \mathbf{B} = \mathbf{i} + \mathbf{k}$$

Find **A** + **B**, **A** − **B**, and 2**A** + 3**B**.

Solution

$$\mathbf{A} + \mathbf{B} = (3\mathbf{i} + \mathbf{j} - \mathbf{k}) + (\mathbf{i} + \mathbf{k})$$
$$= 4\mathbf{i} + \mathbf{j}$$
$$\mathbf{A} - \mathbf{B} = (3\mathbf{i} + \mathbf{j} - \mathbf{k}) - (\mathbf{i} + \mathbf{k})$$
$$= 2\mathbf{i} + \mathbf{j} - 2\mathbf{k}$$
$$2\mathbf{A} + 3\mathbf{B} = 2(3\mathbf{i} + \mathbf{j} - \mathbf{k}) + 3(\mathbf{i} + \mathbf{k})$$
$$= 9\mathbf{i} + 2\mathbf{j} + \mathbf{k}$$

2-5 SCALAR PRODUCT

We have introduced a scheme for combining two displacement vectors to find the total displacement. This operation is called *vector addition* because it is analogous to the familiar addition of numbers. Is there an operation analogous to the *multiplication* of two numbers, i.e., a *product* of two vectors? If so, has it any physical significance?

We certainly cannot use the operation by which we multiply two ordinary numbers, for we do not know how to take the directions of the vectors into consideration or even whether the result should be a vector or a number. If there is to be an operation of vector multiplication, it must be a new operation which we must *define*. Because no definition has yet been given, we are at liberty to define the product of two vectors in any way we please. Our chief interest in vectors is their usefulness in representing physical quantities, so we should try to define a product of two vectors which is useful in physical problems.

It turns out that *two* kinds of products are useful. The first is called the *scalar product* of two vectors. The reason for this name is that the result is a scalar, or ordinary number, and not another vector. The formal definition of the scalar product of two vectors **A** and **B**, which we denote by **A · B**, is

$$\mathbf{A} \cdot \mathbf{B} = AB \cos \theta \tag{2-15}$$

where θ is the angle between **A** and **B** when they are drawn from a common point. Stated in words, the scalar product **A · B** of two vectors **A** and **B** is the magnitude of **A** multiplied by the magnitude of **B** multiplied by the cosine of the angle between the directions of the two vectors. Because of the notation **A · B**, the scalar product of **A** and **B** is also called the *dot product* of **A** and **B**, and in speech **A · B** is read "A dot B."

The geometrical significance of the scalar product is easily seen from Fig. 2-16. The quantity $B \cos \theta$ is the magnitude of the component of **B**

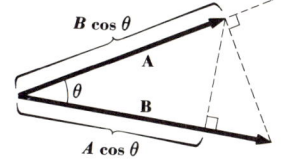

Fig. 2-16 Two vectors **A** and **B** and the angle used in computing the scalar product. The projection of **A** on the direction of **B** and the projection of **B** on the direction of **A** are shown.

in the direction of **A**. It can also be thought of as the *projection* of **B** onto the direction of **A**. Alternatively, we may think of $A \cos \theta$ as the projection of **A** onto the direction of **B**. Thus in general the scalar product of two

40 VECTOR LANGUAGE

vectors can be obtained by projecting either vector onto the direction of the other vector and multiplying the length of this projection by the length of the second vector.

The scalar product **A · B** may be either a positive or negative number. When the angle between **A** and **B** is obtuse, the cosine is negative. In this case the projection of **B** on **A** has sense opposite that of the direction of **A**, and **A · B** is a negative number.

One of the applications of the scalar product arises in calculating the *work* done by a force applied to an object which undergoes a displacement. Work is defined as the component of force in the direction of the displacement multiplied by the magnitude of the displacement. Thus if a force represented by a vector **F** is exerted on a point which undergoes a displacement **d**, the work W done by force during this displacement is $W = $ **F · d**. Other physical applications of the scalar product will be seen later.

Scalar multiplication is *commutative*; the result is independent of the order of factors. The value of the scalar product depends on the cosine of the angle between the two vectors, and this is the same no matter in which sense the angle is measured. Thus for any vectors **A** and **B**,

$$\mathbf{A} \cdot \mathbf{B} = \mathbf{B} \cdot \mathbf{A} \tag{2-16}$$

Are there similar properties? Multiplication of ordinary numbers obeys the *distributive* law:

$$a(b + c) = ab + ac$$

for any numbers a, b, and c. It is not obvious at the outset whether or not it is true that

$$\mathbf{A} \cdot (\mathbf{B} + \mathbf{C}) = \mathbf{A} \cdot \mathbf{B} + \mathbf{A} \cdot \mathbf{C} \tag{2-17}$$

This is a property which must be proved. Reference to Fig. 2-17 shows that the projection of the sum **B + C** on the vector **A** has a length equal to the sum of the projection of **B** on **A** and the projection of **C** on **A**. Since each term in Eq. (2-17) contains one of the projections multiplied by **A**, we see that Eq. (2-17) is in fact correct.

In Fig. 2-17 vectors **A**, **B**, and **C** are drawn in the same plane. This

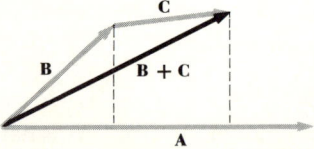

Fig. 2-17 The projection of **B + C** on **A** is the sum of the projections of **B** and **C** on **A**. This fact establishes the validity of Eq. (2-17).

restriction is not necessary. Visualize vector **C** as having a component perpendicular to the page, as well as components in the plane of the page; the sum of the projections of **B** and **C** on **A** is still equal to the projection of the sum **B** + **C**. Thus Eq. (2-17) is true in general for vectors in space, not just for vectors in a plane.

The distributive law of multiplication is useful in computing the scalar product of two vectors from their components. Suppose two vectors **A** and **B** are represented as follows:

$$\mathbf{A} = A_x\mathbf{i} + A_y\mathbf{j} + A_z\mathbf{k}$$
$$\mathbf{B} = B_x\mathbf{i} + B_y\mathbf{j} + B_z\mathbf{k} \tag{2-18}$$

The scalar product is then given by

$$\mathbf{A} \cdot \mathbf{B} = (A_x\mathbf{i} + A_y\mathbf{j} + A_z\mathbf{k}) \cdot (B_x\mathbf{i} + B_y\mathbf{j} + B_z\mathbf{k})$$

Now, using the distributive law, we can multiply out this product, obtaining nine terms as follows:

$$\begin{aligned}\mathbf{A} \cdot \mathbf{B} = &A_x\mathbf{i} \cdot B_x\mathbf{i} + A_x\mathbf{i} \cdot B_y\mathbf{j} + A_x\mathbf{i} \cdot B_z\mathbf{k} \\ + &A_y\mathbf{j} \cdot B_x\mathbf{i} + A_y\mathbf{j} \cdot B_y\mathbf{j} + A_y\mathbf{j} \cdot B_z\mathbf{k} \\ + &A_z\mathbf{k} \cdot B_x\mathbf{i} + A_z\mathbf{k} \cdot B_y\mathbf{j} + A_z\mathbf{k} \cdot B_z\mathbf{k}\end{aligned} \tag{2-19}$$

This rather complicated expression can be simplified greatly. Consider the first term, $A_x\mathbf{i} \cdot B_x\mathbf{i}$. This is a scalar product of two vectors both in the x direction, with magnitudes A_x and B_x, respectively. Because the two vectors are in the same direction, the cosine of the angle between them is unity, and so their scalar product is just the product of their magnitudes:

$$A_x\mathbf{i} \cdot B_x\mathbf{i} = A_xB_x$$

It is readily verified that this product has the correct sign even if A_x or B_x is negative or if both are.

In the second term, $A_x\mathbf{i} \cdot B_y\mathbf{j}$, we have two perpendicular vectors, one in the direction of the x axis, and the other in the direction of the y axis. The angle between these two is 90°; its cosine is zero. Thus

$$A_x\mathbf{i} \cdot B_y\mathbf{j} = 0$$

Proceeding through the nine terms in the same way, we find that all are zero except those in which both vectors are in the same direction. Therefore Eq. (2-19) reduces to

$$\mathbf{A} \cdot \mathbf{B} = A_xB_x + A_yB_y + A_zB_z \tag{2-20}$$

42 VECTOR LANGUAGE

This simple and elegant result gives a method of computing the scalar product of two vectors directly from the components of the vectors.

It should be emphasized once more that although the quantity $\mathbf{A} \cdot \mathbf{B}$ is computed from two vectors, it is not itself a vector quantity because it does not have a direction. It is a scalar quantity, a single number. Its only dependence on the direction of the vectors $\mathbf{A}$ and $\mathbf{B}$ is contained in the angle between the two vectors.

Example
Find the angle between the two vectors $\mathbf{A} = 4\mathbf{i} + 3\mathbf{j}$ and $\mathbf{B} = -2\mathbf{j}$.

Solution
From Eq. (2-20), we find

$$\mathbf{A} \cdot \mathbf{B} = -6$$

But we also know that $\mathbf{A} \cdot \mathbf{B} = AB \cos \theta$, by definition of the scalar product. Now, from Eq. (2-12),

$$A = \sqrt{4^2 + 3^2} = 5 \quad \text{and} \quad B = \sqrt{(-2)^2} = 2$$

so

$$\cos \theta = \frac{|\mathbf{A} \cdot \mathbf{B}|}{AB} = \frac{-6}{(5)(2)} = \frac{-6}{10}$$

$$\theta = 126.9°$$

2-6 VECTOR PRODUCT

We now define a different product of two vectors. Like the scalar product, it involves the product of the magnitudes of the two vectors and the angle between them. Unlike the scalar product, however, the vector product yields a result which is *another vector* instead of a scalar quantity. Furthermore, the dependence on the angle between the two vectors is not the same as in a scalar product.

We use the notation $\mathbf{A} \times \mathbf{B}$ to denote this new product, and we call it the *vector product* of $\mathbf{A}$ and $\mathbf{B}$. Because of this notation, this product is often called the *cross product*; in speech $\mathbf{A} \times \mathbf{B}$ is read "A cross B." The definition of the vector product $\mathbf{A} \times \mathbf{B}$ is as follows: $\mathbf{A} \times \mathbf{B}$ is a vector whose *magnitude* $|\mathbf{A} \times \mathbf{B}|$ is defined to be $AB \sin \theta$, where θ is the angle between the two vectors, measured from the direction of **A** toward that of **B**. The symbolic expression of this statement is

$$|\mathbf{A} \times \mathbf{B}| = AB \sin \theta \tag{2-21}$$

The *direction* of the vector **A** × **B** is perpendicular to the plane containing the vectors **A** and **B** when they are drawn from a common starting point. But there are two opposite directions perpendicular to the plane of **A** and **B**, and we must specify which is to be used. To complete the definition of the direction, we imagine a line perpendicular to the plane of **A** and **B** at the common point. We grasp this line with the right hand, with the fingers curled in the direction we would have to rotate **A** in order to move it into the direction of **B**, always rotating by the smallest possible angle, i.e., an angle less than 180°. We then define the direction of **A** × **B** as the direction in which the thumb of the right hand points. This rule, called the *right-hand rule*, is illustrated in Fig. 2-18.

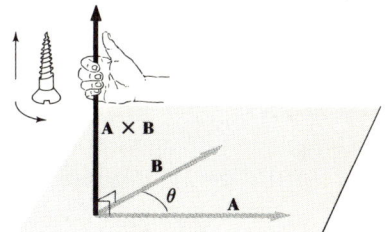

Fig. 2-18 Use of the right-hand rule to find direction of the vector product **A** × **B** of two vectors **A** and **B**.

An alternative and equivalent definition of the direction of **A** × **B** is as follows: Draw **A** and **B** with a common starting point, and imagine a screw with a right-hand thread having its axis along the line normal to the plane of **A** and **B** at this point. Then regard vector **A** as a wrench handle; when the handle is turned by the smaller of the two possible angles until it points in the direction of **B**, the direction in which the screw advances is the direction of **A** × **B**. Those who have difficulty remembering which way to turn a screw to tighten it may prefer the other definition. In either case, it is important to note that the definition specifies turning the first of the two vectors in the product into the second, never the reverse.

The magnitude of the vector product **A** × **B** has a simple geometrical significance, illustrated in Fig. 2-19. $B \sin \theta$ is the height of the parallelogram, and so $AB \sin \theta$ is its area.

An example of the usefulness of the vector product is the representation of *torques*. Torque is the tendency of a force to produce a rotation about

44 VECTOR LANGUAGE

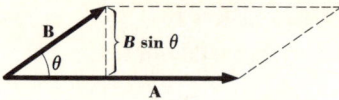

Fig. 2-19 The magnitude of **A** × **B** is the area of the parallelogram.

a specified point. Considering rotations of a body about point O in Fig. 2-20, suppose a force **F** is applied to a point P on the body, whose position with respect to O is described by the vector **r**. The torque τ is defined as $\tau = \mathbf{r} \times \mathbf{F}$. This is consistent with the more elementary definition of torque as the product of the force and the perpendicular *moment arm*. In Fig. 2-20

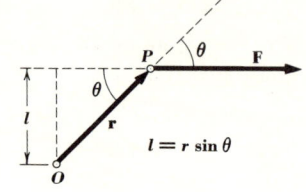

Fig. 2-20 Force **F** applied to point P on a body which rotates about O has a torque with respect to point O given by **r** × **F**. The moment arm of the force is the distance l.

the moment arm is $r \sin \theta$, so the magnitude $|\mathbf{r} \times \mathbf{F}| = Fr \sin \theta$ is just the force times the moment arm. In addition, the *direction* of **r** × **F** designates the direction of the axis about which the body tends to rotate. The concept of torque and its applications are discussed in more detail in Chap. 12.

In the above example, it is legitimate to shift the position of either **r** or **F** in order to apply the right-hand rule, even though in the physical situation they are both actually tied to definite points. The location of the point at which **F** is applied is described by the vector **r,** and this description is not altered by shifting the position of **F** to calculate the vector product. The vector **r** is called a *position vector;* it describes the position of the point P by giving the displacement from the origin O to P, and it is called the position vector of P *with respect to O.*

Example
Find the vector product **A** × **B** for the two vectors lying in the *xy* plane in Fig. 2-21.

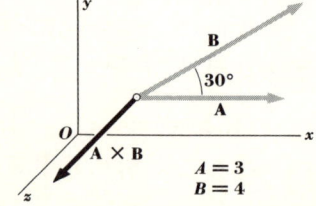

Fig. 2-21 Vector product of two vectors lying in the *xy* plane.

Solution

The magnitude of **A** × **B** is

$$|\mathbf{A} \times \mathbf{B}| = AB \sin 30° = (4)(3)(\tfrac{1}{2}) = 6$$

The direction is perpendicular to the *xy* plane and therefore parallel to the *z* axis. The right-hand rule shows that **A** × **B** is in the positive *z* direction, designated by the unit vector **k**; thus we find

$$\mathbf{A} \times \mathbf{B} = 6\mathbf{k}$$

We now examine some of the properties of the vector product. A natural question to ask is whether the vector product is *commutative*. Is it true that **A** × **B** = **B** × **A**? Referring again to Fig. 2-18, we see that the direction of **B** × **A** is obtained by curling the fingers of the right hand around the line perpendicular to these vectors in a direction we would rotate **B** to bring it into the direction of **A**; in this case, the thumb of the right hand points *downward*. The direction of **B** × **A** is not the same as the direction of **A** × **B**. In fact, it is exactly the opposite direction. The magnitudes of the two quantities are the same, so

$$\mathbf{B} \times \mathbf{A} = -(\mathbf{A} \times \mathbf{B}) \tag{2-22}$$

The vector product of two vectors is *not commutative*.

In view of this result, we may question whether the distributive law of multiplication holds for the vector product. Is it true, or is it not, that

$$\mathbf{A} \times (\mathbf{B} + \mathbf{C}) = (\mathbf{A} \times \mathbf{B}) + (\mathbf{A} \times \mathbf{C}) \tag{2-23}$$

This question may be investigated geometrically by a procedure similar to that which was used to establish the distributive law for the scalar product. The analysis is somewhat more involved than for the scalar product, and the details need not concern us. The result is that the distributive law given by Eq. (2-23) *is* obeyed, provided that the order of factors is preserved as in Eq. (2-23).

If the components of the vectors **A** and **B** are known, the components of the vector product **C** = **A** × **B** can be calculated; to do this we proceed as with the scalar product:

$$\mathbf{A} \times \mathbf{B} = (A_x\mathbf{i} + A_y\mathbf{j} + A_z\mathbf{k}) \times (B_x\mathbf{i} + B_y\mathbf{j} + B_z\mathbf{k}) \tag{2-24}$$

Using the distributive law just discussed, we expand this product:

$$\begin{aligned}\mathbf{A} \times \mathbf{B} = &\; A_x\mathbf{i} \times B_x\mathbf{i} + A_x\mathbf{i} \times B_y\mathbf{j} + A_x\mathbf{i} \times B_z\mathbf{k} \\ &+ A_y\mathbf{j} \times B_x\mathbf{i} + A_y\mathbf{j} \times B_y\mathbf{j} + A_y\mathbf{j} \times B_z\mathbf{k} \\ &+ A_z\mathbf{k} \times B_x\mathbf{i} + A_z\mathbf{k} \times B_y\mathbf{j} + A_z\mathbf{k} \times B_z\mathbf{k}\end{aligned} \tag{2-25}$$

46 VECTOR LANGUAGE

We now consider the terms one at a time. The first term is $A_x\mathbf{i} \times B_x\mathbf{i}$. This is a vector product of two vectors which point in the same direction. The angle between these two vectors is zero, the sine of this angle is zero, and therefore this term vanishes. Similarly, the terms $A_y\mathbf{j} \times B_y\mathbf{j}$ and $A_z\mathbf{k} \times B_z\mathbf{k}$ vanish.

The second term is $A_x\mathbf{i} \times B_y\mathbf{j}$; the angle between the two vectors in this product is 90°, and sin 90° = 1. Therefore the magnitude of this term is $A_x B_y$. What is its direction? Referring to the coordinate system shown in Fig. 2-22a, we notice that if we grasp the z axis with the fingers curled

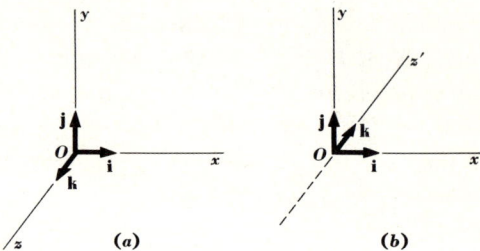

Fig. 2-22 (a) System $Oxyz$ is a right-handed coordinate system. (b) System $Oxyz'$ is left-handed.

in the direction of rotation from $\mathbf{i}$ to $\mathbf{j}$, the thumb points in the position z direction, in the direction of the unit vector $\mathbf{k}$. Thus the second term is

$$A_x\mathbf{i} \times B_y\mathbf{j} = A_x B_y \mathbf{k}$$

There is another term in Eq. (2-25) containing $\mathbf{i}$ and $\mathbf{j}$, namely $A_y\mathbf{j} \times B_x\mathbf{i}$. A similar consideration shows that the magnitude of this term is $A_y B_x$, and its direction is opposite that of the unit vector $\mathbf{k}$. That is,

$$A_y\mathbf{j} \times B_x\mathbf{i} = -A_y B_x \mathbf{k}$$

The sum of the two terms just considered can be written

$$(A_x B_y - A_y B_x)\mathbf{k}$$

The remaining terms in Eq. (2-25) may be discussed in a similar way. The final result is

$$\mathbf{C} = \mathbf{A} \times \mathbf{B} = (A_y B_z - A_z B_y)\mathbf{i} + (A_z B_x - A_x B_z)\mathbf{j} + (A_x B_y - A_y B_x)\mathbf{k} \quad (2\text{-}26)$$

The components of vector $\mathbf{C}$ are thus expressed in terms of the components of $\mathbf{A}$ and $\mathbf{B}$ as

$$\begin{aligned} C_x &= A_y B_z - A_z B_y \\ C_y &= A_z B_x - A_x B_z \\ C_z &= A_x B_y - A_y B_x \end{aligned} \quad (2\text{-}27)$$

For the student learning to work with vector products, it is useful to write out a multiplication table for vector products of the various pairs of unit vectors, as suggested in Prob. 2-15. The vector product of two vectors can also be calculated by means of a determinant:

$$\mathbf{A} \times \mathbf{B} = \begin{vmatrix} \mathbf{i} & \mathbf{j} & \mathbf{k} \\ A_x & A_y & A_z \\ B_x & B_y & B_z \end{vmatrix} \tag{2-28}$$

The student should verify that this formula gives the same result as Eq. (2-26) or (2-27).

Example

A force of magnitude 5 newtons (N), parallel to the positive z axis, is applied at a point with coordinates (0, 2 m, 2 m). What is the torque of this force, with respect to the origin?

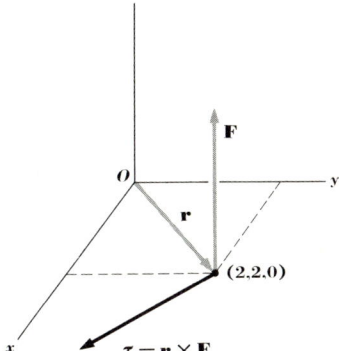

Fig. 2-23 The torque associated with a force is the vector product of the position vector of the point of application of the force and the force.

Solution

The situation is shown in Fig. 2-23. The position vector $\mathbf{r}$ is given by

$$\mathbf{r} = (2 \text{ m})\mathbf{i} + (2 \text{ m})\mathbf{j}$$

and the force vector $\mathbf{F}$ by

$$\mathbf{F} = (5 \text{ N})\mathbf{k}$$

The torque, given by $\boldsymbol{\tau} = \mathbf{r} \times \mathbf{F}$, is

$$\begin{aligned} \boldsymbol{\tau} &= [(2 \text{ m})\mathbf{i} + (2 \text{ m})\mathbf{j}] \times (5 \text{ N})\mathbf{k} \\ &= (10 \text{ N-m})\mathbf{i} \times \mathbf{k} + (10 \text{ N-m})\mathbf{j} \times \mathbf{k} \\ &= (-10 \text{ N-m})\mathbf{j} + (10 \text{ N-m})\mathbf{i} \end{aligned}$$

where 10 N-m is read "10 newton-meters." The torque vector lies in the xy plane, makes 45° angles with the $+x$ and $-y$ axes, and is therefore perpendicular to both **r** and **F**, as required by the definition of the vector product.

We now observe a rather peculiar feature of the vector product. What would have happened if in Fig. 2-22 we had chosen the z axis to point in the opposite direction, as indicated by the dotted line labeled z' in Fig. 2-22b? Then the direction of the unit vector **k** would be reversed. Thus in the first coordinate system **i** $\times$ **j** = **k**, while in the second **i** $\times$ **j** = $-$**k**. Thus Eqs. (2-26) and (2-27) give us the correct components for the vector product of two vectors for certain particular kinds of coordinate systems but not for others.

There are two ways to reconcile this difficulty. One is simply to recognize that there are two kinds of coordinate systems. If we use the right-hand rule throughout, then in systems of the type $Oxyz$ in Fig. 2-22a we have **i** $\times$ **j** = **k**, **j** $\times$ **k** = **i**, and **k** $\times$ **i** = **j**; in systems of the type $Oxyz'$ in Fig. 2-22b, we have **i** $\times$ **j** = $-$**k**, **j** $\times$ **k** = $-$**i**, and **k** $\times$ **i** = $-$**j**. In this case, Eqs. (2-26) and (2-27) are valid in systems of the first type, usually called *right-handed* coordinate systems, but not in the other type, called *left-handed* systems.

Another possibility would be to modify the basic definition of the vector product so that we use the right-hand rule in right-handed systems, but the *left-hand rule* in left-handed systems. For our purposes, this whole problem can be avoided by simply deciding to use *only* right-handed coordinate systems; this choice is adhered to uniformly in the remainder of this book. Thus in all calculations of vector products which use components of vectors, care must be taken to use only right-handed coordinate systems, in which, when the right-hand rule is used, **i** $\times$ **j** = **k**.

2-7 REPRESENTATION OF PHYSICAL QUANTITIES BY VECTORS

Our discussion of vectors has been concerned mostly with displacement vectors. These have a very direct geometrical significance, and combining displacements leads naturally to the general definition of addition of vectors. Many other physical quantities having both magnitude and direction can be conveniently represented by vectors. A few examples of physical quantities which have direction as well as magnitude are velocity, force, and magnetic field.

Any of these quantities can be represented by a directed line segment having the direction of the physical quantity described and a length proportional to the magnitude of the physical quantity. For example, to represent a force, we can use a scale diagram in which each inch on the scale represents 10 lb of force. A force of 30 lb in a given direction is represented on such a scale diagram by a directed line segment whose direction is that of the force and whose length is 3 in.

The fact that a physical quantity can be represented by a vector does not, however, prove that it obeys the laws of vector addition. For example, when two forces are applied simultaneously to the same point on a body, it may be surmised that the effect is the same as when a single force equal to the vector sum of these two forces is applied. This turns out to be correct, but to establish its correctness we must experiment with various combinations of forces and compare the results. In general, the question whether a physical quantity with magnitude and direction obeys the laws of vector addition must be tested experimentally. In the remainder of this book we refer to quantities which can be represented by vectors and which obey the law of vector addition for displacements as *vector quantities*.

Problems

2-1 The coordinates of a point in the *xy* plane are (3 cm, 4 cm).
   a What is its distance from the origin?
   b What are the polar coordinates of the point?

2-2 The polar coordinates of a point in a plane are (10 m, 45°). Find the *x* and *y* coordinates of the point. What is its distance from the origin?

2-3 Find the rectangular coordinates of a point whose polar coordinates are:
   a 10 m, 225°
   b 5 m, 360°
   c 6 ft, 120°

2-4 Consider a coordinate system in which a point *P* is determined by the distances (u,v) from two fixed points, as shown in Fig. P2-4. Do these quantities specify the position uniquely? Explain. Derive expressions for the rectangular coordinates *x* and *y* in terms of *u* and *v*.

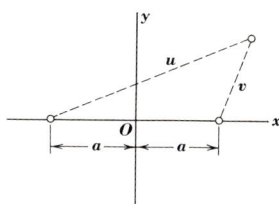

Fig. P2-4

2-5 Vector **A** is parallel to the $+x$ axis and is 4 in. long, and vector **B** makes an angle of 120° counterclockwise from the $+x$ axis and is 4 in. long.
   a Find the vector sum **A** + **B** by drawing a full-scale vector diagram. (Find the magnitude and direction of the vector sum by measuring the diagram.)
   b Calculate the components A_x, A_y, B_x, and B_y, and calculate the components of the vector sum. Compare with the results of part *a*.

2-6 For two vectors **A** and **B** whose directions differ by an angle θ (when they are drawn from a common point), use the product $(\mathbf{A} + \mathbf{B}) \cdot (\mathbf{A} + \mathbf{B})$ to show that the magnitude of the vector sum **A** + **B** is

$$\sqrt{A^2 + B^2 + 2AB \cos \theta}$$

By a similar method, show that the magnitude of **A** − **B** is

$$\sqrt{A^2 + B^2 - 2AB \cos \theta}$$

Is this always equal to the magnitude of **B** − **A**? Explain.

2-7 Find the components of the vector representing the displacement from point $(3,4,-5)$ to point $(1,-3,0)$. What is the *distance* between these points?

2-8 For two vectors **A** = 3**i** − 4**j** and **B** = −**i** −**j**, find the magnitude and direction of **A**, **B**, **A** + **B**, **A** − **B**, and **B** − **A**.

2-9 An xy coordinate system is located in the plane of a clock face, with the origin at the center, the y axis at 12, and the x axis at 3; the minute hand is 3 in. long. Obtain expressions for the displacement of the end of the minute hand:
   a From 12:00 to 12:30
   b From 4:00 to 4:15
   c From 9:00 to 10:00

2-10 By use of the scalar product, find the angle between the vectors **i** + **j** + **k** and **i** − **j** − **k**.

2-11 If a cube whose edges have length $2a$ has its center at O and its sides parallel to the coordinate planes of an xyz coordinate system, what are the position vectors of the corners? How many *different* angles are formed by these vectors? What are the angles?

2-12 For a vector **A** = 3**i** + 2**j** − **k**, find the magnitude of the vector product **A** × **A**.

2-13 Find a *unit* vector perpendicular to the two vectors **i** − **k** and **i** + **j**.

2-14 Construct a multiplication table for the scalar products of all possible pairs of unit vectors **(i,j,k)** in a rectangular coordinate system.

2-15 Construct a multiplication table for the vector products of all possible pairs of unit vectors **(i,j,k)** in a rectangular coordinate system. Is this table valid for both right-handed and left-handed systems? Explain.

51 PROBLEMS

2-16 In the example following Eq. (2-28), show that the torque vector $\boldsymbol{\tau}$ is perpendicular to both $\mathbf{r}$ and $\mathbf{F}$ by computing the scalar products $\boldsymbol{\tau} \cdot \mathbf{r}$ and $\boldsymbol{\tau} \cdot \mathbf{F}$ and showing that they are both zero.

2-17 For the three vectors

$$\mathbf{A} = 3\mathbf{i} + 2\mathbf{j} + 4\mathbf{k}$$
$$\mathbf{B} = -2\mathbf{i} + \mathbf{j} - 2\mathbf{k}$$
$$\mathbf{C} = \mathbf{i} - 2\mathbf{j} + 3\mathbf{k}$$

compute the quantities $(\mathbf{A} + \mathbf{B}) \cdot \mathbf{C}$ and $(\mathbf{A} \cdot \mathbf{C}) + (\mathbf{B} \cdot \mathbf{C})$, and show that the distributive law of scalar multiplication is obeyed.

2-18 For the vectors $\mathbf{A}$, $\mathbf{B}$, and $\mathbf{C}$ given in Prob. 2-17, compute the quantities $(\mathbf{A} + \mathbf{B}) \times \mathbf{C}$ and $(\mathbf{A} \times \mathbf{C}) + (\mathbf{B} \times \mathbf{C})$, and show that the distributive law of vector multiplication is obeyed.

2-19 Show that Eq. (2-28) leads to the same result for the vector product as Eq. (2-26).

2-20 Find the area of the parallelogram formed by the vectors $\mathbf{A}$ and $\mathbf{B}$ in Prob. 2-17 when they are drawn from a common point. Can you think of a way to find this area without using vector language?

2-21 Are the two expressions $\mathbf{i} \times (\mathbf{i} \times \mathbf{j})$ and $(\mathbf{i} \times \mathbf{i}) \times \mathbf{j}$ equal? Generalize this result.

2-22 Under what circumstances is $(\mathbf{A} \times \mathbf{B}) \times \mathbf{C}$ equal to $\mathbf{A} \times (\mathbf{B} \times \mathbf{C})$?

2-23 For any three vectors $\mathbf{A}$, $\mathbf{B}$, and $\mathbf{C}$, prove that $(\mathbf{A} \times \mathbf{B}) \cdot \mathbf{C} = \mathbf{A} \cdot (\mathbf{B} \times \mathbf{C})$.

2-24 For any three vectors $\mathbf{A}$, $\mathbf{B}$, and $\mathbf{C}$, prove that $\mathbf{A} \times (\mathbf{B} \times \mathbf{C}) = \mathbf{B}(\mathbf{A} \cdot \mathbf{C}) - \mathbf{C}(\mathbf{A} \cdot \mathbf{B})$.

Describing Motion | 3

Many branches of physics are concerned with motion, particularly the relation between the motion of a particle or a body and the physical quantities, such as forces, which influence and change this motion. In developing these relations, we need efficient methods for *describing* motion quantitatively. The simplest possible situation is a point moving along a straight line. We define velocity and acceleration for such a situation and consider some examples. The concepts of velocity and acceleration are then generalized to describe the motion of a point in a plane or in space. Finally, we discuss how the velocity and acceleration depend on the motion of the observer; this discussion leads to the concepts of relative velocity and relative acceleration.

3-1 VELOCITY

We consider first the motion of a point along a straight line. The position of the point P is described by its distance x from a fixed point O, called the *origin of coordinates*. The arrow on the righthand end of the line in Fig. 3-1 indicates that x is positive when P is to the right of O and negative when to the left.

We now proceed to measure x at various times. We can determine the value of x with a meterstick and measure the time with a stopwatch, or perhaps take a multiple-flash photograph, in which a light is flashed at regular intervals to show the position of the object at a succession of times. Suppose we observe that at time t_1 the point is at position x_1 and at t_2 it is at x_2. The change Δx in the value of x is given by $\Delta x = x_2 - x_1$, and the corresponding time interval Δt during which this change took place is given by $\Delta t = t_2 - t_1$.

Fig. 3-1 Coordinate system for the motion of a point P along a straight line. The positive direction for x is shown.

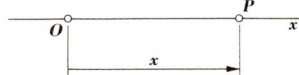

The *average velocity* during the time interval Δt, denoted by v_{av}, is defined as the ratio of the displacement Δx to the time interval during which the displacement occurred:

$$v_{av} = \frac{\Delta x}{\Delta t} \tag{3-1}$$

The subscript "av" indicates that it is an *average* value.

Example
A runner on a slow track runs a hundred-yard dash in 12 s. His average velocity is

$$v_{av} = \frac{100 \text{ yd}}{12 \text{ s}} = 8\tfrac{1}{3} \text{ yd/s}$$

The above result can be expressed in various systems of units. To obtain the average velocity in feet per second, we convert the distance into feet:

$$v_{av} = \frac{300 \text{ ft}}{12 \text{ s}} = 25 \text{ ft/s}$$

We can also convert feet into miles (1 mi = 5,280 ft) and the seconds into hours (1 h = 3,600 s)

$$v_{av} = \frac{300/5{,}280 \text{ mi}}{12/3{,}600 \text{ h}} = 17.0 \text{ mi/h}$$

A useful conversion which is easily derived is

$$60 \text{ mi/h} = 88 \text{ ft/s}$$

In the above example we have not specified whether or not the runner's speed was constant during the 12-s interval. He may have started from rest at the beginning, reached a maximum velocity somewhere during the interval, and then slowed down somewhat near the end. But these considerations do not enter the calculations of the *average* velocity; it is determined entirely by the *total* distance and the *total* time.

54 DESCRIBING MOTION

Implicit in these remarks is the assumption that we can somehow define a velocity at *each instant* during the motion; otherwise the phrases "maximum velocity during the interval" and "velocity at the end" would have no meaning. To define instantaneous velocity at a point at one instant of time, we compute the *average* velocity of a point in a succession of smaller and smaller time intervals, with correspondingly smaller and smaller displacements, always starting at the same point. In general these average velocities may all be slightly different; if they approach a definite value as the time interval shrinks down to zero, we call this value the *instantaneous velocity* of the point. This quantity is denoted simply by v, without a subscript, because it is no longer an average but an instantaneous value.

Here is an example: Suppose the point is observed at a succession of times, with the results shown in the first two columns of Table 3-1. We use these figures and Eq. (3-1) to compute the average velocity in a number of intervals, using $x = 14.00$ ft and $t = 2.00$ s for the beginning of the interval in each case. The calculations are summarized in the remaining columns. It can be seen that as the intervals become smaller and smaller, the velocity comes closer and closer to 64 ft/s; we thus conclude that the instantaneous velocity at $t = 2$ s is $v = 64$ ft/s.

The mathematical terminology for the above process is "finding the *limit* of the ratio of Δx to Δt as Δt approaches zero." The result of this limit is called the *derivative* of x with respect to t:

$$v = \lim_{\Delta t \to 0} v_{av} = \lim_{\Delta t \to 0} \frac{\Delta x}{\Delta t} = \frac{dx}{dt} \tag{3-2}$$

If x is a known function of t, an expression for v can be derived.

Example
Suppose the distance is known to be related to time by the equation

Table 3-1

x, ft	t, s	Δx, ft	Δt, s	$v_{av} = \dfrac{\Delta x}{\Delta t}$, ft/s
14.00	2.00			
14.64	2.01	0.64	0.01	64.0
17.23	2.05	3.23	0.05	64.6
20.56	2.10	6.56	0.10	65.6
50.00	2.50	36.0	0.50	72.0
94.00	3.00	80.0	1.00	80.0

55 VELOCITY [3-1]

$$x = \tfrac{1}{2}gt^2 \tag{3-3}$$

where g is a constant. What is the instantaneous velocity?

Solution

At time t_1 the position x_1 is given by

$$x_1 = \tfrac{1}{2}gt_1^2$$

The position $x_2 = x_1 + \Delta x$ at a later time $t_2 = t_1 + \Delta t$ is obtained by inserting the quantity $t_1 + \Delta t$ in place of t_1 in Eq. (3-3). The result is

$$x_2 = x_1 + \Delta x = \tfrac{1}{2}g(t_1 + \Delta t)^2$$
$$= \tfrac{1}{2}gt_1^2 + gt_1 \Delta t + \tfrac{1}{2}g(\Delta t)^2$$

To find Δx, we subtract the previous equation from this, obtaining

$$x = gt_1 \Delta t + \tfrac{1}{2}g(\Delta t)^2$$

The average velocity during the time Δt is

$$v_{\text{av}} = \frac{\Delta x}{\Delta t} = gt_1 + \tfrac{1}{2}g \Delta t$$

The instantaneous velocity at time t_1 is given by the limit of this expression as Δt approaches 0. In this limit the second term vanishes, and the result is

$$v = \lim_{\Delta t \to 0} \frac{\Delta x}{\Delta t} = gt_1$$

More generally, if the position is given as a function of time by Eq. (3-3), the instantaneous velocity at time t is given by

$$v = gt \tag{3-4}$$

This result can also be obtained directly from Eq. (3-3) by taking the derivative of x with respect to t. If c and n are constants, the derivative of the function $x = ct^n$ is cnt^{n-1}. In this case, the derivative of $\tfrac{1}{2}gt^2$ is $\tfrac{1}{2}g(2t) = gt$, in agreement with Eq. (3-4). Although there is nothing wrong with our first method of obtaining Eq. (3-4) by using the definition of the derivative given by Eq. (3-2), it is usually easier to use formulas for derivatives, once they have been derived and learned.

Example

The position of a point on a line is given by the equation

$$x = bt^2 + 2ct^3 - 6dt^4 \qquad (3\text{-}5)$$

where b, c, and d are constants, each with the appropriate units so x is measured in meters and t in seconds. What is the instantaneous velocity at time t?

Solution

A general theorem about derivatives states that the derivative of a sum of functions is equal to the sum of the derivatives of the functions, so we can differentiate the terms separately and add the results. We obtain

$$v = \frac{dx}{dt} = b(2t) + 2c(3t^2) - 6d(4t^3)$$

$$= 2bt + 6ct^2 - 24dt^3 \qquad (3\text{-}6)$$

It is often useful to represent the variations of the position and velocity of a point with time graphically by plotting x as a function of t, as in Fig. 3-2a. In this figure, the average velocity between time t_1 and time t_2 is the slope of the line which intersects the curve at the points (x_1,t_1) and (x_2,t_2). As the time interval Δt shrinks to zero, this line becomes tangent to the curve at t_1, so the instantaneous velocity is just the slope of the curve. The height of the v-versus-t curve at any value of t is equal to the slope of the x-versus-t curve at the same value of t, as shown in Fig. 3-2b.

3-2 ACCELERATION

If the instantaneous velocity of a point changes with time, it is of interest to find the *rate of change* of the velocity. The time rate of change of velocity is called *acceleration*. This quantity is of particular importance because, as we shall see in Chap. 4, the acceleration of a body is the quantity most directly related to the forces acting on it.

Let the instantaneous velocities of a point at time t_1 and t_2 be v_1 and v_2, respectively; then in exact analogy to our definition of average velocity, we define the *average acceleration* during this time interval as

$$a_{\text{av}} = \frac{v_2 - v_1}{t_2 - t_1} = \frac{\Delta v}{\Delta t} \qquad (3\text{-}7)$$

57 ACCELERATION [3-2]

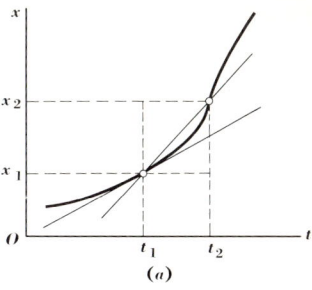

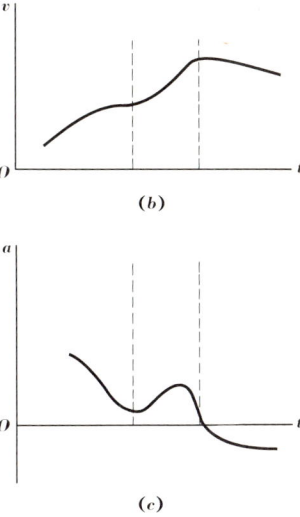

Fig. 3-2 (*a*) Position of a point as a function of time. The average velocity between t_1 and t_2 is represented by the slope of the line joining the two corresponding points on the curve; the instantaneous velocity is represented by the slope of the line tangent to the curve at point (x_1,t_1). (*b*) Velocity as a function of time. The velocity for any t is given by the slope of the *x*-versus-*t* curve at the corresponding *t*. (*c*) Acceleration as a function of time. The acceleration for any t is given by the slope of the *v*-versus-*t* curve at the corresponding *t*. When this slope is negative, the velocity is decreasing and the acceleration negative.

where Δv is the change in velocity during the time interval Δt. We define the *instantaneous acceleration* as the limit of the average acceleration as the time interval Δt approaches zero:

$$a = \lim_{\Delta t \to 0} a_{\mathrm{av}} = \lim_{\Delta t \to 0} \frac{\Delta v}{\Delta t} = \frac{dv}{dt} \tag{3-8}$$

The instantaneous acceleration is the derivative of the instantaneous velocity with respect to time.

The velocity is itself the derivative of the position x with respect to time. In the usual calculus notation, the symbol d/dt represents the operation of taking the derivative of a function with respect to t, so we may write

$$a = \frac{d}{dt}\left(\frac{dx}{dt}\right)$$

58 DESCRIBING MOTION

The derivative of a derivative is called the *second derivative*, usually denoted as d^2x/dt^2. Thus we may express the instantaneous acceleration in any of the following forms:

$$a = \frac{dv}{dt} = \frac{d}{dt}\left(\frac{dx}{dt}\right) = \frac{d^2x}{dt^2} \tag{3-9}$$

Example
If the position is given by Eq. (3-3), what is the instantaneous acceleration?

Solution
From Eq. (3-4), the velocity at time t_1 is

$$v_1 = gt_1$$

and the velocity v_2 at a slightly later time $t_2 = t_1 + \Delta t$ is

$$v_2 = v_1 + \Delta v = g(t_1 + \Delta t)$$

The change in velocity Δv during Δt is just $g\,\Delta t$, and the average acceleration for any interval is

$$a_{\text{av}} = \frac{\Delta v}{\Delta t} = g$$

In this particular case the instantaneous acceleration and the average acceleration for any interval Δt are equal. The instantaneous acceleration is constant: $a = g$. In many types of motion a is *not* constant; this example is a particularly simple one.

Example
If the position is given by Eq. (3-5), find the instantaneous acceleration.

Solution
The instantaneous velocity is given in this case by Eq. (3-6), and the instantaneous acceleration is just the derivative of this equation. Thus

$$a = \frac{dv}{dt} = 2b(1) + 6c(2t) - 24d(3t^2)$$

$$= 2b + 12ct - 72dt^2$$

We note that in this case the instantaneous acceleration is *not* constant but is a function of time.

Just as the instantaneous velocity of a particle at any time is the slope of the graph of x versus t at that time, the instantaneous acceleration is the slope of the graph of v as a function of t, as shown in Fig. 3-2c.

Another kind of information can be obtained from the graph of v versus t. Consider the small time interval Δt in Fig. 3-3; the distance traveled

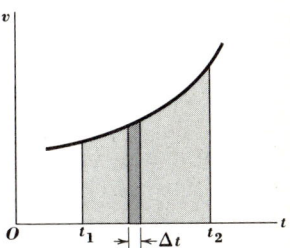

Fig. 3-3 Velocity of a point as a function of time. The area of the strip of width Δt represents approximately the distance traveled during that interval, and the area under the curve between two times t_1 and t_2 represents the total displacement during that interval.

during this interval is $v \, \Delta t$, which is represented graphically by the *area* of the strip of width Δt and height v on the graph. Similarly, the *total* distance traveled between two times t_1 and t_2 is represented graphically by the total area bounded by the vertical lines at t_1 and t_2 and the v-versus-t curve.

3-3 MOTION WITH CONSTANT ACCELERATION

The acceleration of a point is a particularly important concept because it is directly related to the *forces* producing the acceleration. In particular, when an object is acted on by a *constant* force, its acceleration is constant. Let us investigate in more detail the motions which are possible when the acceleration of an object is constant.

We have just seen that if the velocity is proportional to time, $v = gt$, the acceleration a is constant and is in fact equal to the constant g. But if $v = at$, the velocity must be zero when $t = 0$. One can easily imagine types of motion for which this is not true. Suppose that at time $t = 0$, when we start measuring time, the point has some initial velocity, denoted by v_0, and that from then on it undergoes constant acceleration a. Then the velocity changes at a uniform rate and at any later time is represented by

$$v = v_0 + at \qquad (3\text{-}10)$$

We can easily check Eq. (3-10), to see that it does represent motion with constant acceleration, by taking its derivative with respect to time. Since v_0 is a constant, its derivative is zero and $dv/dt = a$, a constant.

If the velocity of a point varies with time according to Eq. (3-10), the position x can also be found as a function of time. One way to do this is by use of the graphical procedure mentioned in Sec. 3-2. Figure 3-4 shows

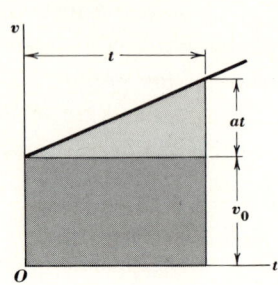

Fig. 3-4 Velocity versus time for a point moving with constant acceleration; the velocity changes uniformly with time, and the slope is constant. The total displacement between time $t = 0$ and a later time t is represented by the total area shown; the area of the rectangular portion is $v_0 t$, and that of the triangular portion $\frac{1}{2}at^2$.

a graph of Eq. (3-10). The displacement of the point between time $t = 0$ and a later time t is represented by the area shown in the figure; by breaking this area into a rectangle and a triangle, we easily find it to be $v_0 t + \frac{1}{2}at^2$. If the initial position of the point (at $t = 0$) is denoted by x_0, the position at any time t is

$$x = x_0 + v_0 t + \tfrac{1}{2}at^2 \tag{3-11}$$

One can take the second derivative of this expression with respect to t to show that it does represent motion with constant acceleration.

Sometimes a problem arises in which we want to know the velocity of an object at a particular location but do not need to know *when* it is at this location. We can eliminate t from Eqs. (3-10) and (3-11) to obtain a relationship between v and x which does not involve t. This is most easily done by solving Eq. (3-10) for t and substituting the expression for t into Eq. (3-11). The details of this calculation are left as an exercise; the result is

$$v^2 - v_0^2 = 2a(x - x_0) \tag{3-12}$$

In summary, when a point moves along a straight line with constant acceleration a, with initial position x_0 and initial velocity v_0 at time $t = 0$, the position x and velocity v at any time t are related by Eqs. (3-10) to (3-12). It should not be necessary to memorize these equations; it is better to understand their derivation sufficiently well to be able to derive them whenever needed.

Example

A car accelerates from 0 to 60 mi/h in 8 s. What is the average acceleration, and what distance does the car travel during this time if the acceleration is constant?

Solution

The velocity at the end of the time interval is $60 \text{ mi/h} = 88 \text{ ft/s}$. To acquire this velocity in 8 s the velocity must have changed, on the average, 11 ft/s each second; the average acceleration is therefore 11 ft/s^2. If we place the car at the origin ($x = 0$) at time $t = 0$ and start it with no initial velocity, then x_0 and v_0 are both zero. In this case the distance covered is given by Eq. (3-11):

$$x = \tfrac{1}{2}at^2 = \tfrac{1}{2}(11 \text{ ft/s}^2)(8 \text{ s})^2 = 352 \text{ ft}$$

Alternatively, we may use Eq. (3-12):

$$x = \frac{v^2}{2a} = \frac{(88 \text{ ft/s})^2}{(2)(11 \text{ ft/s}^2)} = 352 \text{ ft}$$

Example

Objects moving vertically under the action of the earth's gravity move with constant acceleration if the effect of air resistance can be neglected. This acceleration, the same for all objects, is approximately 32 ft/s^2 downward. Suppose a ball is thrown vertically upward with an initial velocity of 64 ft/s. Find the time it takes to reach maximum height, the maximum height reached, the total time elapsed until it returns to the ground, and its velocity when it hits the ground.

Solution

We consider first a simplified model of a physical situation. Rather than trying to describe precisely the motion of a ball, which may be spinning as well as moving up and may deviate somewhat from its straight path as a result of air currents and so forth, we idealize this situation by representing the ball as a point that moves precisely in a straight line with a constant acceleration of magnitude 32 ft/s^2.

We take the origin of coordinates to be at the ground and the positive direction to be upward. Then $a = -32 \text{ ft/s}^2$, $v_0 = 64 \text{ ft/s}$, and $x_0 = 0$. The acceleration is negative because we have chosen the positive direction as upward, so the acceleration corresponds to greater and greater negative velocities with increasing time. According to Eq. (3-10), the velocity at any time t is given by

$$v = v_0 + at = 64 \text{ ft/s} - (32 \text{ ft/s}^2)t$$

At maximum height the ball stops and starts back down, so at this instant

62 DESCRIBING MOTION

$v = 0$. We see that this occurs at $t = 2$ s. (The acceleration at this instant is *not* zero!) At this time the height, according to Eq. (3-11), is

$$x = v_0 t + \tfrac{1}{2}at^2$$
$$= (64\text{ ft/s})(2\text{ s}) + \tfrac{1}{2}(-32\text{ ft/s}^2)(2\text{ s})^2 = 64\text{ ft}$$

The time at which it reaches the ground is the time when $x = 0$. The height at any time is

$$x = (64\text{ ft/s})t + \tfrac{1}{2}(-32\text{ ft/s}^2)t^2$$

Setting this equal to zero and solving the resulting quadratic equation for t, we find the two roots $t = 0$ and $t = 4$ s. The first, of course, is the time the ball *left* the ground; the other is the time of return. We note that the times of ascent and descent are equal. Finally, the velocity at $t = 4$ s is

$$v = 64\text{ ft/s} + (-32\text{ ft/s}^2)(4\text{ s}) = -64\text{ ft/s}$$

The velocity has the same magnitude as the velocity with which the ball was thrown; the direction has reversed, as indicated by the negative sign.

3-4 VELOCITY IN SPACE

We now generalize the concepts of velocity and acceleration so they can be applied to a point whose motion is *not* confined to a straight line. It is convenient to describe the position of the point with a *position vector*, which is a displacement vector from the origin of coordinates to the point. We denote this vector by **r**. If we know the x, y, and z coordinates of the point in a particular coordinate system, the position vector can be represented as

$$\mathbf{r} = x\mathbf{i} + y\mathbf{j} + z\mathbf{k} \qquad (3\text{-}13)$$

We see that x, y, and z are the components of the vector **r**. The distance from the origin to the point is the magnitude of the vector **r**; this is given by

$$r = |\mathbf{r}| = \sqrt{x^2 + y^2 + z^2}$$

Suppose the point moves from P_1, described by position vector $\mathbf{r}_1$, to P_2, described by $\mathbf{r}_2$, during a time interval Δt. Since we have described the position with the vector **r**, it is natural to call the change in position $\Delta \mathbf{r}$. This is the *displacement* of the point during Δt. We see from Fig. 3-5 and the definition of vector addition that $\mathbf{r}_2 = \mathbf{r}_1 + \Delta \mathbf{r}$, so that the displacement $\Delta \mathbf{r}$ (the change in **r**) is given simply by

Fig. 3-5 Two points P_1 and P_2 and the corresponding position vectors $\mathbf{r}_1$ and $\mathbf{r}_2$. The displacement from P_1 to P_2 is described by the vector $\Delta \mathbf{r}$; the average velocity during this displacement has the same direction as $\Delta \mathbf{r}$.

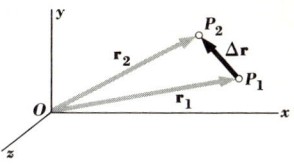

$$\Delta \mathbf{r} = \mathbf{r}_2 - \mathbf{r}_1 \tag{3-14}$$

It is now natural to define the average velocity $\mathbf{v}_{av}$ during Δt as the ratio $\Delta \mathbf{r}/\Delta t$, the displacement per unit time. Because $\Delta \mathbf{r}$ is a vector, $\Delta \mathbf{r}/\Delta t$ is also a vector quantity. It has the same direction as $\Delta \mathbf{r}$, and its magnitude is the magnitude of $\Delta \mathbf{r}$ divided by Δt. If the components of the displacement $\Delta \mathbf{r}$ in a particular coordinate system are known, the corresponding components of the average velocity can be found simply by dividing each component by Δt. These components are, of course, simply the changes Δx, Δy, and Δz in the x, y, and z coordinates, respectively, so we can write the average velocity as

$$\mathbf{v}_{av} = \frac{\Delta x}{\Delta t}\mathbf{i} + \frac{\Delta y}{\Delta t}\mathbf{j} + \frac{\Delta z}{\Delta t}\mathbf{k} \tag{3-15}$$

It is then natural to define the instantaneous velocity as the limit of the average velocity as the time interval Δt approaches zero. Symbolically, we write

$$\mathbf{v} = \lim_{\Delta t \to 0} \mathbf{v}_{av} = \lim_{\Delta t \to 0} \frac{\Delta \mathbf{r}}{\Delta t} = \frac{d\mathbf{r}}{dt} \tag{3-16}$$

This, in fact, defines what we mean by the derivative of the vector $\mathbf{r}$ with respect to time, $d\mathbf{r}/dt$.

If a point moves along a curve (Fig. 3-6) so that it is at P_1 with position

Fig. 3-6 Velocity of a point moving along a curved path. The average velocity $\mathbf{v}_{av}$ between P_1 and P_2 is directed along the line joining these points; the instantaneous velocity $\mathbf{v}_1$ at P_1 is tangent to the curve at P_1.

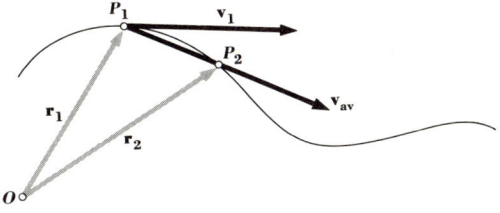

vector $\mathbf{r}_1$ at time t_1 and at P_2 with position vector $\mathbf{r}_2$ at time t_2, the direction of the average velocity during this time interval is, by definition, the direction from P_1 to P_2, as shown. The direction of the instantaneous velocity at time

64 DESCRIBING MOTION

t_1 is *tangent* to the curve at P_1. As we have seen, velocity is in general a vector quantity.

If the position of a point at any time is described by the coordinates x, y, and z, then the *components* of the instantaneous velocity in the x, y, and z directions are, respectively,

$$v_x = \frac{dx}{dt} \qquad v_y = \frac{dy}{dt} \qquad v_z = \frac{dz}{dt} \tag{3-17}$$

These results can also be obtained directly by differentiating Eq. (3-13) with respect to t. Since the unit vectors do not change with time, we have

$$\mathbf{v} = \frac{d\mathbf{r}}{dt} = \frac{dx}{dt}\mathbf{i} + \frac{dy}{dt}\mathbf{j} + \frac{dz}{dt}\mathbf{k} \tag{3-18}$$

This single *vector equation* is equivalent to the three equations for the components, Eqs. (3-17).

The magnitude of the velocity, which we denote by v, is called the *speed*. It is, of course, a scalar and not a vector quantity. The speedometer on a car is properly named; it measures the magnitude of velocity but not its direction. If the components of velocity are known, the speed is given by

$$v = \sqrt{v_x^2 + v_y^2 + v_z^2} \tag{3-19}$$

Example

A point moves in the xy plane; its coordinates vary with time as follows:

$$x = 2t \qquad y = 19 - 2t^2$$

where x and y are measured in meters and t in seconds. Find its distance from the origin and the magnitude and direction of its velocity at time $t = 2$ s. At what time is the position vector perpendicular to the velocity vector?

Solution

The distance from the origin is given by

$$r = \sqrt{x^2 + y^2} = \sqrt{(2t)^2 + (19 - 2t^2)^2}$$

At $t = 2$ s, this gives

$$r = \sqrt{(2 \times 2)^2 + (19 - 2 \times 2^2)^2}$$
$$= \sqrt{137} \text{ m} \simeq 11.7 \text{ m}$$

The components of velocity are

$$v_x = \frac{dx}{dt} = 2 \text{ m/s}$$

$$v_y = \frac{dy}{dt} = -4t \text{ m/s}$$

At $t = 2$ s, $v_x = 2$ m/s and $v_y = -8$ m/s; the magnitude of v at this time is

$$v = |\mathbf{v}| = \sqrt{v_x^2 + v_y^2} = \sqrt{(2 \text{ m/s})^2 + (-8 \text{ m/s})^2} = \sqrt{68} \text{ m/s}$$
$$\simeq 8.24 \text{ m/s}$$

The direction is downward and to the right; the velocity at $t = 2$ s makes an angle ϕ with the $+x$ axis given by

$$\phi = \tan^{-1} \frac{v_y}{v_x} = \tan^{-1} \frac{-8 \text{ m/s}}{2 \text{ m/s}} = \tan^{-1}(-4) \simeq -76°$$

The velocity vector $\mathbf{v}$ is perpendicular to the position vector $\mathbf{v}$ whenever the scalar product $\mathbf{v} \cdot \mathbf{r}$ is zero. In terms of components,

$$\mathbf{v} \cdot \mathbf{r} = v_x x + v_y y = (2)(2t) + (-4t)(19 - 2t^2) \quad \text{m}^2/\text{s}$$

When $\mathbf{v} \cdot \mathbf{r} = 0$,

$$4t - 76t + 8t^3 = 0$$

Rearranging and factoring,

$$8t(t^2 - 9) = 0$$

The roots of this equation are clearly $t = 0, \pm 3$ s.

By eliminating t from the x and y expressions, we can also obtain an equation for the path of the particle. We have

$$y = 19 - 2t^2 = 19 - 2\frac{x^2}{2} = 19 - \tfrac{1}{2}x^2$$

This is the equation of a parabola with its axis on the y axis opening downward, as shown in Fig. 3-7, which also shows the positions of the particle when $\mathbf{v} \cdot \mathbf{r} = 0$. How are the coordinates of these positions obtained?

This example illustrates the fact that once the coordinates are given as functions of time, we know everything there is to know about the motion. We can find the particle's position at any time, the magnitude and direction of its velocity, and the shape of its path.

66 DESCRIBING MOTION

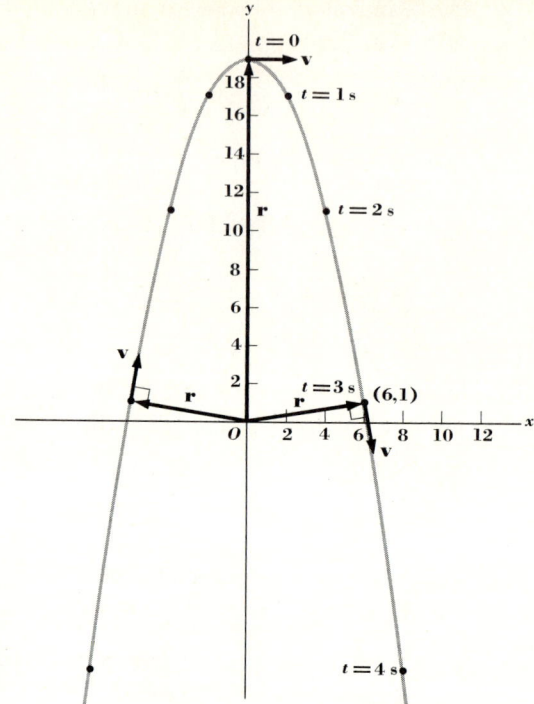

Fig. 3-7 Trajectory of a particle whose x and y coordinates are given functions of time as in the example.

3-5 ACCELERATION IN SPACE

Next we consider the *acceleration* of a point moving in space. Suppose that at time t_1 the point is at P_1 and has instantaneous velocity $\mathbf{v}_1$; at time t_2 it is at P_2 and has velocity $\mathbf{v}_2$. Each of these velocities is tangent to the path at the corresponding position. Acceleration is defined as the time rate of change of velocity. Drawing the vectors $\mathbf{v}_1$ and $\mathbf{v}_2$ from a common point as in Fig. 3-8, we see that the change in velocity is the quantity $\Delta\mathbf{v}$ shown in this figure. That is, $\Delta\mathbf{v}$ is the quantity which must be added to $\mathbf{v}_1$ to obtain $\mathbf{v}_2$, so $\Delta\mathbf{v} = \mathbf{v}_2 - \mathbf{v}_1$. We now define the *average acceleration* during this time interval as

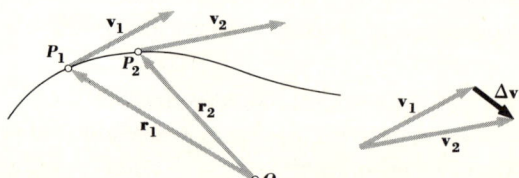

Fig. 3-8 Change in velocity of a point moving along a curved path. The average acceleration during the interval Δt is found from the change in velocity $\Delta\mathbf{v}$, obtained from the vector diagram shown.

$$\mathbf{a}_{av} = \frac{\Delta \mathbf{v}}{\Delta t} \tag{3-20}$$

by which we mean a vector with the direction of $\Delta \mathbf{v}$ and a magnitude which is that of $\Delta \mathbf{v}$ divided by Δt. Finally, we define the instantaneous acceleration as

$$\mathbf{a} = \lim_{\Delta t \to 0} \frac{\Delta \mathbf{v}}{\Delta t} = \frac{d\mathbf{v}}{dt} \tag{3-21}$$

If the position of the point is described by cartesian coordinates x, y, and z, all of which vary with time, then the components of acceleration are given by

$$a_x = \frac{dv_x}{dt} = \frac{d^2x}{dt^2} \qquad a_y = \frac{dv_y}{dt} = \frac{d^2y}{dt^2} \qquad a_z = \frac{dv_z}{dt} = \frac{d^2z}{dt^2} \tag{3-22}$$

Example
Find the acceleration in the previous example.

Solution
We found $v_x = 2$ m/s and $v_y = -4t$ m/s. Thus

$$a_x = \frac{dv_x}{dt} = 0 \qquad a_y = \frac{dv_y}{dt} = -4 \text{ m/s}^2$$

The x component of velocity is constant; hence the x component of acceleration is zero. In this example, the acceleration is constant in both magnitude and direction; it always points in the negative y direction, and it is *never* in the direction of **v**.

In general the acceleration is different from zero when *either* the magnitude or direction of the velocity changes. For example, a particle moving in a circle with constant speed has an acceleration, even though the magnitude of the velocity is constant, because the *direction* of the velocity changes continuously.

We also note that although the velocity is always tangent to the path, there is no reason to expect that the acceleration will be tangent to the path. A ball thrown straight up has, after it leaves the hand, an upward velocity but a downward acceleration. In the parabolic-trajectory example above, the

68 DESCRIBING MOTION

acceleration is always vertical and is nowhere tangent to the path. For uniform circular motion it turns out that the acceleration is always perpendicular to the velocity; we return to this problem in Chap. 6.

Example
The minute hand on a clock is 6 in. long. Find the average velocity of its end between 12:00 and 12:15.

Solution
The displacement during this interval is $6\sqrt{2}$ in. in the direction shown in Fig. 3-9. The average velocity has this direction and a magnitude of

$$\frac{6\sqrt{2} \text{ in.}}{15 \text{ min}} = \frac{2\sqrt{2}}{5} \text{ in./min} = 0.56 \text{ in./min}$$

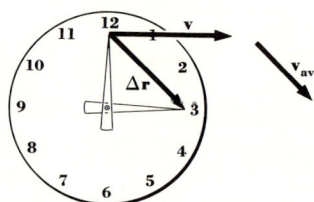

Fig. 3-9 Displacement of the end of the minute hand between 12:00 and 12:15 and average velocity in this interval. The vector **v** is the instantaneous velocity at 12:00.

Example
Find the instantaneous velocity of the end of the minute hand at 12:00.

Solution
First we find the speed, which is constant, from the fact that in 60 min the hand travels

$$(2\pi)(6 \text{ in.}) = 12\pi \text{ in.}$$

The speed is thus 12π in./60 min $= \pi/5$ in./min $\simeq 0.63$ in./min. The velocity is to the right in Fig. 3-9, and its magnitude is 0.63 in./min.

Example
Find the average acceleration between 12:00 and 12:15.

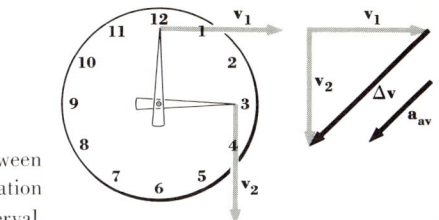

Fig. 3-10 Change in velocity between 12:00 and 12:15 and average acceleration during this interval.

Solution
The velocities at the two points are shown in Fig. 3-10. The change in velocity, $\Delta \mathbf{v} = \mathbf{v}_2 - \mathbf{v}_1$, is shown in the diagram. Its magnitude is

$$\frac{\pi\sqrt{2}}{5} \text{ in./min} \simeq 0.89 \text{ in./min}$$

The magnitude of the average acceleration is

$$\frac{\pi\sqrt{2}/5 \text{ in./min}}{15 \text{ min}} = 0.059 \text{ in./min}^2$$

and its direction is as shown in Fig. 3-10. What is the direction of the *instantaneous* acceleration at 12:00?

3-6 RELATIVE MOTION

Although we have not explicitly said so, the velocity discussed up to now has always described motion as it appears to an observer at rest with respect to the *origin O* of our coordinate system. This is because the velocity of a point is defined in terms of the change in the position vector of this point with respect to the origin.

It sometimes happens, however, that we want to know how the motion looks to an observer who is himself moving with respect to O. If the observer is in a car traveling at 50 mi/h, and if the car ahead of him is traveling at 60 mi/h, that car is moving away from the observer at the rate of 10 mi/h. The velocity of the car ahead *relative to this observer* is 10 mi/h. In general, if the point and observer are both moving, the velocity as it appears to the moving observer is called the *velocity relative to the moving observer*.

We can easily work out a general relationship for relative velocities. Let O be the origin of our coordinate system, as usual, and let B be the position of a moving observer. The point whose motion is being observed is A. The position of this point is described by the vector $\mathbf{r}_A$, and its velocity with respect to O is $\mathbf{v}_A = d\mathbf{r}_A/dt$. Similarly, let the position B of the moving observer be given by the vector $\mathbf{r}_B$; his velocity with respect to O is $\mathbf{v}_B = d\mathbf{r}_B/dt$. The position vectors are shown in Fig. 3-11.

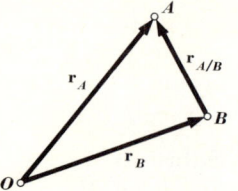

Fig. 3-11 Describing the position of a point A by giving its position vector $\mathbf{r}_{A/B}$ relative to a moving observer B. This observer's position is described by the position vector $\mathbf{r}_B$.

The displacement from the moving observer at B to the point A is given by the vector $\mathbf{r}_{A/B}$ in the figure. This is related to the other position vectors by

$$\mathbf{r}_A = \mathbf{r}_B + \mathbf{r}_{A/B} \tag{3-23}$$

The motion of A relative to B, which is the motion seen by the observer at B, is determined entirely by the change in the *relative position vector* $\mathbf{r}_{A/B}$, the position vector of A relative to B. Hence the velocity of A as observed from B, which we denote by $\mathbf{v}_{A/B}$, is $\mathbf{v}_{A/B} = d\mathbf{r}_{A/B}/dt$. The relation of $\mathbf{v}_{A/B}$ to $\mathbf{v}_A$ and $\mathbf{v}_B$ is obtained by taking the time derivative of Eq. (3-23). We find

$$\mathbf{v}_A = \mathbf{v}_B + \mathbf{v}_{A/B} \tag{3-24}$$

This equation says that the velocity $\mathbf{v}_A$ of A with respect to O can be thought of as the vector sum of two quantities: the velocity $\mathbf{v}_{A/B}$ of A relative to the moving observer at B and the velocity $\mathbf{v}_B$ of that observer with respect to O.

Example
An airplane flying straight north at 400 mi/h observes a second plane 3 mi straight east, which appears to the navigator of the first plane to have a velocity straight west with a magnitude (determined with doppler radar) of 300 mi/h. What is the second plane's velocity with respect to the ground? Are the two planes on a collision course? If so, how soon will they collide?

Solution
Let B be the first plane. Then $\mathbf{v}_B$ is 400 mi/h north. Let A be the second plane; its velocity $\mathbf{v}_{A/B}$ with respect to B is 300 mi/h west. The velocity

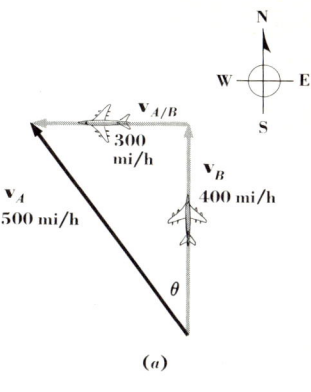

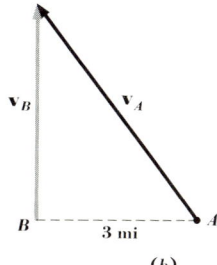

Fig. 3-12 The ground velocity of the second plane is the vector sum of its velocity relative to the first plane and the first plane's ground velocity.

$\mathbf{v}_A$ of the second plane with respect to the ground is then given by Eq. (3-24). The vector sum is easily obtained from a vector diagram, as shown in Fig. 3-12a; the triangle is a 3-4-5 right triangle, so the magnitude of $\mathbf{v}_A$ is 500 mi/h, in a direction 36.8° west of north, where $36.8° = \tan^{-1} \frac{3}{4}$. This vector diagram does not show the actual positions of the planes; this is shown in Fig. 3-12b, which also shows the paths as viewed from the ground. The planes are indeed on a collision course, and unless one plane changes course, they will collide after 36 s, the time required for A to travel 3 mi relative to B at a relative velocity of 300 mi/h. The collision will occur 4 mi north of the initial position of B.

One can also discuss acceleration measured by a moving observer. Taking the time derivative of Eq. (3-24), we find

$$\mathbf{a}_A = \mathbf{a}_B + \mathbf{a}_{A/B} \tag{3-25}$$

where $\mathbf{a}_B$ is the acceleration of the moving observer B relative to the origin O and $\mathbf{a}_{A/B}$ is the rate of change of the relative velocity of A with respect to B.

In the particular case where the moving observer B moves relative to the origin with constant velocity, we have $\mathbf{v}_B = $ constant and $\mathbf{a}_B = 0$. In this special case $\mathbf{a}_A = \mathbf{a}_{A/B}$; that is, the observer at O and the one at B both measure the same acceleration for A. Since acceleration is directly related to force, this is a significant result, and its implications will be discussed in Chap. 4.

The essential point in this discussion is the central role of the observer or, more generally, of the frame of reference used for the description of motion. The position, velocity, and acceleration of a point depend on the frame of reference, and for any specific description of motion it is essential to understand the nature of the particular frame of reference being used.

In discussing relative velocities, we have implicitly assumed that the *time intervals* used to define the velocity of point A are the same whether measured by the observer at O or the moving observer at B. That is, we have assumed that time has an absolute meaning, independent of any particular frame of reference. This seems obvious; we are accustomed to communicating information about time at very great speeds via radio and other means. But is it really so obvious? The position and velocity of a point depend on the motion of the observer; if a time interval between two events at this point is observed, why should it not also depend on the observer's motion?

The point of view of newtonian mechanics, which forms the subject matter of two-thirds of this book, is that time and time intervals *do* have a significance independent of motion of the observer. This point of view is, however, inconsistent with the experimental observation that the velocity of light is independent of the motion of its source. That is, Eq. (3-24) is not correct for light. The modifications in the principles of mechanics necessitated by this discovery are discussed in Chaps. 14 and 15. It turns out that the assumption of a time scale which is the same for all observers is valid only when all the velocities are very small in magnitude compared to the speed of light; when this is not the case, the formulations must be modified. The necessary modifications are embodied in the special theory of relativity, developed by Einstein in 1905.

Problems

3-1 For a point moving on a straight line, at a distance x from a fixed point, what is the displacement between each of the following pairs of initial and final points?
    a $x = 10$ cm to $x = 5$ cm
    b $x = 10$ cm to $x = -5$ cm
    c $x = 5$ cm to $x = 10$ cm
    d $x = -5$ cm to $x = -10$ cm

73 PROBLEMS

3-2 An object's position x along a straight line was observed at a succession of times t, with the following results:

x, m	t, s
3.00	1.0
3.14	1.5
3.29	2.0
3.42	2.5
3.57	3.0

    a   Plot a graph showing x as a function of t.
    b   Find the velocity of the object. Is it constant? Discuss.
    c   Where was the object at time $t = 0$?

3-3 If the position of an object is given by $x = (2 \text{ cm/s}^3)t^3$, where x is measured in centimeters and t in seconds, find:
    a   The average velocity between $t = 1$ s and $t = 2$ s
    b   The instantaneous velocities at $t = 1$ s and $t = 2$ s
    c   The average acceleration between $t = 1$ s and $t = 2$ s
    d   The instantaneous accelerations at $t = 1$ s and $t = 2$ s

3-4 In Eq. (3-5), what units must the constants b, c, and d have for dimensional consistency? Using these units, show that Eq. (3-6) is also dimensionally consistent.

3-5 In Fig. 3-3 and the accompanying discussion, the area of the small strip is not precisely $v \, \Delta t$ because of the small triangular area at the top. Does this omission affect the validity of the final result? Discuss.

3-6 A particle moves along a line, with velocity varying with time according to

$$v = 6t^2 - 12t^3$$

If the particle is at the origin ($x = 0$) at time $t = 0$, find its position as a function of time. *Hint:* What function of t must x be in order for its derivative to equal the given function for v?

3-7 In Prob. 3-6, suppose the particle is at the point $x = 3$ at time $t = 0$. Find its position as a function of time.

3-8 A particle moves along a straight line, with the distance from the origin given by

$$x = A \sin \omega t$$

where A and ω are constants.
    a   What is the maximum distance of the point from the origin?
    b   What is the particle's velocity as a function of time?
    c   What is the particle's acceleration as a function of time?

Show that at each instant $a = -\omega^2 x$, that is, that the acceleration is proportional to the distance from the origin. The following formulas for derivatives can be used even if you have not yet studied their derivation.

$$\frac{d}{dt}\sin \omega t = \omega \cos \omega t \qquad \frac{d}{dt}\cos \omega t = -\omega \sin \omega t$$

3-9 Derive Eq. (3-12) from Eqs. (3-10) and (3-11).

3-10 If a jet airliner must attain a speed of 200 mi/h on the runway for takeoff, and if the runway is 1 mi long, what acceleration must it have (assuming it is constant)? How does this compare with the acceleration of a 1974 Pontiac starting from rest?

3-11 If an automobile starts from rest, accelerates uniformly, and reaches 60 mi/h in 8 s, what is its acceleration? What distance does it travel in this time? How much time would be required to travel $\frac{1}{4}$ mi if the acceleration remained constant? Is it reasonable to assume that the acceleration is constant? Discuss.

3-12 A ball is thrown vertically upward. When it has reached one-half its maximum height, its velocity is 32 ft/s.
 a What is the maximum height?
 b What was the initial velocity?
 c How much time is required to reach maximum height? One-half maximum height?

3-13 A juvenile delinquent wants to throw a baseball through a glass skylight in the ceiling of a store, 32 ft overhead. If the glass will not break unless the baseball has a speed of at least 16 ft/s, with what speed must it be thrown from the floor to achieve the desired result?

3-14 A ball is thrown vertically upward with an initial speed of 64 ft/s.
 a After how much time has the speed decreased to 32 ft/s?
 b How much time is required for the ball to fall back to its starting point?
 c What is its acceleration at the instant it reaches its maximum height?

3-15 An automobile traveling 60 mi/h passes a radar checkpoint in a 40 mi/h speed zone. As he passes, a motorcycle policeman starts from rest at this point, accelerates uniformly, reaching 60 mi/h in 10 s, and continues to accelerate until he overtakes the automobile.
 a How much time elapses during the chase?
 b How far from the checkpoint is the motorist apprehended?
 c How fast is the motorcycle going when it passes the automobile?

3-16 A ball is dropped from a building 128 ft above the ground, and at the same time another ball is thrown vertically upward from the ground with an initial velocity of 64 ft/s. Where and after how much time do the balls collide? Note that the time is independent of the acceleration of gravity; is this surprising?

75 PROBLEMS

3-17 A particle moves in the xy plane, with coordinates given as functions of time by

$x = 3t + 5$
$y = \frac{1}{2}t^2 + 3t - 4$

   a Write an expression for the position vector of the particle as a function of time.
   b Sketch the path described by the particle.
   c Find the velocity (x and y components) as a function of time. In particular, find the magnitude and direction of the velocity at $t = 4$.
   d Find the acceleration (x and y components) as a function of time. Find the magnitude and direction of the acceleration at $t = 4$.

3-18 In Fig. 3-7, find the minimum distance from the origin to the particle and the times when this distance occurs by writing an expression for the square of the distance $x^2 + y^2$ and setting the time derivative of this quantity equal to zero to find maxima and minima. Do your results confirm predictions based on geometric reasoning?

3-19 A particle moves in the xy plane, with coordinates given as functions of time as follows:

$x = R \cos \omega t$
$y = R \sin \omega t$

where R and ω are constants.
   a Show that the distance of the particle from the origin is constant and equal to R, so that the particle moves in a circle of radius R. Does it move clockwise or counterclockwise?
   b Find the x and y components of velocity, using the formulas for the derivatives of trigonometric functions given in Prob. 3-8. Show that the magnitude of the velocity is constant and equal to ωR.
   c By computing the scalar product $\mathbf{v} \cdot \mathbf{r}$ show that $\mathbf{v}$ is always perpendicular to $\mathbf{r}$. Could you have predicted this on geometric grounds?
   d Find the x and y components of acceleration. Show that the magnitude of the acceleration is constant and given by $a = \omega^2 R$. Combine this with the relation $v = \omega R$ from part *b* to obtain the alternate form $a = v^2/R$.
   e Write expressions for the position vector and the acceleration vector, and show that $\mathbf{a} = -\omega^2 \mathbf{r}$. Hence show that the acceleration is always directed toward the center of the circle.

3-20 Is the average speed of a point always the same as the magnitude of its average velocity? Explain.

3-21 Consider a simple pendulum made of a ball on the end of a string. Find the *directions* of its velocity and acceleration at each end of the swing and at the midpoint of the swing.

76 DESCRIBING MOTION

3-22 Show that an object moving in a circular path with constant speed has an acceleration directed toward the center of the circle.

3-23 A canoeist who can paddle 4 mi/h in still water paddles upstream on a river whose current flows 2 mi/h. After reaching a point $\frac{1}{2}$ mi upstream, he turns around and comes back to the starting point. How much time is required for the round trip?

3-24 A canoeist who can paddle 10 ft/s in still water paddles across the Greenbrier River, which is 200 ft wide. It is summer, and the Greenbrier has a weak current flowing 6 ft/s.
a If he heads straight across, how far downstream does he reach the shore? How much time elapses?
b In what direction should he head in order to land at a point directly across from the starting point? How much time elapses during this trip?

3-25 A jet airplane whose airspeed is 500 mi/h flies from Miami to New York in the aftermath of hurricane Thelma, during which the wind blows from the east at 50 mi/h.
a What is meant by airspeed?
b In what direction must the plane head if New York is 1,000 mi straight north from Miami?
c What is the groundspeed? Will the plane arrive significantly late?

3-26 A man on a train moving in the $+x$ direction with constant velocity $v_x = 5$ m/s observes a ball being tossed up in the air by a boy standing on the station platform. With respect to a coordinate system on the platform, the ball's motion is described by

$$x = 0$$
$$y = -\tfrac{1}{2}gt^2 + t$$

where g is a constant.
a If the man on the train describes the motion with respect to coordinate axes x' and y' moving with the train, what are x' and y' as functions of time if the origins of the two coordinate systems coincide at time $t = 0$?
b What geometric figure is traced out by the ball with respect to the $x'y'$ system?
c What is the acceleration of the ball as seen by the man on the train? As seen by the boy on the platform? Comment on your answer.

PERSPECTIVE I

In the three opening chapters two principal themes have been discussed in considerable detail. The first of these is the general framework within which physics is constructed. Physics is an empirical science; we believe that only by observing the physical world quantitatively is it possible to arrive at physical truth. Furthermore, an essential part of any such investigation is the simplification and idealization of the physical situation by construction of a model which permits analysis of the essential features while stripping away unimportant or inconsequential details.

The second main theme has been the development of mathematical language suitable for efficient description of physical phenomena. In particular, we have seen that vectors are useful for describing the position of a particle and its motion, and we have hinted at the usefulness of vectors in describing other physical quantities. The language of vectors has been used to define quantities such as velocity and acceleration which are useful in describing motion precisely. The description of motion constitutes the subject of kinematics.

In developing the techniques of kinematics, we have relied heavily on geometry and geometric intuition but relatively little on detailed observation of motion and its causes. Thus it is not surprising that our study of the motion of a particle is not yet complete; it says nothing yet about the causes of motion. We do not yet have any general guiding principles to connect a simplified model of a situation to the motion which can be described by our kinematic developments.

For example, we observe that the acceleration of a freely falling body is constant and nearly independent of the size or structure of the object.

We can use this observation to make detailed calculations of the motion, but we are not yet in a position to understand why the acceleration should be constant. More generally, if we know the acceleration of a body, we can obtain its velocity and position, and conversely, but we cannot yet relate these quantities to the causes of the motion.

Thus the next essential task is to obtain, through experiments and generalizations from them, some principles relating the motion of a particle to the causes of this motion. These principles, embodied in Newton's laws of motion, are introduced, along with some elementary applications, in the next three chapters.

Principles of Dynamics | 4

We have discussed various methods of describing the motion of a point but have not yet considered the *causes* of motion. In this chapter we discuss the relationship between motion and the influences which produce and change it. The study of such relationships is called *dynamics*.

4-1 INERTIA

The word inertia is used often in everyday conversation. When we say that a person has a "lot of inertia," we usually mean that it is difficult to prod him into action or to movement of any sort; he tends to remain motionless. The term is also used in another sense. We speak of a loaded truck knocking down a row of telephone poles because of its great inertia. In this sense inertia refers to a property by which a body once set in motion is difficult to stop.

In mechanics, the term *inertia* has an exact meaning related to both the common meanings discussed above. The property of a body which we call its inertia is its tendency to remain at rest or if set into motion to maintain this state of motion.

It is a matter of common observation that an object at rest remains at rest unless acted upon by some external influence. Your pencil does not of its own accord jump up from your desk and fly across the room. To make it do so you must pick it up and throw it. The idea that an object set into motion continues to move uniformly unless something makes it stop is not quite so obvious, and examples of such motion are less easy to find. A billiard ball rolling on a smooth table may roll for a long distance, but it eventually comes to a stop. A car coasting along a level road with the gears in neutral finally rolls to a stop.

On the basis of such observations, Aristotle (384–322 B.C.) concluded that the "natural state" of all bodies is a state of *rest*. He believed that to make a body move, or to keep it moving once it is started, it is necessary to exert some external influence on the body. The heavenly bodies, incidentally, were excluded from this generalization. Aristotle thought that the planets, the sun, and the moon revolved around the earth in circular paths and that this was the "natural state" for these bodies.

These teachings of Aristotle were accepted for many centuries and were not seriously questioned until the time of Galileo. In the following imaginary dialogue Galileo compares his theories, represented by Salviati, with the older aristotelian ideas, represented by Simplicio. The two men have been discussing the motion of a hard ball rolling on a smooth plane. They have agreed that if the ball rolls uphill, its motion is *retarded*, while if it rolls downhill, its motion is *increased*. Then Salviati continues:

Now tell me what would happen to the same movable body placed upon a surface with no slope upward or downward.

Simplicio *Here I must think a moment about my reply. There being no downward slope, there can be no natural tendency toward motion; and there being no upward slope, there can be no resistance to being moved, so there would be indifference between the propensity and the resistance to motion. Therefore it seems to me that it ought naturally to remain stable. But I forgot; it was not so very long ago that Sagredo gave me to understand that this is what would happen.*

Salviati *I believe it would do so if one set the ball down firmly. But what would happen if it were given an impetus in any direction?*

Simplicio *It must follow that it would move in that direction.*

Salviati *But with what sort of movement? One continually accelerated, as on the downward plane, or increasingly retarded as on the upward one?*

Simplicio *I cannot see any cause for acceleration or deceleration, there being no slope upward or downward.*

Salviati *Exactly so. But if there is no cause for the ball's retardation, there ought to be still less for its coming to rest; so how far would you have the ball continue to move?*

Simplicio *As far as the extension of the surface continued without rising or falling.*

Salviati *Then if such a space were unbounded, the motion on it would likewise be boundless? That is, perpetual?*

Simplicio *It seems so to me, if the movable body were of durable material.*

Salviati *That is of course assumed, since we said that all external and accidental impediments were to be removed, and any fragility on the part of the moving body would in this case be one of the accidental impediments.*

Galileo discussed the same observations as Aristotle, yet he reached quite different conclusions. The most striking difference is that while Aristotle confined his discussion entirely to phenomena which were directly observed, Galileo used these observations as a springboard to consideration of an *idealized model*, about which he could make generalizations. He argued that if the conditions of his idealized experiment are *approximately* realized, then the motion takes place *approximately* as he describes it.

In fairness, we must admit that Aristotle's conclusion was not wrong. Any familiar object, when set in motion, eventually comes to rest unless something keeps it moving. But Aristotle's conclusion did not lead to deeper understanding of the phenomena, while Galileo's invited further investigation of the causes of change in the velocity of an object and the conditions under which an object *does* move with constant velocity. The usefulness of Galileo's point of view results in large part from his use of a model as a generalization and idealization of phenomena observed in the real world.

Galileo's conclusion regarding the "natural state of motion" of a body was restated by Sir Isaac Newton (1642–1727). Paraphrasing Newton's words slightly, we make the following statement:

A body initially at rest continues in a state of rest, and a body initially in motion continues to move with uniform velocity in a straight line unless compelled by some external agency to alter these states of motion.

This generalization is now called *Newton's first law of motion*. It is important to notice that the state of uniform motion is just as "natural" as the state of rest.

Suppose an object is seen by two different observers, one at rest with respect to the object, the other in a frame of reference moving with uniform velocity relative to the first. To the second observer the object seems to be

82 PRINCIPLES OF DYNAMICS

in uniform motion; to *both* observers the object seems to be in a "natural" state of motion, in accordance with Newton's first law. Thus both observers may use the same basic principles to understand the motion, even though they use different coordinate systems. We shall see that this is also true for the other laws of motion, with some exceptions which will be discussed.

4-2 FUNDAMENTAL EXPERIMENTS

In formulating the principles of dynamics, as in all of physics, we are guided by experimental observations. In this chapter we consider several simple yet crucial experiments. Although the precise details of these experiments are not crucial, it is useful to think of the motion of disks or pucks on an air table, supported by jets of air blown through holes in the table, as in Fig. 4-1. Such an object moves with extremely small friction because it is supported by a cushion of air and does not actually touch the surface.

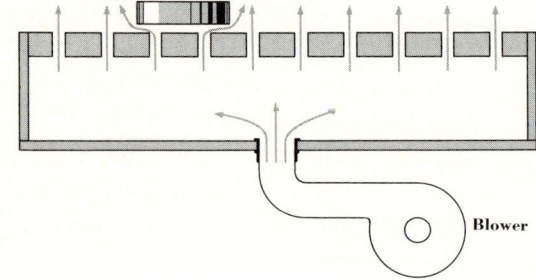

Fig. 4-1 Cross section of an air table, showing a puck supported by jets of air from small holes in the table. Air pressure is supplied by a blower.

To study the motion of an air-supported puck we need to record its position at various times; a convenient technique is to take a multiple-flash photograph using a strobe light flashing at regular intervals, thus giving a series of images on the film.

Suppose we give the puck a certain initial velocity and observe its motion by means of the multiple-flash photograph technique, obtaining a picture similar to Fig. 4-2. Careful measurements show that the puck traverses equal

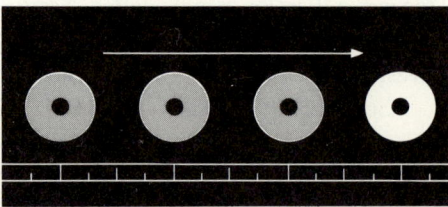

Fig. 4-2 Sketch made from a multiple-flash photograph of an air-supported puck moving with constant velocity.

distances in equal times; its velocity is constant. Thus, according to Newton's law, this object moves without external influence, at least within the precision of this experiment.

Now we apply an external influence to the puck. We choose as an easily controllable influence a stretched spring attached to the puck, which exerts a *force* on the puck. We assert tentatively that when the spring is stretched a constant amount, the force on the puck is constant. We attach one end of the spring to the puck and pull the other end just hard enough to keep the spring stretched a constant amount; the resulting motion can be observed with the multiflash photograph technique.

The result of such an experiment is shown in Fig. 4-3. It is evident

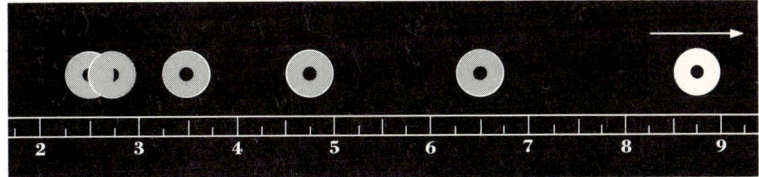

Fig. 4-3 Sketch made from a multiple-flash photograph of an air-supported puck accelerating under the action of a stretched spring.

from this photograph that the puck is *accelerated* by the spring. To describe the motion more fully, we calculate the acceleration at various points along the path of the puck by finding the changes in average velocity from interval to interval, just as in Sec. 3-2. The calculations are shown in Table 4-1. We find that, within the limitations of experimental precision, the acceleration is *constant*. This is a significant result. We have exerted a constant force, and the result has been a constant acceleration.

The experiment may be repeated with various modifications. Suppose we stretch the spring by a larger amount than previously. The acceleration then turns out to be greater than before, but if the spring is stretched a *constant* amount, the acceleration is again *constant*. We may apply this spring to different pucks made of different materials and having different dimensions. Again, although the actual value of the acceleration is different, a constant acceleration results from a constant force. We also note that within each experiment the acceleration does not change as the velocity of the puck increases.

It may also be observed that if material is added to the puck, the acceleration resulting from a given force is decreased. That is, if the same force is applied to two objects, one light and one heavy, the heavy object

Table 4-1

t, s (1)	x, cm (2)	Δx, cm (3)	v_{av}, cm/s (4)	Δv, cm/s (5)	a, cm/s^2 (6)
0.0	2.50				
		0.24	2.4		
0.1	2.74			5.1	51
		0.75	7.5		
0.2	3.49			5.1	51
		1.26	12.6		
0.3	4.75			5.0	50
		1.76	17.6		
0.4	6.51			4.9	49
		2.25	22.5		
0.5	8.76			4.7	47
		2.72	27.2		
0.6	11.48				

accelerates more slowly than the light object. In the language with which we began this chapter, we say that the heavy object has more inertia than the light one.

4-3 INERTIAL MASS

We are now ready to define a physical quantity corresponding to the intuitive idea of inertia, namely *inertial mass* or simply *mass*. We first specify a procedure for comparing the inertial masses of two bodies. The same force (as determined by stretching a spring a fixed amount) is applied to two different bodies, and the accelerations are measured. These are denoted by a_1 and a_2 and the inertial masses by m_1 and m_2. If the accelerations are equal, the masses are equal; if $a_1 = a_2$, then $m_1 = m_2$, by definition. If the first body is accelerated *more* than the second, it has *less* inertia and therefore less mass. Thus we define the ratio of the masses to be the inverse of the ratio of the accelerations:

$$\frac{a_1}{a_2} = \frac{m_2}{m_1} \tag{4-1}$$

For example, if the first body has an acceleration of 10 m/s^2 under a given force while the second has an acceleration of 1 m/s^2 under the same force, $a_1/a_2 = 10$. Then $m_2/m_1 = 10$; the second body has 10 times as much inertial mass as the first, since it accelerates only one-tenth as much under a given force.

Having specified a method for comparing the inertial masses of any two bodies, we can choose a standard body to use as a unit of mass for comparison with all other objects. The internationally accepted unit of mass is the kilogram (kg). The standard is a piece of platinum-iridium alloy kept at the International Bureau of Weights and Measures in Sèvres, France. The kilogram is the basic unit of mass in the mks system. In the cgs system of units, the basic unit of mass is the gram (g). The relation between the two is that 1 kg = 1,000 g.

Our definition of inertial mass is an *operational* one in that it specifies a method which can be used for actually *measuring* the inertial mass of any body. We now outline a crucial experiment which will help to demonstrate the usefulness of this concept. Suppose we measure the masses of two bodies using the methods outlined above, obtaining the results m_1 and m_2. Then without adding any additional material we fasten the two bodies together so that they move together as one. What is the mass of this composite body?

We suspect immediately that the composite mass is $m_1 + m_2$; that is, the total mass of a composite body is the sum of the masses of its constituent parts. But this conclusion does not follow from the observations made thus far, and so it must be regarded as a hypothesis subject to experimental confirmation. We must *measure* the mass of the composite body to see whether or not it is equal to $m_1 + m_2$.

Such experiments have been performed under a wide variety of conditions; and the total mass is indeed equal to the sum of the masses of the constituent parts, confirming our hypothesis. Conversely, if a piece of material is cut up into small pieces, the sum of the masses of the individual pieces always equals the original mass of the object. The *total* mass is not changed in the processes of putting together or taking apart.

This principle also holds for chemical reactions. When a combustible material is burned and all the combustion products are collected, including gases, the final total mass is the same as the original mass. Hence, even in a process which involves rearrangement of the atoms and molecules of the material, the total mass is constant.

Generalizing from these and many similar observations, we state the principle of conservation of mass:

In any physical system, if no matter is either added to or taken from the system, the total mass is constant.

This is the first of several *conservation principles* in mechanics. Such principles are very powerful because they enable us to find relations between the initial and final states of a system even if, between these states, the system goes through a very complicated process whose details may not be known.

What about nuclear reactions? In a nuclear reactor, the nucleus of a uranium atom may undergo a process called *fission*, in which it splits into two or more fragments. When one measures the masses of the original nucleus and of the fragments, one finds that the mass of the original nucleus is measurably larger than the sum of the masses of the fragments. Does this violate the principle?

There are two possible points of view. One says simply that the law of conservation of mass is violated. The other, more useful point of view says that we have stated the principle of conservation of mass in too restricted a form and that we should try to generalize the concept of mass, if possible, so that the generalized principle of conservation of mass is *not* violated by this process. Such a generalization is provided by the special theory of relativity, which regards mass and energy as two manifestations of a single basic physical quantity. If a quantity of mass (in our restricted sense) disappears, a proportional amount of energy must appear. This energy, in fact, is what makes it possible to generate power using nuclear reactions. The relationship of mass and energy is discussed in detail in Chap. 15.

Strictly speaking, there are also mass changes associated with chemical reactions in which heat is absorbed or given off. These mass changes are ordinarily of the order of 10^{10} times smaller than the mass of the reactants, so small that they have never been measured. In nuclear fission, on the other hand, the fraction is more like 1 part in 10^5, large enough to be observed experimentally with considerable precision.

In conclusion, the principle of conservation of mass is strictly correct only when it is generalized to include mass changes associated with energy changes in the system. In all cases of motion which do not involve nuclear reactions or creation or destruction of fundamental particles, however, the mass changes are so small that we can describe the situation very precisely with a model in which the total mass does not change.

4-4 FORCE AND MOTION

We have observed that the harder a spring pulls on a body, the greater is its acceleration and that, for a given force, the acceleration of the body is inversely proportional to its mass. Thus it is reasonable to make a quantitative definition of force such that

$$\mathbf{a} = \frac{\mathbf{F}}{m} \tag{4-2}$$

The acceleration **a** of the body is directly proportional to the force **F** acting on it and inversely proportional to its mass. This equation, rewritten as

$$\mathbf{F} = m\mathbf{a} \tag{4-3}$$

defines force quantitatively. This relation is often called *Newton's second law of motion*. It is a *vector* equation; $\mathbf{F}$ and $\mathbf{a}$ have the same direction.

Equation (4-3) also defines a *unit* of force. In the mks system, the force necessary to give a body of mass 1 kg an acceleration of 1 m/s^2 is called 1 newton (N). The corresponding unit of force in the cgs system is the dyne (abbreviated dyn): 1 dyn produces an acceleration of 1 cm/s^2 when exerted on a body of mass 1 g. In the British system the fundamental units of force, mass, and acceleration are, respectively, the pound (lb), the slug, and the foot per second per second (ft/s^2).

We are now ready for another basic experiment. By using springs stretched different amounts, we can exert several different forces on the same body. In each case we measure the acceleration and thereby determine the force. Force $\mathbf{F}_1$ produces acceleration $\mathbf{a}_1$, $\mathbf{F}_2$ acceleration $\mathbf{a}_2$, and so forth. What happens if we attach two such springs to the same body at the same time so that they pull simultaneously, one spring exerting force $\mathbf{F}_1$ and the other $\mathbf{F}_2$?

It seems *reasonable* that the force on the body should be $\mathbf{F}_1 + \mathbf{F}_2$, so that the acceleration is $\mathbf{a}_1 + \mathbf{a}_2$. This conjecture must be tested by experiment; we must *measure* the acceleration resulting from the simultaneous application of the two forces. The results of such experiments show, in fact, that the acceleration *is* the sum of the two individual accelerations. This conclusion illustrates the *principle of superposition of forces*, discussed in more detail in the next section.

Example

A spring balance is a calibrated spring used to measure forces by observing the amount of stretch of the spring. Suppose a spring balance is used to apply a constant horizontal force of magnitude 6 N to a block having a mass of 2 kg which slides on a smooth, frictionless horizontal plane. What acceleration results?

Solution

The acceleration is given by $a = F/m$. Inserting the values given, we obtain

$$a = \frac{6 \text{ N}}{2 \text{ kg}} = 3 \text{ N/kg}$$

88 PRINCIPLES OF DYNAMICS

This result does not have the usual units of acceleration. To simplify the units, we recall that, according to the definition, $1 \text{ N} = 1 \text{ kg-m/s}^2$. Inserting this in the previous equation, we obtain

$$a = 3 \text{ m/s}^2$$

The direction of **a** is the same as that of **F**.

In solving numerical problems, the *units* of each physical quantity should always be carried along. This procedure ensures that a consistent set of units is used and helps guard against obvious mistakes in the calculation.

Example
Suppose two spring balances are attached to the 2-kg block of the previous example. One pulls to the right with a constant force of 6 N, the other to the left with a constant force of 4 N. What is the acceleration of the block?

Solution
First we must decide which direction is positive and which negative and indicate this on a diagram. In Fig. 4-4 the positive direction is to the right.

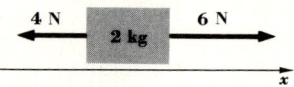

Fig. 4-4 Two forces acting simultaneously on a block on a smooth surface. The acceleration is in the direction of the vector sum of the forces.

The forces on the block are $+6 \text{ N}$ (to the right) and -4 N (to the left). The total force therefore is

$$F = 6 \text{ N} - 4 \text{ N} = 2 \text{ N}$$

The acceleration produced by this force is

$$a = \frac{2 \text{ N}}{2 \text{ kg}} = 1 \text{ m/s}^2$$

The positive result indicates an acceleration in the $+x$ direction (to the right).

Example
A body of mass 3 kg moves along a straight line with a velocity which varies with time according to the relation

$$v = 4t^2 - 3t + 6$$

where v is measured in meters per second and t in seconds. What force acts on the body?

Solution
The mass is given, and if we knew the acceleration, we could use $F = ma$ to find the force. But with the velocity given, it is a simple matter to obtain the acceleration. We have

$$a = \frac{dv}{dt} = 8t - 3$$

with a in meters per second per second and t in seconds. Thus the force is given by

$$F = ma = (3 \text{ kg}) (8t - 3) \quad \text{m/s}^2$$
$$= 24t - 9 \quad \text{N}$$

In this example the force is not constant but a function of time.

4-5 SUPERPOSITION OF FORCES

Thus far we have considered only problems in which the motion occurs along a straight line and in which the forces are directed along this line. There are many situations where this is not the case; the path may be curved, and the forces may have various directions. The relationship $\mathbf{F} = m\mathbf{a}$ is readily generalized to deal with these situations.

Let us return to the air-supported puck on a smooth, horizontal plane. We can describe the position of the puck by giving the x and y coordinates of a particular point, say the center of the puck, with respect to a pair of coordinate axes on the plane. The motion can then be described with the x and y components of velocity and acceleration of the puck.

We now apply a force F_x in the direction of the x axis. This results in an acceleration a_x along the x axis, given by

$$F_x = ma_x = m\frac{d^2x}{dt^2}$$

Now we return the puck to the starting point and apply a force F_y along the y axis. The resulting acceleration a_y is given by

$$F_y = ma_y = m\frac{d^2y}{dt^2}$$

What happens if F_x and F_y are applied simultaneously?

90 PRINCIPLES OF DYNAMICS

Once again we have a question that must be answered experimentally. Experiment shows that when several forces act on a body simultaneously, the resulting acceleration is the same as that which would result from the application of a single force equal to the *vector sum* of the individual forces. Thus it is established experimentally that force is a *vector quantity* and that forces can be combined using vector addition.

The relationship between the acceleration of the body and the forces which cause it is in general a *vector* relationship, which we represent symbolically

$$\Sigma \mathbf{F} = m\mathbf{a} \tag{4-4}$$

The symbol $\Sigma \mathbf{F}$ means the *vector sum* of all the forces acting on the body.

Because Eq. (4-4) is a vector relation, it can also be expressed in terms of its components in any coordinate system. Stated symbolically, Eq. (4-4) is equivalent to the three scalar equations

$$\Sigma F_x = ma_x \qquad \Sigma F_y = ma_y \qquad \Sigma F_z = ma_z \tag{4-5}$$

where ΣF_x means the algebraic sum of the x components of all the forces, and so forth. Since the components of the vectors in a particular coordinate system are often the most convenient quantities to use in vector calculations, these relationships are very useful.

Example

Two constant forces are applied simultaneously to a 2-kg block on a smooth plane, as shown in Fig. 4-5, which is a top view of the situation. Find the magnitude and direction of the acceleration which results.

Solution

The components of the forces are

$$F_{1x} = 20 \text{ N} \qquad F_{1y} = 0$$
$$F_{2x} = -10 \text{ N} \qquad F_{2y} = 10\sqrt{3} \text{ N}$$

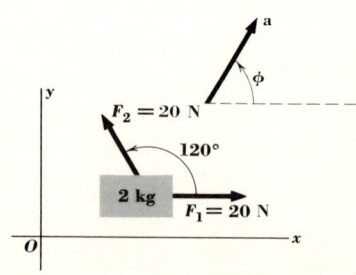

Fig. 4-5 Two forces in different directions applied simultaneously to a block on a smooth plane. The direction of the acceleration is shown.

91 SUPERPOSITION OF FORCES [4-5]

From Eqs. (4-5), the components of acceleration are given by

$$20 \text{ N} - 10 \text{ N} = (2 \text{ kg})a_x$$
$$10 \sqrt{3} \text{ N} = (2 \text{ kg})a_y$$

from which we obtain

$$a_x = 5 \text{ m/s}^2 \qquad a_y = 5 \sqrt{3} \text{ m/s}^2$$

The magnitude of the acceleration is

$$a = (a_x^2 + a_y^2)^{1/2} = [5^2 + (5 \sqrt{3})^2]^{1/2} \text{ m/s}^2 = 10 \text{ m/s}^2$$

and the angle ϕ in Fig. 4-5, which describes the direction of **a**, is given by

$$\phi = \tan^{-1} \frac{a_y}{a_x} = \tan^{-1} \sqrt{3} = 60°$$

Example

A 2-kg body moving in the xy plane has x and y coordinates given as functions of time by

$$x = 2t^2 + 1$$
$$y = \tfrac{1}{2}t^3 - 4$$

where x and y are measured in meters and t in seconds. Sketch the path of the body during the first 3 s, and find the magnitude and direction of the force at $t = 2$ s. Is **F** in the same direction as **a**? As **v**?

Solution

The path is shown in Fig. 4-6, which can be obtained by computing the x and y coordinates at various times, such as $t = 0$, 1 s, 2 s, and 3 s, as shown. To find the force, we first differentiate to find the components of acceleration, obtaining the components of velocity along the way:

$$v_x = \frac{dx}{dt} = 4t \qquad a_x = \frac{dv_x}{dt} = \frac{d^2x}{dt^2} = 4$$

$$v_y = \frac{dy}{dt} = \tfrac{3}{2}t^2 \qquad a_y = \frac{dv_x}{dt} = \frac{d^2y}{dt^2} = 3t$$

92 PRINCIPLES OF DYNAMICS

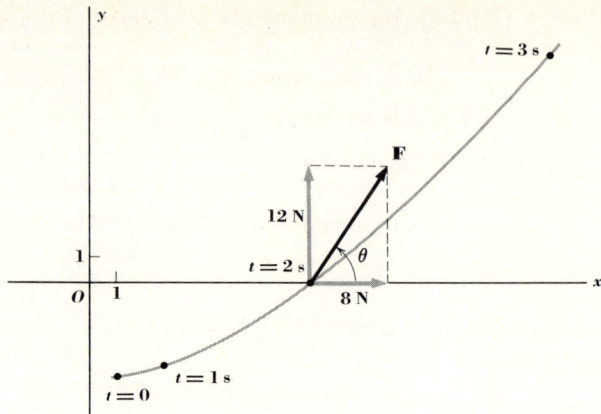

Fig. 4-6 Trajectory of a body whose x and y coordinates are given functions of time, as in the example. The position of the body at time $t = 2$ s and the components of force at this time are shown. From these components the magnitude and direction of the force vector can be obtained.

Then, using $\mathbf{F} = m\mathbf{a}$,

$$F_x = ma_x = (2 \text{ kg}) (4 \text{ m/s}^2) = 8 \text{ N}$$
$$F_y = ma_y = (2 \text{ kg}) (3t \text{ m/s}^2) = 6t \text{ N}$$

At $t = 2$ s, $F_x = 8$ N, $F_y = 12$ N;

$$|F| = \sqrt{F_x^2 + F_y^2} = \sqrt{(8 \text{ N})^2 + (12 \text{ N})^2} = \sqrt{208} \text{ N} \simeq 14.4 \text{ N}$$

$$\theta = \tan^{-1} \frac{12 \text{ N}}{8 \text{ N}} = \tan^{-1} 3/2 \simeq 55.3°$$

Because of the vector relation $\mathbf{F} = m\mathbf{a}$, $\mathbf{a}$ *always* has the same direction as $\mathbf{F}$, as may be verified by finding the direction of $\mathbf{a}$ at $t = 2$ s. We have

$$a_x = 4 \text{ m/s}^2 \qquad a_y = 6 \text{ m/s}^2 \qquad \theta = \tan^{-1} \frac{6 \text{ m/s}^2}{4 \text{ m/s}^2} = \tan^{-1} 3/2$$

The components of $\mathbf{v}$ at $t = 2$ s are

$$v_x = 8 \text{ m/s} \qquad v_y = 6 \text{ m/s}$$

and the direction of $\mathbf{v}$ at this time is given by $\tan^{-1}[(6 \text{ m/s})/(8 \text{ m/s})] \simeq 36.9°$. We note that the velocity does *not* have the same direction as the force, and there is no reason it should. The velocity is, however, tangent to the path at each instant.

The force can also be obtained using vector language throughout. The position vector of the object is given by

SUPERPOSITION OF FORCES [4-5]

$$\mathbf{r} = x\mathbf{i} + y\mathbf{j} = (2t^2 + 1)\,\mathbf{i} + (\tfrac{1}{2}t^3 - 4)\,\mathbf{j}$$

From this we obtain by successive differentiations

$$\mathbf{v} = \frac{d\mathbf{r}}{dt} = 4t\mathbf{i} + \tfrac{3}{2}t^2\mathbf{j}$$

$$\mathbf{a} = \frac{d\mathbf{v}}{dt} = \frac{d^2\mathbf{r}}{dt^2} = 4\mathbf{i} + 3t\mathbf{j}$$

Thus

$$\mathbf{F} = m\mathbf{a} = (2\text{ kg})(4\mathbf{i} + 3t\mathbf{j}) \quad \text{m/s}^2$$
$$= 8\mathbf{i} + 6t\mathbf{j} \quad \text{N}$$

and at $t = 2$ s, $\mathbf{F} = 8\mathbf{i} + 12\mathbf{j}$ N.

Example
A body of mass 5 kg is acted upon by a force with components

$$F_x = 10 \text{ N} \qquad F_y = 5 - 10t \quad \text{N}$$

If it is at rest at the origin at time $t = 0$, find the components of velocity and the coordinates as functions of time.

Solution
This example is the previous one in reverse. First we were given the coordinates as functions of time (and thus the motion) and asked to find the force; this involved taking derivatives to obtain the acceleration. Now we are given the force and asked to find the motion; mathematically speaking this involves the process of *integrating*, which we may think of as *antidifferentiating*. First we use $\Sigma\mathbf{F} = m\mathbf{a}$ in component form to find the components of acceleration:

$$a_x = \frac{dv_x}{dt} = \frac{F_x}{m} = 2 \text{ m/s}^2$$

$$a_y = \frac{dv_y}{dt} = \frac{F_y}{m} = 1 - 2t \quad \text{m/s}^2$$

Next we ask what the functions v_x and v_y must be in order to give the above results when differentiated. The reader can easily verify that if the derivative of a function is given by At^n, where a and n are constants, the function itself must be $At^{n+1}/(n+1) + C$, where C is any constant (since the

94 PRINCIPLES OF DYNAMICS

derivative of a constant is zero. Thus we find for the velocity components

$$v_x = 2t + C_1 \quad \text{m/s}$$
$$v_y = t - t^2 + C_2 \quad \text{m/s}$$

where C_1 and C_2 are some unknown constants. But we also know that at time $t = 0$ we have $v_x = v_y = 0$, and this can be true only if $C_1 = C_2 = 0$. Thus

$$v_x = 2t \text{ m/s} = \frac{dx}{dt}$$

$$v_y = t - t^2 \text{ m/s} = \frac{dy}{dt}$$

Now we repeat the same process to find x and y as functions of time:

$$x = t^2 + C_3 \quad \text{m}$$
$$y = \frac{t^2}{2} - \frac{t^3}{3} + C_4 \quad \text{m}$$

But at time $t = 0$, $x = y = 0$, so C_3 and C_4 must be zero, and

$$x = t^2 \quad \text{m}$$
$$y = \frac{t^2}{2} - \frac{t^3}{d} \quad \text{m}$$

We note that if the initial velocity or position had been different from zero, the C's would not all be zero. For example, the reader may verify that if the particle is initially at the point (3 m, 7 m), the coordinates are

$$x = t^2 + 3 \quad \text{m}$$
$$y = t^2 - \frac{t^3}{d} + 7 \quad \text{m}$$

Thus we see that the "given the motion, find the force" problem is closely related to the operation of differentiation, and the "given the force, find the motion" problem to integration. In later chapters we shall see that the second type of problem often requires solution of a differential equation; this is basically an integration process, but many specialized techniques are available.

4-6 ACTION AND REACTION

In our general discussion of the relations of force and motion, one question remains to be discussed. How are the forces exerted *on* a body related to the forces which the body exerts on *other* bodies? As an example, consider a locomotive pulling a train up a hill. The locomotive exerts a certain force **F** on the coupler of the first car of the train, and the car correspondingly pulls backward on the locomotive with a force **F′**, as shown in Fig. 4-7. The

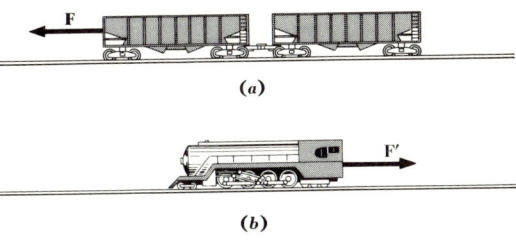

Fig. 4-7 (a) Force exerted on the first car of a train by the locomotive. (b) Force exerted on the locomotive by the first car. The locomotive pulls as hard on the car as the car pulls on the locomotive, but the two pulls are in opposite directions.

two forces appear to be in opposite directions, and it seems reasonable that their magnitudes should be equal. But this hypothesis, like several preceding ones, must be tested experimentally before it is accepted.

Consider the following experiment: Two cars with unequal masses roll on a straight smooth track, as shown in Fig. 4-8. Between the cars we place

Fig. 4-8 Two cars on a smooth track with a compressed spring which pushes them apart when the string holding them is cut.

a spring. We compress the spring and tie the two cars together (with the spring compressed). Then we burn the string; the cars fly apart, and we can measure their final velocities.

If the forces which the cars exert on each other are opposite in direction but equal in magnitude, we can predict how their final velocities should be related.

Let the masses of the two cars be m_1 and m_2, the forces acting on them, $\mathbf{F}_1$ and $\mathbf{F}_2$, and their accelerations, $\mathbf{a}_1$ and $\mathbf{a}_2$. If the forces are equal in magnitude but opposite in direction, then $\mathbf{F}_1 = -\mathbf{F}_2$, and, correspondingly,

$$m_1 \mathbf{a}_1 = -m_2 \mathbf{a}_2 \tag{4-6}$$

That is, the ratio of the accelerations of the two objects is constant at all times during the motion. Since the acceleration of each object is the rate of change of its velocity, the two final velocities, $\mathbf{v}_1$ and $\mathbf{v}_2$, must be in the same ratio:

$$m_1 \mathbf{v}_1 = -m_2 \mathbf{v}_2 \tag{4-7}$$

This is reasonable; the more massive car should have the smaller final velocity.

Experiment shows that the final velocities *are* related in the manner predicted by assuming that $\mathbf{F}_2 = -\mathbf{F}_1$. Furthermore, more detailed observations in which the accelerations of the cars are measured at various instants show directly that Eq. (4-6) is satisfied.

The generalization resulting from these observations is *Newton's third law of motion*, which may be stated as follows: *If body 1 exerts a force on body 2, body 2 simultaneously exerts a force on body 1 which is equal in magnitude but opposite in direction to the force on it.* Both forces may vary with time, but at each instant $\mathbf{F}_2 = -\mathbf{F}_1$.

This is often summarized by saying "action equals reaction" or "for every action there is an equal and opposite reaction," in which *action* is to be interpreted as force.

Example

A locomotive with mass m pulls a train with total mass $5\,m$ out of a station with constant acceleration a. What force is required to accelerate the train, how is this force applied, and what is the tension in the coupler linking the locomotive to the train?

Solution

In solving this problem we introduce an important and useful problem-solving technique, the use of *free-body diagrams*. The technique is to isolate the system under discussion, or a part of the system, draw a diagram showing only the isolated part, and indicate the interactions with everything else by showing the forces acting on the isolated part. In the present problem we need consider only horizontal forces. First, taking the train as a whole, we note that the total mass is $6m$, and the horizontal force must therefore have magnitude $F = (6m)a = 6ma$. If the cars roll freely without friction, this force must be applied to the locomotive drive wheels by the track, as shown in Fig. 4-9a. If this direction of force seems wrong, recall that a sprinter starting off pushes backward with his shoes on the track, and the corresponding reaction force exerted *on* the shoes by the track must be forward. The present situation is analogous.

To find the coupler forces we must take the train apart. Figure 4-9b is a free-body diagram of the locomotive, showing the force $F = 6ma$ just

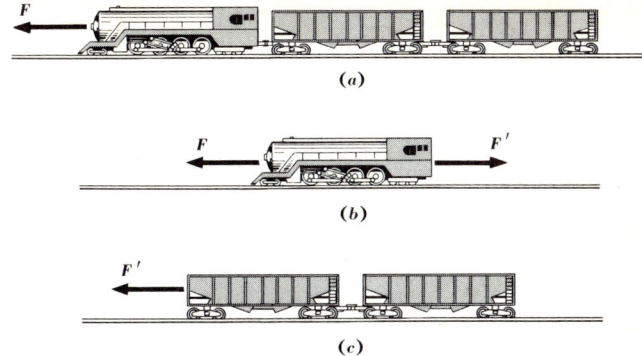

Fig. 4-9 Free-body diagrams showing horizontal forces on train: (*a*) forces on train considered as a system, (*b*) forces on locomotive, (*c*) forces on cars considered as a system.

obtained and the backward force F' at the coupler. Note that these are the forces *on* the locomotive, not the forces it exerts on anything else. Applying $\Sigma \mathbf{F} = m\mathbf{a}$ to this system, we find

$$F - F' = 6ma - F' = ma$$
$$F' = 5ma$$

An alternative procedure is to use a free-body diagram for the cars of the train, as in Fig. 4-9*c*. The only horizontal force on the cars is F', so we find immediately

$$F' = (5m)a$$

in agreement with the previous result.

Newton's third law of motion has a number of important consequences, one of which is indicated by Eq. (4-7). The product $m\mathbf{v}$ of the mass of a body and its velocity is called the *momentum* of the body. This is a vector quantity having the same direction as $\mathbf{v}$. Equation (4-7) shows that the vector sum of the momenta of the two cars in our experiment is zero after the two cars have separated, despite the fact that each has a definite velocity. At the beginning of the experiment, both cars were at rest, so the total momentum was zero. Thus the total momentum of the system is the same at the end of the process as at the beginning. This is a very simple illustration of another conservation principle, the principle of *conservation of momentum*, which is discussed in detail in Chap. 7.

4-7 FRAMES OF REFERENCE

It has been pointed out in Chap. 3 that the velocity and acceleration of a body depend on the motion of the observer. To say that a body has a certain velocity or acceleration has meaning only when we specify the coordinate system in which the position is measured. The position, velocity, and acceleration of an object depend on the *frame of reference* used to measure these quantities.

Motion of the frame of reference is an important consideration in the formulation of Newton's first and second laws of motion. Suppose a body is moving with constant velocity (as observed by one observer) with no forces acting on it, in accordance with Newton's first law. If another observer measures the velocity, does it always appear to him to be constant? The answer, quite clearly, is no. If this observer is *accelerated* with respect to the first, then the body appears to him to have a nonzero acceleration, even though no forces act on it. If Newton's first law is obeyed for one observer or in one frame of reference, it is *not* obeyed for another observer who is accelerated relative to the first.

Similar statements can be made with regard to Newton's second law. If a body has an acceleration with respect to one frame of reference but none with respect to another, then the equation $\Sigma \mathbf{F} = m\mathbf{a}$ cannot be valid in both. One could, of course, take the point of view that the forces are different when viewed in the two frames of reference and contrive the definition of force for each system so as to achieve agreement with $\Sigma \mathbf{F} = m\mathbf{a}$ in both systems. But this complicates the whole situation considerably. It is simpler, as well as more reasonable physically, to take the point of view that the forces are the same in all coordinate systems but that Newton's laws are valid only in certain special ones.

Given an observer in a particular frame of reference, how do we know whether or not the laws of motion will be obeyed? One is tempted to answer this question simply by saying that Newton's laws are obeyed if the frame of reference is not accelerated. But this immediately suggests the question, *accelerated with respect to what?* And then we are back where we started!

The point is that the only possible way to establish that Newton's laws hold in a particular frame of reference is by experiment. A coordinate system in which Newton's laws are found to be valid is called an *inertial frame of reference*. Our simple experiments with air-supported pucks used the earth's surface as a frame of reference. There Newton's laws are obeyed, at least within the precision of these experiments. The earth, then, is at least approximately an inertial frame of reference.

Very precise observations show that the earth is *not quite* an inertial frame of reference. This can be understood at least in part by remembering that the earth rotates about its axis and travels in an orbit about the sun. If the sun, rather than the earth, is an inertial frame of reference, a coordinate system on the earth has small accelerations relative to the coordinate system on the sun, which make it not quite inertial. Similarly, if the sun is accelerated relative to other stars, it may deviate slightly from being an inertial frame. We should be able to observe these deviations, at least in principle, by observation of the motions of the sun and other stars.

Thus, we are certain that inertial frames of reference exist, and in practice we can approximate one as closely as the accuracy of our observations. For motion of objects on the earth, it is often sufficient to consider the earth as an inertial frame of reference. In discussing the motion of the earth around the sun, this is clearly not sufficient; but it may be sufficient to consider the sun as an inertial frame. The choice of a coordinate system and the assumption that a particular frame of reference is inertial for purposes of a particular problem become part of the simplified *model* which we construct for the problem.

This discussion may have suggested that there is in the universe only one truly inertial frame of reference. Fortunately, nature is not so miserly as this. We recall that in the discussion of relative velocity in Sec. 3-6 it was shown that if two observers (located at points O and B in that discussion) move relative to each other with *constant relative velocity* $\mathbf{v}_B$, they obtain the same measurement of the acceleration of an object at point A. Now suppose that in the system O, Newton's second law $\mathbf{\Sigma F} = m\mathbf{a}_A$ is valid. Since $\mathbf{a}_A = \mathbf{a}_{A/B}$, we may also write $\mathbf{\Sigma F} = m\mathbf{a}_{A/B}$. We have assumed that the forces and mass are the same in both systems; thus Newton's second law is also valid for the observer at B.

Therefore any frame of reference which moves with constant velocity with respect to an inertial frame is also inertial. If we can find one inertial frame of reference, we can also find infinitely many others, each moving with constant velocity relative to the original. There is no single frame of reference in the universe which is favored over all others for describing motion; all inertial frames of reference are equivalent.

Another way of saying the same thing is that the laws of motion are *invariant*, or unchanging, with respect to a transformation from one inertial frame of reference to another.

The concept of the invariance of physical principles with respect to transformations from one coordinate system to another has been of tremendous importance in the development of theoretical physics. One of the problems which led to the development of the special theory of relativity was the discovery that the equations of electromagnetism are *not* invariant under transformations of the type we have discussed. A study of the modifications which must be made on the transformation in order for the laws of electrodynamics to have the same form in two coordinate systems led to the Lorentz transformations and later to the special theory of relativity. These transformations are discussed in detail in Chap. 14.

Problems

4-1 A car whose weight is 3,200 lb (mass 100 slugs) and traveling initially at 60 mi/h is brought to rest in a distance of 320 ft in a panic stop. What force is exerted on the car during this stop?

4-2 A bullet whose mass is 5 g leaves the muzzle of a rifle with a speed of 300 m/s. If the barrel is 1.0 m in length, what force was exerted on the bullet while it was in the barrel? State any assumptions made in constructing a simplified model of the situation.

4-3 A car whose weight is 4,000 lb (mass 125 slugs) accelerates from rest to 90 ft/s (about 60 mi/h) in 10 s. What average force is necessary to produce this acceleration? How is the force applied to the car?

4-4 A small European car has a mass of 600 kg. It can accelerate from 10 to 15 m/s in a distance of 50 m. What average force is exerted on the car?

4-5 A body whose mass is 2 kg lies on a smooth horizontal plane to which is attached a rectangular coordinate system.
   a If a force of 6 N in the $+x$ direction is applied, find the acceleration.
   b If a force of 8 N in the $+y$ direction is applied, find the acceleration.
   c If both forces are applied, find the acceleration (magnitude and direction).
   d Find the x and y components of the acceleration in part *c*.

4-6 If the units of force, length, and time were regarded as fundamental units, rather than mass, length, and time, what would be the units of mass in terms of the fundamental units?

4-7 A subway train consists of four cars, each having mass m. Only the first car contains motors. The train starts from rest with constant acceleration a.
   a Find the tension in each coupler between cars.
   b Describe *all* the forces acting on the first car.
   Hint: Free-body diagrams are indispensable!

4-8 A child's toy circus train consists of three wagons coupled together, with a pull string on the first wagon. Each wagon has a mass of 0.5 kg.
   a With what force must the child pull to give the train an acceleration of 0.2 m/s^2?
   b With this acceleration, find the forces in the couplers joining the first and second cars and the second and third.

4-9 An arrow having a mass of 50 g is shot from a bow. The bowstring is drawn back until each half makes an angle of 5° with its original direction; at this time, the tension in the bowstring is 400 N. Find the initial acceleration of the arrow when it is released.

4-10 A boy pulls a four-wheel cart of mass 15 kg along a horizontal surface with a rope, pulling upward at 30° to the horizontal.
   a How much force is required to maintain a uniform velocity?
   b How much force is required for a constant acceleration of 3.0 m/s^2?

4-11 A particle of mass m moves along a straight line, and its displacement x from the origin is given by

$$x = A \sin \omega t$$

where A and ω are constants.
   a Find the force on the particle as a function of time.
   b Show that at each instant the force and displacement are related by $F = -m\omega^2 x$; that is, the force is directly proportional to the displacement and in the opposite direction.

4-12 A particle of mass 2 kg is initially at rest at the origin. Starting at time $t = 0$, it experiences a force directed along the x axis, given by $F = 4 - t$ (in newtons).
   a Find the velocity and position of the particle as functions of time.
   b During what time intervals are the force and velocity in opposite directions? The force and displacement?
   c Draw graphs showing acceleration, velocity, and displacement as functions of time.

4-13 A box whose mass is 10 kg rests on the flat floor of a truck. The truck starts with an acceleration of 1 m/s^2. The floor is fairly smooth, and the box begins to slide; its motion is opposed by a small frictional force of 5 N.
   a If the box is initially 5 m away from the back edge of the truck, how much time elapses before it falls out?
   b What distance does the truck travel before the box falls out?

4-14 In Prob. 3-19, find the x and y components of force on the particle. Show that the magnitude of the force is constant and is given by $F = m\omega^2 R$ and that the force always acts in a direction from the particle toward the origin.

4-15 A truck moves with constant acceleration **A** relative to the earth. In the back of the truck is a billiard table. If a billiard ball of mass m is subjected to several forces, and if its acceleration **a** relative to the table is measured, is it true that $\Sigma \mathbf{F} = m\mathbf{a}$? If not, formulate a correct equation relating the forces, the mass, and the acceleration.

4-16 A man is riding in an elevator at the top of a very tall building. The cables have all broken, and the safety devices have all failed, so the elevator is falling freely. In his excitement, the man lets go of his briefcase. Describe its motion with respect to the elevator. Does this behavior illustrate a violation of Newton's second law? Explain.

4-17 A body of mass 5 kg moves in the xy plane under the action of a force given by $\mathbf{F} = 4t\mathbf{i} + 6t^2\mathbf{j}$ N.
a If the body is initially at rest at the origin, obtain expressions for the velocity and position vectors as functions of time.
b Sketch the path of the body.
c How do your results change if the initial position is (2 m, 3 m) instead of (0,0)?
d If the initial value of y (at $t = 0$) is -8 m, what must be the initial value of x in order for the path to pass through the origin, assuming zero initial velocity as before?

4-18 Suppose a given cartesian coordinate system is known to be an inertial frame of reference. Another set of coordinate axes has its origin at the same point as the first but is rotating with respect to it about one of the coordinate axes. Is this an inertial frame of reference? Explain.

4-19 Two blocks rest on a perfectly smooth surface and are connected by a spring. The blocks are pulled apart, stretching the spring, and are then released without any initial velocity.
a Will the blocks always collide? Explain.
b At what point does the collision occur?

4-20 A locomotive engineer reads an excerpt from a freshman physics text and then decides to quit his job. His reason is that, according to Newton's third law, the train always pulls backward on the locomotive with a force just as great as that which the locomotive exerts on the train, and therefore the train can never move. As personnel supervisor, you are assigned the task of explaining the situation. Explain it.

4-21 A simple accelerometer for a car can be made by attaching one end of a spring to the dash panel and the other end to a mass which rests on a nearly frictionless surface. When the car accelerates, the spring stretches by an amount which depends on the acceleration. Suppose the mass is 2.0 kg and the spring force is directly proportional to the distance it is stretched, a force of 1.0 N producing a stretch of 0.05 m. In a particular trial the spring stretched 0.15 m. What was the acceleration?

4-22 A body of mass m slides without friction on a smooth horizontal surface with velocity **v**. It is proposed to change this motion so that the velocity is perpendicular to its original direction but has the same magnitude. This change is to be produced by applying a constant force **F** in a direction at an angle of 135° to the original velocity. Show that such a force will produce the desired effect, and find the time the force must act to produce the change.

Dynamics of a Particle I | 5

We consider in this chapter several applications of the principles of dynamics developed in Chap. 4. Using a point as a model for a small body, we discuss various situations in which the relationship between force and acceleration is useful. Motion under the influence of gravity is one of the most familiar phenomena in everyday experience, so it is natural to start with several examples of motion of this kind.

5-1 MASS AND WEIGHT

The acceleration of a freely falling body is constant and is the same for all bodies. This statement was discussed in Sec. 1-4; it was observed that it is exactly correct only in idealized situations in which air resistance and other extraneous effects are completely absent. Nevertheless, there are many cases where a reasonably precise description of motion can be obtained from a model in which the acceleration is assumed to be constant.

The acceleration of falling objects on the earth, called the *acceleration of gravity* and denoted by g, has been measured by many different methods with various degrees of precision. The numerical value of g varies slightly at different points near the earth's surface, as shown in Fig. 5-1. In this text, we usually use, for convenience, the values

$$g = 9.80 \text{ m/s}^2$$

in mks units, and

$$g = 32.2 \text{ ft/s}^2$$

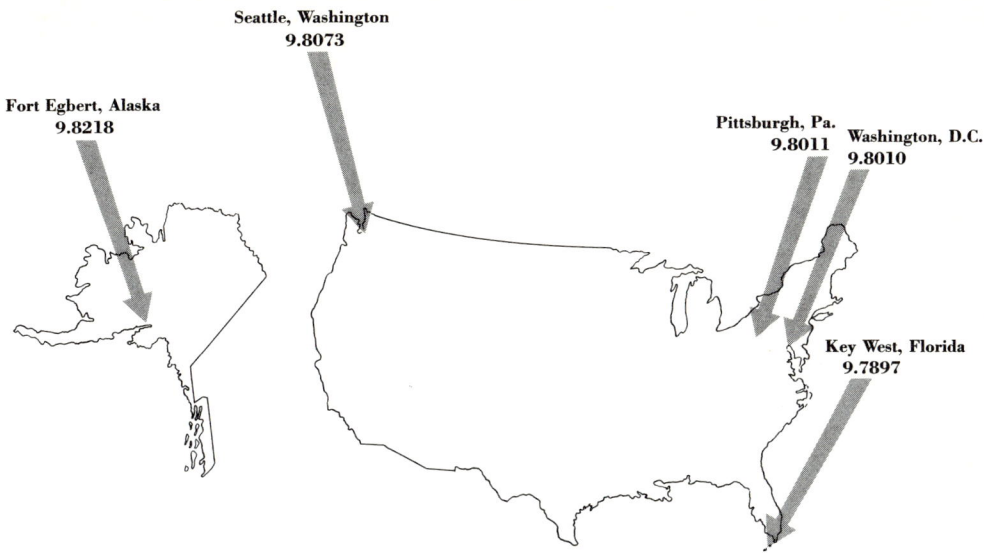

Fig. 5-1 Values of the acceleration of gravity g at various locations.

in the British system. These are in error by not more than 0.2 percent for points in North America.

The force responsible for the acceleration of a falling body is the earth's gravitational attraction, which we call the *weight* of the body. The weight of any body is therefore a *force* which acts in the same direction as the acceleration it would have if permitted to fall freely. According to Newton's second law, the force F required to give a mass m an acceleration g is

$$F = mg$$

But this force is just what we have called the weight of the body. Introducing the symbol W for weight, we write

$$W = mg \qquad (5\text{-}1)$$

The weight of a body is proportional to its mass. For example, a body having a mass of 2.0 kg has a weight of

$$W = (2.0 \text{ kg})(9.8 \text{ m/s}^2) = 19.6 \text{ kg·m/s}^2 = 19.6 \text{ N}$$

Weight and mass are two distinctly different physical quantities. The mass of a body describes its inertial properties and depends on the quantity of matter contained in the body, while weight is a force resulting from the earth's attraction for a body. The units in which the two quantities are measured are different. The fact that mass and weight are proportional is an experimental fact deduced from the observation that all bodies fall with the same acceleration. It is not something which could have been predicted without this essential observation.

Strictly speaking, Eq. (5-1) should be written as a *vector* equation. Since the weight of a body is a force, it is a vector quantity; the acceleration of gravity is also a vector quantity, pointing in the direction of free fall. Thus we may write

$$\mathbf{W} = m\mathbf{g} \tag{5-2}$$

It has already been pointed out that the acceleration of gravity is somewhat different at various points on the earth. On other planets or on the moon its value may be *much* different than on earth. Thus the weight of a body is not simply a property of the body but is determined jointly by its mass and the characteristics of the planet or other aggregation of matter to which it is attracted. On the surface of the moon, for example, the acceleration of gravity is about 1.67 m/s^2, so that a body whose mass is 1 kg has a weight on the moon of 1.67 N, rather than 9.80 N as on earth. The *mass* of this body, however, is the same on earth as on the moon. Close to the surface of the sun, $g = 274$ m/s^2!

We often say, in common speech, that a certain quantity of material "weighs 1 kg." This is incorrect; what we mean is that the *mass* of material is 1 kg. But because the weight of a quantity of matter at a given location on the earth's surface is always proportional to its mass, one can measure masses indirectly by measuring weights. When we place two bodies on a chemical balance, we are comparing not their *masses* but their *weights;* if two bodies have equal weights in the same gravitational field, they also have equal masses. The inscriptions on standard bodies ("weights") used for weighing, such as 50 g or 10 kg, are always units of mass.

5-2 FREE FALL

When a body moves in a vertical line under the influence of no force other than its weight, the acceleration is directed downward and has constant magnitude g. It is easy to calculate the details of the motion of such a body under these idealized conditions. We performed such a calculation in Chap. 3 when we computed various quantities related to the motion of a ball thrown straight up.

Example

With what initial velocity must a standing high jumper leave the ground if he jumps to a maximum height h? How much time elapses between the time he leaves the ground and the time he reaches his maximum height?

Solution

We take the origin of coordinates at ground level and denote the vertical distance of the jumper above the ground at any time by y. This coordinate choice means that velocity is positive when it is upward and negative when downward, and similarly for acceleration. The acceleration is constant and is given by

$$\frac{d^2y}{dt^2} = -g \qquad (5\text{-}3)$$

The minus sign appears because the force of gravity acts downward, while we have chosen the positive direction of y as upward. The velocity at any time is obtained by integrating Eq. (5-3) to obtain

$$v = \frac{dy}{dt} = -gt + v_0 \qquad (5\text{-}4)$$

where v_0 is the velocity at time $t = 0$. Integrating again, we obtain the position as a function of time:

$$y = -\tfrac{1}{2}gt^2 + v_0 t + y_0 \qquad (5\text{-}5)$$

where y_0 is the position at time $t = 0$. If the coordinate system is chosen so that $y = 0$ at $t = 0$, then the initial position is $y_0 = 0$. For the reader who is not yet familiar with integration, Eqs. (5-4) and (5-5) can be obtained directly from Eqs. (3-10) and (3-11) with appropriate symbol changes.

When the jumper reaches the top of his jump, his velocity is instantaneously zero. From Eq. (5-4) it can be seen that $dy/dt = 0$ at a time t_1 given by

$$t_1 = \frac{v_0}{g}$$

Inserting this value of t in Eq. (5-5) for the position (with $y_0 = 0$), we find that the height h at this time is

$$h = -\frac{g}{2}\left(\frac{v_0}{g}\right)^2 + v_0\frac{v_0}{g} = \frac{v_0{}^2}{2g}$$

108 DYNAMICS OF A PARTICLE I

The initial velocity, the maximum height, and the acceleration of gravity are related by this equation. To reach any given height h, the jumper must have an initial velocity

$$v_0 = \sqrt{2gh}$$

On the moon, where the acceleration of gravity has only about one-sixth its value on earth, one can jump about six times as high as on the earth for a given value of v_0.

When other forces in addition to gravity act on a falling body, the motion is more complicated. In practical situations, the retarding force of air resistance is often important. This force is not constant but increases with the velocity of the object relative to the air. In discussing this problem we again make use of the idea of a free-body diagram introduced in Sec. 4-6. A free-body diagram for a body moving downward under the action of gravity and air resistance is shown in Fig. 5-2. The force F in this diagram represents

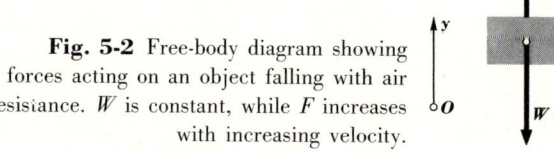

Fig. 5-2 Free-body diagram showing forces acting on an object falling with air resistance. W is constant, while F increases with increasing velocity.

air resistance opposing the motion. The choice of coordinates is always arbitrary; this time we take the positive direction of y to be *downward*, so forces, velocities, and accelerations are positive when downward, negative when upward.

The force of air resistance is not constant but depends on the velocity; therefore the acceleration of the object is not constant. If we know how the force of air resistance varies with velocity, we can still write an equation of motion based on $\Sigma \mathbf{F} = m\mathbf{a}$. As a simple model to describe air resistance approximately, we may assume it to be *directly proportional* to velocity. That is,

$$F = -kv = -k\frac{dy}{dt} \tag{5-6}$$

In Eq. (5-6), k is a proportionality constant. The minus sign denotes that F and v are in opposite directions; F always *opposes* the motion. The weight is mg, acting downward (the negative direction).

Applying $\Sigma \mathbf{F} = m\mathbf{a}$ to this problem, we find

$$-mg - k\frac{dy}{dt} = m\frac{d^2y}{dt^2} \tag{5-7}$$

This equation does not give y as a function of t but is a *differential equation* relating various derivatives of y. To complete the solution we must find a function $y(t)$ which is consistent with this equation and with whatever information is available about the initial position and velocity. This process, called *solving* the differential equation, involves specialized techniques with which the reader may not yet be familiar. Even without these techniques, however, we can see several interesting features of the motion.

Suppose first the body is released from rest at the origin. Initially the term $k\,dy/dt$ in Eq. (5-7) corresponding to air resistance is zero, and the accelerating force is mg. As the velocity increases, $k\,dy/dt$ becomes larger, the net accelerating force is *less* than mg, and the acceleration decreases. Finally, a speed is reached at which $-mg = k\,dy/dt$, at which point the acceleration becomes zero and the object moves with constant speed. The final velocity, called the *terminal velocity*, is given by $-mg/k$. If the body starts from rest, it never exceeds this velocity, no matter how long it falls. At terminal velocity, the downward force of gravity is just balanced by the upward force of air resistance, so $\Sigma \mathbf{F} = \mathbf{0}$.

Figure 5-3 shows graphs of the acceleration, velocity, and position of

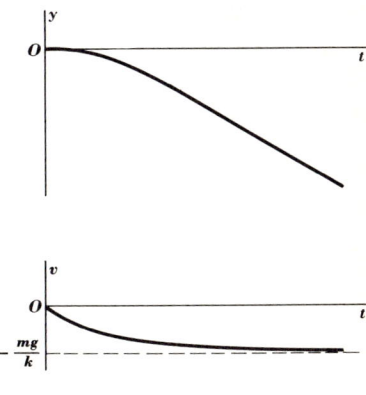

Fig. 5-3 Position, velocity, and acceleration as functions of time for an object falling with air resistance proportional to velocity. The terminal velocity is shown.

the body as functions of time. As time increases, the acceleration decreases from its initial value g to zero. The velocity increases from zero to its asymptotic terminal value, mg/k. The graph of y as a function of t shows at first a proportionality to t^2, since for small values of t the retarding force of air resistance is not important. At larger values it approaches a straight line as the velocity approaches the terminal velocity and becomes nearly constant.

The constant k in Eq. (5-6) depends on the size and shape of the body, the temperature of the air, and many other factors. A man falling through the air has a terminal velocity on the order of 50 to 100 m/s; a feather may have as little as 0.1 m/s. For the reader with some acquaintance with exponential functions, we quote the solution of Eq. (5-7). For a body initially at rest at $y = 0$,

$$y = -\frac{mg}{k}t + \frac{m^2 g}{k^2}(1 - e^{-(k/m)t}) \tag{5-8}$$

Several questions about this solution are raised in the problems for this chapter.

The assumption that the force of air resistance is directly proportional to velocity is not very realistic; in actual fact, it usually depends on velocity in a more complicated manner. Nevertheless, the simple model in which air resistance is proportional to velocity is sometimes a useful approximation.

5-3 BALLISTIC TRAJECTORIES

In the examples of the preceding section, the forces and therefore the acceleration were always along the line of the initial velocity, and consequently the object moved in a straight line. But what happens when the body has a component of initial velocity which is *not* in the same direction as the acceleration? Suppose, for example, that a ball is thrown with an initial velocity of magnitude v_0 inclined to an angle α to the horizontal and that after it is released, no forces except gravity act on it. In this case the motion is confined to a vertical plane determined by the initial velocity; there is no force perpendicular to this plane. We can also include in this model the assumption that the earth is an inertial frame of reference.

A convenient coordinate system is shown in Fig. 5-4, in which the y axis is vertical and the x axis is horizontal; the motion then lies in the xy plane. We assume for simplicity that the ball starts at time $t = 0$ at the origin. Our task now is to find its x and y coordinates as functions of time; these functions constitute a complete description of the motion.

The governing principle, as always, is **ΣF = ma.** In our model the

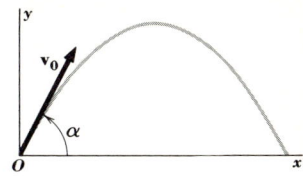

Fig. 5-4 Convenient coordinate system for describing the trajectory of a ballistic missile. The missile is assumed to be launched from point O at time $t = 0$. The magnitude and direction of the initial velocity are shown.

only force is the downward force of gravity. Thus there is no horizontal acceleration, and the vertical acceleration is constant and equal to $-g$. That is, in our coordinate system

$$a_x = \frac{d^2x}{dt^2} = 0 \qquad a_y = \frac{d^2y}{dt^2} = -g \tag{5-9}$$

The initial velocity $\mathbf{v}_0$ can also be represented by its x and y components:

$$v_{0x} = v_0 \cos \alpha \qquad v_{0y} = v_0 \sin \alpha \tag{5-10}$$

The crucial point in the analysis of this problem is that by representing $\mathbf{\Sigma F} = m\mathbf{a}$ in terms of components we can treat the horizontal and vertical components of the motion independently. Because there is no horizontal force, $a_x = 0$; and thus v_x is constant and equal at all times to its initial value $v_0 \cos \alpha$. We have chosen the coordinate system so $x = 0$ when $t = 0$. Thus at any later time t, $x = v_0 \cos \alpha \, t$.

A similar analysis may be made for the y components. The vertical motion is not altered by the presence of a horizontal component; the vertical acceleration is still $-g$, and v_y and y vary just as in the case of vertical free fall. The vertical velocity changes uniformly with time, starting with the initial value $v_{0y} = v_0 \sin \alpha$. The formulas of Sec. 3-3 for constant acceleration are applicable; making the appropriate substitutions, and collecting the formulas for the x and y components, we obtain

$$v_x = v_0 \cos \alpha \qquad v_y = v_0 \sin \alpha - gt \tag{5-11}$$
$$x = v_0 \cos \alpha \, t \qquad y = v_0 \sin \alpha \, t - \tfrac{1}{2}gt^2 \tag{5-12}$$

These expressions can also be obtained by integrating Eqs. (5-9) directly, using the appropriate initial values of velocity and position.

Equations (5-11) are illustrated in Fig. 5-5, sketched from a multiple-flash photograph of the trajectory of a small steel ball. The x component of velocity is constant. The y component, initially positive, becomes smaller, eventually reaches zero, and then becomes negative and larger. For comparison, the figure also shows the vertical velocity of an object projected straight up with the same value of v_{0y}, but with $v_{0x} = 0$.

A variety of useful information can be obtained from Eqs. (5-11) and

112 DYNAMICS OF A PARTICLE I

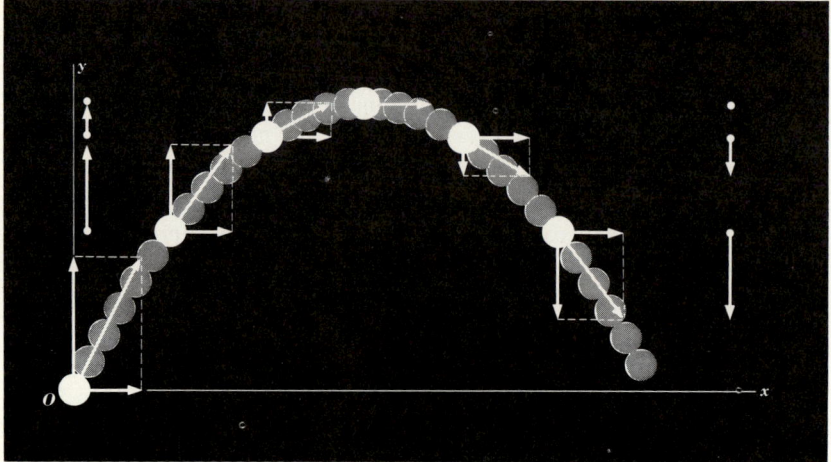

Fig. 5-5 Components of velocity of a ballistic missile at various points in its trajectory. The points are separated by equal time intervals. The vertical arrows at the sides of the figure represent the velocity of a missile fired straight up with initial velocity v_{y0} at corresponding times.

(5-12). For example, to find the maximum height reached by the ball, we find the time at which the vertical component of velocity is zero; this is the time it stops rising and begins to descend. From the second of Eqs. (5-11), we see that this occurs when

$$t = \frac{v_0 \sin \alpha}{g} \tag{5-13}$$

The corresponding height h is found by inserting Eq. (5-13) into the second of Eqs. (5-12):

$$h = +v_0 \sin \alpha \frac{v_0 \sin \alpha}{g} - \frac{g}{2}\left(\frac{v_0 \sin \alpha}{g}\right)^2 = \frac{v_0^2 \sin^2 \alpha}{2g} \tag{5-14}$$

Similarly, we can find the horizontal *range* of the ball. To do this, we first find the time at which it returns to earth by setting $y = 0$ in the second of Eqs. (5-12). We find that $y = 0$ when

$$t = 0 \qquad t = \frac{2v_0 \sin \alpha}{g} \tag{5-15}$$

The first is the time the ball *leaves* the ground; the second is the time it returns. The range is simply the value of x at this time, obtained by inserting this value of t in the first of Eqs. (5-12). Denoting the range by R, we find

$$R = v_0 \cos \alpha \, \frac{2v_0 \sin \alpha}{g} = \frac{v_0^2 \sin 2\alpha}{g} \tag{5-16}$$

It can be shown that the *maximum* range occurs when $\alpha = 45°$. Proof of this statement is left as an exercise.

The equation describing the shape of the trajectory can be found from Eqs. (5-12) by eliminating the variable t between them. This is most easily accomplished by solving the x equation for t and substituting the result into the y equation. The result is

$$y = -\frac{gx^2}{2v_0^2 \cos^2 \alpha} + \tan \alpha \, x \tag{5-17}$$

The general *form* of this relationship is

$$y = Ax - Bx^2 \tag{5-18}$$

where A and B are constants which depend on the initial velocity, the angle, and the acceleration of gravity. The curve described by Eq. (5-18) is a *parabola*. We have thus shown that an object moving through the air under the action of gravity alone moves in a *parabolic* path.

The results contained in Eqs. (5-12) can be obtained somewhat more simply by using vector language. The initial velocity is $\mathbf{v_0}$, and the acceleration is $-g\mathbf{j}$. If the position of the projectile at any time t is described by the vector $\mathbf{r}$ from the origin O to the projectile, then

$$\mathbf{a} = \frac{d^2\mathbf{r}}{dt^2} = -g\mathbf{j} \tag{5-19}$$

Integrating this expression with respect to time, and using the fact that at $t = 0$ the velocity is $\mathbf{v_0}$, we obtain

$$\mathbf{v} = \frac{d\mathbf{r}}{dt} = -gt\mathbf{j} + \mathbf{v_0} \tag{5-20}$$

Finally, integrating again and using the fact that at time $t = 0$ the position is $\mathbf{r} = \mathbf{0}$ gives

$$\mathbf{r} = -\tfrac{1}{2}gt^2\mathbf{j} + \mathbf{v_0}t \tag{5-21}$$

Expressing v_0 in terms of its components,

$$\mathbf{v_0} = v_0 \cos \alpha \, \mathbf{i} + v_0 \sin \alpha \, \mathbf{j}$$

we obtain

$$\mathbf{r} = (v_0 \cos \alpha \; t)\mathbf{i} + (-\tfrac{1}{2}gt^2 + v_0 \sin \alpha \; t)\mathbf{j} \tag{5-22}$$

This is a vector equation whose components are the two scalar equations (5-12).

We again consider the effect of air resistance. For example, if the force of air resistance is proportional to the velocity and in the opposite direction $(\mathbf{F} = -k\mathbf{v})$, the total force on the body is

$$\mathbf{F} = -mg\mathbf{j} - k\mathbf{v} = m\mathbf{a} \tag{5-23}$$

Equation (5-19) is then replaced by

$$\mathbf{a} = \frac{d^2\mathbf{r}}{dt^2} = -g\mathbf{j} - \frac{k}{m}\frac{d\mathbf{r}}{dt} \tag{5-24}$$

which is equivalent to the two scalar equations

$$\frac{d^2x}{dt^2} = -\frac{k}{m}\frac{dx}{dt} \qquad \frac{d^2y}{dt^2} = -g - \frac{k}{m}\frac{dy}{dt} \tag{5-25}$$

These equations, like those for the one-dimensional motion of a falling object, can be solved for x and y as functions of t.

The y equation is identical to Eq. (5-7) for straight-line motion, except for a sign difference resulting from our choice of the upward direction for y as positive. Thus the y component of velocity eventually reaches a terminal value independent of the initial value. The x velocity decreases continuously toward zero; the horizontal force always tends to reduce this component of velocity.

5-4 EXAMPLES

In this section we discuss two situations which can be represented by a model in which a particle moves under the action of a constant force. Several general comments about problem-solving strategy are also included. We consider first a car rolling down a long, straight hill, inclined at an angle θ with the horizontal, without the use of brakes or engine. What is its acceleration?

First we identify the forces by means of a free-body diagram, as discussed previously. We remove the road and all other surroundings from the picture and draw only the car itself, with all the forces acting on it, as in Fig. 5-6. The forces are the weight $\mathbf{W}$ of the car, acting downward, and the force $\mathbf{N}$ exerted by the road on the car. To determine the direction of $\mathbf{N}$, we note

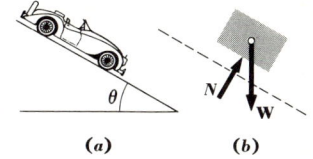

Fig. 5-6 (*a*) Physical problem. (*b*) Free-body diagram from a simplified model of the physical situation.   (*a*)   (*b*)

that if the car rolls without any help from the engine or hindrance from the brakes, it might just as well be sliding down a perfectly frictionless hill. That is, in this model the contact between the tires and the road has no effect either in slowing the car or in accelerating it. Hence, **N** cannot have any component parallel to the surface of the road, and it must be perpendicular to the surface. In fact, the letter **N** stands for *normal*, a synonym for perpendicular.

It is important to understand what forces are *not* included in the free-body diagram. The car pushes on the road with a force equal in magnitude but opposite in direction to **N**; that is, −**N**. This force is *not* included in the free-body diagram because it is not a force exerted *on* the car but a force which the car exerts on something else. Such forces are *never* included in a free-body diagram. In applying the principle $\Sigma \mathbf{F} = m\mathbf{a}$, only the forces acting *on* the body under study are included.

Next we introduce a coordinate system. The choice of directions for our coordinate axes is arbitrary, but since we are interested in acceleration along the road, it is natural to take this as one of the coordinate-axis directions, as shown in Fig. 5-7. In this figure, the vector **W** has been replaced by

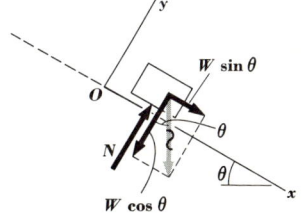

Fig. 5-7 Coordinate system used to analyze the simplified model of a car rolling down a hill. The weight of the car is represented by its *x* and *y* components in this coordinate system.

its components. It is always advisable to cross out the original vector on the diagram, as shown; otherwise there is danger of including the same force twice in the calculations.

It is clear that there is no acceleration in the *y* direction; hence $\Sigma F_y = ma_y = 0$. That is,

$$N - W \cos \theta = 0$$

116 DYNAMICS OF A PARTICLE I

The force N with which the road pushes up on the car is *less than* the weight of the car.

The only force in the x direction is $W \sin \theta$. The acceleration a in this direction can then be determined from

$$W \sin \theta = ma$$

But $W = mg$, so

$$a = g \sin \theta \tag{5-26}$$

The acceleration is independent of the mass of the car. This should not be astounding; if the mass, which characterizes the inertia, is increased, then the weight, which causes the acceleration, also increases proportionately. Therefore, just as for a freely falling body, the acceleration is independent of the mass.

It is useful to check Eq. (5-26) for some particular values of the angle θ. First, when $\theta = 0$, the acceleration becomes zero, as we should expect for a flat road. Also, if $\theta = 90°$, the hill is vertical, and the acceleration becomes simply g. Thus this result seems reasonable for the two particular angles we have examined. In problems of this type, it is always a good idea to check a result for some particular values of an angle or other quantity to see whether it is reasonable.

The next problem involves some new features. The situation is shown in Fig. 5-8. We are asked to find the tension in the string connecting the two bodies and their accelerations.

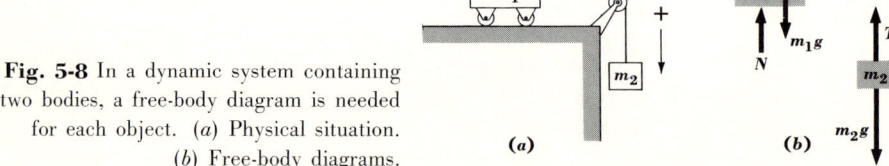

Fig. 5-8 In a dynamic system containing two bodies, a free-body diagram is needed for each object. (*a*) Physical situation. (*b*) Free-body diagrams.

A simplified model of the situation includes the following idealizations: The car with mass m_1 rolls without friction on the smooth horizontal plane, the pulley over which the string runs has no friction, air resistance is negligible, the string has negligible mass and does not stretch, and each body moves precisely in a straight line, m_1 in a horizontal line, m_2 in a vertical line.

Before proceeding with the details of the solution, we observe two important features. First, because the string does not stretch, the magnitudes of the accelerations of the two bodies are the same. To verify this statement, note that in any time interval t the two bodies move equal distances; hence their average velocities in any interval must have the same magnitude. Therefore, the *instantaneous* velocities are equal in magnitude at every instant, so the rates of change of velocity must also be equal in magnitude. The magnitude of each acceleration may be denoted by a.

Second, the string exerts a force on each body. It pulls to the right on m_1 and upward on m_2. But the *magnitudes* of these two forces are equal. If the pulley were not present, this would follow immediately from Newton's third law, which states that the force which m_1 exerts on m_2 is equal in magnitude but opposite in direction to that which m_2 exerts on m_1. Since the pulley has no friction in its bearings, its only effect is to change the direction of the force; it cannot change its magnitude. Therefore the two forces must have the same magnitude. We denote this magnitude by T, an abbreviation for *tension*.

There may be some temptation to look at Fig. 5-8 and jump immediately to the conclusion that the tension is equal to the weight of the body of mass m_2. This would be a rash act, and incorrect as well. If T were equal to $m_2 g$ (which it is not), then m_2 would be in equilibrium and would not accelerate. This consideration should make it clear that in this problem T must be *less* than $m_2 g$.

Now we are ready to proceed with the details. It is essential to remember that $\mathbf{\Sigma F} = m\mathbf{a}$ refers to the forces acting on one body and its acceleration. In this problem we have two bodies; we must apply the laws of motion to *each one separately*. For this purpose, we draw a free-body diagram for each body and write an equation of motion for each. The appropriate free-body diagrams are shown in Fig. 5-8b.

We must also choose a coordinate system for each body. Because we are treating them separately, the coordinate systems need not be the same, but it is essential to define a system for each body and then use it consistently. In this problem m_1 moves along a horizontal line, and it is natural to define its position in terms of the distance from some origin, which may be taken as the initial position. We arbitrarily designate the positive direction to the right, as shown in Fig. 5-8a. Thus any quantity (velocity, acceleration, or force) pertaining to m_1 is positive if directed to the right, negative if to the left. Similarly, we arbitrarily designate the positive direction for m_2 as downward.

Because mass m_1 has no vertical acceleration,

$$N - m_1 g = 0$$

Furthermore its horizontal acceleration a_1 is given by

$$T = m_1 a_1 \qquad (5\text{-}27)$$

There are two forces acting on m_2: its weight in the positive direction and the tension of the string in the negative direction. Therefore,

$$m_2 g - T = m_2 a_2 \qquad (5\text{-}28)$$

Because the string does not stretch, we have the additional relationship discussed above, $a_1 = a_2$, and from now on we denote this quantity simply as a.

Now, considering Eqs. (5-27) and (5-28), we note that each of these contains the same two unknown quantities, T and a. They can therefore be solved simultaneously; the easiest way to do this is simply to add the two equations to eliminate T. We then solve the resulting equation for a and find

$$a = \frac{m_2}{m_1 + m_2} g \qquad (5\text{-}29)$$

Is this a reasonable result? First, if $m_1 = 0$, m_2 should drop as a freely falling body; this is what Eq. (5-29) predicts, since if $m_1 = 0$, then $a = g$. Also, if $m_2 = 0$, we expect the acceleration to be zero, and it is. So Eq. (5-29) gives sensible results in these two special cases.

Inserting Eq. (5-29) into Eq. (5-27), we find for the tension

$$T = \frac{m_1 m_2}{m_1 + m_2} g \qquad (5\text{-}30)$$

Again, we can check this expression for some particular cases. For example, if m_1 is very much larger than m_2, the acceleration is very small. In this case, we may neglect the second term in the denominator in Eq. (5-30), and the tension is almost equal to $m_2 g$.

To illustrate the arbitrariness of choice of coordinate systems, suppose we had taken the positive direction for m_1 to be up rather than down. Then instead of Eq. (5-28) we would have $T - m_2 g = m_2 a_2$. The relationship between the two accelerations would then be $a_2 = -a_1$. The reader may complete the solution and verify that Eq. (5-29) still gives the acceleration of m_1 and Eq. (5-30) the tension. Acceleration a_2 now is the negative of Eq. (5-29), but its physical meaning is the same as before.

It is well to review the strategy used in solving this problem. We first noted that the bodies do not move independently; because they are connected by an inextensible string, their accelerations are related. We also noted a relationship between the forces acting on the two bodies. We drew a free-body diagram for each body, showing all the forces acting on that body but no others. We then chose a coordinate system for each body, applied $\Sigma \mathbf{F} = m\mathbf{a}$ to each body, obtaining two equations. Finally, we solved the two equations simultaneously to obtain the two desired quantities, T and a.

Occasionally in the analysis of a system in which several different masses are in motion, we may end up with more unknown quantities than equations. This usually means that we have forgotten or overlooked or specified incompletely one or more relationships among the various quantities, in which case other relationships must be sought.

5-5 CONTACT FORCE

In the examples of Sec. 5-4 several kinds of forces were considered. One was the force of gravity; another was the pull of a string; a third was the force exerted by contact between two surfaces, which we call a *contact* force. In both examples, the contact force happened to be perpendicular to the surface of contact. Whenever there is friction between two surfaces, however, the contact force also has a component parallel to the surface; this component is usually called the *force of friction*.

The *total* contact force on a body can always be represented as the vector sum of two forces: a force perpendicular (or normal) to the surface, called the *normal force*, and one parallel to the surface, the force of friction. These can be conveniently represented on a free-body diagram for the body, as shown in Fig. 5-9.

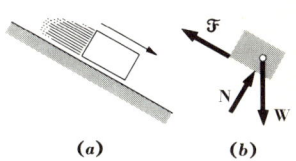

Fig. 5-9 (*a*) Block sliding down an inclined plane. (*b*) Free-body diagram showing the components of the contact force exerted on the block by the plane. The normal force **N** is perpendicular to the plane; the force of friction $\mathcal{F}$ is parallel to it. The weight of the block is also shown.

The behavior of the friction force is in general rather complicated. Some of its characteristics are, however, quite familiar. For example, if there is relative motion between the two surfaces, the force of friction always acts in such a direction as to oppose the relative motion. If the block in Fig. 5-9 is sliding down the plane, the force of friction on the block acts *up* the plane as shown.

The magnitude of the force of friction is related to that of the normal force. The traction of a car on icy streets may be improved by putting several hundred pounds of concrete blocks in the trunk of the car to increase the normal force between tires and pavement. In some situations, the force of friction is somewhat greater if there is no relative motion between the two objects than if there is relative motion. More complicated still, the force of friction may depend on the relative velocity of the surfaces. This is the case, for example, if the friction is associated with the viscosity of oil or other lubricant between two surfaces.

Thus it is very hard to make any quantitative statement about the behavior of the force of friction which is general enough to be of much importance. Nevertheless, there is a rule of thumb which is sometimes useful in practical problems. When there is relative motion between two objects, it sometimes turns out that the magnitude of the force of friction $\mathcal{F}$ is approximately *proportional* to that of the normal force N and independent of the relative velocity. Even when this is not exactly true, it may be a useful approximation. This approximation then becomes part of the idealized model for the problem under consideration.

If the force of friction does behave in this particularly simple manner, we express the relationship symbolically as

$$\mathcal{F} = \mu N \tag{5-31}$$

The constant μ is known as the *coefficient of friction* for the two surfaces; it expresses the proportionality between the magnitudes of $\mathcal{F}$ and N. $\mathcal{F}$ is always perpendicular to N, and their vector sum is the total contact force.

Example

How much force is required to push a 50-kg box across the floor at constant speed if the coefficient of friction between floor and box is 0.5?

Solution

A free-body diagram for the problem is shown in Fig. 5-10, where F is the

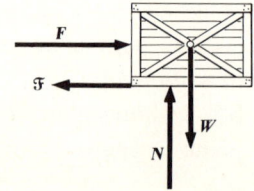

Fig. 5-10 Free-body diagram for a box pushed arcoss a rough floor by a horizontal force **F**.

unknown force. In this case, the normal force is just equal to the weight of the box, or (50kg) (9.8 m/s²) = 490 N, since there is no vertical acceleration. The force of friction when the box is moving is (0.5) (490 N) = 245 N. A force of 245 N is required to keep the box moving at constant speed. If a force greater than 245 N is exerted, the box accelerates.

It is nearly always best to work problems of this general nature as far as possible using algebraic symbols for the various physical quantities. This method has several advantages. It saves arithmetic, it facilitates examination of the final result to see whether it is reasonable, and it leads to results that have more general usefulness than a numerical answer. The problems at the end of this chapter should be worked first using symbols, even if numerical values are given. Then the numbers may be inserted into the final symbolic result.

Example

Instead of pushing the box of the previous example, suppose we pull it with a rope acting upward at an angle θ with the horizontal as shown in Fig. 5-11a. What minimum force F is necessary, and what acceleration results if the actual force is greater than this minimum?

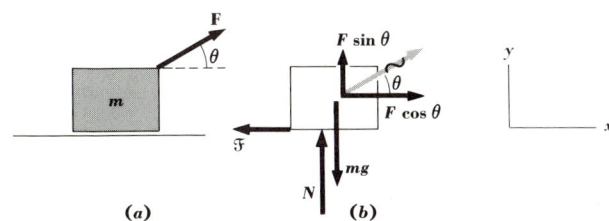

Fig. 5-11 (a) Box pulled across a rough floor by a rope pulling at an angle θ with the horizontal. (b) Free-body diagram for this problem. The force **F** exerted by the rope is represented by its horizontal and vertical components.

Solution

A free-body diagram for the body is shown in Fig. 5-11b, in which **F** is represented by its components. The new feature in this problem is that the normal force is *not* equal to the weight of the box, because **F** has a vertical component, which tends to lift the box, thus reducing the normal force.

It is easy to find N, however. If we do not pull hard enough on the rope to lift the box off the floor entirely, there is no vertical acceleration, so the sum of the vertical components of the forces must be zero. Taking the upward direction as positive, we obtain

$$F \sin \theta + N - mg = 0 \tag{5-32}$$

from which we can find the normal force N in terms of the other quantities. Next, the horizontal acceleration a is determined from the equation

$$F \cos \theta - \mathcal{F} = ma \tag{5-33}$$

Now if we assume that the force of friction $\mathcal{F}$ is proportional to the normal force as represented by Eq. (5-31) and use the expression for N given by Eq. (5-32), we obtain $\mathcal{F} = \mu(mg - F \sin \theta)$, and

$$F \cos \theta - \mu(mg - F \sin \theta) = ma \tag{5-34}$$

from which a can be determined. The minimum force needed to keep the box moving with no acceleration is obtained by setting the left-hand side of this equation equal to zero. The result is

$$F_{\min} = \frac{\mu mg}{\cos \theta + \mu \sin \theta} \tag{5-35}$$

When the rope is horizontal, $\theta = 0$, the denominator of Eq. (5-35) becomes unity, and the result reduces to $F = \mu mg = \mu W$, in agreement with the previous situation.

Problems involving frictional forces are full of pitfalls. If there is no relative motion between the surfaces, Eq. (5-31) does not necessarily give the correct force of friction but only an upper limit for the possible values. It may be anywhere between zero and this maximum. In some cases the maximum static friction force is somewhat *larger* than that given by Eq. (5-31), corresponding to the fact that it is sometimes more difficult to start an object sliding than to keep it sliding once it is started. In this case the maximum force of static friction is sometimes represented by $\mathcal{F}_{\max} = \mu_s N$, where μ_s, the *coefficient of static friction*, has a value somewhat larger than μ, which is then called the *coefficient of kinetic friction*. Even then, this description of the behavior of friction forces is at best an approximation which represents a complicated phenomenon in a simple but inexact way.

5-6 EQUILIBRIUM

When a body is at rest or is moving with constant velocity in an inertial frame of reference, it is said to be in *equilibrium*. In both cases, the acceleration is zero. Sometimes a distinction is made between *static* equilibrium, in which the object is at rest in a particular coordinate system, and *dynamic* equilibrium, in which it moves with constant velocity in that system. This is a somewhat artificial distinction, since both situations can be treated with the same principle of dynamics. For equilibrium, $\Sigma \mathbf{F} = m\mathbf{a}$ becomes simply $\Sigma \mathbf{F} = 0$.

Example

In mountaineering, one occasionally has use for a technique known as a *Tyrolean traverse*. This technique, illustrated in Fig. 5-12a, is useful in

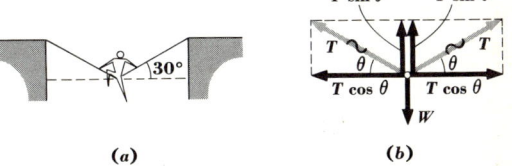

Fig. 5-12 (*a*) Mountaineer crossing a glacier crevasse on a rope anchored at both ends (Tyrolean traverse). (*b*) Free-body diagram for the climber in this situation.

crossing crevasses on glaciers or deep notches on rocky mountain ridges. A climber weighing 180 lb hangs in the middle of the rope, each side of which makes an angle of 30° with the horizontal as shown. Find the tension in the rope; in particular, if its tensile strength is 2,000 lb, will it break?

Solution

We work the problem symbolically as far as possible. The climber's weight is W, the tension in the rope T, and the angle the ropes make with the horizontal θ. The climber is acted upon by three forces: the two rope forces on the two sides and his weight. These forces are shown in a free-body diagram in Fig. 5-12b, in which the rope forces are shown in terms of their components. Because of the symmetry of the situation, it is clear that the tensions in the two sides of the rope are equal. Since there is no acceleration, the sum of the vertical components of force must be zero:

$$T \sin \theta + T \sin \theta - W = 0$$

Thus the tension in each rope is

$$T = \frac{W}{2 \sin \theta}$$

When $\theta = 90°$, the rope hangs straight down. In this case $T = W/2$; each side supports half the climber's weight, as we should expect. When θ is very small, the tension becomes very large, expressing the fact that it is impossible to keep a rope stretched exactly horizontal with a weight at its center.

If the climber's weight is 180 lb and $\theta = 30°$, then

$$T = \frac{180 \text{ lb}}{2 \sin 30°} = 180 \text{ lb}$$

If the angle is *smaller* than 30°, the tension is *greater* than the climber's weight. If the rope is stretched very tightly, so $\theta = 2°$, then $T = 2{,}570$ lb, which is about the breaking strength of a ½-in. manila rope. In actual practice nylon ropes are used, which are more elastic than manila. In this situation a nylon rope would stretch enough to make θ sufficiently large to avoid danger of breakage. Even so, anyone who aspires to be a mountaineer needs some knowledge of physics.

5-7 NATURE OF FORCES

In some respects, the discussion of forces in this chapter has been somewhat superficial. The concepts of *weight* and *contact force* are very useful for many problems, even without any further inquiry into their natures. A more fundamental understanding of these forces is necessary, however, if we are to see their relationship to other kinds of physical phenomena.

A case in point is gravitational interaction. Two problems which superficially seem to be quite independent are the attraction of bodies on the earth toward its center (weight) and the forces responsible for the motion of planets around the sun. Newton perceived that *both* these problems suggest the existence of a general interaction in which *every* body attracts every other body with a force proportional to the product of their masses. The resulting physical principle, now known as *Newton's law of universal gravitation*, accounts for both the weight of bodies on the earth and the forces responsible for the motion of planets around the sun and satellites around the planets. The subject of gravitation is discussed in detail in Chap. 8.

The contact force provides another example of the application of a basic principle to several different kinds of phenomena. Since matter is composed of atoms and molecules, the contact force must result from the individual forces which atoms and molecules exert on each other. One may ask whether these forces are similar in nature to those which hold atoms together in a molecule or a solid body.

Again the answer is affirmative. In this case the forces are basically electromagnetic in nature. Although they cannot be discussed in much detail without some background in electromagnetic theory, several qualitative observations can be made. Atoms are composed of electrically charged particles; the positively charged nucleus of an atom is surrounded by electrons having negative charges. In a neutral atom, the *total* electric charge is zero. If two such neutral atoms are very far apart, they exert no electric forces on each other, since each appears to the other simply as a neutral particle. But as they come together, the positive charges may come closer together, on the average, than the negative charges, or vice versa, and the atoms may exert electric forces on each other. Such forces are responsible for chemical binding, the definite structure and coherence of solids, the contact force between two solid objects, the adhesive force of glue, and many other phenomena.

It is reasonable to ask whether *all* forces have either a gravitational or electromagnetic basis or whether there are additional fundamentally different kinds of forces. A major problem in the early days of nuclear physics (1930–1940) was that electromagnetic and gravitational forces could not account for the very strong attraction of protons and neutrons in atomic nuclei. It is now known that these forces are neither electromagnetic nor gravitational but of a different nature; they are usually called *nuclear forces*. They are extremely short-range forces, normally insignificant at distances greater than 10^{-14} m; but within their range they are very much stronger than electromagnetic forces. Thus two positively charged protons can be held together in a nucleus despite the fact that the electric force between them is repulsive. Even today, the details of nuclear forces are not completely understood. Still other kinds of forces have been discovered in fundamental-particle interactions. These are important chiefly in interactions involving the creation or destruction of particles.

When one examines the fundamental nature of these interactions more deeply, the very concept of force becomes somewhat dubious. We have discussed forces in connection with the accelerations of particles and their inertial properties. There are situations, however, in which it is not adequate to picture a particle as a point having a definite position in space. It is known, for example, that in some respects electrons do not behave as point particles at all but exhibit behavior characteristic of *waves*. In situations of this sort, it is not so easy to define the acceleration of the particle, and the concept of force loses some of its usefulness.

For this reason, it can be argued with some justification that on this very fundamental level the concept of force is an empty and unnecessary idea. It is, in fact, possible to describe the interaction between any two systems and the observable consequences of this interaction without ever using the idea of force. Nevertheless, the concept of force is extremely useful in analyzing a great variety of mechanical situations, especially macroscopic phenomena, and it is used extensively in the remainder of this text.

Problems

5-1 A spring balance measures the *weight* of a body hung from it, but by the addition of a scale giving the corresponding masses it can be made to measure *mass*. If this spring balance is now taken to the moon, will it still measure weights correctly? Masses?

5-2 A plumb bob is hung from the ceiling of an automobile. As the automobile accelerates away from a green traffic light, it is observed that the plumb is displaced 10° from its straight-down position.
  a What is meant by "straight down"?
  b What is the acceleration of the car?

5-3 In Prob. 5-2, suppose it is desired to calibrate the angular scale on the plumb bob to read accelerations directly. Explain how this might be done. If the car is now driven on the moon instead of the earth, will the calibration still be correct? Explain. Does the same problem arise with the accelerometer of Prob. 4-21? Explain.

5-4 An elevator has a mass of 500 kg. Find the tension in the supporting cable when the elevator is:
  a At rest
  b Accelerating upward at 2.0 m/s²
  c Accelerating downward at 2.0 m/s²

5-5 A worn-out elevator cable can support a maximum static load of 500 kg. If the actual mass of elevator and occupants is 400 kg, what maximum upward acceleration can the elevator have without breaking the cable? If a girl with mass 40 kg stands in the elevator on a bathroom scale calibrated in kilograms, what will the scale read while the elevator has this maximum acceleration?

5-6 In the discussion of free fall with air resistance in Sec. 5-2, suppose a body is given an initial velocity equal to the terminal velocity; that is, $v_0 = -mg/k$. Find its velocity and position as functions of time.

5-7 If a spherical object moves through a viscous fluid, the viscous force opposing the motion is proportional to both the speed v and the radius r of the sphere. Specifically, the force is given by $F = 6\pi\eta rv$, where η is a constant characterizing the viscosity of the fluid. Two spheres of diameters 1.0 and 2.0 cm, both homogeneous and made of the same material, are dropped in a viscous fluid. It is found that the terminal velocity of the 1.0-cm sphere is 10 cm/s. What is the terminal velocity of the 2.0-cm sphere?

5-8 A certain fluid has the property that the resisting force it exerts on a sphere of radius r moving through it with speed v is given by $F = Crv^2$, where C is a constant.

a What dimensions or units must C have?
b If a solid sphere of radius 1.0 cm falls through the fluid with a terminal velocity of 10 cm/s, what will be the terminal velocity of a sphere of the same material having a radius of 2.0 cm?

5-9 Verify that Eq. (5-8) is a solution of the problem of free fall with air resistance. That is, show that it satisfies Eq. (5-7), and show that it gives a velocity of zero at time $t = 0$.

5-10 In the discussion of free fall with air resistance in Sec. 5-2, suppose the body is given an initial downward velocity *greater* than the terminal velocity. Draw graphs for acceleration, velocity, and position as functions of time, similar to those in Fig. 5-3 for the case of zero initial velocity.

5-11 When a missile is fired from level ground, show that the maximum range for any given initial speed occurs when it is aimed 45° upward from the horizontal.

5-12 Prove that when a missile is aimed at an angle $45° + \delta$, its range (on level ground) is the same as when it is aimed at $45° - \delta$, where δ is any angle between 0 and 45°. This interesting result was obtained by Galileo, long before the formal development of newtonian mechanics.

5-13 In the kickoff at a football game, the ball is kicked at an angle of 30° with the horizontal, and it returns to the ground 50 yd downfield. With what velocity did it leave the ground initially? How long was it in the air?

5-14 A rock climber stands at the edge of a cliff 200 ft high. It drops straight down for 100 ft, then there is a horizontal ledge 10 ft wide, and then another vertical drop of 100 ft. He drops a rock in such a way that it just misses the edge of the ledge halfway down.

a If the rock is thrown horizontally, what was its initial velocity?
b Find the magnitude and direction of its velocity when it passes the ledge and when it strikes the ground.
c How far from the base of the cliff does it land?

5-15 A mortar placed on the edge of a cliff 50 m high is fired 45° above the horizontal, away from the cliff, with an initial speed of 100 m/s. How far from the base of the cliff does the shell land?

128 DYNAMICS OF A PARTICLE I

5-16 An electron in an oscilloscope tube is accelerated from rest to a speed of 1.0×10^7 m/s in a distance of 5 cm in the electron gun of the tube. The mass of the electron is 9.1×10^{-31} kg.
   a What force was exerted on the electron?
   b During the acceleration, the force of gravity also acts on the electron. Does this have a significant effect on the motion? Explain.

5-17 After the electron in Prob. 5-16 leaves the electron gun, it passes between two electrostatic deflection plates 5 cm long, where it experiences a constant transverse force of magnitude 2.0×10^{-15} N. After passing these, it travels freely to the screen 10 cm away.
   a At what angle to the axis of the tube is the electron moving as it leaves the deflection plates?
   b How much is it deflected from the point at which it would strike the screen if the plates were not present?

5-18 A body whose mass is 5 kg is dropped from a bridge. A strong wind blows under the bridge, exerting a constant horizontal force of 25 N on the body. Describe the trajectory of the body in detail.

5-19 Equation (5 8) describes the motion of a falling body with air resistance starting at $y = 0$ and with no initial velocity. If the body is given an initial velocity v_0, the motion is described instead by the function

$$y = -\frac{mg}{k}t - \frac{m}{k}\left(v_0 - \frac{mg}{k}\right)\left(1 - e^{-(k/m)t}\right)$$

   a Show that when $v_0 = 0$, this expression reduces to Eq. (5-8).
   b Show that this expression satisfies Eq. (5-7).
   c Show that this expression gives a velocity v_0 at time $t = 0$.

5-20 Show that the solution of Eq. (5-25a) corresponding to $x = 0$ and $v_x = v_0 \cos \alpha$ at time $t = 0$ is

$$x = \frac{v_0 \cos \alpha \, m}{k}\left(1 - e^{-(k/m)t}\right)$$

Using this result together with the result quoted in Prob. 5-19, show that the trajectory of a body moving under gravity and air resistance is described by

$$x = \frac{v_0 \cos \alpha \, m}{k}\left(1 - e^{-(k/m)t}\right)$$

$$y = \frac{mg}{k}t + \frac{m}{k}\left(v_0 \sin \alpha + \frac{mg}{k}\right)\left(1 - e^{-(k/m)t}\right)$$

5-21 A technical report on a certain airplane stated that the plane is designed so that in pulling out of a dive it is able to "withstand a force of $5g$." Explain what this means. Describe carefully the forces that act on a pilot during such a pullout.

5-22 By flying along a certain path, an airplane pilot can, for a limited time, achieve a state of motion in which objects in the plane seem to be weightless. Explain how this is possible. What must the shape of the path be?

5-23 A bunch of bananas is tied to one end of a rope. The rope passes over a pulley, and a monkey hangs from its other end. His weight is exactly equal to that of the bananas, which initially hang just out of reach above him. Naturally, he starts to climb the rope in an effort to reach them. Describe what happens. Does he ever reach them?

5-24 The device shown in Fig. P5-24 is called *Atwood's machine*; every freshman physics text contains at least one problem on this machine.

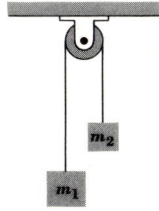

Fig. P5-24

   a Describe in detail a simplified *model* which can be used to predict the motion of this machine, and specify a suitable coordinate system for each body.
   b Find the acceleration of each body and the tension in the string.
   c If a spring balance is inserted in each side of the string, what will these balances read?
   d Check the results of part *b* for the following special cases: $m_1 = m_2$; $m_2 = 0$. Do the results agree with your intuitive expectations?

5-25 For the situation shown in Fig. P5-25, find the accelerations of the two bodies and the tension in the string. Is the tension the same or different in the three

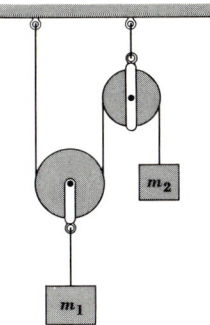

Fig. P5-25

sections of string passing over the pulleys? Explain. Check your general results for the following special cases: $m_1 = 0$; $m_1 = 2m_2$. Do the results agree with your intuitive predictions?

5-26 In Prob. 4-7, what must the minimum coefficient of friction between wheels and rails be for the train to start without slipping?

5-27 The coefficient of friction of a block on an inclined plane can be determined by raising the plane just enough for the block to slide with uniform speed and measuring the angle of inclination of the plane.
   a Derive a relationship between the angle and the coefficient of friction for these conditions.
   b Using this method, measure the coefficients of friction of a penny and of a rubber eraser on a book cover.
   c Are you measuring the coefficient of *static* or *kinetic* friction? Explain.

5-28 In the situation of Fig. 5-10, suppose the force F is not constant but instead increases uniformly with time: $F = At$, where A is a constant.
   a Assuming that the coefficients of static and kinetic friction are equal, draw a graph of the frictional force as a function of time and a graph of acceleration as a function of time.
   b Obtain expressions for the velocity and position as functions of time.

5-29 A man pushes a board of mass m through a circular saw by pushing down on it with a stick as shown in Fig. P5-29.

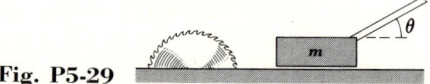

Fig. P5-29

   a Find the force F which must be exerted to make the board slide with constant speed, in terms of the angle θ, the coefficient of friction, and the mass of the board.
   b Show that if the angle is too steep, the board cannot be made to slide, no matter how great a force is applied. Find this critical angle.

5-30 A block of mass m rests on a rough plane inclined at an angle θ to the horizontal. If it is pushed by a force $\mathbf{F}$ applied horizontally as shown in Fig. P5-30, how

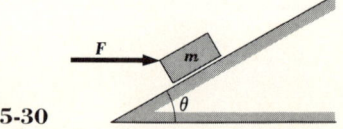

Fig. P5-30

much force is required to make it move with constant speed? Show that, for a given block and plane, if the plane is steeper than a certain critical angle, the block cannot be pushed up by a horizontal force, no matter how great. Find the critical angle in terms of the other quantities.

5-31 In Fig. 5-8, suppose the body of mass m_1 slides on the horizontal surface instead of rolling and that the coefficient of friction is μ.
    a In terms of m_1 and μ, find the minimum value m_2 must have for motion to occur if the system is initially at rest.
    b Assuming m_2 is large enough for motion to occur, find the acceleration of each body and the tension in the string.
    c If m_2 is initially a distance h above the floor, how much time is required for it to reach the floor if it starts from rest?

5-32 For the situation shown in Fig. P5-32, find the accelerations of both bodies and the tension in the string, assuming that the inclined plane is frictionless.

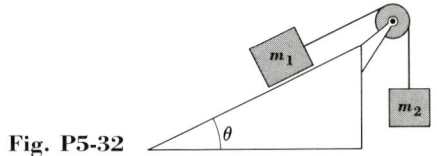

Fig. P5-32

    a Show that the direction of the acceleration depends on the relative magnitudes of the masses.
    b Show that when $\theta = 0$, your result reduces to that found in Sec. 5-4. Also check the special cases $m_1 = 0$ and $m_2 = 0$ (with $\theta \neq 0$).

5-33 In the situation of Prob. 5-32, if the plane is not frictionless but has $\mu = 0.1$, find the accelerations when $\theta = 30°$ and
    a $m_1 = 10$ kg, $m_2 = 1$ kg
    b $m_1 = 1$ kg, $m_2 = 10$ kg

5-34 A man sits in a bosun's chair supported by a rope. The rope passes over a pulley, and he holds its end in his hands. How much force must his hands support?

5-35 A steel girder of mass 2,000 kg is supported by a cable from the boom of a crane.
    a If the cable is 50 m long, how much horizontal force must be applied to the girder to pull it sideways 2 m?
    b If the ultimate strength of the cable is 30,000 N, how far sideways can the girder be pulled (by a larger force) before the cable breaks?

Dynamics of a Particle II | 6

The relations between force and motion for a body moving in a circular path are considered in this chapter. The concepts of centripetal acceleration and centripetal force are introduced. The laws of motion for a body moving in a plane are formulated using polar coordinates, and expressions for the components of velocity and acceleration in polar coordinates are derived.

6-1 CIRCULAR MOTION

Circular motion is important for a variety of reasons. Historically, this motion was considered in an attempt to understand the motions of the planets, which move in nearly circular orbits around the sun. Many practical problems in the dynamics of machinery involve rotational motion in which bodies or parts of bodies move in circular paths. One elementary model of the structure of simple atoms (the Bohr theory) treats the electrons as moving in circles around the nucleus. Furthermore, the concepts of angular velocity and related ideas in circular motion are essential background for the development of the concept of angular momentum in Chap. 12; this in turn provides the key to analysis of rotational motion of solid bodies.

Consider a particle (which may be a model for a small body) moving in a circular path of radius R whose center is at point O. The position of any point P on this circle can be described by the angle θ between the radius line at this point and some fixed line, as shown in Fig. 6-1. In effect, we are simply using the *polar coordinates* of P, introduced in Chap. 2. In this case, the radial coordinate is R (a constant), and so once the radius of the circle is known, one additional coordinate θ is sufficient to specify the position of the point.

Fig. 6-1 Particle at point P moves on a circle of radius R. Its position is described by the angle θ between the radius line OP and a fixed line. Conventionally, θ is positive when measured counterclockwise, as shown, and negative when measured clockwise. Similarly, the rate of change of θ with time describes the motion of P. If θ is increasing (counterclockwise motion), the angular velocity is positive; if decreasing (clockwise motion), it is negative.

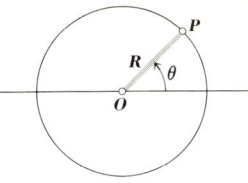

When the point moves in a circle, the angle θ changes; the motion can be described by means of the time rate of change of the angle. This rate of change is called the *angular velocity* of the point, denoted by ω:

$$\omega = \frac{d\theta}{dt} \tag{6-1}$$

Usually, for reasons which will soon become apparent, the angle θ is measured in radians (rad) rather than in degrees. In this case, the units of ω are radians per unit time. Clearly, ω may be either positive or negative. If the point P in Fig. 6-1 moves in a clockwise direction, then θ decreases with time and ω is a negative quantity. As defined above, ω is a scalar quantity; hence the term "angular velocity" is something of a misnomer. Nevertheless, this term is frequently used, so we use it here. Later in this chapter we shall define an angular velocity which is a *vector* quantity.

When the angular velocity changes with time, it is useful to define an angular acceleration, denoted by α:

$$\alpha = \frac{d\omega}{dt} = \frac{d^2\theta}{dt^2} \tag{6-2}$$

There is a simple relation between the speed of the particle and its angular velocity. To obtain this relationship, we consider Fig. 6-2. In a time

Fig. 6-2 Relationship between a small change $\Delta\theta$ and the resulting displacement for a point moving on a circle of radius R. From this diagram, the relationship between the angular velocity and the speed of point P can be found.

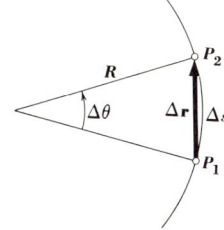

interval Δt the particle moves from point P_1 to P_2. During this time, the angle θ changes by an amount $\Delta\theta$, and the particle moves a distance Δs along an arc of the circle. If θ is measured in radians, then by definition of radian measure,

$$\Delta\theta = \frac{\Delta s}{R} \quad \text{or} \quad \Delta s = R\,\Delta\theta \tag{6-3}$$

The displacement during this interval is $\Delta\mathbf{r}$, and the average velocity is $\Delta\mathbf{r}/\Delta t$. When $\Delta\theta$ is small, $|\Delta\mathbf{r}|$ and Δs are very nearly equal, and in the limit as $\Delta\theta \to 0$ they become exactly equal. Thus the magnitude of the average velocity is approximately

$$v_{\text{av}} = \frac{\Delta s}{\Delta t} = R\,\frac{\Delta\theta}{\Delta t} \tag{6-4}$$

In the limit when $\Delta\theta$ and Δt both approach zero, this equation becomes exact. Then the left side of Eq. (6-4) becomes simply the instantaneous speed, and the right side becomes R multiplied by the instantaneous angular velocity. That is,

$$v = R\omega \tag{6-5}$$

In reviewing the above derivation, we note that θ must be measured in radians, otherwise Eq. (6-3) is not correct. Thus the units of ω must be radians per unit time. Also, in this and the following discussion, it is essential to distinguish clearly between velocity and angular velocity. The two are quite different physical quantities, with different units and different properties. Sometimes, in order to emphasize this distinction, we use the term *linear velocity* for velocity.

Example
A satellite moves in a circular orbit of radius 4,000 mi around the earth. It makes 1 revolution (r) in approximately 1.5 h. Find its angular velocity and speed.

Solution
Since 1 r is 2π rad, the angular velocity is $\omega = 2\pi$ rad/1.5 h $= 4\pi/3$ rad/h $= 4.19$ rad/h, or $\pi/2{,}700$ rad/s. The speed can be found from Eq. (6-5):

$$v = (4{,}000 \text{ mi})\left(\frac{4\pi}{3} \text{ rad/h}\right) = 1.68 \times 10^4 \text{ mi/h}$$

Equation (6-5) gives the *magnitude* of the linear velocity of point P; the *direction* of the velocity is always tangent to the circle. Thus, even when the magnitude of the velocity is constant, its direction continuously changes as the point moves around the circle. For this reason *the velocity is not constant;* the object is *accelerated*. This should also be apparent from the fact that a particle has zero acceleration only when it is at rest or moving with constant velocity in a straight line; such is not the case here. In the next section we determine the magnitude and direction of the acceleration in order to find the force required to produce this motion.

6-2 CENTRIPETAL ACCELERATION

We now calculate the acceleration of a body moving in a circle, considering first the special case where the speed is constant (usually called *uniform circular motion*). Again we consider motion from point P_1 to point P_2 during a time interval Δt. Figure 6-3 shows the velocity at these two points. Since

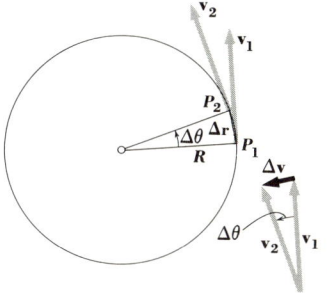

Fig. 6-3 Point moving in a circle of radius R with uniform speed is accelerated because the direction of the velocity changes. The velocities at two points P_1 and P_2 are shown. The vector diagram at the right shows the calculation of the change in velocity $\Delta \mathbf{v}$ between these two points.

the object moves with constant speed, $\mathbf{v}_1$ and $\mathbf{v}_2$ have the same magnitude; their directions differ by an angle $\Delta \theta$, as shown. Also shown in the diagram is the change of velocity $\Delta \mathbf{v} = \mathbf{v}_2 - \mathbf{v}_1$ between the two points.

The speed of the object, which is constant, is denoted simply by v. How is $\Delta \mathbf{v}$ related to the other quantities? The triangle formed by the velocity vector diagram in Fig. 6-3 is *similar* to that formed by the points O, P_1, and P_2. Each triangle is isosceles, and in each case the angle between the two equal sides is $\Delta \theta$. Thus

$$\frac{|\Delta \mathbf{v}|}{v} = \frac{|\Delta \mathbf{r}|}{R} \tag{6-6}$$

If Δt is the time required for the particle to move from P_1 to P_2, the magnitude of the average acceleration is given by

$$a_{\text{av}} = \frac{|\Delta \mathbf{v}|}{\Delta t} = \frac{v}{R}\frac{|\Delta \mathbf{r}|}{\Delta t} \tag{6-7}$$

The quantity $|\Delta \mathbf{r}|/\Delta t$ is the magnitude of the average velocity during Δt, as before.

To find the instantaneous acceleration, we let Δt approach zero. In this limit, $|\Delta \mathbf{r}|/\Delta t$ again becomes the instantaneous speed v of the particle, so we can write

$$a = \frac{v}{R}v = \frac{v^2}{R} \tag{6-8}$$

The magnitude of the instantaneous acceleration is proportional to the *square* of the speed and inversely proportional to the radius of the circle. This dependence on R should not be surprising; the smaller R is, the more sharply the object must curve. The fact that a is proportional to the square of v may seem a little surprising at first, but a little thought shows that it is a reasonable result. If we double v, the velocity has twice as great magnitude, and its direction changes twice as fast; hence the acceleration increases by a factor of 4.

What is the *direction* of this acceleration? Referring to Fig. 6-3 again, we see that when $\Delta \theta$ becomes very small, the change in velocity $\Delta \mathbf{v}$ is very nearly perpendicular to both $\mathbf{v}_1$ and $\mathbf{v}_2$. In the limit as $\Delta t \to 0$, $\Delta \mathbf{v}$ is perpendicular to a tangent to the circle at the point we are considering. Thus, in this limit $\Delta \mathbf{v}$, hence also $\Delta \mathbf{v}/\Delta t$, points toward the *center* of the circle. *The instantaneous acceleration is directed toward the center of the circle.*

It is sometimes useful to represent the magnitude of this acceleration in terms of the angular velocity. This is easily accomplished by combining Eqs. (6-5) and (6-8):

$$a = \frac{(R\omega)^2}{R} = \omega^2 R \tag{6-9}$$

Because the acceleration in uniform circular motion is always directed toward the center of the circle, it is called *centripetal* acceleration. The word centripetal is derived from two Latin words meaning "seeking the center." We summarize what we have discovered thus far: When a body moves in a circular path of radius R with uniform speed v and angular velocity ω, these quantities are related by

$$v = R\omega$$

and the body has an acceleration, directed toward the center of the circle, of magnitude

$$a = \frac{v^2}{R} = \omega^2 R$$

In these equations ω must be measured in radians per unit time.

The velocity and acceleration relations for uniform circular motion, Eqs. (6-5), (6-8), and (6-9), can be obtained by an alternate derivation using derivatives of sines and cosines. Although this derivation is in no way superior to the one presented above, studying it may enhance understanding of the various kinematic relationships. A reader who is not yet familiar with derivatives of trigonometric functions may omit this discussion and proceed to the examples at the end of this section.

We begin with the relation between rectangular and polar coordinates introduced in Sec. 2-1:

$$x = R \cos \theta \qquad y = R \sin \theta \tag{6-10}$$

In this case R, the radius of the circle, is constant. Thus the position vector of a point on the circle can be expressed as

$$\mathbf{r} = R \cos \theta \, \mathbf{i} + R \sin \theta \, \mathbf{j} \tag{6-11}$$

Suppose that at time $t = 0$ the particle is crossing the x axis and that it moves counterclockwise with constant angular velocity ω. Then $\theta = 0$ at $t = 0$, and at any later time θ is given simply by $\theta = \omega t$. Thus Eq. (6-11) becomes

$$\mathbf{r} = R \cos \omega t \, \mathbf{i} + R \sin \omega t \, \mathbf{j} \tag{6-12}$$

To find the velocity of the particle we take the time derivative of Eq. (6-12):

$$\mathbf{v} = \frac{d\mathbf{r}}{dt} = -R\omega \sin \omega t \, \mathbf{i} + R\omega \cos \omega t \, \mathbf{j} \tag{6-13}$$

In terms of components,

$$v_x = -R\omega \sin \omega t \qquad v_y = R\omega \cos \omega t$$

The magnitude of the velocity is given by

$$v = \sqrt{v_x^2 + v_y^2} = [R^2 \omega^2 (\sin^2 \omega t + \cos^2 \omega t)]^{1/2} = R\omega \tag{6-14}$$

in agreement with Eq. (6-5). We also note that the velocity at each instant is perpendicular to the position vector at that instant, hence tangent to the circle. This is easily established by taking the scalar product $\mathbf{v} \cdot \mathbf{r}$:

$$\mathbf{v} \cdot \mathbf{r} = (R \cos \omega t \, \mathbf{i} + R \sin \omega t \, \mathbf{j}) \cdot (-R\omega \sin \omega t \, \mathbf{i} + R\omega \cos \omega t \, \mathbf{j}) = 0$$

Since the scalar product is zero, and since neither vector has zero length, the cosine of the angle between them must be 90°.

The acceleration is found by differentiating Eq. (6-13):

$$\mathbf{a} = \frac{d\mathbf{v}}{dt} = -R\omega^2 \cos \omega t \, \mathbf{i} - R\omega^2 \sin \omega t \, \mathbf{j} \tag{6-15}$$

In terms of components,

$$a_x = -R\omega^2 \cos \omega t \qquad a_y = -R\omega^2 \sin \omega t$$

The magnitude of the acceleration is given by

$$a = (a_x^2 + a_y^2)^{1/2} = [(-R\omega^2)^2(\cos^2 \omega t + \sin^2 \omega t)]^{1/2} = R\omega^2 \tag{6-16}$$

Using Eq. (6-14), we can express this as

$$a = \frac{v^2}{R}$$

These results agree with Eq. (6-9) and (6-8), respectively. Also, comparing Eqs. (6-15) and (6-12), we obtain the vector relation

$$\mathbf{a} = -\omega^2 \mathbf{r} \tag{6-17}$$

showing that the direction of $\mathbf{a}$ is always opposite that of $\mathbf{r}$, that is, toward the center of the circle.

Example
Find the speed and acceleration of a bug perched on the edge of a 12-in. $33\frac{1}{3}$ r/min record.

Solution
We first find the angular velocity in radians per second. Since $1 \text{ r} = 2\pi$ rad, we find

$$\omega = (33\frac{1}{3} \text{ r/min})(2\pi \text{ rad/r}) \frac{1 \text{ min}}{60 \text{ s}} = 3.49 \text{ rad/s}$$

Since $R = 6$ in., we find

$$v = R\omega = (6 \text{ in.})(3.49 \text{ rad/s}) = 20.9 \text{ in./s}$$

The acceleration is

$$a = \omega^2 R = (3.49 \text{ rad/s})^2 (6 \text{ in.}) = 73.0 \text{ in./s}^2$$

In mks units,

$$a = (73.0 \text{ in./s}^2)(0.0254 \text{ m/in.}) = 1.86 \text{ m/s}^2$$

Example
If the turntable, initially at rest, reaches its final angular velocity 2 s after being turned on, what is its average angular acceleration?

Solution

$$\alpha_{av} = \frac{\Delta \omega}{\Delta t} = \frac{3.49 \text{ rad/s}}{2 \text{ s}} = 1.74 \text{ rad/s}^2 = 1.74 \text{ s}^{-2}$$

Note that in carrying the units through these examples, we have sometimes omitted radians. This is permissible because the radian measure of angles is a ratio of two lengths. The number expressing the size of an angle in radians is a ratio of two quantities having the same units and is thus a *pure number* without any units, as discussed in Sec. 1-2. Despite the fact that we usually refer to an angle as "2 radians," and so forth, it would be equally correct to say simply $\theta = 2$. Similarly, an angular velocity may be designated as "4 s^{-1}" and an angular acceleration as "6 s^{-2}."

6-3 CENTRIPETAL FORCE

As for every accelerated motion, the acceleration of an object in uniform circular motion must be produced by a force related to the acceleration by $\mathbf{\Sigma F} = m\mathbf{a}$, assuming that the acceleration is measured in an inertial frame of reference. Thus, when an object undergoes uniform circular motion, the vector sum of the forces acting on it must always be directed toward the center of the circle and have magnitude

$$F = ma = m\left(\frac{v^2}{R}\right) = m(\omega^2 R) \tag{6-18}$$

140 DYNAMICS OF A PARTICLE II

where F is understood to mean the magnitude of the vector sum of the forces. In the two forms of Eq. (6-18), the quantity in parentheses is the magnitude of the acceleration of the object. The force responsible for the centripetal acceleration is sometimes referred to as the *centripetal force*.

The best way to attack a problem involving uniform circular motion is first to draw a free-body diagram showing all the forces acting on the body and then to compute the vector sum of these forces and equate the result to the mass times the centripetal acceleration. This technique is illustrated in the following examples.

Example
Consider again the bug on the phonograph turntable in the example of Sec. 6-2. If the bug has a mass of 2.0×10^{-5} kg (0.02 g), what force is being applied to it? How is this force applied?

Solution
Figure 6-4 shows a free-body diagram for the bug. The phonograph turntable is shown in cross section, and we may imagine that the bug is moving into

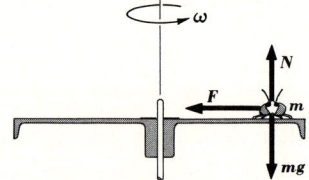

Fig. 6-4 Free-body diagram for a bug on the edge of a phonograph record. The turntable is shown in cross section; the spindle about which it rotates is vertical, in the plane of the page.

the plane of the paper. The forces acting on the bug are its weight and the two components of the contact force exerted by the record. Because there is no vertical acceleration, the sum of the vertical forces must be zero, and thus the normal force N has magnitude

$$N = mg = (2.0 \times 10^{-5} \text{ kg})(9.8 \text{ m/s}^2) = 1.96 \times 10^{-4} \text{ N}$$

The force F, which is the horizontal (frictional) component of the contact force, is responsible for the centripetal acceleration. We have already found in Sec. 6-2 that the acceleration has magnitude $a = 1.86$ m/s². The magnitude of the friction force must therefore be

$$F = ma = (2.0 \times 10^{-5} \text{ kg})(1.86 \text{ m/s}^2) = 3.72 \times 10^{-5} \text{ N}$$

Example
What must the minimum coefficient of friction μ_s between the bug and the record be if it is not to slide off?

Solution
The *maximum* force of friction which can be exerted is given by

$$F = \mu_s N$$

if the frictional component of the contact force can be represented in this simple manner. Therefore the *minimum* value of μ_s is

$$\mu_{min} = \frac{F}{N} = \frac{3.72 \times 10^{-5} \text{ N}}{1.96 \times 10^{-4} \text{ N}} = 0.19$$

If the coefficient of friction is smaller than this, the bug will slide off.

Example
What force on the bug opposes this force of friction and keeps the bug in equilibrium so that it does not slide off?

Solution
This is a loaded question, intended as a trap for sleepy or gullible readers. The answer, which we state with the strongest possible emphasis, is that there is *no* force opposing the force of friction. *The bug is not in equilibrium;* it accelerates toward the center of the circle, and there *must* be an unbalanced force acting toward the center. Circular motion is *not* equilibrium, and in circular motion **ΣF** is *never* zero!

One sometimes encounters the term *centrifugal force* in discussions of circular motion. What does this mean? First, the term itself is derived from two Latin words meaning "to flee from the center." The implication is that there is a force tending to pull the object away from the center of the circle. This actually is not the result of a force at all but simply the tendency of the body to move in a straight line. It is true that the body exerts a force of *reaction* which is directed outward from the circle, and this force may properly be called centrifugal force. But centrifugal force is never exerted *on* the body undergoing circular motion. In solving problems involving circular motion, it is best not to use this term at all: it is unnecessary and potentially confusing.

Example

A passenger in a car making a right turn seems to be thrown toward the left side of the car. Is this not an example of centrifugal force on the passenger?

Solution

No! The passenger is thrown to the left *relative to the car* because he tends to continue moving in a straight line relative to an inertial frame of reference while the car is deviating from its straight-line path. As he leans against the left side of the car, he feels himself pushing to the left against the left side of the car, and the car exerts a force on him *toward the right* to produce his centripetal acceleration toward the right. The force with which the passenger pushes on the car may be called centrifugal force; it is not a force exerted *on* the passenger, however, but rather a force which *he* exerts *on* the car.

Example

A passenger rides around in a ferris wheel of radius 20 m which makes 1 r every 10 s. What force does he exert on the seat when he is at the bottom of the circle if his mass is 75 kg?

Solution

The forces on the rider are his weight, acting downward, and the force the seat exerts on his posterior, acting upward. The vector sum of these must equal the mass times the acceleration. The acceleration has a magnitude

$$a = \omega^2 R = \left(\frac{2\pi \text{ rad}}{10 \text{ s}}\right)^2 (20 \text{ m}) = 7.90 \text{ m/s}^2$$

When he is at the bottom of the circle, his acceleration is directed upward. Therefore there must be a net upward force on the rider of

$$F = ma = (75 \text{ kg})(7.90 \text{ m/s}^2) = 592 \text{ N}$$

That is, the upward force exerted by the seat must be greater in magnitude than the man's weight by 592 N. Since his weight is $(75 \text{ kg})(9.8 \text{ m/s}^2) = 735$ N, the seat must push up with a total force of $592 \text{ N} + 735 \text{ N} = 1{,}327$ N on the passenger.

Example
If the passenger in the above example is sitting on a bathroom scale, what does it read?

Solution
The reading of a bathroom scale is equal to the amount of force it exerts upward on the object resting on its top surface. Therefore, the bathroom scale reads 1,327 N (assuming it is calibrated in newtons!).

Example
A satellite moves in a circular orbit around the earth, close enough to earth for its weight to be very nearly equal to its value on earth. How fast must it move for its weight to provide just the right centripetal acceleration to maintain the circular orbit? What is the time required for one revolution?

Solution
The only force acting on the satellite is its weight, which we assume to be mg. Equating this to the product of its mass and centripetal acceleration gives

$$mg = m\frac{v^2}{R}$$

Thus the necessary speed is given by

$$v = \sqrt{gR}$$

where R, the radius of the orbit, is slightly greater than the earth's radius, which is about 6.4×10^6 m. Thus

$$v = \sqrt{(9.8 \text{ m/s}^2)(6.4 \times 10^6 \text{ m})} = 7.9 \times 10^3 \text{ m/s}$$

which is equal to about 17,700 mi/h. The angular velocity ω is given by

$$\omega = \frac{v}{R} = \frac{7.9 \times 10^3 \text{ m/s}}{6.4 \times 10^6 \text{ m}} = 1.2 \times 10^{-3} \text{ s}^{-1}$$

144 DYNAMICS OF A PARTICLE II

The time required for one revolution (called the *period* of the motion) is the time for the angle to increase by 2π. This is given by

$$T = \frac{2\pi}{\omega} = \frac{2\pi}{1.2 \times 10^{-3} \text{ s}^{-1}} = 5.1 \times 10^3 \text{ s} = 85 \text{ min}$$

Example

A ball of mass m is attached to two strings which are in turn attached to a vertical rod rotating with constant angular velocity ω. Thus the ball describes a horizontal circle of radius R, as shown in Fig. 6-5a. Find the tension in each string.

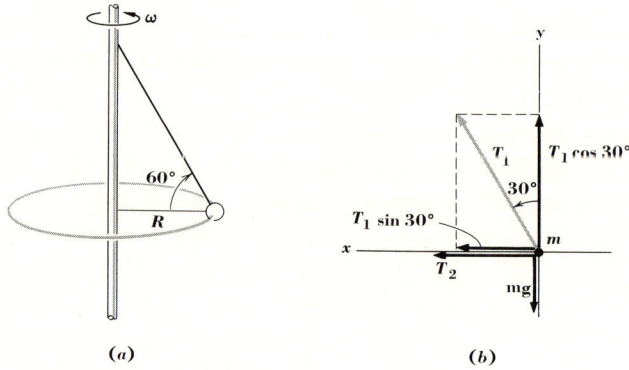

Fig. 6-5 (a) A ball travels in a circular path under the action of gravity and two strings shown. (b) A free-body diagram showing vertical and horizontal components of the force acting on the ball.

Solution

The first step is to remember that we are going to use $\Sigma \mathbf{F} = m\mathbf{a}$, so a free-body diagram is essential. Figure 6-5b shows a free-body diagram, and an axis system to be used in computing components of vectors. The force T_1 is represented in terms of its components. The components of acceleration can be obtained immediately: $a_x = \omega^2 R$ and $a_y = 0$. Applying $\Sigma \mathbf{F} = m\mathbf{a}$ for the x and y components respectively, we obtain

$$T_1 \sin 30° + T_2 = ma_x = m\omega^2 R$$
$$T_1 \cos 30° - mg = ma_y = 0$$

The second equation gives $T_1 = 2mg/\sqrt{3}$, showing that T_1 is always *greater* than the weight of the ball by the factor $2/\sqrt{3}$ and is independent of angular velocity. Substituting this expression for T_1 in the first equation and solving for T_2, we obtain

$$T_2 = m\left(\omega^2 R - \frac{g}{\sqrt{3}}\right)$$

As might be expected, T_2 increases with ω. When $\omega^2 = g/(R\sqrt{3})$, $T_2 = 0$, and when ω is less than this critical value, T_2 becomes negative. Physically this means the string force would have to change direction, which is of course impossible; the string cannot push on the ball; it can only *pull*. If we replace the lower string by a light rigid rod, this result makes sense. In particular, when $\omega = 0$, the rod must push *outward* with a force $T_2 = mg/\sqrt{3}$ to keep the ball in equilibrium. When it is moving, of course, the ball is *not* in equilibrium.

6-4 VECTOR ANGULAR VELOCITY

Although the concept of vector angular velocity is not absolutely necessary to the topics of this chapter, it is introduced here for two reasons. First, it provides an elegant derivation of Eqs. (6-8) and (6-9). Second, it paves the way for analysis of more complicated examples of rotational motion which will appear in later chapters.

In considering angular velocity as a vector quantity, the central idea is to think of the object in a circular motion as rotating about an *axis*. In Fig. 6-6, for example, point P may be thought of as a point on a wheel rotating

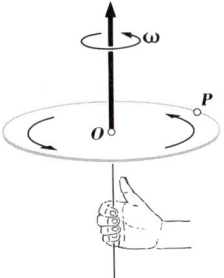

Fig. 6-6 Point moving in a circle whose center is at O may be thought of as rotating about an axis through O perpendicular to the plane of the circle as though it were a point on the rim of an imaginary wheel. The vector angular velocity lies in the direction of this axis. The right-hand rule shows that in this example the angular velocity points upward as shown.

about an axis through the center of the circle, perpendicular to its plane, as shown. We now define the angular velocity $\boldsymbol{\omega}$ as a vector quantity whose direction is that of the axis of rotation and whose magnitude is the angular

146 DYNAMICS OF A PARTICLE II

velocity previously defined, $d\theta/dt$. To define the direction of the velocity vector completely, we use the same definition as for the vector product of two vectors, namely the *right-hand rule*. We wrap the fingers of the right hand around the axis in the direction of rotation; the direction of the angular velocity is the direction in which the thumb points. Thus in Fig. 6-6, $\boldsymbol{\omega}$ points upward.

The position of point P can be described by its position vector $\mathbf{r}$ with respect to an origin O which lies along the axis of rotation but not necessarily at the center of the circle (Fig. 6-7). We now assert that the *velocity* of point P is given by

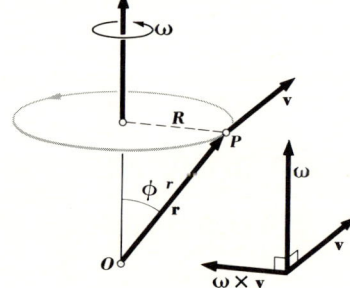

Fig. 6-7 Relationship among the velocity of point P, its position, and the angular velocity. The radius R of the circle is related to the magnitude of r of the position vector $\mathbf{r}$ by $R = r \sin \phi$.

$$\mathbf{v} = \boldsymbol{\omega} \times \mathbf{r} \tag{6-19}$$

To verify Eq. (6-19), we first note that according to Eq. (6-5) the magnitude of the velocity is given by

$$v = R\omega = (r \sin \phi)\omega$$

which is equal to the magnitude of the vector product $\boldsymbol{\omega} \times \mathbf{r}$. Furthermore, consideration of Fig. 6-7 shows that the direction of $\mathbf{v}$ is perpendicular to the plane containing the vectors $\boldsymbol{\omega}$ and $\mathbf{r}$ and is, in fact, in the direction of the vector product $\boldsymbol{\omega} \times \mathbf{r}$. Thus we have verified Eq. (6-19) with respect to both magnitude and direction.

Having established Eq. (6-19), we may permit ourselves the luxury of using it to find the *acceleration* of point P by taking the time derivative:

$$\mathbf{a} = \frac{d\mathbf{v}}{dt} = \frac{d}{dt}(\boldsymbol{\omega} \times \mathbf{r}) \tag{6-20}$$

In uniform circular motion, $\boldsymbol{\omega}$ is constant, so this can be rewritten

$$\mathbf{a} = \boldsymbol{\omega} \times \frac{d\mathbf{r}}{dt} = \boldsymbol{\omega} \times \mathbf{v} \tag{6-21}$$

Any concern about the legitimacy of this procedure for differentiating the vector product can be resolved by going back to the definition of a derivative. The velocity at time t is

$$\mathbf{v} = \boldsymbol{\omega} \times \mathbf{r}$$

The velocity at a somewhat later time $t + \Delta t$ is

$$\mathbf{v} + \Delta\mathbf{v} = \boldsymbol{\omega} \times (\mathbf{r} + \Delta\mathbf{r}) = \boldsymbol{\omega} \times \mathbf{r} + \boldsymbol{\omega} \times \Delta\mathbf{r}$$

Thus $\Delta\mathbf{v} = \boldsymbol{\omega} \times \Delta\mathbf{r}$; dividing by Δt and letting $\Delta t \to 0$, we obtain Eq. (6-21). Incidentially, if $\boldsymbol{\omega}$ is *not* constant, Eq. (6-21) contains an additional term $(d\boldsymbol{\omega}/dt) \times \mathbf{r}$.

Referring to Fig. 6-7 again, we note that the vectors $\boldsymbol{\omega}$ and $\mathbf{v}$ are always perpendicular, and so the magnitude of $\boldsymbol{\omega} \times \mathbf{v}$ is simply the product of the magnitudes of the vectors, ωv. Thus, from Eq. (6-21) we find

$$a = \omega v$$

Using Eq. (6-5), we have

$$a = \omega^2 R = \frac{v^2}{R}$$

in agreement with our previous results. Reference to Fig. 6-7 also shows that the direction of $\boldsymbol{\omega} \times \mathbf{v}$ is toward the center of the circle, as expected.

6-5 VELOCITY AND ACCELERATION IN POLAR COORDINATES

For a point moving in a plane, especially in problems involving rotational motion, polar coordinates are often more convenient than cartesian coordinates. For example, the path of a body moving in a circle is described simply by stating that the polar coordinate r has the constant value R. To describe this path in cartesian coordinates, we should have to say that the coordinates x and y vary according to the equation

$$x^2 + y^2 = R^2$$

Also, there are many situations in which the force acting on a body depends only on its distance from a fixed point. For example, if the sun is at the origin, its gravitational force on the earth at point (r,θ) is proportional simply to $1/r^2$.

For such problems, it is useful to represent vector quantities, such as force, velocity, and acceleration, by means of components parallel and perpendicular to the line joining O and P, rather than by the more familiar x and y components. Such components are called the *polar* components of a vector, since they make use of a polar coordinate system. When point P in Fig. 6-8 moves with velocity **v** as shown, the velocity can be represented

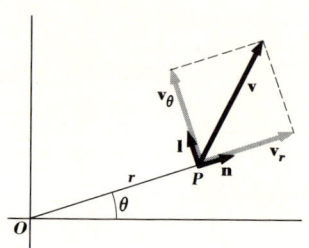

Fig. 6-8 Representation of the velocity **v** of point P by means of its radial and tangential component vectors $\mathbf{v}_r$ and $\mathbf{v}_\theta$, respectively. The unit vectors **n** and **l**, whose directions change as P moves, are shown for this position.

by means of the component vector $\mathbf{v}_r$ along the line from O to P and the component vector $\mathbf{v}_\theta$ perpendicular to this direction.

The component v_r, also called the *radial* or r component of **v**, corresponds to motion in which r changes; v_θ, called the *tangential* or θ component of **v**, corresponds to motion in which θ changes. It is easy to find these components if the coordinates r and θ and their time derivatives are known. When only r changes with time, the velocity has only a radial component, with magnitude dr/dt. When only θ changes, the particle's speed is $r\,d\theta/dt$, as we have previously seen. Thus, in general, when both r and θ vary with time, the polar components of velocity are

$$v_r = \frac{dr}{dt} \qquad v_\theta = r\frac{d\theta}{dt} \tag{6-22}$$

Just as with rectangular coordinates, it is convenient to introduce unit vectors to specify the directions of components in this scheme. We introduce the unit vectors **n**, pointing in the direction of increasing r, and **l**, pointing in the direction of increasing θ, as illustrated in Fig. 6-8. These unit vectors are different from the unit vectors **i**, **j**, and **k** used with rectangular coordinates in that they do not have fixed directions in space but change direction when P moves. Thus the directions of **n** and **l** depend on the coordinate θ of the point P. Using the unit vectors **n** and **l** and the components of **v** given by Eqs. (6-22), we can express **v** as

$$\mathbf{v} = \frac{dr}{dt}\mathbf{n} + r\frac{d\theta}{dt}\mathbf{l} \tag{6-23}$$

This result may also be derived in a more formal manner. First, the position vector of point P with respect to O is given by

149 VELOCITY AND ACCELERATION IN POLAR COORDINATES [6-5]

$$\mathbf{r} = r\mathbf{n} \tag{6-24}$$

The velocity is the time derivative of **r**. In general both r and θ vary with time, and as a result the unit vector **n** also varies. Thus the time derivative of Eq. (6-24) must be written

$$\mathbf{v} = \frac{d\mathbf{r}}{dt} = \frac{d}{dt}(r\mathbf{n}) = \frac{dr}{dt}\mathbf{n} + r\frac{d\mathbf{n}}{dt} \tag{6-25}$$

Now, what is $d\mathbf{n}/dt$? To evaluate this quantity, we make use of the fact that **n** depends on θ and write

$$\frac{d\mathbf{n}}{dt} = \frac{d\mathbf{n}}{d\theta}\frac{d\theta}{dt} \tag{6-26}$$

To evaluate $d\mathbf{n}/d\theta$, consider Fig. 6-9, which shows the unit vector **n** corresponding to angle θ and also **n** corresponding to a slightly different angle

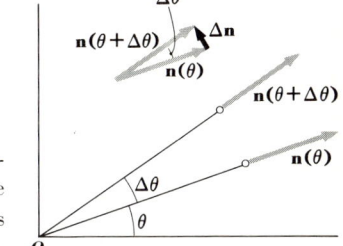

Fig. 6-9 Change in the unit vector **n** corresponding to a small change in the angle θ. When $\Delta\theta \longrightarrow 0$, $\Delta\mathbf{n}$ becomes perpendicular to **n**.

$\theta + \Delta\theta$. Because **n** is a unit vector, its magnitude is unity. Therefore, by the definition of radian measure of angles, the magnitude of $\Delta\mathbf{n}$ is approximately $|\Delta\mathbf{n}| = \Delta\theta$, and in the limit as $\Delta\theta \to 0$, $|\Delta\mathbf{n}/\Delta\theta| = 1$. Also, as $\Delta\theta \to 0$, $\Delta\mathbf{n}$ becomes perpendicular to **n** and assumes the direction of **l**. Therefore

$$\frac{d\mathbf{n}}{d\theta} = \mathbf{l} \tag{6-27}$$

In precisely the same way, we obtain the relation

$$\frac{d\mathbf{l}}{d\theta} = -\mathbf{n} \tag{6-28}$$

Finally, using Eq. (6-26) and the corresponding equation for **l**,

$$\frac{d\mathbf{n}}{dt} = \frac{d\theta}{dt}\mathbf{l} \quad \text{and} \quad \frac{d\mathbf{l}}{dt} = -\frac{d\theta}{dt}\mathbf{n} \tag{6-29}$$

Inserting these results in Eq. (6-25), we find

$$\mathbf{v} = \frac{dr}{dt}\mathbf{n} + r\frac{d\theta}{dt}\mathbf{l} \tag{6-30}$$

in agreement with Eqs. (6-22) and (6-23). Point P has a component of velocity in the radial direction of magnitude dr/dt and a component in the θ direction equal to the instantaneous radius r multiplied by the instantaneous angular velocity $d\theta/dt$.

The polar components of acceleration can be found in precisely the same way. We differentiate Eq. (6-30) with respect to t, observing that each term is a product of several factors, all of which may vary with time. We find

$$\mathbf{a} = \frac{d^2r}{dt^2}\mathbf{n} + \frac{dr}{dt}\frac{d\mathbf{n}}{dt} + \frac{dr}{dt}\frac{d\theta}{dt}\mathbf{l} + r\frac{d^2\theta}{dt^2}\mathbf{l} + r\frac{d\theta}{dt}\frac{d\mathbf{l}}{dt}$$

Using Eqs. (6-29) and grouping the r and θ components together gives

$$\mathbf{a} = \left[\frac{d^2r}{dt^2} - r\left(\frac{d\theta}{dt}\right)^2\right]\mathbf{n} + \left[r\frac{d^2\theta}{dt^2} + 2\frac{dr}{dt}\frac{d\theta}{dt}\right]\mathbf{l} \tag{6-31}$$

This is a very powerful result because it provides a means of computing the polar components of acceleration for any possible motion of a point in a plane in terms of the polar coordinates and their derivatives.

It is instructive to see how Eq. (6-31) reduces to familiar results in special cases. For example, for uniform circular motion r is constant, and its derivatives are zero. Also, $d\theta/dt$ is constant, so $d^2\theta/dt^2 = 0$. In this case, Eq. (6-31) becomes simply

$$\mathbf{a} = -r\left(\frac{d\theta}{dt}\right)^2\mathbf{n}$$

The magnitude of this expression agrees with Eq. (6-9); it also gives the direction of the centripetal acceleration. Another familiar case occurs when θ is constant and the point moves along a radial line. In this case all the derivatives of θ are zero, and we obtain

$$\mathbf{a} = \frac{d^2r}{dt^2}\mathbf{n}$$

It may happen that a particle moves in a circle with nonconstant angular velocity. In this case all the derivatives of r are zero, but both $d\theta/dt$ and $d^2\theta/dt^2$ are in general different from zero. In this case Eq. (6-31) becomes

$$\mathbf{a} = -r\left(\frac{d\theta}{dt}\right)^2\mathbf{n} + r\frac{d^2\theta}{dt^2}\mathbf{l}$$

The first term is the centripetal acceleration, as before, corresponding to the change of direction of **v**; the second, which is *parallel* to the instantaneous velocity, corresponds to a change in the *magnitude* of **v** which must occur if $d\theta/dt$ is not constant.

Finally, when r changes (so that the path is not circular), there are in general a radial acceleration d^2r/dt^2 and a tangential acceleration containing dr/dt. This, the last term in Eq. (6-31), results from the fact that even when the angular velocity of a particle is constant, its speed increases as r increases. This contribution to the acceleration is called the *Coriolis* acceleration.

Example
A bug walks with constant speed u outward along a radial line on a phonograph record which turns with constant angular velocity ω. Find the polar components of the velocity and acceleration of the bug.

Solution
In this situation $dr/dt = u$, and $d\theta/dt = \omega$. Both these quantities are constant, so $d^2r/dt^2 = 0$, and $d^2\theta/dt^2 = 0$. From Eq. (6-30), we find

$$\mathbf{v} = u\mathbf{n} + r\omega\mathbf{l}$$

and, from Eq. (6-31),

$$\mathbf{a} = -r\omega^2\mathbf{n} + 2u\omega\mathbf{l}$$

If the bug's mass is m, the force necessary to produce this acceleration is

$$\mathbf{F} = -mr\omega^2\mathbf{n} + 2mu\omega\mathbf{l}$$

The magnitude of the force is

$$F = m\sqrt{(r\omega^2)^2 + (2u\omega)^2}$$

Knowing the components of **a** and **F**, we can find the directions of these vectors relative to the line OP.

Problems

6-1 A grindstone of diameter 8 in. turns at 1,800 r/min. Find the speed of a point on its rim. What is the angular velocity of the grindstone?

6-2 If a motor requires 3 s to attain its full speed of 1,800 r/min after being turned on, what is its angular acceleration? How many revolutions does it make during this interval?

6-3 When an object moves in a circle with constant speed, is its acceleration constant? Explain.

6-4 A body of mass m_1 slides on a frictionless horizontal table. A string attached to it passes through a hole in the table, and a body of mass m_2 hangs vertically from the other end of the string. If the first body moves in a circle of radius R, what must its speed be?

6-5 A conical pendulum is made from a ball of mass m attached to the end of a string of length l, the other end of which is tied to the ceiling. The ball is set into motion in a circular path, as shown in Fig. P6-5. What must its speed be for the string to make a given angle θ with the vertical?

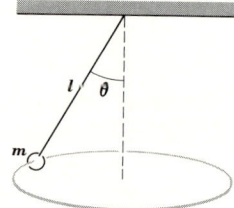

Fig. P6-5

6-6 In Prob. 6-5, show that the angular velocity of the ball is given by $\omega^2 = g/h$, where h is the vertical distance from the point of suspension to the plane of the path.

6-7 The radius of the earth is approximately 6.40×10^6 m.
   a What is the earth's angular velocity in radians per second?
   b What is the speed of a point on the earth's surface at the equator?
   c What is the centripetal acceleration of a point at the equator?

6-8 Suppose we stop the rotation of the earth and weigh a 1-kg body at the equator, using a spring balance, and find its weight to be precisely 9.80 N. Now we let the earth rotate and weigh the body again. What result do we obtain? Would a similar result occur if we used a platform balance? Explain, and comment on the phrase "no springs—honest weight," sometimes seen at the meat counter of the supermarket.

6-9 In Prob. 6-8, suppose the earth rotated so fast that a body at the equator appeared to have no weight at all. What would be the length of the day?

6-10 A pilot flying a jet plane in a vertical circle observes that just as he reaches the top of the circle, objects in the plane, including himself, appear weightless. If his speed at this point is 400 m/s, what is the radius of his path?

PROBLEMS

6-11 A familiar backyard trick is to put some water in a bucket and then whirl the bucket in a vertical circle fast enough to prevent its falling out. If the radius of the circle is 1.0 m, what minimum speed must the bucket have? What minimum angular velocity?

6-12 A curve in a road has a radius of 500 ft. At what angle should it be banked so that a car can negotiate it at 60 mi/h without any tendency to slip sideways? Does this depend on the mass of the car? Explain.

6-13 Find the angle at which a curve of radius R in a road should be banked so that a car can negotiate it at speed v without any tendency to slip sideways.

6-14 In Prob. 6-13, if the coefficient of friction between road and tires is μ, and if the curve is banked for speed v, what is the maximum speed (in terms of v, μ, and R) at which the curve can be negotiated without the car's actually sliding off the road?

6-15 A ball of mass m is connected to two strings attached to a rotating vertical rod as shown in Fig. P6-15. Assuming that the angular velocity ω of the rod

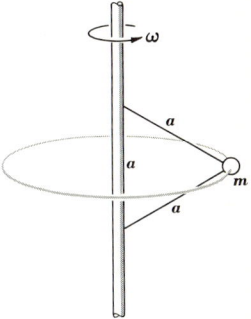

Fig. P6-15

is the same as that of the ball, which travels in a circular path around the rod, and that ω is great enough for both strings to be taut, find the tension in each string in terms of the relevant quantities. What is the minimum value of ω for the lower string to remain taut?

6-16 A model airplane whose mass is 0.5 kg flies in a horizontal circle at the end of a control wire 50 m long (the bottom end of which is held by the controller) with a speed of 50 m/s. The wire makes a constant angle of 30° with the horizontal.
    a Describe a simplified model of this situation which can be used to compute the tension in the control wire.
    b Compute the tension in the wire.

6-17 A clown in the circus drives a small car around the inside surface of a cylinder having a vertical axis and a radius of 20 m. The coefficient of friction between this surface and the car tires is 1.0, and the total mass of car and driver is 400 kg.

a Explain carefully, using a free-body diagram, how it is possible to drive in a horizontal circle inside the cylinder without falling.

b What minimum speed is necessary to prevent falling?

6-18 Because of the rotation of the earth, a plumb bob does not hang exactly in the direction of its weight (the earth's gravitational attraction). Assuming that the radius of the earth is approximately 6×10^6 m and your latitude is $40°$, find how much the plumb bob deviates from the direction of the force of gravity.

6-19 A particle of mass m experiences a force described by $\mathbf{F} = \mathbf{v} \times \mathbf{B}$, where $\mathbf{v}$ is its velocity and $\mathbf{B}$ is a constant vector. If the particle is given an initial velocity $\mathbf{v}_0$ perpendicular to $\mathbf{B}$, describe its subsequent motion in detail.

6-20 In Prob. 6-19 suppose the particle is given a velocity which has a component parallel to $\mathbf{B}$ in addition to the component perpendicular to $\mathbf{B}$. Describe the path of the particle.

6-21 An astronaut in a circular orbit around the earth in a space capsule describes the phenomenon of *weightlessness*, in which an object released in the capsule just floats instead of falling to the floor. Does this mean that the earth's gravity does not act on such an object? Discuss. What is meant by weightlessness in this situation?

6-22 Find the vector angular velocity (magnitude and direction) which describes the rotation of the earth about its axis.

6-23 A bug whose mass is 0.05 g walks with a constant speed of 5 cm/s along a radial line painted on a phonograph record turning at $33\frac{1}{2}$ r/min. What minimum coefficient of friction is necessary to prevent slipping when the bug is 10 cm from the center?

6-24 In Prob. 6-23, if the turntable has an angular acceleration of 1.0 rad/s^2, how is the result changed?

6-25 A phonograph turntable starts from rest and accelerates uniformly to a final angular velocity of 5 rad/s in 2 s.
  a Find the polar components of the acceleration of a point on the rim, 0.10 m from the axis.
  b Find the magnitude and direction of this acceleration.
  c If a bug is perched at this point, what minimum coefficient of friction is necessary to prevent its slipping off?

PERSPECTIVE II

We have now completed the development of the general principles relating force and motion for a particle. Newton's laws of motion can be used to solve two general classes of problems which can be summarized concisely as "given the forces, find the motion" and "given the motion, find the forces." In particular, we can describe in detail the possible motions of a particle under the earth's gravitational force and in a number of other situations in which the forces are constant. We have also considered a few simple cases in which the forces are not constant. One particular case of considerable practical as well as theoretical interest is that in which a particle moves in a circle with constant speed; the laws of motion, together with the kinematics techniques developed in earlier chapters, provide a sufficient basis for a complete analysis of this problem.

There are, however, many situations in which it is not feasible to apply the laws of motion directly, either because the forces are not known or because the motion is too complicated to be readily described in detail. For example, in a collision between two bodies, large and variable forces may act for a very short time, and it may be difficult to describe these forces and the resulting accelerations in detail. We are not yet prepared to solve problems such as these. We should like to have principles which enable us to relate the motion at one time with that at a later time, even if the details of what happened in between are not known.

Two such principles are those of conservation of momentum and conservation of energy. These may be thought of as integral principles, in that they consider the total change of motion between two times, in contrast to Newton's second law, which gives the rate of change of velocity and which may therefore be called a differential principle. Still, these conservation principles are not new generalizations from experiment but can be derived from Newton's laws of motion. They are discussed, along with a number of applications, in the next four chapters.

Momentum   7

The concept of momentum plays an important role in the analysis of collisions between particles or bodies. A few simple collisions, in which the total momentum of the system does not change, are discussed. We define the total momentum of a system of particles, state the principle of conservation of momentum, and demonstrate its relationship to Newton's laws of motion. We then generalize the principle to include collisions involving particles which have no mass and introduce the concept of impulse to aid in the analysis of forces involved in collisions.

7-1 SIMPLE COLLISIONS

In Sec. 4-6 we discussed a simple example of the motion of two cars, each under the action of forces exerted by the other. Analysis of experiments with these cars led to the conclusion that the two forces were equal in magnitude but opposite in direction. The generalization based on this and other similar observations is Newton's third law of motion. Defining the *momentum* of each car as the product of its mass and velocity, we found that in these experiments the total momentum of the system was the same before and after the interaction.

Now let us investigate the concept of momentum in more detail. We begin with another example. Suppose a rifle bullet is fired at a target mounted on a car which is free to roll on a nearly frictionless track; in an idealized model it may be assumed that the net force on the car is zero, except during the bullet's impact (Fig. 7-1). The car and target are at rest before they are struck by the bullet, and afterward the bullet is embedded in the target. The problem is to find the velocity of the car after the bullet strikes.

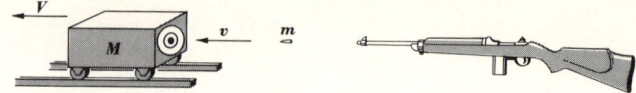

Fig. 7-1 Rifle bullet shot into a target mounted on a car which rolls on a frictionless track. The initial velocity of the bullet and the final velocity of the car can be measured, and the total momentum of the system before and after the bullet strikes the target can be compared experimentally.

In principle, we can find the force on the car, compute its acceleration, and find the final velocity. In practice, this is extremely difficult to do, because the impact occurs in a very short interval of time. Large and variable forces are exerted during this interval, and there are large accelerations, but it is difficult to obtain detailed information about the forces. Thus it is desirable to try to relate the initial and final motions without having to go into the details of what happened *during* the collision.

The situation can be investigated experimentally. Let the mass of the bullet be m, that of the target and car M, the initial velocity of the bullet v, and the final velocity of the car, target, and bullet V. The velocities v and V can be measured by multiflash photography or some other technique, and we can also measure the masses. As might be expected, V turns out to be smaller than v (assuming the car and target were initially at rest). Furthermore, the ratio of final to initial velocity turns out to be just the reciprocal of the ratio of the total mass of target, car, and bullet to the mass of the bullet itself. That is,

$$\frac{V}{v} = \frac{m}{M + m} \tag{7-1}$$

It is easy to think of situations involving similar collisions. Suppose a boy sitting in a wagon catches a football. We can measure the football's speed just before it is caught and the speed of the wagon (with the boy and the football in it) just after the ball is caught. In this situation, the wagon may not be such an ideal, frictionless system as the car on a frictionless track, and so the wagon eventually rolls to a stop. But if we measure the velocity of the ball *just before* it is caught and the velocity of the wagon *just after*, the results are consistent with Eq. (7-1).

This relation may be written in a more illuminating form as follows:

$$mv = (M + m)V \tag{7-2}$$

Before the collision, a mass m moves with velocity v; after the collision, a total mass $M + m$ moves with velocity V. Equation (7-2) says that the product of mass and velocity is the same before and after the collision. As we have already mentioned, the product of mass and velocity is called *momentum*. What we have discovered is that in a collision of this sort the total momentum before the collision is the same as that after; i.e., momentum is *conserved*. Before the collision, an object with relatively small mass has all the momentum and therefore has to be moving relatively fast. After the collision, the momentum is shared with a much larger mass, and the velocity is correspondingly smaller.

On the basis of such experiments, we make the tentative hypothesis that in any collision of two bodies, when no forces act on either except the force each exerts on the other, the total momentum of the system is the same before and after the collision. In the following sections this principle, called the *principle of conservation of momentum*, is generalized and related to the principles of mechanics which we have already learned.

7-2 MOMENTUM

When a body with mass m moves with velocity $\mathbf{v}$, its *momentum*, denoted by $\mathbf{p}$, is defined as

$$\mathbf{p} = m\mathbf{v} \tag{7-3}$$

Momentum, like velocity, is a vector quantity. Its magnitude is the mass of a particle times its speed, and its direction is the same as the direction of the velocity. Strictly speaking, Eq. (7-3) defines momentum only for a point particle having no extension in space. But as we have already seen in many instances, a point may be used as a model of a body which does have spatial extension so long as internal motion and rotational motion are of no importance. When they *are* important, a more detailed discussion is necessary. Momentum of a system of particles is treated in more detail in Chap. 11.

The units of momentum are those of mass times velocity. The unit of momentum in the mks system is the kilogram-meter per second. In the British system it is the slug-foot per second.

Newton's second law of motion ($\mathbf{\Sigma F} = m\mathbf{a}$) may be stated in terms of momentum. When we take the time derivative of Eq. (7-3), we find

$$\frac{d\mathbf{p}}{dt} = \frac{d}{dt}(m\mathbf{v}) = m\frac{d\mathbf{v}}{dt} = m\mathbf{a} \tag{7-4}$$

Therefore Newton's second law of motion may be written

$$\mathbf{\Sigma F} = \frac{d\mathbf{p}}{dt} \tag{7-5}$$

That is, the time rate of change of momentum of a particle is equal to the vector sum of all the forces acting on it.

In Eq. (7-4) we assume that m is constant. Later we shall consider systems in which it may appear that mass is variable, but we shall see that it is always possible to divide the system into pieces such that the mass of each piece is constant. Even in the theory of relativity, where one sometimes speaks of relativistic mass increase, the idea of variable mass is not necessary. Hence it is never necessary to deal with a particle having variable mass, and we shall not attempt to do so.

Newton's laws of motion may used to *predict* the experimental results found in Sec. 7-1. Consider two bodies having masses m_1 and m_2, completely free except for their mutual-interaction forces. Let the force on the first body (exerted by the second) be $\mathbf{F}_1$ and the force on the second body (exerted by the first) be $\mathbf{F}_2$. Denoting the respective momenta as $\mathbf{p}_1 = m_1 \mathbf{v}_1$ and $\mathbf{p}_2 = m_2 \mathbf{v}_2$, we find, from Eq. (7-5),

$$\mathbf{F}_1 = \frac{d\mathbf{p}_1}{dt} \quad \text{and} \quad \mathbf{F}_2 = \frac{d\mathbf{p}_2}{dt} \tag{7-6}$$

But according to Newton's third law, the two forces $\mathbf{F}_1$ and $\mathbf{F}_2$ must be equal in magnitude and opposite in direction: $\mathbf{F}_1 = -\mathbf{F}_2$. Thus

$$\frac{d\mathbf{p}_1}{dt} = -\frac{d\mathbf{p}_2}{dt} \tag{7-7}$$

or

$$\frac{d}{dt}(\mathbf{p}_1 + \mathbf{p}_2) = 0 \tag{7-8}$$

Although $\mathbf{p}_1$ and $\mathbf{p}_2$ may change, the sum $\mathbf{p}_1 + \mathbf{p}_2$ is constant, as shown by the fact that its time derivative is zero. Thus the sum of the two momenta is constant.

This is the same conclusion reached in Sec. 7-1 as a generalization from experimental observations, but now we have *derived* this result from Newton's laws of motion. Thus the principle of conservation of momentum is not a new fundamental principle but one already contained in Newton's second and third laws of motion. This does not mean, of course, that the experiments are without value. They supply additional corroboration of Newton's laws of motion, especially the third law, and they supply the motivation for introducing the concept of momentum and demonstrate its usefulness.

We now define the *total momentum* of a collection of particles as the *vector sum* of the individual momenta. Denoting the total momentum of a system by **P,** we define it as

$$\mathbf{P} = \sum_i \mathbf{p}_i = \sum_i m_i \mathbf{v}_i \tag{7-9}$$

Using this definition, we can state the principle of conservation of momentum somewhat more concisely. *In any isolated system the total momentum is constant.* By "isolated system" we mean a system which does not interact with its surroundings, although parts of the system may interact with each other. Thus far, we have considered systems containing only two bodies. But it is clear that this principle can be generalized immediately to a system containing any number of particles provided that the internal forces obey Newton's third law. When one part of the system exerts a force on another part, the second part reacts on it with an equal and opposite force. Thus the two rates of change of momentum resulting from these forces are equal and opposite, and the *total* momentum does not change.

Two notes of caution are necessary regarding the principle of conservation of momentum. First, it is applicable only to an *isolated* system. In applications of the principle, it is necessary to decide on the boundaries of the system being considered and to make sure that this system does not interact with anything outside it. If it does, the total momentum of the system is in general *not* constant. The concept of an isolated system is, of course, an idealization. In practice we sometimes deal with systems which are *nearly* isolated. In applying the principles of mechanics to such systems, one may use a simplified model in which small interactions with the surroundings are omitted, so that in the model the system *is* isolated.

Second, the principle is valid only when the momenta are computed from velocities measured in an inertial frame of reference. The reason for this limitation becomes clear when we consider the derivation of the principles from Newton's laws. $\mathbf{\Sigma F} = m\mathbf{a}$ is valid only in an inertial frame of reference, so any results derived from it can be valid only in an inertial frame. Furthermore, it is necessary to use the *same* inertial frame of reference for all the velocities, since in two different frames of reference the momenta are in general different. Thus in applications of the principle it is essential to specify not only the system but also the *frame of reference* used.

Example
A small car whose mass is 800 kg moves north at 15 m/s. It collides at a very icy intersection with a car whose mass is 1,600 kg and whose velocity

Fig. 7-2 (a) Two cars collide at a very icy intersection. After the collision, they move off as one mass. (b) Momentum vector diagram showing the total momentum **P** as the vector sum of the momenta of the two cars. The total momentum of the system is the same before and after the collision.

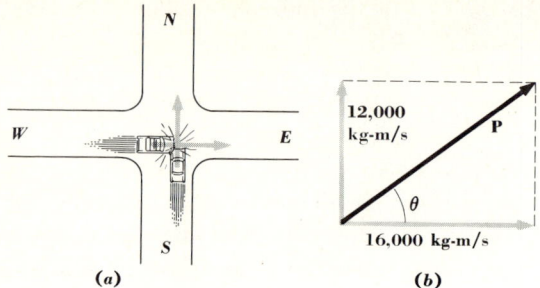

is 10 m/s east, as shown in Fig. 7-2a. After the collision the cars are tangled up and move on as one mass. How can the insurance adjuster find the magnitude and direction of the final velocity?

Solution

We can idealize the situation by assuming that the streets are so slippery that there are no horizontal forces exerted on the cars except those which they exert on each other; the two cars then constitute an isolated system. We draw a *momentum vector diagram* (Fig. 7-2b) showing the momentum of each car and the vector sum. According to the principle of conservation of momentum, the total momentum after the collision, represented by **P** in the diagram, must be equal to the vector sum of the two momenta before the collision. The momentum vector diagram happens to be a 3-4-5 right triangle, so the magnitude of the total momentum is 20,000 kg-m/s, and the angle θ is about 37°. The total mass of the wreckage after collision is 2,400 kg, so its speed is

$$\frac{20{,}000 \text{ kg-m/s}}{2{,}400 \text{ kg}} = 8.33 \text{ m/s}$$

It is important to understand that Fig. 7-2b is *not* a velocity vector diagram. It would not be correct to find the final velocity by simply taking the vector sum of the two velocities; there is no physical basis for such a procedure.

7-3 VARIABLE MASS

In discussing the relation of momentum to Newton's laws of motion, we stated somewhat dogmatically that particles with variable mass were not to be considered. There are, however, many mechanical situations which are most readily described in terms of a system whose mass changes. The most obvious example is a rocket, which becomes lighter as it burns up and ejects its fuel. How can the principles of mechanics be applied to such a system?

Rather than trying to discuss this question in general, we concentrate on a particular example, calculating the acceleration of a rocket. For simplicity, we consider a rocket in outer space, sufficiently far away from all other matter for it to be considered as an isolated system with no external forces. We denote the mass of the rocket at some instant of time by m and its speed at this time, relative to some inertial frame, by v. Now suppose that in an interval of time Δt a quantity of fuel Δm is burned and expelled from the rear end of the rocket with a speed u *relative to the rocket* (Fig. 7-3). As

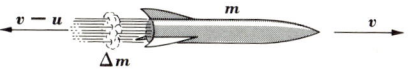

Fig. 7-3 Rocket accelerates by expelling a quantity of mass Δm with a speed u relative to the rocket in a time Δt.

a result, the rocket's speed increases by an amount Δv. If we can find Δv in terms of the other quantities, we can relate the acceleration of the rocket to the rate at which fuel is burned.

At time t the momentum is simply mv. At time $t + \Delta t$ the rocket's velocity has increased to $v + \Delta v$, and its mass has decreased to $m + \Delta m$. The plus sign in this expression is correct because Δm, the change in mass of the rocket, is a negative quantity, so $m + \Delta m$ is *less than* m. The momentum of the rocket at time $t + \Delta t$ is $(m + \Delta m)(v + \Delta v)$. The burned fuel also has momentum. Its speed relative to the rocket is u; its velocity relative to the inertial frame in which v is measured is therefore $v - u$. The mass of the burned fuel is $-\Delta m$, a positive quantity, and its momentum is $-\Delta m(v - u)$.

The rocket and this fuel, taken together, constitute an isolated system, so according to the principle of momentum conservation the *total* momentum is the same at time t as at time $t + \Delta t$. That is,

$$mv = (m + \Delta m)(v + \Delta v) - \Delta m(v - u) \tag{7-10}$$

In applying conservation of momentum, we have considered a system which includes the rocket and the fuel burned during the interval Δt but *not* the fuel already burned before this time.

Simplifying Eq. (7-10), we find

$$m\,\Delta v = -\Delta m(u + \Delta v)$$

Dividing by $m\, \Delta t$, we have

$$\frac{\Delta v}{\Delta t} = -\frac{\Delta m}{\Delta t}\frac{u + \Delta v}{m}$$

which gives the *average* acceleration during the interval Δt. Finally, to find the instantaneous acceleration, we take the limit of this expression as $\Delta t \to 0$ and find

$$a = \frac{dv}{dt} = -\frac{dm}{dt}\frac{u}{m} \qquad (7\text{-}11)$$

This result makes sense; it says that the acceleration is proportional to the rate at which fuel is burned (dm/dt) and to the relative velocity u with which burned fuel is expelled from the rocket. Also, the acceleration is inversely proportional to the instantaneous mass of the rocket. Since m decreases as fuel is burned, dm/dt is negative. If dm/dt is constant, a increases as m decreases.

Equation (7-11) is a differential equation which can be solved to find the velocity of the rocket as a function of time if the rate of fuel consumption dm/dt, which may vary with time, is known. The *total* change in velocity corresponding to burning of a given mass of fuel turns out to be independent of dm/dt, provided u is constant. Letting v_0 and v_f be the initial and final velocities, $M + M_f$ the total initial mass of rocket plus fuel, and M the remaining mass of rocket after the fuel M_f is burned, we integrate both sides of Eq. (7-11):

$$\int_{v_0}^{v_f} dv = -\int_{M+M_f}^{M} u\frac{dm}{m} \qquad (7\text{-}12)$$

The result of this operation, the details of which are left as an exercise, is

$$v_f - v_0 = u \ln\left(1 + \frac{M_f}{M}\right) \qquad (7\text{-}13)$$

This equation shows that if the ratio of fuel mass M_f to mass of the "empty" rocket M is large enough, the final velocity v_f may be larger than the exhaust speed u, even if the initial velocity v_0 is zero or negative. In such a case, there is an interval during which the final velocity of some of the burned fuel is in the *same* direction as that of the rocket.

The above analysis can be generalized to the situation where the rocket is in a gravitational field. Suppose the rocket is aimed vertically upward from the surface of the earth, where it experiences a gravitational force of magnitude mg. The system is no longer isolated, and its momentum no longer constant, but the change in total momentum during the interval Δt can easily be found. The *rate* of change of total momentum is equal to the only external force; if the positive direction is upward, this is $-mg$, and the total momentum change during Δt is simply $-mg\,\Delta t$. That is, the total upward momentum at $t + \Delta t$ is *less* than that at t by $mg\,\Delta t$, and Eq. (7-10) becomes

$$mv - mg\,\Delta t = (m + \Delta m)(v + \Delta v) - \Delta m(v - u)$$

We simplify this, divide by Δt, and take the limit as $\Delta t \to 0$, obtaining

$$a = \frac{dv}{dt} = -\frac{dm}{dt}\frac{u}{m} - g \tag{7-14}$$

which reduces to Eq. (7-11) when $g = 0$. Recalling that dm/dt is negative, we see from Eq. (7-14) that if $-u\,dm/dt = mg$, the acceleration is instantaneously zero and the *thrust* of the rocket, of magnitude $u\,dm/dt$, just balances its weight.

When a particle moves relative to an observer at a speed which is not small compared with the speed of light, the equations of motion used by that observer must be modified, as we have already remarked. It has been found that the correct relativistic generalization of Newton's second law is

$$\mathbf{F} = \frac{d}{dt}\frac{m\mathbf{v}}{\sqrt{1 - v^2/c^2}} \tag{7-15}$$

If v is much smaller than c, this reduces to Eq. (7-5).

The form of Eq. (7-15) suggests a corresponding generalization of the definition of momentum:

$$\mathbf{p} = \frac{m\mathbf{v}}{\sqrt{1 - v^2/c^2}} \tag{7-16}$$

Thus at speeds approaching that of light, the momentum of a particle is no longer proportional to its velocity but is related to the velocity in a more complicated way.

In Chap. 14 we return to relativistic mechanics for a more complete discussion, but even while developing the principles of newtonian mechanics we should recognize their limitations. It must be made clear, however, that the principles of mechanics which arise from the theory of relativity are *generalizations* of Newton's laws of motion and not *contradictions* of them.

7-4 MASSLESS PARTICLES

A further generalization of the concept of momentum is necessary in dealing with particles which do not have mass in the ordinary sense. It has been well established by experiment that electromagnetic radiation of wavelength λ transports *energy* in indivisible bundles, or *quanta*, with the energy of each quantum related to the wavelength by

$$E = \frac{hc}{\lambda} \qquad (7\text{-}17)$$

In this expression, E is the energy of one quantum, c is the speed of light, and h is a universal constant called *Planck's constant*. A quantum of electromagnetic radiation is usually called a *photon*. In some respects it may be thought of as a particle, although it has no mass in the ordinary sense.

We now state without proof a relationship from the theory of relativity. This result is derived in Chap. 15 after a discussion of the foundations of this theory, but we introduce it here because of its relevance and importance to the concept of momentum. It is this: Any particle which has energy must also have momentum, whether it has mass or not. Furthermore, for a particle with zero mass which moves with the speed of light, as do photons of electromagnetic radiation, the energy E and magnitude of momentum p are related by

$$E = pc \qquad (7\text{-}18)$$

This momentum cannot be expressed in terms of a mass, since the particle has no mass; it is associated instead with the *energy* of the particle.

Thus each photon of electromagnetic radiation must have associated with it a momentum with magnitude given by

$$p = \frac{E}{c} = \frac{h}{\lambda} \qquad (7\text{-}19)$$

For photons of visible light, this is a very small amount of momentum. The value of Planck's constant is

$$h = 6.62 \times 10^{-34} \text{ kg-m}^2/\text{s}$$

The yellow light from a sodium-vapor lamp has a wavelength of about

$$\lambda = 589 \times 10^{-9} \text{ m}$$

Hence one photon of this light has a momentum of

$$p = \frac{h}{\lambda} = 1.12 \times 10^{-27} \text{ kg-m/s}$$

If a sodium atom initially at rest emits a photon of yellow light, it must recoil in a direction opposite the direction of emission in order for the total momentum of the system to be the same after the emission as before, namely zero (Fig. 7-4). To calculate the recoil velocity, observe that the sodium atom

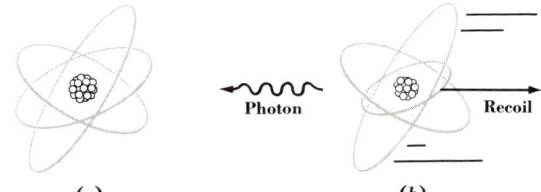

Fig. 7-4 Sodium atom initially at rest (*a*) emits a photon and recoils in the direction opposite the direction of emission (*b*).

must have a momentum of magnitude 1.12×10^{-27} kg-m/s. The mass of a sodium atom is about 3.8×10^{-26} kg, so its speed after emission is 0.029 m/s. This is a very small speed, much smaller than the speed arising from random thermal motion, and it is not readily observable. But in similar processes a *nucleus* may emit a quantum of electromagnetic radiation (called a *γ ray*) which may have on the order of a million times as much momentum as a photon of visible light. Recoils arising from these emissions are easily observed.

Another kind of massless particle, the *neutrino*, appears in collisions in which elementary particles are created or destroyed. Like the photon, it has energy and momentum related by Eq. (7-18). Other characteristics, particularly the angular momentum and the possible kinds of interactions, are different from those of the photon, and the two are distinctly different particles.

The above remarks are not intended to be any kind of thorough exposition of the properties of photons and neutrinos. Such an explanation must wait until more fundamental groundwork in mechanics and electricity has been laid. But just as with the relativistic generalization of the concept of momentum, the essential point is that the principle of conservation of momentum is applicable to situations more general than those with which we began the chapter, in this case to collisions involving massless particles.

7-5 COLLISIONS

Although conservation of momentum is a very important principle in the analysis of collisions, it is not in itself sufficient for a complete analysis;

additional information is usually needed. The following examples will illustrate the point.

A student experimenting with a linear air track observes collisions with two gliders of equal mass m. In one situation the first glider is given an initial velocity v while the second is at rest. After the collision the first glider has stopped, and the second has acquired velocity v. Momentum is conserved; the total momentum is mv both before and after the collision. In a second experiment with the same initial conditions the two gliders stick together and move away with a common final velocity $v/2$. Again momentum is conserved; the initial momentum is mv, the final momentum $(2m)(v/2) = mv$. Thus conservation of momentum by itself cannot predict the final state of motion from given initial conditions.

As these experiments suggest, collisions can be classified in terms of the *relative velocity* before and after the collision. In the first example the relative velocity is the same in magnitude (but opposite in direction) before and after; in the second example it is zero after the collision. One might imagine a collision in which an explosive charge between the bodies is detonated by the impact, driving them apart with a relative velocity *greater* in magnitude than that before the collision.

The ratio of final to initial relative-velocity magnitudes is called the *coefficient of restitution*, denoted by e. The following names are customarily used:

$e = 0$: inelastic collision

$0 < e < 1$: semielastic collision

$e = 1$: elastic collision

$e > 1$: superelastic collision

As we shall see in later chapters, the value of e is closely related to energy relations. When $e = 1$, the total kinetic energy after the collision is the same as before; i.e., kinetic energy is conserved. When $e < 1$, the final kinetic energy is less than the initial, and when $e > 1$, it is greater.

Example
A 0.2-kg glider on an air track with an initial velocity of 0.5 m/s collides elastically ($e = 1$) with a 0.6-kg glider initially at rest. Find the final velocities of both gliders.

Solution

Let the unknown velocities be v_1 and v_2, and let the positive direction be that of the initial velocity of the 0.2-kg glider. From momentum conservation we obtain the relation

$$(0.2 \text{ kg})(0.5 \text{ m/s}) + (0.6 \text{ kg})(0) = (0.2 \text{ kg})v_1 + (0.6 \text{ kg})v_2$$

Since the collision is elastic, $e = 1$, and the relative velocity has the same magnitude after the collision as before. We arbitrarily consider the velocity of the 0.6-kg glider relative to the 0.2-kg one, rather than the reverse. The equality of the two relative-velocity magnitudes gives the relation

$$|0 - 0.5 \text{ m/s}| = |v_2 - v_1|$$

It does not necessarily follow that $v_2 - v_1 = -0.5$ m/s. In fact, a little thought reveals that the relative velocity as defined above must be positive before the collision and negative after. Thus the correct equation without the absolute-value bars is

$$v_2 - v_1 = 0.5 \text{ m/s}$$

This and the momentum-conservation equation can be solved simultaneously to yield the values

$$v_1 = -0.25 \text{ m/s}$$
$$v_2 = +0.25 \text{ m/s}$$

The reader may substitute these values back into the equations to verify that they do indeed satisfy momentum conservation and the condition that the collision be *elastic*.

When the velocities in a collision problem are not all confined to a single line, as they were in the above air-track example, we have an additional complication. In this case, stating the value of e is still not sufficient to define the collision completely; more information must be given. Consider, for example, a collision of two bodies whose motion is confined to a plane, such as two billiard balls. When the initial velocities are given, four scalar quantities are required to describe the final state of motion. These might be the x and y components of the two final velocities or their magnitudes and directions. Each component of momentum must be conserved, so momentum conservation gives two equations relating these quantities. A third relation comes from the coefficient of restitution, but there are still more unknowns than equations. If the magnitude or direction of one of the final velocities is given, the problem becomes soluble.

Example

A billiard ball with mass m and initial speed v collides elastically with a second ball of mass m initially at rest, as shown in Fig. 7-5. The first ball moves off at 45° to its original direction. Find the final velocities of both balls.

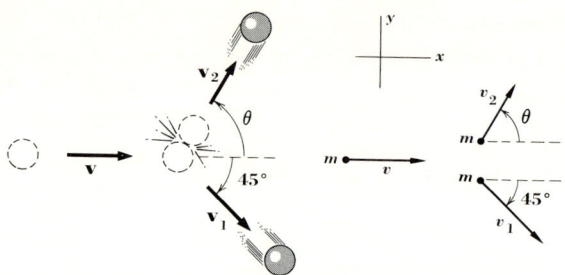

Fig. 7-5 Collision of two billiard balls when one ball is initially at rest.

Solution

We choose coordinate axes as shown. Conservation of the x and y components of momentum gives the relations

$$mv = mv_1 \cos\theta + mv_2 \cos 45°$$
$$0 = mv_1 \sin\theta - mv_2 \sin 45°$$

The relative-velocity relation gives

$$v = \sqrt{(v_1 \cos\theta - v_2 \cos 45°)^2 + (v_1 \sin\theta + v_2 \sin 45°)^2}$$

These three relations are sufficient to solve for the three unknowns v_1, v_2, and θ. Actually carrying out the solution requires some acrobatics, the details of which are left as a problem. The results are $v_1 = v_2 = v/\sqrt{2}$ and $\theta = 45°$. That is, the final velocities are perpendicular, a fact known to every good billiard player. In fact, it can be shown that the two final velocities are perpendicular even if the angle of the first is different from 45°. This problem will be discussed in more detail in Chap. 11, where we shall also learn a much simpler way to obtain this result using the center-of-mass coordinate system.

7-6 IMPULSE

The principle of conservation of momentum makes it possible to relate the initial and final velocities in a collision without knowing the details of the forces exerted during the collision. When Newton's third law holds and there are no external forces, the total momentum is the same at every instant before, during, and after the collision. The momentum of each individual body participating in the collision is, of course, not constant. The total change in momentum of an object provides some general information about the forces exerted during the collision. For this purpose, it is useful to introduce the concept of *impulse*.

When a force **F** is applied to an object, its momentum changes at a rate given by

$$\frac{d\mathbf{p}}{dt} = \mathbf{F}$$

In a small time interval Δt, the momentum changes by an amount $\Delta \mathbf{p}$ approximately equal to $\mathbf{F}\,\Delta t$. To relate the total momentum change during the collision to the force, the time of the collision (say t_1 to t_2) can be divided into N intervals, the momentum change in each calculated, and the results added:

$$\sum_{i=1}^{N} \Delta \mathbf{p}_i = \sum_{i=1}^{N} \mathbf{F}_i\,\Delta t_i = \mathbf{p}_2 - \mathbf{p}_1$$

In the limit as all the intervals approach zero and the number of intervals becomes very large, the sums become integrals, and we have

$$\mathbf{p}_2 - \mathbf{p}_1 = \int_{\mathbf{p}_1}^{\mathbf{p}_2} d\mathbf{p} = \int_{t_1}^{t_2} \mathbf{F}\,dt \qquad (7\text{-}20)$$

where p_1 and p_2 are the initial and final values of p. The time integral of the force on the right-hand side of Eq. (7-20) is called the *impulse* of this force. With this definition, we see that the total change of momentum of an object during any time interval equals the impulse of the force applied to the object during that interval.

If the force is constant, the impulse is equal to the product of the force and the time interval during which it acts. Otherwise, the integral in Eq. (7-20) can be evaluated only when **F** is known as a function of t. Like momentum, impulse is a vector quantity, and the two quantities have the same units. In general, both the magnitude and direction of **F** may vary during a collision.

In the particular case where the force acts always in the same direction, its behavior can be represented graphically by plotting the magnitude of force as a function of time, as in Fig. 7-6. The impulse of the force for any time

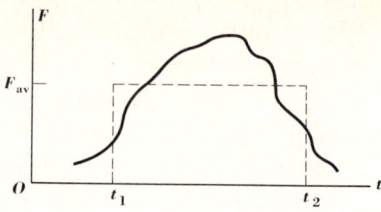

Fig. 7-6 The impulse of a force in a specified time interval is represented graphically by the area under the curve of force as a function of time. The average force is represented by the height of the rectangle having the same area as that under the curve.

interval, say t_1 to t_2, is then represented by the area under the curve in this interval, as shown. A rectangle having the same area can be constructed as illustrated; the height of this rectangle then represents the *average* force, and thus the impulse is equal to the product of the average force and the time interval.

Forces during collisions often have large magnitudes and act for short times. Such forces are often called *impulsive forces*, and the concept of impulse is useful in discussing them.

Example

A tennis ball with negligible initial velocity is struck by a racket. By means of high-speed photography, we ascertain that as the ball leaves the racket, it has a speed of 30 m/s and that the ball and racket are in contact for 0.02 s. What average force acts on the ball?

Solution

The mass of the ball is about 0.060 kg, so when it leaves the racket, it has a momentum of 1.8 kg·m/s. Since the initial momentum was zero, this must be equal to the impulse delivered by the racket. This impulse could be achieved by a *constant* force of 90 N acting for 0.02 s or by any other force which varies with time in such a way that the *time average* over the 0.02-s interval is 90 N. Even though the force is in general not constant during such a collision, the total impulse of the force is always equal to the *average* force multiplied by the time interval. That is,

$$\Delta \mathbf{p} = \int_{t_1}^{t_2} \mathbf{F}\, dt = \mathbf{F}_{av}(t_2 - t_1) = \mathbf{F}_{av}\, \Delta t \tag{7-21}$$

If the impulsive force changes direction during the time it acts, Eq. (7-20) is somewhat more difficult to evaluate. Often the most straightforward method of solution if the force is known is to calculate the components of momentum change from the corresponding components of the force. An example of such a calculation appears in Chap. 12 in connection with the scattering of α particles from atomic nuclei by electrical repulsion.

7-7 GENERALIZATIONS

In reviewing the development of the principle of conservation of momentum, two points must be emphasized. First, if we accept Newton's laws of motion, the principle of conservation of momentum is not new. Rather, it can be derived from Newton's laws using theoretical considerations, without the need for any additional experimental evidence.

Second, the principle of conservation of momentum depends for its validity on *both* the second and third laws of motion. For two interacting bodies, the changes in momentum are equal in magnitude and opposite in direction only if the corresponding forces are equal in magnitude and opposite in direction. We may well question whether this is always the case. What is the status of the third law? Is it a universal law, obeyed under all circumstances, or does it have limitations?

There is certainly good experimental evidence the third law is obeyed in collisions of macroscopic bodies, where contact forces are involved. There is also good experimental evidence to indicate that the *gravitational* force (Chap. 8) of one body on another is equal and opposite to the force of the second body on the first. Coulomb's law of electrostatics, which is based on experimental observations, also states that the forces between two charged particles at rest are equal and opposite and are directed along a line joining the two particles, just as with gravitational forces. For all these forces the third law of motion is obeyed.

When two electrically charged particles are in relative motion, the situation becomes more interesting. Electromagnetic fields are not transmitted instantaneously from one point in space to another but travel with the speed of light. By the time the field of the first particle reaches the position of the second, the first particle has moved to a new position. In computing the force on one particle at a particular instant of time we must use the position and motion of the other particle at a somewhat earlier time; in this case action and reaction are not *simultaneous*. Thus the rates of change of momentum for the two particles are *not* exactly equal and opposite, and the total momentum of such a system is *not* constant.

This is a somewhat disconcerting development. It reminds us of the situation which arose in Chap. 4 in the discussion of the conservation of mass for a nuclear-fission process. In that case there was an apparent violation of the principle of conservation of mass; here we have an apparent violation of conservation of momentum.

As with conservation of mass, we may take two different points of view. One is simply to say that the third law of motion is not a universally valid law and let it go at that. But a much more productive and useful attitude is to ask whether the momentum which seems to appear or disappear during the interaction of two charged particles may be associated with some additional momentum somewhere else in the system, perhaps due to the electromagnetic field.

It turns out that it *is* possible to define a momentum associated with the electromagnetic field. When this additional momentum is taken into consideration, the *total* momentum of the system, including the *mechanical* momentum (calculated from the masses and velocities of the charged particles) and the *electromagnetic* momentum (associated with the fields), is *always* conserved. Thus once more we see an example of a physical principle which, originally developed in a rather specific situation, can be generalized to take in wider and wider classes of phenomena.

Problems

7-1 What is the momentum of a car whose weight is 15×10^3 N and whose speed is 100 km/h?

7-2 Two men, each having a mass of 70 kg, sit at the ends of a 40-kg canoe. They try to propel the canoe by throwing or rolling a 5-kg bowling ball back and forth between them.
   a Comment on the success or failure of this scheme. What procedure do you recommend? Assume that the canoe glides through the water without friction.
   b How is your answer changed if the canoe's interaction with the water is included?

7-3 A car with mass 1,500 kg is standing at a traffic light when another car with mass 2,000 kg strikes it from the rear with speed 20 m/s. After the collision all the wreckage moves away together. What is the speed of the wreckage just after the collision?

7-4 A rifle bullet whose mass is 5 g is shot into a block of wood mounted on a car which rolls without friction on a smooth track. The mass of block and car is 2 kg. The car is initially at rest, but after the bullet lodges in the block, it has a speed of 1 m/s. What was the initial speed of the bullet?

PROBLEMS

7-5 A car with mass 2,000 kg and speed 10 m/s collides head on with a car with mass 1,000 kg. What speed must the second car have for the velocity of the wreckage after the collision to be zero?

7-6 A girl sitting in a canoe throws a rock out the stern end, horizontally, with a speed of 5 m/s. The girl's mass is 50 kg, that of the canoe is 30 kg, and that of the rock is 2 kg. If the canoe is initially at rest, what speed does it have after the rock is thrown?

7-7 A ball of mass m and initial velocity $\mathbf{v}$ collides head on with another ball of equal mass, initially at rest. If the collision is elastic, find the velocities of both balls after the collision.

7-8 A ball of mass m_1 and initial velocity $\mathbf{v}$ collides head on with a ball of mass m_2, initially at rest. If the collision is elastic, find the velocities of both balls after the collision.

7-9 A croquet ball collides with an identical ball initially at rest. After the collision, the first moves off at 45° to its original direction. If the collision is elastic, find the velocities of both balls after the collision.

7-10 Complete the solution of the billiard-ball example in Sec. 7-5 to obtain the values of v_1, v_2, and θ.

7-11 In a billiard game a player wishes his cue ball to strike another ball initially at rest so that the second ball moves off at 45° to the initial direction of the cue ball. Assuming the ball surfaces to be perfectly smooth and the collision elastic, describe how the cue ball must be aimed. What is the final direction of the cue ball?

7-12 A nucleus of radium 226, containing 88 protons and 138 neutrons, undergoes a radioactive decay in which it emits an α particle, containing 2 protons and 2 neutrons. The masses of proton and neutron are very nearly equal. If the nucleus is initially at rest, the α particle leaves with a speed of 1.5×10^7 m/s. What is the final speed of the residual nucleus?

7-13 A shell is fired upward at 45° to the horizontal with an initial speed of 300 m/s. At the highest point in its trajectory it explodes into two equal fragments. If one fragment falls vertically after the explosion with zero initial vertical velocity, what is the velocity of the other fragment just after the explosion?

7-14 If a ball collides with a perfectly smooth flat stationary surface, and if its speeds before and after the collision are equal, prove that the angle the velocity makes with the surface is the same before and after the collision; i.e., the angle of incidence is equal to the angle of rebound.

7-15 A truck of mass M can roll without friction along a long straight horizontal road. At the beginning the truck is at rest and N men, each having mass m, stand in the truck.

a Suppose all the men run with speed v_0 (relative to the truck) toward the rear of the truck and jump off. What is the final speed of the truck?

b Suppose the men run one at a time and jump off, again each with speed v_0 relative to the truck. What is the final speed of the truck? Compare your result to that for part *a*.

7-16 A machine gun fires 10-g bullets at a target with a speed of 300 m/s at a rate of 480 bullets per minute. If the target is supported so that it does not move, what average force must the supports exert to keep it in position?

7-17 A uniform cable hangs vertically with the bottom end just touching the floor. The upper end is then released, allowing the cable to fall and to coil up on the floor. Show that the total force exerted by the cable on the floor at any time while the cable is falling is equal to three times the weight of the coiled-up part of the cable already on the floor. Assume the cable coils up smoothly without bouncing.

7-18 Water from a fire hose is aimed horizontally at a brick wall. The hose is 10 cm in diameter, and the water flows at a speed of 10 m/s. If the water falls straight down after striking the wall, what force does it exert on the wall?

7-19 A jet airplane traveling 200 m/s takes air into its engines at the rate of 50 kg/s. This air is mixed with 2 kg of fuel (per second), burned, and ejected from the rear of the engine at a speed *relative to the airplane* of 400 m/s. How much thrust does the engine provide?

7-20 A railroad coal car has an empty mass of 20×10^3 kg. It rolls under a coal chute with an initial speed of 10 m/s. Coal is dropped vertically into it at the rate of 5×10^3 kg/s. Find the acceleration and velocity of the car as functions of time.

7-21 A rocket whose total mass is M is pointed vertically upward near the surface of the earth. If burned fuel is ejected at speed u relative to the rocket, at what initial rate must fuel be burned in order for the rocket to remain motionless without external support, i.e., to lift its own weight?

7-22 Derive Eq. (7-13) from Eq. (7-12). For what ratio of fuel mass M_f to "empty mass" M is the final velocity exactly equal to the exhaust speed u if $v_0 = 0$?

7-23 A rocket with empty mass M and fuel mass M_f burns fuel at a constant rate and with constant exhaust speed u, burning the total fuel in a time T.
a Obtain an expression for the total mass $m = M + M_f$ of rocket plus fuel as a function of time.
b Use this result together with Eq. (7-11) to find the velocity as a function of time. From this, find the position as a function of time.

7-24 What must be the ratio of fuel mass to empty mass for a rocket if it starts from rest and attains a final velocity twice as great as the exhaust speed u?

PROBLEMS

7-25 The mass of an electron is 9.11×10^{-31} kg, and the speed of light c is 3.00×10^8 m/s. Comparing the classical definition of momentum given by Eq. (7-3) with its relativistic generalization [Eq. (7-16)], by how much is the classical expression in error if:
   a $v = 0.01c$
   b $v = 0.5c$
   c $v = 0.9c$

7-26 A radioactive isotope of cobalt, ^{60}Co, emits an electromagnetic photon (γ ray) whose wavelength is about 0.932×10^{-12} m. The nucleus contains 27 protons and 33 neutrons, each with a mass of about 1.66×10^{-27} kg. If the nucleus is at rest before emission, what is its speed afterward? Is it necessary to use the relativistic generalization of momentum to find this speed?

7-27 In Prob. 7-26, suppose that the cobalt atom is in a metallic crystal containing 0.01 mol of cobalt (about 6.02×10^{21} atoms) and that the crystal recoils as a whole, rather than just the single nucleus. Find the recoil velocity. (This recoil of the whole crystal rather than a single nucleus is called the *Mössbauer effect*, in honor of the man who discovered it in 1958.)

7-28 A tennis ball approaches a player's racket at 10 m/s in a direction 45° down from the horizontal. It is struck hard and leaves at 20 m/s, 30° up from the horizontal. What impulse was given to the ball? If the collision with the racket lasted 0.01 s, what average force did the racket exert? The mass of a tennis ball is about 60 g.

7-29 If the player holding the racket in Prob. 7-28 is standing on a frictionless ice-covered court and has no velocity before hitting the ball, what is his velocity after hitting it if his mass is 80 kg?

7-30 An automobile whose mass is 2,000 kg is moving 30 m/s when the brakes are applied hard. The road is clean and dry, and the coefficient of friction is $\mu = 1.0$.
   a Using the impulse-momentum principle, find the minimum time required to stop.
   b Find the acceleration of the car; compute the stopping time from this and the initial speed. Compare your result with that of part a.

7-31 A billiard ball of mass 0.2 kg bounces off a cushion on the table. Both the initial and final velocities are at 45° to the table, and the initial and final speeds are both 2 m/s.
   a Using a suitable coordinate system, find the change in momentum of the ball.
   b What impulse was given to the ball?
   c What impulse did the ball give to the table?

7-32 A force applied to a certain body is constant in direction but increases in magnitude at a uniform rate from 0 to 20 N during an interval of 0.1 s. It is then constant for 0.2 s and finally decreases at a uniform rate to zero in another 0.1 s.
    a   Graph the force as a function of time; find the impulse and the average force.
    b   If this force is applied to a 3-kg body having an initial velocity of 1 m/s in the direction of the force, what is the final velocity of the body?

Dynamics of a Particle III | 8

A general formulation of the principle of gravitational attraction leads to the concept of a *force field*. The motion of planets and satellites is discussed, and then the more general problem of a particle moving under the action of a central force is introduced. Consideration of the motion of a charged particle in electric and magnetic fields leads to the concept of a velocity-dependent force. Finally, we discuss briefly how information can be obtained about the forces of interaction among elementary particles by observing collisions of such particles.

8-1 GRAVITATION

Every body of finite mass near the surface of the earth experiences a force which pulls it toward the earth; we call this force the *weight* of the body. We noted in Chap. 5 that the weight of a body at a given location is always proportional to its mass, as shown by the observation that the acceleration of a freely falling body is independent of its mass.

Nearly 300 years ago Newton speculated that the force which attracts bodies to the earth might be a particular example of a more general kind of interaction. In particular, Newton believed that the motion of the planets around the sun should obey the same laws of motion he had formulated for motion of bodies on earth. If so, there must be *forces* acting on the planets to make them move in their nearly circular orbits around the sun. What is the nature of these forces?

In a truly brilliant flash of insight, Newton saw that the forces responsible for both planetary motion and the phenomenon of weight could be understood if one assumed that *every* body in the universe attracts every other body in the universe with a force dependent on their masses. This hypothesis forms the basis for Newton's law of gravitation, and the forces are called *gravitational forces* or *gravitational interactions*. Newton was even able to calculate how this attractive force must depend on the masses of the bodies and the distance between them, on the basis of information concerning the orbits of planets. We discuss in Sec. 8-3 some of the details of his analysis. First, however, we discuss some more direct evidence of the gravitational interaction.

The first direct observations of gravitational attraction between bodies in a laboratory were made by Cavendish in 1798. One form of the apparatus he used, shown in Fig. 8-1, is called a *Cavendish balance*. A dumbbell-shaped

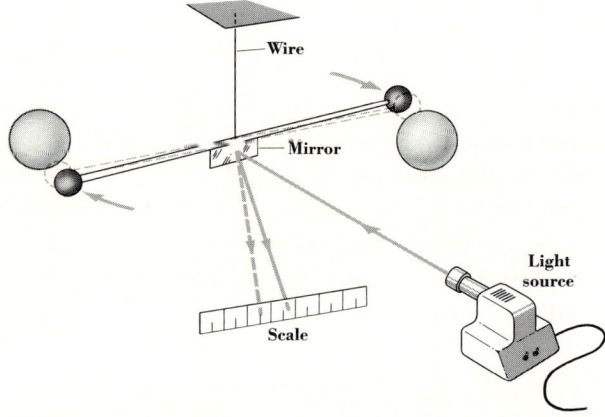

Fig. 8-1 Schematic diagram of a Cavendish balance. When the large spheres are brought close to the smaller spheres on the rod suspended by the thin wire, a slight deflection of this rod is observed. The angular deflection can be measured precisely by reflecting a beam of light from a mirror attached to the rod onto a ground-glass scale.

object is suspended by a very thin wire or fiber as shown. When massive bodies, such as lead spheres, are brought close to the ends of the dumbbell, a slight twisting of the dumbbell is observed, indicating an attractive force. Furthermore, it is possible to calibrate the stiffness of the suspending wire by independent means, to find the amount of force corresponding to a given angular deflection. Thus the gravitational force between two known masses, although very weak, can be *measured* directly.

The observations of Cavendish corroborated Newton's deductions from measurements of planetary orbits, which may be stated as follows: When two bodies with masses m_1 and m_2 are separated by a distance r, each exerts on the other an *attractive force,* directed along the line between the two bodies, of magnitude

$$F = G\frac{m_1 m_2}{r^2} \tag{8-1}$$

where G is a universal constant which has the same value for all masses. Gravitational forces obey Newton's third law; the forces which two bodies exert on each other act along the same line and have equal magnitudes but opposite directions.

Because the gravitational forces between bodies of ordinary size are extremely small and the apparatus rather delicate, it is difficult to obtain a precise value of G. The presently accepted value is

$$G = 6.67 \times 10^{-11} \text{ N-m}^2/\text{kg}^2 \tag{8-2}$$

This constant is called the *universal gravitational constant.* Newton was not able to deduce the actual value of G from astronomical observations, for reasons to be discussed in Sec. 8-3, but the method of Cavendish measures G directly.

It may seem somewhat strange that an attractive *force* between two bodies should be proportional to quantities (their masses) originally introduced to characterize *inertial* properties. One can, in fact, define two different kinds of masses: an *inertial* mass, used in $\Sigma \mathbf{F} = m\mathbf{a}$, and a *gravitational* mass, used in Eq. (8-1). As already pointed out, evidence indicates that these two masses may be used interchangeably. Many very precise observations have been made in an effort to detect any small difference between inertial mass and gravitational mass. The results of these investigations have shown that these two are, in fact, equivalent, at least to within 1 part in 10^9.

As formulated above, the law of gravitation can be applied only to particles having no extension in space or to bodies sufficiently small compared to the distance between them to permit their representation by a model in which they become points. The law is easily generalized to bodies of finite size, however. We imagine each body to be divided into a very large number of small elements, calculate the force which each element of one body exerts on each element of the other, and calculate the vector sum of all these forces by integration. In general this can be a very complex problem, but in many cases of practical interest it is actually quite simple.

First, when the distance between two bodies is large compared with the size of either, only a small error is introduced by assuming that all the mass of each body is concentrated at a single point. Also, since Newton's third law holds for the interaction between two points, it also holds for the forces between two extended objects, because the total force which each body exerts on the other is the vector sum of forces between individual particles. Thus the gravitational forces between *any* two bodies, no matter what their shape, are equal in magnitude and opposite in direction.

When a body has spherical symmetry, it can be shown that its gravitational interaction is the same as though all the mass were concentrated at its center. This statement can be proved by starting with Eq. (8-1) and using a theorem of vector analysis known as *Gauss' theorem*. The proof will not be given in detail here, but Fig. 8-2 illustrates its application. An irregularly

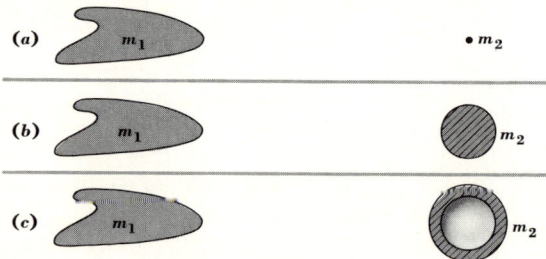

Fig. 8-2 Gravitational interaction between an irregularly shaped body of mass m_1 and another body of total mass m_2. In (a), m_2 is a point mass. In (b), it is a homogeneous sphere with the same total mass. When the center of the sphere coincides with the position of the point mass in (a), the force exerted by m_1 is the same in both cases; the force on m_1 is also the same in the two cases. In (c), m_2 is a spherical shell (shown in cross section) with its center at the same point as the point mass in (a). In all three cases, the forces exerted by one body on the other are the same.

shaped body of mass m_1 exerts a gravitational force on three different bodies, all spherically symmetric and all with the same total mass m_2. Because of the spherical symmetry, the total force is the same in each case.

This is a very useful result. In the Cavendish experiment, for example, it shows that the force between a large lead sphere and a small sphere suspended by a fiber is the same as though all the mass of each sphere were concentrated at its center. This result was first established by Newton, and Cavendish used it in interpreting his experimental data to obtain the value of G. This principle is used again in discussing the orbits of planets and satellites in Sec. 8-3.

One can easily obtain a relation between the universal gravitational constant G and the acceleration of gravity g on the surface of the earth. We model the earth as a uniform sphere of total mass M and radius R. Then a body of mass m at the surface of the earth is attracted toward its center just as though the earth's mass were all concentrated there. The magnitude of the attractive force, according to Eq. (8-1), is

$$F = G\frac{Mm}{R^2} \tag{8-3}$$

But this force is exactly the same as the weight of the body, mg. Thus

$$mg = G\frac{Mm}{R^2}$$

and

$$g = \frac{GM}{R^2} \tag{8-4}$$

The quantities g and G can be measured in the laboratory. It is also possible to measure the radius of the earth, so this relation gives us a means of calculating the *mass* of the earth! This calculation is left as a problem.

8-2 FORCE FIELDS

Discussion of the gravitational interaction is facilitated by introducing the concept of a *force field*. To illustrate this concept, we consider a distribution of mass, as in Fig. 8-3. When a particle of mass m, which we may call a

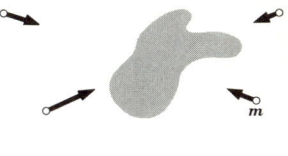

Fig. 8-3 Distribution of mass shown produces a gravitational field **g** at every point in space. Arrows representing the field **g** at several points are shown. A particle of mass m experiences a force $\mathbf{F} = m\mathbf{g}$, where **g** is the gravitational field at the position of the particle.

test particle, is brought into the vicinity of this mass distribution, the test particle experiences a gravitational force, the vector sum of the forces due to individual elements in the distribution.

The force on the test particle depends on its mass and its position in space. Such a position-dependent force is called a *force field*; we may visualize a vector at each point in space representing the force on a test particle at that point.

Because the gravitational force on a test particle is proportional to its mass m, description of the gravitational force field is simplified further by representing the force $\mathbf{F}$ on the particle at any point as the product of m and another vector $\mathbf{g}$, which we call the *gravitational field*. That is,

$$\mathbf{F} = m\mathbf{g} \tag{8-5}$$

The advantage of this scheme is that if we know $\mathbf{g}$ at every point in space, we can find the gravitational force exerted on *any* point mass by simply multiplying $\mathbf{g}$ by its mass.

To summarize this new point of view: Instead of regarding the distribution of mass as exerting a gravitational force directly on the test body of mass m, we regard the distribution as creating a *gravitational field* $\mathbf{g}$. The field then exerts a force, given by Eq. (8-5), on m. This new point of view does not introduce any new experimental information, and we cannot make any calculations with it which could not, at least in principle, be made by using Eq. (8-1) directly. It does, however, simplify calculations in many practical cases. Furthermore, there is good reason to think that the field itself has a real physical significance and is not simply a device to simplify calculations.

We have used the same symbol for the gravitational field as was used in Chap. 5 for the acceleration of gravity. This is no accident. If a body of mass m is in a gravitational field $\mathbf{g}$, and if no other forces act on it, then we see immediately that the acceleration of the body is just $\mathbf{g}$. This is a special property of the gravitational field because the force is proportional to the *mass* of the particle. In Sec. 8-5 we consider two different force fields in which the force is proportional to the *electric charge* of the particle. In such cases, we must first determine the force on the particle and then use Newton's laws of motion to find the acceleration.

As an example of the use of the gravitational-field concept, consider a situation in which a particle of mass M is located at the origin of a polar coordinate system. We consider the gravitational field for points in the plane of this coordinate system. Suppose a particle m is located at point P with coordinates (r,θ). The gravitational force on m is then

$$\mathbf{F} = -\frac{GmM}{r^2}\mathbf{n} \tag{8-6}$$

where the unit vector $\mathbf{n}$ and the minus sign indicate that the force is directed toward the origin, where M is located. Then according to Eq. (8-5), the gravitational field $\mathbf{g}$ at point P is given by

$$\mathbf{g} = -\frac{GM}{r^2}\mathbf{n} \tag{8-7}$$

One may ask whether m must always be positive; there is nothing in the principles discussed above which would prohibit *negative* values. No particle with negative inertial or gravitational mass has been observed, but it has been proposed that the *antiparticles* such as negative protons and antineutrons may have positive inertial masses but negative gravitational masses. There are theoretical arguments on both sides of this proposal, but no experimental evidence to support or refute the speculation.

There are other kinds of fields in addition to force fields. In any situation where a vector quantity is associated with each point in space, we describe those vectors collectively as a *vector field*. A familiar example is the velocity of air when the wind is blowing. The air may have a different velocity at each point in space; thus there is a velocity vector associated with each point in space. We refer to all these as the *velocity field*. Similarly, electric and magnetic fields, to be described in Sec. 8-5, are examples of vector fields.

There are other fields which are not vector fields but which associate a *scalar* quantity with each point in space. An example is temperature; every point in a room has a certain temperature, and we refer to the temperature at each point in space as the *temperature field*. Such a field associates with each point a scalar quantity rather than a vector quantity, and so we call it a *scalar field*. There are also fields which associate with each point in space a more complicated mathematical quantity called a *tensor*, having more than three components. These are called *tensor fields*; they are of considerable importance in the quantum theory of elementary particles, as well as in the areas of elasticity and fluid mechanics.

8-3 PLANETARY MOTION

The problem which originally led Newton to the discovery of the law of gravitation was that of the motion of planets around the sun. Even earlier, Johannes Kepler, whose death preceded the birth of Newton by 12 years, had discovered three rules concerning the motions of the planets, now called *Kepler's laws*. These discoveries were in turn based on astronomical observations made by Kepler's teacher, the Danish astronomer Tycho Brahe (1546–1601).

The first of Kepler's laws states that each planet moves in an elliptical path, with the sun located at one focus of the ellipse. The second, to be discussed in more detail in Chap. 12, states that a line drawn from the sun to any planet sweeps out area at a uniform rate. The third states that the average distance R from the sun to any planet is related to the time T required for one revolution about the sun. Specifically,

$$T^2 = KR^3 \tag{8-8}$$

where K is a constant with the same value for all planets. T is called the *period* of the motion.

Now what has all this to do with Newton's laws of motion and the law of gravitation? To see the relationship, we construct a simple model of the motion of a planet around the sun. Let the mass of the sun be M and that of the planet m. It has been observed that the elliptical orbits of the planets are very nearly circular, so we assume that our idealized planet moves in a circle with radius R. We suppose that M is very much larger than m, corresponding to the fact that the sun is much more massive than any planet; then the center of the sun may be regarded as stationary in an inertial frame of reference. In addition, we neglect the gravitational forces exerted by the other planets and stars and consider only the gravitational attraction of the sun. Thus in this model the planet moves under the action of a force of constant magnitude, directed always toward a fixed point (the sun) as shown in Fig. 8-4.

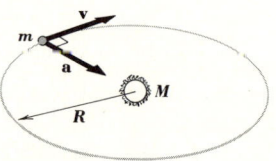

Fig. 8-4 Idealized model for the motion of a planet around the sun. The sun, with mass M, is assumed to be stationary; the planet, with mass m, moves in a circle of radius R with the sun at its center. The speed v of the planet is constant; its acceleration is always directed toward the sun.

If the planet moves in a circular path with the sun at its center, the force and therefore the acceleration are directed toward the center and are always perpendicular to the velocity. As a result, the velocity changes only in direction, not in magnitude; the *speed* of the planet is constant. The planet's acceleration, as discussed in Sec. 6-2, is given by

$$a = \frac{v^2}{R} \tag{8-9}$$

The force which produces this acceleration is the gravitational attraction of the sun. Equating this force to the mass of the planet times its acceleration, we find

$$\frac{GMm}{R^2} = ma = m\frac{v^2}{R} \tag{8-10}$$

The speed of the planet and the radius of its orbit are not independent but are related by Eq. (8-10). Furthermore, the mass m of the planet can be divided out of this relation, so that the speed which a planet must have in order to move in an orbit of given radius is $v^2 = GM/R$, independent of its mass. For instance, if we could saw the earth in two, each piece would continue in the same orbit (neglecting effects due to rotation of the earth).

The speed of the planet is simply related to its period T, the time required for one revolution. The circumference of the orbit is $2\pi R$, so $v = 2\pi R/T$. Substituting this in Eq. (8-10) and canceling m, we find

$$\frac{GM}{R^2} = \left(\frac{2\pi R}{T}\right)^2 \frac{1}{R}$$

which can be rewritten

$$T^2 = \frac{4\pi^2}{MG} R^3 \qquad (8\text{-}11)$$

Here, lo and behold, is Kepler's third law! This equation has the same form as Eq. (8-8); furthermore, we have found that the constant K is related simply to the gravitational constant G and the mass M of the sun. Since the radius of a planetary orbit and its period can be measured by astronomical observations, this equation can be used to compute the mass of the sun. The calculation is left as a problem.

As previously mentioned, we have reversed the historical order in the development of Kepler's third law by starting with the law of gravitation and deriving Eq. (8-11). Newton started with Kepler's law and deduced the law of gravitation from it by assuming that the gravitational force is given by an equation

$$F = G\frac{m_1 m_2}{r^n} \qquad (8\text{-}12)$$

with G and n as unknown constants, and then computing the relation between T and R for planetary orbits on this basis. Then the value of n could be chosen to make the result agree with Kepler's third law. This calculation, which is an interesting exercise, is left as a problem.

Although we have discussed orbits in terms of the motion of planets around the sun, it should be evident that this discussion can easily be adapted to the motion of satellites around the earth. The earth then plays the role of the sun in the solar-system discussion, and the satellites are analogous to the planets. As pointed out in Sec. 8-1, the gravitational force exerted by the earth on a satellite is the same as it would be if all the mass of the earth were concentrated at its center. To be correct, we should say "almost the same," because the earth is not exactly spherical in shape. In fact, precise observations of the motion of satellites have shown deviations from the motion which would occur if the earth were really a homogeneous sphere; these deviations have been used to compute the extent to which it differs from spherical shape. It turns out that the earth is flattened by a few miles near the poles and is slightly pear-shaped. To the extent that the earth's shape may be approximated as a uniform sphere, however, the force acts as though all the mass were concentrated in the center (Fig. 8-5).

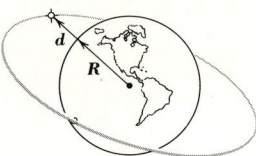

Fig. 8-5 One possible orbit of an artificial satellite around the earth. The radius of the earth is R, and the satellite moves in a circular orbit a distance d above the earth, which is considered to be spherical. The gravitational force exerted by the earth on the satellite is the same as it would be if all the mass of the earth were concentrated at its center.

Example
Find the period of a satellite which moves in a circular orbit about the earth at an altitude of 500 mi above its surface.

Solution
We use Eq. (8-11) to solve this problem. It is convenient to let R be the radius of the earth; then when the satellite is a distance d above the earth, the radius of its orbit is $R + d$. Equation (8-11) becomes

$$T^2 = \frac{4\pi^2}{MG} (R + d)^3$$

The mass M of the earth can be found from the value of g at the surface, as discussed in Sec. 8-1. From Eq. (8-4), we have

$$\frac{1}{M} = \frac{G}{gR^2}$$

Thus the equation for T becomes

$$T^2 = \frac{4\pi^2}{G} \frac{G}{gR^2} (R + d)^3 = 4\pi^2 \frac{R}{g} \left(1 + \frac{d}{R}\right)^3$$

or

$$T = 2\pi \left(\frac{R}{g}\right)^{1/2} \left(1 + \frac{d}{R}\right)^{3/2}$$

For an approximate numerical value we may take $R = 4{,}000$ mi and $d = 500$ mi, so $d/R = \frac{1}{8}$. Then $R \simeq 6.44 \times 10^6$ m.

$$T = 2\pi \sqrt{\frac{6.44 \times 10^6 \text{ m}}{9.8 \text{ m/s}^2}} \left(1 + \frac{1}{8}\right)^{3/2} = 7{,}270 \text{ s} = 121 \text{ min}$$

The period increases with increasing d, as we expect. For a satellite close to earth, $d \simeq 0$, we find $T = 85$ min, which is the *minimum* period for an earth satellite. For the moon, $d = 240{,}000$ mi, approximately, and $T = 27.3$ d.

When the orbit is not circular, the calculations become somewhat more involved. In the next sections, we outline briefly some general methods which can be used to compute noncircular orbits of planets and satellites.

8-4 CENTRAL-FORCE PROBLEMS

Let us consider the general class of problems in which a particle with mass m experiences a force which always acts along the line joining this particle to a fixed point, called the *center of force*. Planetary motion, in our idealized model, is an example of such motion, the center of force in this case being the center of the sun. A force which always acts toward or away from a fixed point, with magnitude depending only on the distance from the fixed point, is called a *central force*.

The gravitational force exerted on a particle by another particle which is stationary in an inertial frame of reference is an example of a central force. Another example is the force exerted on an electrically charged particle by another stationary charged particle. This situation arises in the simplest model of the hydrogen atom when one considers the motion of the electron in the force field produced by the nucleus, which is much more massive than the electron and may be considered at rest. A third example is a particle attached to one end of a spring, the other end of which is stationary in an inertial frame of reference. In this case, the spring always pulls toward the fixed end or pushes away from it. Thus there are several kinds of problems which fall into the general category of *central-force problems*.

An important characteristic of central-force problems is that the *motion always takes place in one plane* containing the center of force and the initial position and velocity vectors; there is never a component of acceleration perpendicular to this plane, and so there can never be any component of velocity perpendicular to it. In the special case in which $\mathbf{v}_o$ and $\mathbf{a}$ lie along the same line, there is no unique plane determined by these vectors, but in this case the motion is confined to a single line through O.

We use the polar coordinate system introduced in Sec. 6-5 and place the origin O of the coordinate system at the center of force. Any force which acts always along the position vector $\mathbf{r}$ joining O and P and whose magnitude depends only on the distance between them can be represented as

$$\mathbf{F} = f(r)\mathbf{n} \tag{8-13}$$

where $f(r)$ is the magnitude of the force, a function only of the distance r, and the unit vector $\mathbf{n}$ indicates that the direction of the force is always toward or away from the center of force O (Fig. 8-6).

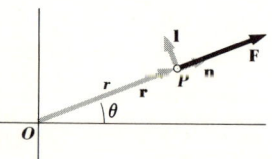

Fig. 8-6 Polar coordinate system used for central-force problems. The unit vectors $\mathbf{n}$ and $\mathbf{l}$ are shown. In a central-force field, the force on a particle at point P is always in the direction of $\mathbf{n}$ if the center of force is at the origin O.

The equation of motion is $\mathbf{F} = m\mathbf{a}$. We now use Eq. (6-31) to express the acceleration in terms of the polar coordinates and their derivatives, obtaining

$$f(r)\mathbf{n} = m\left[\frac{d^2r}{dt^2} - r\left(\frac{d\theta}{dt}\right)^2\right]\mathbf{n} + m\left[r\frac{d^2\theta}{dt^2} + 2\frac{dr}{dt}\frac{d\theta}{dt}\right]\mathbf{l} \tag{8-14}$$

This is a vector equation relating the force to the acceleration. Writing separately the r and θ components of this equation, we obtain the two scalar equations

$$f(r) = m\left[\frac{d^2r}{dt^2} - r\left(\frac{d\theta}{dt}\right)^2\right]$$
$$0 = m\left[r\frac{d^2\theta}{dt^2} + 2\frac{dr}{dt}\frac{d\theta}{dt}\right] \tag{8-15}$$

The second of these equations becomes more interesting when divided by m and multiplied by r:

$$0 = r^2\frac{d^2\theta}{dt^2} + 2r\frac{dr}{dt}\frac{d\theta}{dt} \tag{8-16}$$

This is simply the derivative with respect to time of the quantity $r^2\, d\theta/dt$. That is,

$$\frac{d}{dt}\left(r^2\frac{d\theta}{dt}\right) = 0 \quad \text{or} \quad r^2\frac{d\theta}{dt} = \text{constant} \tag{8-17}$$

For *any* central-force motion, the angular velocity $d\theta/dt$ is always inversely proportional to r^2.

For reasons discussed in detail in Chap. 12, it is customary to make the following abbreviation:

$$L = mr^2 \frac{d\theta}{dt} \tag{8-18}$$

where L is a constant. Equation (8-18) can be used to eliminate $d\theta/dt$ from the first of Eqs. (8-15) as follows:

$$f(r) = m\left[\frac{d^2r}{dt^2} - r\left(\frac{L}{mr^2}\right)^2\right] \tag{8-19}$$

The advantage of this procedure is that it reduces the equation to one containing only the coordinate r, and not θ.

There is no point in carrying the general analysis much further than this. In a specific problem, when $f(r)$ is known, Eq. (8-19) can be solved to find the coordinate r as a function of time. This function can then be inserted in Eq. (8-17) to obtain a differential equation for θ as a function of time. The solution of this equation then completes the solution of the problem, since the polar coordinates r and θ are then known functions of time.

It is instructive to see that the first of Eqs. (8-15) reduces to something familiar in the special case in which the orbit is a circle. Then r is constant; thus $d^2r/dt^2 = 0$, and we obtain

$$f(r) = -mr\left(\frac{d\theta}{dt}\right)^2 \tag{8-20}$$

That is, for a circular orbit the force $f(r)$ must be just the mass m times the centripetal acceleration $r(d\theta/dt)^2$. Of course, we already know this from the considerations of Chap. 6. The minus sign in Eq. (8-20) indicates that the force must be directed toward O, *opposite* to the unit vector **n**; a circular orbit is possible only with an attractive force.

It is often helpful to represent the behavior of the force graphically. In Fig. 8-7, we show $f(r)$ for two different kinds of interactions. Figure 8-7a shows a force proportional to $1/r^2$, always negative corresponding to the

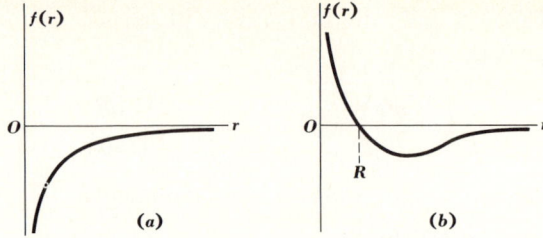

Fig. 8-7 Two different central-force fields. In (a) the force is always attractive, as in the case of gravitational interaction. In (b), the force is negative (attractive) at large values of r, but at sufficiently small values of r it becomes positive (repulsive). At R the force is zero.

direction of an attractive electric or gravitational force. Figure 8-7b shows a force which is very small for very large values of r, becomes larger and negative (corresponding to an attraction toward the center of force) for smaller r, but becomes positive (repulsive) at still smaller r. Forces which hold atoms together in a molecule exhibit this general behavior. The distance R at which the force becomes zero is related to the equilibrium separation between the atoms in the molecule.

It turns out that in the case of a force which is proportional to $1/r^2$, such as a gravitational attraction or an electric force between two charged particles, the possible orbits always have the shapes of conic sections; they are always circular, elliptical, parabolic, or hyperbolic. The circular and elliptical orbits are called *bound* orbits, since in these cases the particle attains some maximum distance from the center of force and then returns. The parabolic and hyperbolic orbits, on the other hand, are *unbound* in the sense that the particle comes in toward the center of force from a great distance, attains a certain distance of closest approach, and then goes infinitely far away again. Some comets have such paths, but the orbits of recurring comets, such as Halley's, are flat ellipses.

8-5 ELECTRIC AND MAGNETIC FORCES

We have remarked several times that electric and magnetic field forces on electrically charged particles are of primary importance in the microscopic structure of matter. It would not be appropriate at this point to try to define exactly what electric charge is, but we may discuss how electromagnetic forces on charged particles can be described in terms of electric charge and electric and magnetic fields.

Consider the situation shown in Fig. 8-8, in which two parallel metal

Fig. 8-8 When two flat parallel metal plates (shown in cross section) are connected to a battery, an excess of positive charge accumulates on one plate and an excess of negative charge on the other. As a result, an electric field **E** is established in the region between the plates. If the plates are very large compared with their separation, the electric field is nearly uniform in magnitude and direction between the plates, except near the edges.

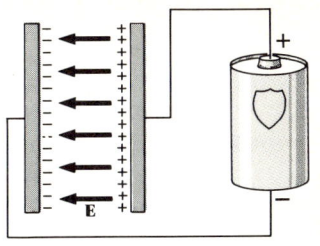

plates are connected to the terminals of a battery. When a charged particle, such as an electrified pith ball or an electron, is introduced between these plates, it experiences a force. Furthermore, it is found that this force is proportional to the magnitude of the *charge* of the particle, which in turn can be defined in terms of the force exerted on the particle by a standard charge at a specified distance.

We describe this situation precisely by saying that an *electric field* exists between the metal plates. To define the field, we specify that the force on the particle is equal to the product of the charge (a scalar) and the electric field (a vector). Denoting the force on the particle as **F**, the charge (which may be either positive or negative) as q, and the electric field as **E**, we have

$$\mathbf{F} = q\mathbf{E} \tag{8-21}$$

In the mks system the unit of charge is the *coulomb* (C), and the magnitude of **E** may be expressed in newtons per coulomb.

This situation is very similar to that in which we defined the gravitational field **g** in Sec. 8-2. The electric field is produced by an excess of positive charge on one plate and an excess of negative charge on the other, just as the gravitational field was produced by a certain distribution of mass. The details of the computation of the electric field from a given distribution of charge need not concern us here; the essential point is that in the presence of the electric field **E**, a particle of charge q experiences a force given by Eq. (8-21).

As an example, suppose that a charged particle moving with an initial velocity $\mathbf{v}_0$ enters a region between two plates (Fig. 8-9). For simplicity, we assume that the electric field between the plates is uniform in both magnitude and direction. What trajectory does the particle describe?

In answering this question, let us suppose that the electric field force on the particle is so large that all other forces acting on it, such as gravity,

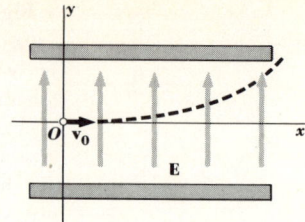

Fig. 8-9 Charged particle with initial velocity $\mathbf{v}_0$ enters a uniform electric field **E**. The parabolic path of the particle is shown.

may be neglected. The constant force $\mathbf{F} = q\mathbf{E}$ on the particle produces a constant acceleration determined by Newton's second law:

$$q\mathbf{E} = m\mathbf{a} \tag{8-22}$$

Taking axes as shown in Fig. 8-9, with the origin of the coordinate system at the initial position of the particle, we find that the x and y coordinates as functions of time are given by

$$x = v_0 t \qquad y = \frac{1}{2} \frac{qE}{m} t^2 \tag{8-23}$$

The x velocity is constant, since there is no force in the x direction to produce an acceleration. The y acceleration is constant; hence y is proportional to t^2. This is exactly the situation which occurred in our discussion of the trajectory of a particle under the force of gravity. Here, as before, the path is parabolic in shape.

This scheme is commonly used to deflect the electron beam in an oscilloscope tube. When q is a negative quantity, as in the case of electrons, the path of the particle curves off below the x axis instead of above it. Thus, in principle, this is a means of determining whether a charge is positive or negative. If the electric field magnitude E is known, the quantity q/m can be determined, although it is not possible by this means to measure either q or m separately.

Now we consider a different situation, the motion of a charged particle in a *magnetic* field **B**. Magnetic forces on charged particles are employed widely in research on fundamental particles. The magnetic field may be produced by a material which is permanently magnetized or by an electromagnet. A typical laboratory magnet has the general shape shown in Fig. 8-10. It consists of a soft-iron yoke with a gap in one side and one or more coils of wire. A magnetic field is produced in the gap when an electric current is passed through the coil. For simplicity, we assume that the magnetic field is uniform in magnitude and direction in the region between the pole faces

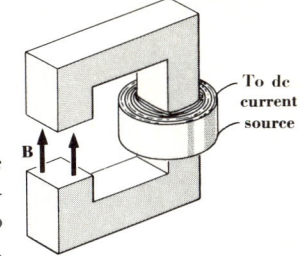

Fig. 8-10 Laboratory electromagnet. The soft-iron core is designed so that the magnetic field **B** is produced chiefly in the gap in the yoke.

and that its direction is as shown in the figure. A small compass can be used to determine the direction of a magnetic field.

We now investigate what happens when a particle with charge q moves through the field. A number of interesting features emerge. First, when the particle is at rest or moves parallel to the direction of the magnetic field **B**, *no force* is exerted on it, and it moves in a straight line with constant velocity, again assuming that all forces other than the magnetic field forces may be neglected. When the particle moves perpendicular to the field direction, it experiences a force proportional to q. This force turns out to be *perpendicular* to both **B** and the velocity **v** of the particle. Furthermore, the *magnitude* of this force is found to be proportional to the *speed* of the particle! Thus the effect of this force on the particle is quite different from that of the electric field force.

It has been found experimentally that the magnitude of the magnetic field force on a charged particle is proportional to the sine of the angle between the velocity and the magnetic field. Thus we may say that

$$F = qvB \sin \theta \tag{8-24}$$

Now $vB \sin \theta$ is just the magnitude of the vector product $\mathbf{v} \times \mathbf{B}$, as defined in Sec. 2-6. Furthermore, since **F** is perpendicular to both **v** and **B**, the *direction* of **F** is the same as that of $\mathbf{v} \times \mathbf{B}$. Thus we may write

$$\mathbf{F} = q\mathbf{v} \times \mathbf{B} \tag{8-25}$$

If, in Fig. 8-10, the magnetic field **B** is always vertical and the particle enters the region between the poles with an initial velocity **v** which is horizontal, then **v** and **B** are perpendicular. The force is horizontal, with magnitude

$$F = qvB \tag{8-26}$$

Figure 8-11 shows a top view of this situation. The circles with center dots

Fig. 8-11 Force on a charged particle moving in a magnetic field. The circles with dots at their centers indicate that the magnetic field **B** is directed out of the plane of the paper. If the charged particle q has a velocity **v** in the plane of the paper, the force **F** exerted on it is also in this plane, as shown. If the magnetic field is uniform, the particle moves in a circular path. The figure shows the path of a positively charged particle. How is it different if q is negative?

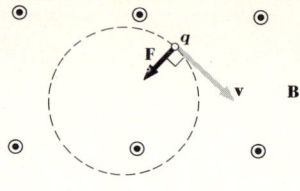

indicate the "heads" of the magnetic field vectors, showing that the field is directed out of the page. Application of Eq. (8-25) and the right-hand rule to the situation of Fig. 8-11 shows that the force acting on the particle is in the direction shown. This force thus produces an acceleration *perpendicular* to the direction of the instantaneous velocity. But this is precisely what happens when a particle moves in a circular path. If its speed is constant, the acceleration is always perpendicular to the velocity and always has the same magnitude. Thus, as long as the charged particle remains in the region in which **B** is uniform, it moves in a *circular* path.

The radius R of the path can be found by equating the magnitude of the force on the particle, given by Eq. (8-26), to the mass times the centripetal acceleration v^2/R. This gives

$$qvB = m\frac{v^2}{R}$$

From this, we find the radius of the circle:

$$R = \frac{mv}{qB} \tag{8-27}$$

In a cyclotron, charged particles such as protons move in circular paths in a field created by the cyclotron magnet. During each revolution, they are given a push by an electric field produced by charged metallic plates in the cyclotron, called *dees*. As a result, each revolution is made in a slightly larger circle than the previous one, and the particles accumulate greater and greater energies as they move in larger and larger circles.

Equation (8-27) shows that the radius of curvature of the path of a particle in a magnetic field is proportional to the magnitude of the momentum mv of the particle; hence a measurement of the radius of curvature of a particle in a magnetic field can be used to determine its momentum. This relationship, or a relativistic generalization of it, is widely used in mass spectrometry, research in high-energy physics, and in design of special-purpose vacuum tubes such as magnetrons.

When both electric and magnetic fields are present, the total force exerted on a charged particle is the sum of the contributions of the two. That is,

$$\mathbf{F} = q(\mathbf{E} + \mathbf{v} \times \mathbf{B}) \qquad (8\text{-}28)$$

This equation is called the *Lorentz law of force for charged particles*, after its discoverer, H. A. Lorentz, a Dutch theoretical physicist, who first formulated the electron theory of matter about 1900. There are many practical situations in which Eq. (8-28) can be used to study the motion of charged particles in electric and magnetic fields. A few examples of such situations are included in the problems for this chapter.

8-6 SCATTERING

In many of the examples of this chapter, the problem has been to predict the motion of a particle when the forces acting on it are known. But there are situations, especially in the investigation of fundamental-particle interactions, in which the forces are *not* known. For example, in investigating the forces which act between nuclear particles at close range, one useful technique is to fire a beam of protons at a target containing other protons, perhaps in the form of liquid hydrogen. As a result of these close-range interactions, some of the protons are *scattered* away from their original direction and emerge from the target in directions different from that in which they entered. By various techniques it is possible to measure the number of particles emerging in each of several directions. Then the problem is to reconstruct the kind of force which must have been responsible for this particular motion.

This is often a considerably more difficult task than the problem of predicting the motion when the force is known. Yet experiments of this kind have produced nearly all the knowledge we now have regarding the structure of nuclei, nuclear forces, and interactions among the various fundamental particles. In these situations the problem is analogous to that with which Newton was confronted when he started with Kepler's laws expressing the observed motion of the planets and set out to deduce the law of force responsible for this motion. A particular scattering experiment, Rutherford scattering, is discussed in detail in Chap. 12. This experiment was of crucial importance in the early days of investigation of the internal structure of atoms, about 1910.

Problems

Note: Dimensions and masses of bodies in the solar system are summarized in Appendix C.

8-1 In the Cavendish balance shown in Fig. 8-1, suppose that the large spheres are of lead (density 11.3 g/cm^3) and are 6 cm in diameter and that the small spheres on the suspended rod are of brass (8.40 g/cm^3) 1.0 cm in diameter. If the distance between centers of the large sphere and the nearer small one is 5 cm, find:
 a The gravitational force on the small sphere due to the nearby large one;
 b The gravitational force on the large sphere due to the small one.

8-2 Consider a Cavendish balance constructed as in Prob. 8-1, with a suspension fiber which exerts a force on each small sphere proportional to its displacement from the equilibrium position. When the spheres are displaced 1.0 cm, the restoring force is 1.5×10^{-8} N on each. If now the lead spheres are placed as shown in Fig. 8-1, with centers 10 cm from the equilibrium positions of the small spheres, how much are the small spheres deflected? Through what angle does the dumbbell turn?

8-3 Two automobiles, each with total mass 2,000 kg, are on a horizontal frictionless road and are separated by 100 m.
 a What is the magnitude of the gravitational interaction force?
 b What is the initial acceleration of each car?

8-4 Considering the earth to be a solid uniform sphere, find the density it must have in order to account for the observed acceleration of gravity at the surface.

8-5 Using the data of Appendix C, compute the densities of the moon and the first five planets, and compare your results.

8-6 The asteroid Toro, discovered in 1964, has been proposed as a likely target for manned spacecraft landings. Its diameter is about 5 km. If the density of Toro is the same as that of earth, find the acceleration of gravity at its surface.

8-7 If all the planets in the solar system had the same density, how would the acceleration of gravity on the surface of a planet depend on its radius? Derive a general expression.

8-8 Compare the magnitudes of the gravitational forces exerted on the earth by the sun and by the moon.

8-9 A spacecraft launched from earth is aimed directly toward the moon. At what distance from earth does the net gravitational force become directed toward the moon rather than earth?

8-10 Suppose that particles with negative mass exist. Consider a particle with positive mass m and one with equal negative mass m initially separated by a distance d and at rest. If there are no forces on either except their mutual interaction, describe the subsequent motion in detail. Is the principle of conservation of momentum violated by this motion?

8-11 An airplane flies over Chicago at an altitude of 32,000 ft. Compare its weight with the weight of the same plane on the ground at the same location. Express your result as a percent change in weight.

8-12 Is the acceleration of gravity on earth altered significantly by the fact that the moon and sun, as well as the earth, exert gravitational forces on an object at the surface of earth?

8-13 A thin rod has length $2a$ and total mass M. Find the gravitational field due to this rod at a point P on its perpendicular bisector, a distance y from the center of the rod. *Hint:* Divide the rod into elements of length dx; find the x and y components of gravitational field at P due to this element, and integrate on x to find the total components. One component of the total field is zero, and this can be shown using symmetry, without actually integrating.

8-14 A body in the shape of a thin ring of radius R, with total mass M, exerts a gravitational force on a point mass m which lies somewhere on a line perpendicular to the plane of the ring through its center. Find the gravitational force on m as a function of its distance from the plane of the ring. *Hint:* Divide the ring into small elements; find the components of force parallel and perpendicular to the axis from each element. The perpendicular components must add to zero because of the symmetry of the situation.

8-15 A body is dropped out of a spaceship a distance R away from the surface of the earth, where R is equal to earth's radius. If it has negligible initial velocity and air resistance can be neglected, find the speed of the body when it strikes the earth. An approximate result can be obtained by estimating an average gravitational force during the fall, but the problem can also be solved exactly. Is the neglect of air resistance reasonable?

8-16 It has been asserted that it is impossible to distinguish experimentally between an inertial frame of reference with a gravitational field $\mathbf{g}$ and an acceleration of the frame of reference of $-\mathbf{g}$ with respect to an inertial frame if it is not known in advance whether or not the frame of reference is inertial. Explain why this is true. Does the validity of this statement depend on the equivalence of inertial and gravitational masses? Explain.

8-17 The earth's orbit is approximately circular, with a radius of 93×10^6 mi. Mars travels around its orbit once in 687 d. Using only this information, find the radius of the orbit of Mars.

8-18 Using Eq. (8-11) and the known data on the radius and period of the earth's orbit, compute the mass of the sun.

8-19 Find the period of an artificial satellite which moves in a circular orbit 2,000 mi above the surface of the earth.

8-20 Some communications satellites have a circular orbit with a period of exactly 1 d and thus have the same rotational motion about the earth's axis as the earth itself. Then, from a given point on earth, such a satellite appears to be stationary. What must the radius of its orbit be?

8-21 Find the minimum period for a satellite in a circular orbit around a spherical planet of radius R and uniform density ρ.

8-22 What is the minimum period for a satellite of the planet Jupiter? The data in Appendix C will be useful.

8-23 If Kepler's third law had turned out to be $T = KR^2$ instead of Eq. (8-8), what form would the law of gravitation have to have in order to be consistent with this result?

8-24 If the gravitational force were proportional to $1/r^3$ instead of $1/r^2$, what form would Kepler's third law have?

8-25 If the gravitational force were given by $F = Gm_1m_2/r^n$ (where n is an integer) instead of Eq. (8-1), what form would Kepler's third law have?

8-26 A particle of mass m is attracted to the origin of coordinates with a force whose magnitude is given by $F = kr$, where k is a constant and r is the distance from the origin. Find the relation between radius and speed in order for the particle to move in a circular path.

8-27 A particle of mass m is attached to one end of a spring whose unstretched length is l and when stretched exerts a force of magnitude $F = kx$, where k is a constant and x is the amount of stretch. The other end of the spring is fastened to a fixed point. Under what circumstances can the particle move in a circular path? How are its speed and radius related for such a path?

8-28 The mass of the planet Jupiter has been deduced from observations of the periods of its moons and the radii of their orbits. Explain in detail how this is done.

8-29 A certain central-force field, sometimes used to describe interactions between atoms in a diatomic molecule, is given by

$$f(r) = -\frac{A}{r^6} + \frac{B}{r^{12}}$$

Draw a graph of this function. At what value of r does the maximum negative force occur? What is the force at this point? At what value of r is the force zero?

8-30 A particle moves with constant speed along a line parallel to the y axis in an xy coordinate system. In terms of polar coordinates, show that the quantity $r^2\, d\theta/dt$ is constant for this motion. Is this an example of central-force motion? Explain.

8-31 In Fig. 8-8, the magnitude of the electric field **E** (in mks units) at points not too near the edge is equal to V/d, where V is the battery voltage (in volts) and d is the distance between the plates (in meters). If an electron leaves the left-hand plate with no initial velocity, and if $V = 100$ V and $d = 0.01$ m, with what velocity does it strike the right-hand plate? $q = 1.6 \times 10^{-19}$ mks units.

8-32 In Prob. 8-31, show that the final velocity is independent of the distance between the plates.

8-33 Derive Eqs. (8-23) in detail from Eq. (8-22) and the given initial conditions.

8-34 In the situation shown in Fig. 8-9, if the particle is an electron with initial speed 1.5×10^7 m/s, find the electric field necessary to deflect the particle 0.01 m from the x axis at the point at which it emerges from the plates. The horizontal distance traveled between plates is 0.1 m.

8-35 A charged particle moves in a uniform magnetic field. Describe the motion qualitatively if:
  a The particle is initially at rest.
  b The initial velocity is parallel to the field.
  c The initial velocity is perpendicular to the field.
  d The initial velocity makes an angle θ with the field.

8-36 An electrical engineer wishes to create a magnetic field that will make an electron move in a circular orbit of radius 1 cm with an angular velocity of 10^9 rad/s in a vacuum. The mass of the electron is 9.1×10^{-31} kg and its charge 1.6×10^{-19} C.
  a Find the speed of the electron and the magnitude of the magnetic field.
  b In one simple model of the hydrogen atom, the electron moves in a circular orbit of radius 0.5×10^{-10} m with angular velocity 4×10^{16} rad/s. If a magnetic field were responsible for this motion (which it is not), find the magnitude of the necessary field.

8-37 A proton moves in a magnetic field between the poles of a large electromagnet in a cyclotron. The particle can move in a circle of maximum radius 0.5 m. If the magnetic field is 2 Wb/m^2 (the mks unit of magnetic field), what speed must the proton have to move in such a circular path?

8-38 Show that when a charged particle moves in a magnetic field, its speed is always constant. Is this true even if the magnetic field is not uniform? Explain.

8-39 If a charged particle moves in a region containing uniform electric and magnetic fields which are perpendicular to each other, it is possible for the particle to move in a straight line with constant speed. Explain how this is possible. Find the velocity with which the particle must move in terms of **E, B,** and other relevant quantities.

8-40 A charged particle in a uniform magnetic field is given an initial velocity which is neither parallel nor perpendicular to the magnetic field. Describe its subsequent motion. *Hint:* Represent the velocity in terms of components parallel and perpendicular to the field, and find the force arising from each component.

Mechanical Energy | 9

The concept of energy is one of the most important unifying principles underlying all the various branches of physics. In this chapter we introduce the concepts of *work* and *kinetic energy* and develop the relationship between them. We then generalize the concept of energy to include potential energy, the energy associated with position rather than motion. This leads naturally to the principle of conservation of energy and to a discussion of conservative force fields. Finally, we discuss nonmechanical forms of energy and generalizations of the principle of conservation of energy.

9-1 WORK

The word *work* as used in everyday conversation carries the idea of exertion or effort. Consideration of examples of physical labor or exertion shows that the most tiring operations are those in which a force is exerted on an object while it undergoes a displacement. Pushing a car with an empty gas tank with a force of 50 lb while running behind the car is much more tiring than applying the same force for the same amount of time to a stationary car. This is not to imply, of course, that a person pushing with a force of 50 lb against a brick wall will never get tired, but pushing a moving object is much more tiring.

In physics, *work* is given a precise meaning related to these observations about a force acting through a displacement. Suppose a constant horizontal force F acts on an object which undergoes a displacement Δx in the direction of the force. The *work* done by the force during this displacement, which we denote by W, is defined to be

$$W = F \Delta x \tag{9-1}$$

204 MECHANICAL ENERGY

By this definition, a man pushing a moving car does an amount of work proportional to the *force* he exerts and to the *displacement* of the car. A man pushing a stationary car does no work. The essential idea of work is that an object must be displaced during the application of a force.

The units of work are those of force multiplied by displacement. In the mks system, the unit of work is the newton-meter. This unit appears so frequently that it is given a special name, the joule (J). That is,

$$1 \text{ J} = 1 \text{ N-m} \tag{9-2}$$

In the British system, where the unit of force is the pound and that of displacement the foot, the unit of work is the foot-pound. We note that the units of work, like those of force, are *derived* units which can be expressed in terms of the fundamental units of mass, length, and time.

When the force and displacement are along the same line, they may have opposite directions. In this case, F and Δx have opposite signs, and W is a *negative* quantity. It is illuminating to represent Eq. (9-1) graphically. In Fig. 9-1 we plot the force F on the vertical axis and the position x on

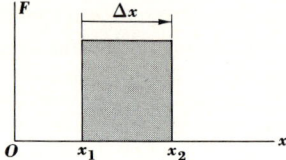

Fig. 9-1 Work done by force **F** during the displacement x_1 to x_2 is represented by the rectangular area.

the horizontal axis. The work done during the displacement Δx is equal to the area under the line representing the force, in the interval $\Delta x = x_2 - x_1$.

What happens if the force is not constant during the displacement? We can still use the definition given in Eq. (9-1) provided Δx is sufficiently small for the force F not to vary appreciably over this interval. That is, we separate the interval from x_1 to x_2 into a series of very small intervals, calculate the work done in each of them, and add all the results to find the *total* work done. This procedure is shown in Fig. 9-2, where the curve shows the

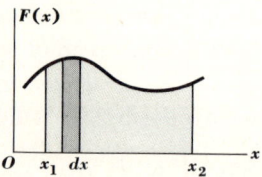

Fig. 9-2 Work done by the force $F(x)$ between x_1 and x_2 is represented by the area under the curve between these limits. The work done in a small displacement dx is represented by the area of the small strip of width dx.

variation of F with x. The work done in each displacement dx is represented by the area of the corresponding thin strip, and the total work is represented by the total *area* under the curve from x_1 to x_2. In the language of integral calculus, the work dW done during an infinitesimal displacement dx is $dW = F\,dx$, and the total work done during the displacement from x_1 to x_2 is

$$W = \int_{x_1}^{x_2} F(x)\,dx \tag{9-3}$$

As an example, consider the amount of work required to stretch a spring from its relaxed state ($x = 0$) to a final state in which it is stretched an amount X. In many cases, the distance the spring stretches is approximately proportional to the force exerted on it; we describe such a spring by a model in which the force F needed to stretch the spring a distance x is

$$F = kx \tag{9-4}$$

where k is a proportionality constant which depends on the construction of the spring, usually called the *spring constant* or *force constant*. The variation of the force with displacement is shown graphically in Fig. 9-3, where the

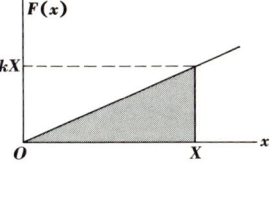

Fig. 9-3 The force exerted on a stretched spring, stretching it from an initial un-stretched position to a final displacement X, is represented by the diagonal line, assuming that the force is proportional to the displacement x. The triangular shaded area under this line represents the work done on the spring during this process; this is equal to $\frac{1}{2}kX^2$.

shaded area denotes the total work done during the extension. We see immediately that the area of this triangle is $\frac{1}{2}kX^2$. Thus the work done on a spring in stretching it from zero stretch to a maximum stretch X is

$$W = \tfrac{1}{2}kX^2 \tag{9-5}$$

Equation (9-5) can also be obtained from Eq. (9-3):

$$W = \int_0^X F(x)\,dx = \int_0^X kx\,dx = \tfrac{1}{2}kX^2 \tag{9-6}$$

It should be pointed out that springs do not always behave in the simple manner indicated by Eq. (9-4), which may be thought of as an idealized *model* describing the behavior of a spring approximately. If the force depends on the displacement in a more complicated way, the integral is correspondingly more complicated.

There are situations where force and displacement do not have the same *direction*. In pushing our car to the gas station, we may exert an upward force on the rear end, as well as a forward force, in order to improve our footing. If a force of magnitude F is exerted in a direction making an angle θ with the direction of the displacement as in Fig. 9-4, how much work

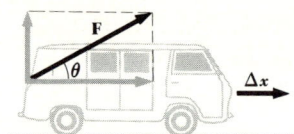

Fig. 9-4 Example of a force applied to a body whose displacement is not in the same direction as the force.

is done by the force? To answer this question, we resolve **F** into a component parallel to the displacement and one perpendicular to it. The component perpendicular to the displacement does no work, since the body has no component of displacement in its direction. Thus the work is associated only with the component of force in the direction of displacement: $F \cos \theta$. Therefore, if the magnitude and direction of **F** are constant, the amount of work done during the displacement Δx is defined as

$$W = F \cos \theta \, \Delta x \tag{9-7}$$

The work is the component of force in the direction of the displacement multiplied by the magnitude of the displacement. Alternatively, we may say that the work is the magnitude of the force multiplied by the magnitude of the displacement multiplied by the cosine of the angle between their directions.

All this should sound familiar. What we have written is simply the *scalar product* of the force and the displacement, as defined in Sec. 2-5. Denoting a general vector displacement by $\Delta \mathbf{r}$ (representing a change in the position vector **r** of a point), we say that when a constant force **F** acts on a body which undergoes a displacement $\Delta \mathbf{r}$, the work done by the force is defined as

$$W = \mathbf{F} \cdot \Delta \mathbf{r} \tag{9-8}$$

This is the most general definition of mechanical work. It can be used, at least in principle, to compute the amount of work done in a situation such as that shown in Fig. 9-5. A body moves from a point P_1 with position

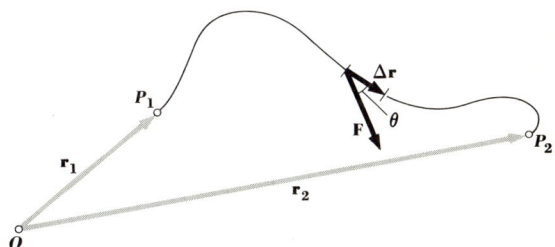

Fig. 9-5 Body displaced from point P_1 to P_2 along the curved path shown. The force on the body varies as it moves. The work done by the force during the small displacement $\Delta \mathbf{r}$ is $\mathbf{F} \cdot \Delta \mathbf{r}$.

vector $\mathbf{r}_1$ to a point P_2 with position $\mathbf{r}_2$ along the curved path shown in the figure. The force on the body may change, both in magnitude and in direction, as the body moves along the path. We write the total amount of work done during this displacement symbolically as

$$W = \int_{P_1}^{P_2} \mathbf{F} \cdot d\mathbf{r} \tag{9-9}$$

This is called a *line integral*. Its meaning is simply this: We divide the path into a very large number of very small displacements $d\mathbf{r}$, and we add all the elements of work $dW = \mathbf{F} \cdot d\mathbf{r}$, a process completely analogous to ordinary integration. Of course, to evaluate Eq. (9-9) for a particular problem, we must have a precise description of the shape of the path and how the force $\mathbf{F}$ varies with position on the path.

Example
The force on a certain particle in the xy plane varies with position according to the relation

$$\mathbf{F} = k[2xy\mathbf{i} + (x^2 + y^2)\mathbf{j}]$$

where k is a constant. The particle moves in a straight line from the origin to point $(0,a)$ and then in another straight line from $(0,a)$ to (a,a). How much work is done?

Solution

In general, $d\mathbf{r} = \mathbf{i}\, dx + \mathbf{j}\, dy$. The first part of the path is along the y axis, so $x = 0$ and $d\mathbf{r} = \mathbf{j}\, dy$. Thus the work done in this part of the path is

$$W = \int \mathbf{F} \cdot d\mathbf{r} = k \int_0^a [2(0)y\mathbf{i} + (0 + y^2)\mathbf{j}] \cdot \mathbf{j}\, dy$$

$$= k \int_0^a y^2\, dy = \tfrac{1}{3}ka^3$$

The second part of the path is parallel to the x axis; we have $y = a$ and $d\mathbf{r} = \mathbf{i}\, dx$. Thus for this part of the path

$$W = k \int_0^a [2xa\mathbf{i} + (a^2 + y^2)\mathbf{j}] \cdot \mathbf{i}\, dx$$

$$= k \int_0^a 2xa\, dx = \tfrac{1}{2}ka^3$$

The total work is

$$W = \tfrac{1}{3}ka^3 + \tfrac{1}{2}ka^3 = \tfrac{5}{6}ka^3$$

Example

In the above example, suppose the particle moves along a straight line from the origin to the point (a,a). How much work is done?

Solution

There are various choices for the integration variable. Suppose we choose to integrate on a variable s measured along the path. Then $x = s/\sqrt{2}$ and $y = s/\sqrt{2}$, and s varies from 0 to $\sqrt{2}a$. The total work is

$$W = k \int_0^{\sqrt{2}a} \left[2\left(\frac{s}{\sqrt{2}}\frac{s}{\sqrt{2}}\right)\mathbf{i} + \left(\frac{s^2}{2} + \frac{s^2}{2}\right)\mathbf{j}\right] \cdot \left[\frac{ds}{\sqrt{2}}\mathbf{i} + \frac{ds}{\sqrt{2}}\mathbf{j}\right]$$

$$= k \int_0^{\sqrt{2}a} \sqrt{2}s^2\, ds = \tfrac{4}{3}ka^3$$

Alternately, we could choose x as the integration variable and set $y = x$:

$$W = k \int_0^a (2x^2\mathbf{i} + 2x^2\mathbf{j}) \cdot (dx\, \mathbf{i} + dx\, \mathbf{j})$$

$$= k \int_0^a 4x^2\, dx = \tfrac{4}{3}ka^3$$

in agreement with the above result. We note that the work for this path is different from that for the other path, even though the end points are the same. In general when the force is position-dependent, as it is here, the work is different for different paths. Only in special cases, to be discussed later, is the work the same for *all* paths between two given points.

The foregoing discussion shows that work, although computed from two vector quantities, is itself a *scalar* quantity; it has magnitude but no direction. We do just as much work pushing a car 1 mi east to the gas station as pushing it 1 mi north to another gas station. The angle between the force and displacement is an essential part of the definition of work, but these directions do not give work a direction. In the two situations in Fig. 9-6, the same amount of work is done, despite the fact that the two forces act in different directions.

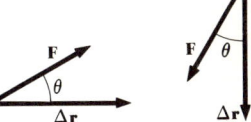

Fig. 9-6 Work done by the force is the same in these two examples, although **F** and **Δr** are in different directions in the two cases.

9-2 WORK AND KINETIC ENERGY

To show the usefulness of the concept of work, we now develop the relationship between work and energy. We consider first the apparatus used in Chap. 4 to develop the principles of mechanics, a frictionless hockey puck on a horizontal surface. Suppose we apply a constant force F to a puck initially at rest, and the puck accelerates. When it has moved a distance x, the force has done an amount of work $W = Fx$. How is this work related to the velocity acquired by the puck?

We use the results from Sec. 3-3 for straight-line motion with constant acceleration. For a body with initial velocity v_0, constant acceleration a, and final velocity v, we found

$$v^2 - v_0^2 = 2a(x - x_0) \tag{9-10}$$

where x_0 is the initial position. In the present case, $v_0 = 0$, we may take $x_0 = 0$, and the acceleration is given by $a = F/m$. Making these substitutions, we find

$$v^2 = 2\frac{F}{m}x \quad \text{or} \quad \tfrac{1}{2}mv^2 = Fx \tag{9-11}$$

which shows that the final speed of the body is simply related to the total work Fx done by the force.

If the initial velocity is not zero, we can still derive a simple relationship. Substituting $a = F/m$ and $x_0 = 0$ in Eq. (9-10) and rearranging, we obtain

$$\tfrac{1}{2}mv^2 - \tfrac{1}{2}mv_0^2 = Fx \tag{9-12}$$

This time we obtain a difference of two terms, each in the form $\tfrac{1}{2}mv^2$. Anticipating the importance of this quantity in later discussions, we give it the name *kinetic energy* and denote it by E. When a particle of mass m moves with speed v, its kinetic energy E is defined as

$$E = \tfrac{1}{2}mv^2 \tag{9-13}$$

We may now reinterpret Eqs. (9-11) and (9-12) using this concept. In Eq. (9-11), the puck started from rest (therefore with zero kinetic energy) and acquired an amount of kinetic energy exactly equal to the work done by the force. When the object started with initial velocity v_0 (thus with initial kinetic energy $\tfrac{1}{2}mv_0^2$), the kinetic energy $\tfrac{1}{2}mv^2$ at the end of the process was greater than that at the beginning by an amount equal to the work done, as in Eq. (9-12). In each case, the *increase* in kinetic energy was equal to the work done by force F.

Thus far, we have considered only situations in which force, displacement, and initial velocity all lie along one straight line. We shall soon see that the conclusion is more general than this, but before proceeding, it is important to note that the relation between work and kinetic energy is not a new discovery; it is not based on any experimental observations other than those which led to Newton's laws. Rather, we have *defined* work and kinetic energy and have then *derived* from Newton's laws the relation between these quantities. It is important to keep in mind the distinction between fundamental laws, based directly on observation, and derived principles based on definitions of new quantities and derivations from principles already established. The relation between work and kinetic energy is in the latter category.

We see from Eq. (9-12) that kinetic energy must have the same units as work. It is easy to verify that this is in fact the case. The mks unit of force is the newton. From Newton's second law, $1 \text{ N} = 1 \text{ kg·m/s}^2$. Thus,

$$1 \text{ N·m} = 1 \text{ J} = 1 \text{ kg·(m/s)}^2$$

Thus the joule is also a unit of kinetic energy.

Like work, kinetic energy is a scalar quantity. In defining it, we have specified that the *speed* of the particle be used. The kinetic energy of a body depends only on its speed and not on the *direction* of its motion. Although kinetic energy superficially resembles momentum in that both contain m and v, they are quite different physical quantities. Momentum is a vector quantity, while kinetic energy is a scalar. Momentum is proportional to **v**, kinetic energy to v^2.

When several forces act on an object at once, then **F** in the previous discussion must be identified as the *total* force. In such cases, it is necessary to distinguish carefully between the work done by *all* the forces and the work done by an individual force. The following example illustrates this point.

Example
A box whose mass is 10 kg is pushed a distance of 10 m across a rough floor with coefficient of friction $\mu = 0.5$ by a force of 100 N. How much work is done by this force? How much work is done by *all* the forces? What is the velocity of the object after it is displaced 10 m?

Solution
The weight is $mg = 98$ N, and so the force of friction opposing the motion is 0.5×98 N $= 49$ N, as shown in Fig. 9-7. The net force on the box is

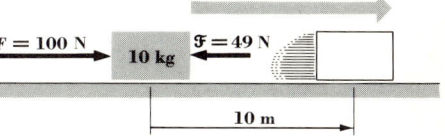

Fig. 9-7 Forces acting on a box being pushed across a rough floor. The work done by each force can be computed.

51 N to the right, so the total work done is 510 N-m $= 510$ J. This is to be equated to the kinetic energy of the box at the end of its 10-m travel. That is,

$$510 \text{ N-m} = \tfrac{1}{2}(10 \text{ kg})v^2$$

from which we find $v = 10.1$ m/s. The work done by F is $(100 \text{ N})(10 \text{ m}) = 1{,}000$ J, and the work done by the force of friction on the box is -490 J. The latter is negative because the force of friction acts to the left, while the displacement is to the right. The force of friction does negative work on the box; i.e., it tends to *decrease* the kinetic energy of the box.

Because of the reaction to the force of friction, the box exerts a forward force on the *floor*, and the point of application of this force moves through

212 MECHANICAL ENERGY

a distance as the box moves. Thus the force of friction does work *on* the floor. This force is, of course, balanced by other forces which hold the floor in place, so the kinetic energy of the floor does not change, but energy is transferred to the floor in the form of heat. Thus, the fact that -490 J of work is done *on* the box corresponds to the fact that the box does 490 J of work on the floor, giving it thermal energy in the process.

The relation between work and kinetic energy expressed by Eq. (9-12) has been derived only for the special case of a constant force acting along the direction of the initial velocity. We now proceed to generalize this result. If F is not constant but is a function of position, say $F(x)$, that is, F is a function of x, we may divide the path into a large number of elements small enough for F to be nearly constant in each interval. Then the change in kinetic energy in each interval, which we may call ΔE, is still given by

$$\Delta E = F \, \Delta x \tag{9-14}$$

and the total change in kinetic energy is the sum of all the quantities $F(x) \, \Delta x$, which in the limit as all the Δx's approach zero is simply the integral of $F(x)$. If the end points of the motion are x_1 and x_2, and if v_1 and v_2 are the speeds at these points, then

$$E_2 - E_1 = \tfrac{1}{2}mv_2^2 - \tfrac{1}{2}mv_1^2 = \int_{x_1}^{x_2} F(x) \, dx \tag{9-15}$$

This rather intuitive derivation can be made more precise by noting that in the limits as $\Delta x \to 0$, Eq. (9-14) becomes

$$F(x) = \frac{dE}{dx} \tag{9-16}$$

This relation, however, can be derived directly from $F = ma$. Using the chain rule for derivatives, we find

$$\frac{dE}{dx} = \frac{d}{dx}(\tfrac{1}{2}mv^2) = \frac{d}{dv}(\tfrac{1}{2}mv^2)\frac{dv}{dx} = mv\frac{dv}{dx} \tag{9-17}$$

But also

$$F = ma = m\frac{dv}{dt} = m\frac{dv}{dx}\frac{dx}{dt} = mv\frac{dv}{dx} \tag{9-18}$$

Combining Eqs. (9-17) and (9-18) and integrating, we obtain

$$\int_{x_1}^{x_2} F(x)\, dx = \int_{E_1}^{E_2} dE = E_2 - E_1 \tag{9-19}$$

which is the same as Eq. (9-15).

To generalize further, suppose the force and the displacement are in different directions. We consider a displacement $\Delta \mathbf{r}$ occurring in a time interval Δt. During this same interval, the velocity changes by $\Delta \mathbf{v}$. If Δt is short enough to permit regarding the force as practically constant during this interval, then the work is $\mathbf{F} \cdot \Delta \mathbf{r}$. This can be related to the average acceleration during the interval as follows:

$$\mathbf{F} \cdot \Delta \mathbf{r} = m \frac{\Delta \mathbf{v}}{\Delta t} \cdot \Delta \mathbf{r} = m\, \Delta \mathbf{v} \cdot \frac{\Delta \mathbf{r}}{\Delta t} = m \mathbf{v}_{av} \cdot \Delta \mathbf{v} \tag{9-20}$$

In Eq. (9-20) we have used the definitions of average velocity and average acceleration and the fact that the scalar product of two vectors is commutative. To find the *total* work and the *total* change in kinetic energy during travel over a finite path from P_1 to P_2, we add all the elemental amounts of work and the corresponding velocity changes, taking the limit as all the $\Delta \mathbf{r}$'s become very small. That is,

$$\int_{P_1}^{P_2} \mathbf{F} \cdot d\mathbf{r} = \int_{v_1}^{v_2} m\mathbf{v} \cdot d\mathbf{v} \tag{9-21}$$

The integrand on the right side is the differential of the quantity $\frac{1}{2} m \mathbf{v} \cdot \mathbf{v} = \frac{1}{2} mv^2$, so that the right-hand side is equal to $\frac{1}{2} m v_2^2 - \frac{1}{2} m v_1^2$. Thus the final result is

$$\int_{P_1}^{P_2} \mathbf{F} \cdot d\mathbf{r} = \tfrac{1}{2} m v_2^2 - \tfrac{1}{2} m v_1^2 = E_2 - E_1 \tag{9-22}$$

which is the most general form of the relation between work done on a body and its change in kinetic energy.

In summary, we have shown that the change in kinetic energy of a particle during any motion is equal to the total work done on it by *all* the forces acting on it during this motion. The performance of work transfers energy from one body to another. Conversely, by virtue of its motion, the particle itself possesses the capacity to do work. When a particle with some initial velocity is stopped, it exerts a force on whatever agent stops it; this force may act through a distance. The agent which stops it, of course, does negative work on the particle. Thus we may regard kinetic energy as associated with a *capacity for doing work*. This is a useful point of view in considering the concept of *potential energy*, introduced in Sec. 9-4.

9-3 POWER

There are situations in which it is important to know not only the work done by a force and the resulting change in kinetic energy but also the time rate at which the processes occur. In such cases, the important question may be not *how much* work is done but rather *how quickly* it is done. This leads to the concept of *power*.

Power is defined as the *time rate of doing work*. If work ΔW is done during a time interval Δt, the *average power* during this interval is defined as

$$P_{av} = \frac{\Delta W}{\Delta t} \tag{9-23}$$

Similarly, we define the instantaneous power P as

$$P = \frac{dW}{dt} \tag{9-24}$$

The units of power are those of work per unit time. In the mks system, the unit of power is the joule per second. This unit appears so frequently that it is given a special name, the watt (W). That is,

$$1 \text{ W} = 1 \text{ J/s}$$

This is the familiar unit of power used in rating electric bulbs. A 100-W bulb converts electric energy into light and heat at the rate of 100 J/s.

The corresponding unit in the English system is the foot-pound per second. Another unit, regrettably still used in engineering work, is the horsepower. By definition,

$$1 \text{ horsepower} = 1 \text{ hp} = 550 \text{ ft-lb/s}$$

A horse can do work at the rate of several horsepower for short periods; this definition originally was an average for a horse working all day. Another useful conversion factor is

$$1 \text{ hp} = 746.0 \text{ W}$$

That is, if a 1-hp electric motor were 100 percent efficient, it would consume electric energy and convert it into mechanical work at the rate of 746 W. Practical motors are always less than 100 percent efficient and therefore consume somewhat more electric power than the mechanical-power output. The difference is lost mainly as heat.

When a force does work on a moving body, the corresponding power may be expressed in terms of the average force and the body's velocity. From Eq. (9-23) we find

$$P_{av} = \frac{\Delta W}{\Delta t} = \frac{\mathbf{F}_{av} \cdot \Delta \mathbf{r}}{\Delta t} = \mathbf{F}_{av} \cdot \mathbf{v}_{av} \qquad (9\text{-}25)$$

Taking the limit as $\Delta t \to 0$ gives

$$P = \mathbf{F} \cdot \mathbf{v} \qquad (9\text{-}26)$$

Example

A man pushes on the rear of a stalled car with a force of 200 N in a direction 30° up from the horizontal, as in Fig. 9-4. If the car moves 1 m/s, what power does the man develop?

Solution

According to Eq. (9-26),

$$\begin{aligned}P = \mathbf{F} \cdot \mathbf{v} &= (200 \text{ N})(1 \text{ m/s})(\cos 30°) \\ &= 173 \text{ N·m/s} = 173 \text{ J/s} \\ &= 173 \text{ W} = 0.232 \text{ hp}\end{aligned}$$

9-4 POTENTIAL ENERGY

The relationship between work and kinetic energy, developed in Sec. 9-2, enables us to relate the motion of a body at one point to the motion at another point without having to describe the details of the intervening motion, provided only that we can calculate the *work* done on the body during the intervening motion. Because the work calculation may well be simpler than a complete description of the motion, the work-energy relation is a powerful and useful one.

Calculation of work can be simplified further when, as in many situations of practical interest, the force on a body is determined by its position. For example, a body attached to a stretched spring experiences a force which depends on the amount of stretch, which of course varies with the position of the body. Similarly, the gravitational force on a body, calculated from Newton's law of gravitation as discussed in Sec. 8-1, depends on the body's position. In such cases, calculating the work done by these forces can be simplified by use of the concept of *potential energy*, which we consider next.

An example will exhibit the basic idea. Suppose we lift a body of mass m a distance h above the floor and release it with zero initial velocity. It falls straight down with constant acceleration g, corresponding to the constant force of gravity mg. The total work done by this force during the fall is $W = mgh$. According to our work-energy theorem, this must equal the kinetic energy $\frac{1}{2}mv^2$ acquired during the fall:

$$\frac{1}{2}mv^2 = mgh \tag{9-27}$$

where v is the body's speed just before it reaches the floor. The farther the body falls, the more work can be done and the greater the kinetic energy when it reaches the floor. This energy which is "available" as a result of the body's initial position but which actually becomes kinetic energy only at the end of the fall is called *potential energy*.

Specifically, when the body begins its fall from height h, we say it has potential energy mgh. As it falls, the potential energy decreases and the kinetic energy increases. In fact, it is easy to show that the *sum* of kinetic and potential energies is constant during the fall. Let v_1 be the speed at height h_1 and v_2 the speed at height h_2; assume for definiteness that $h_2 < h_1$ so $v_2 > v_1$. The work done by gravity between these two points is $mg(h_1 - h_2)$, and according to the work-energy theorem this must equal the change in kinetic energy $\frac{1}{2}mv_2^2 - \frac{1}{2}mv_1^2$. Thus

$$\frac{1}{2}mv_2^2 - \frac{1}{2}mv_1^2 = mg(h_1 - h_2) \tag{9-28}$$

Rearranging this equation, we find

$$\frac{1}{2}mv_1^2 + mgh_1 = \frac{1}{2}mv_2^2 + mgh_2 \tag{9-29}$$

The sum of kinetic and potential energies, the *total* energy, is the same at the two points and hence does not change during the motion. At the beginning the energy is all potential, at the end it is all kinetic, and we speak of a conversion from potential to kinetic energy during the fall. This is a simple example of the principle of *conservation* of mechanical energy, to be discussed in more detail in the next section.

The key step in the above discussion is to represent the work done by the force in terms of a potential energy, a quantity that depends on the *position* of the body. Another familiar example is the work done by a stretched spring, discussed in Sec. 9-1. We found that if the spring force is given by $F = kx$, where x is the amount of stretch, the total work required to stretch the spring a distance X is $\frac{1}{2}kX^2$. Correspondingly, when the spring returns to its unstretched state, it can do an amount of work $\frac{1}{2}kX^2$ on a body attached to it. Thus a spring stretched a distance X has *potential energy* $\tfrac{1}{2}kX^2$.

In both these examples, we have chosen a point where the potential energy is zero. For the falling body it was the floor; for the spring it was the unstretched position. For the falling body this choice is quite arbitrary. If we cut a hole in the floor, the body can fall to negative values of h, corresponding to negative potential energy, but Eq. (9-29) is still valid. We could just as well have taken the potential energy to be zero at the level of the floor beneath, or the roof, and it would not have mattered, since only *changes* in potential energy are used in the work-energy relation. The choice of the zero point for the stretched spring is also arbitrary; that fact is somewhat less obvious, but it will become clear later.

We now proceed to define potential energy more generally. Let a particle move along a straight line, with coordinate x, under the action of a force F which depends only on x and which we denote as $F(x)$. The potential energy is then also a function of x, and we denote it by $V(x)$. Let the zero point for potential energy be x_0; that is, $V(x_0) = 0$. We *define* the potential energy $V(x)$ at any other point x to be the work done by the force when the body moves from the initial point x to the zero point x_0. That is,

$$V(x) = \int_{x}^{x_0} F(x')\,dx' \tag{9-30}$$

We have used the integration variable x' in the work integral to avoid confusing this variable with the limits x_0 and x.

Once $V(x)$ is known, it can be used to find the work done by the force when the body moves from any point x_1 to any other point x_2. We use two fundamental properties of integrals. The first is that for any values of x_0, x_1, and x_2 and for any function $F(x)$,

$$\int_{x_0}^{x_2} F(x)\,dx = \int_{x_0}^{x_1} F(x)\,dx + \int_{x_1}^{x_2} F(x)\,dx \tag{9-31}$$

The other is that interchanging the upper and lower limits of an integral merely changes its sign. Thus Eq. (9-31) can be written

$$\int_{x_1}^{x_2} F(x)\,dx = \int_{x_1}^{x_0} F(x)\,dx - \int_{x_2}^{x_0} F(x)\,dx \tag{9-32}$$

Combining this with Eq. (9-30), we obtain

$$W_{12} = \int_{x_1}^{x_2} F(x)\,dx = V(x_1) - V(x_2) = V_1 - V_2 \tag{9-33}$$

which gives the work done by the force $F(x)$ during the displacement from x_1 to x_2, abbreviated as W_{12}, in terms of the potential energies V_1 and V_2 at the two points. We note that if V is smaller at x_2 than at x_1, the work is positive, and conversely. When the force does positive work, the potential energy decreases; when it does negative work, the potential energy increases.

Example

One end of a spring whose spring constant is 200 N/m is held stationary, and the other is attached to a body which moves along the line of the spring from $x = 0.5$ m to $x = 1.0$ m. How much work does the spring do on the object?

Solution

Here $x_1 = 0.5$ m and $x_2 = 1.0$ m. The corresponding potential energies are

$$V(x_1) = \tfrac{1}{2}(200 \text{ N/m})(0.5 \text{ m})^2 = 25 \text{ J}$$
$$V(x_2) = \tfrac{1}{2}(200 \text{ N/m})(1.0 \text{ m})^2 = 100 \text{ J}$$

According to Eq. (9-33), the work done by the spring is

$$W_{12} = 25 \text{ J} - 100 \text{ J} = -75 \text{ J}$$

The minus sign is to be expected, since the force and displacement are in opposite directions. If the object had moved from 1.0 m to 0.5 m, the spring would have done a positive quantity of work, and the potential energy would have decreased.

It is not always possible to describe work done by a force in terms of a potential-energy function. There are cases, of which frictional forces are an example, in which the force depends on the state of motion as well as on the position. Then the integral in Eq. (9-30), which defines potential energy, cannot be evaluated in an unambiguous way, and there is no such thing as a potential-energy function.

When the motion is not confined to a straight line, the definition of potential energy must be generalized. This is easily accomplished by using the general definition of work given by Eq. (9-9). The force and potential energy are now functions of the position vector of a point P in space, which we call $\mathbf{r}$. We choose a reference point P_0, with position vector $\mathbf{r}_0$, at which the potential energy is zero. That is, $V(\mathbf{r}_0) = 0$. Then we define the potential energy at any other point P (with position vector $\mathbf{r}$) in exact analogy to Eq. (9-30):

$$V(\mathbf{r}) = \int_{P}^{P_0} \mathbf{F} \cdot d\mathbf{r} \tag{9-34}$$

and again the work done by the force as the body moves from P_1 to P_2 is given by

$$\begin{aligned} W_{12} &= \int_{P_1}^{P_2} \mathbf{F} \cdot d\mathbf{r} = \int_{P_1}^{P_0} \mathbf{F} \cdot d\mathbf{r} + \int_{P_0}^{P_2} \mathbf{F} \cdot d\mathbf{r} \\ &= V(\mathbf{r}_1) - V(\mathbf{r}_2) \end{aligned} \tag{9-35}$$

There is, however, an additional problem which was not present in the one-dimensional situation: There are many paths leading from P_1 to P_2; may not the work depend on the path chosen as well as the end points? Indeed it may; we have already seen in Sec. 9-1 an example where the work *does* depend on the path. Thus in general the definition of V given by Eq. (9-34) is ambiguous, and hence V is not uniquely defined.

This result is not as catastrophic as it may appear because, as mentioned in Sec. 9-1, there are important classes of force fields for which the work is actually *independent* of the path. A familiar example is a uniform gravitational field. The force of gravity on a falling body is the same whether it falls straight down or in a parabolic path; in each case the work done by gravity is determined only by the *vertical* displacement. A horizontal component of displacement cannot contribute to the work, since the force is always perpendicular to this component. Thus the work on a body moving in a uniform gravitational field depends only on its change in height, and the potential energy V for such a body is always given by $V = mgh$, where h is the height above the reference level (the level where V is zero), even if the body moves horizontally as well as vertically.

The uniform gravitational field is the simplest example of a *conservative force field*, which is a force field having the property that the work during any displacement is independent of the path and depends only on the end points. We return to this important concept in Sec. 9-7, but first we discuss some examples of the usefulness of potential energy.

9-5 CONSERVATION OF ENERGY

We are now ready to complete the discussion of the relationship between kinetic and potential energies. We have already found that the change in kinetic energy of a body which undergoes a displacement is equal to the total work done by the forces acting on it. This result is summarized in Eq. (9-22):

$$E_2 - E_1 = \int_{P_1}^{P_2} \mathbf{F} \cdot d\mathbf{r}$$

For cases in which the total work can be expressed in terms of a change in a potential-energy function, Eq. (9-35) combined with Eq. (9-22) gives $E_2 - E_1 = V_1 - V_2$, or

$$E_1 + V_1 = E_2 + V_2 \tag{9-36}$$

This important result states that when the work can be expressed in terms of a potential energy, the sum of kinetic and potential energies, $E + V$, does not change during the displacement even though E and V change individually. The sum $E + V$ is called the *total mechanical energy* and is denoted by H. Thus the total mechanical energy is constant throughout the process. This result is the *principle of conservation of energy*.

This general principle is illustrated by the example at the beginning of Sec. 9-4. Here is another example. Consider again a spring with one end fixed and a mass m at the other end. Suppose that when the spring is in its relaxed position, we give the mass an initial velocity v_0 in such a direction as to stretch the spring. As the mass moves, the spring does *negative* work on it since, as the spring is being stretched, it is pulling in the direction of less stretch. In doing negative work, the spring decreases the kinetic energy of the mass from $\frac{1}{2}mv_0^2$ to zero. The position X at which the mass comes to rest is determined by equating the loss of kinetic energy to the work done on the spring:

$$\tfrac{1}{2}mv_0^2 = \tfrac{1}{2}kX^2 \tag{9-37}$$

The mass stops, reverses its direction of motion, and returns to the starting point. Since the spring force depends only on the position and not on the direction of motion, during the return trip the spring does an amount of work equal in *magnitude* to that which it did in the first half of the motion but this time positive rather than negative. Correspondingly, when the mass returns to the starting point, it must have the same kinetic energy, and thus the same speed, as when it started, although its velocity is now reversed in direction.

We now describe this motion in terms of potential energy. Initially, the potential energy is zero, since the spring is not stretched. As the mass moves, its kinetic energy decreases, and its potential energy increases by a corresponding amount, so the sum of the two quantities is constant, equal to the initial kinetic energy:

$$H = \tfrac{1}{2}mv_0^2$$

Now let x be any position of the mass between 0 and X, and let v be the velocity at this point. Then

$$H = \tfrac{1}{2}mv^2 + \tfrac{1}{2}kx^2 = \tfrac{1}{2}mv_0^2 \tag{9-38}$$

This expression may be used to find the speed v of the mass for any value of x. It does not give the *direction* of the velocity but only its *magnitude*, because kinetic energy depends only on the speed (magnitude of the velocity).

Figure 9-8 shows how the kinetic and potential energies of the system

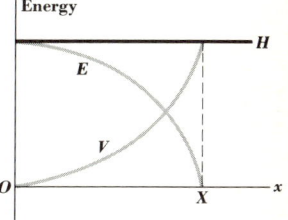

Fig. 9-8 As the mass attached to the spring moves from its initial position ($x = 0$) to its final position ($x = X$), the kinetic energy E decreases from its initial value to zero, and the potential energy V increases from zero to its final value. The sum of these two, equal to the total energy H, is constant.

depend on the coordinate x. We have used the same vertical scale for both, and we have drawn a horizontal line corresponding to the constant total energy H. The kinetic energy is maximum when the potential energy is zero, and conversely. For every value of x, however, the sum of the two is equal to the total mechanical energy. Thus this problem illustrates the *conservation* of mechanical energy.

Two points need to be emphasized. First, the conservation of energy in this situation is not a new physical principle but a result which follows from definitions of kinetic and potential energies and from Newton's laws of motion. Second, the *reversible* nature of the work is important. The work done by this spring during the return trip must be exactly the negative of that done during the first half of the motion. It is so in this example because the force depends only on x and not on the direction of motion. As a result, all the kinetic energy which is converted to potential energy can be reconverted into kinetic energy on the return trip.

Example

A ballistic missile is launched with an initial speed v_0 at an angle α to the horizontal. What maximum height h does the missile attain assuming that gravity is the only force acting on it during its flight?

Solution

This problem has already been solved in Sec. 5-3 using Newton's laws of motion directly. The result obtained there was

$$h = \frac{v_0^2 \sin^2 \alpha}{2g}$$

The energy principle can be used to obtain this result more easily. Initially, the missile has kinetic energy $\frac{1}{2}mv_0^2$. When it reaches the highest point on its path, it still has the same horizontal component of velocity, $v_0 \cos \alpha$, as at the beginning, since there has been no horizontal acceleration; but its vertical component of velocity has been reduced to zero. There has been a corresponding decrease in kinetic energy from $\frac{1}{2}mv_0^2$ to $\frac{1}{2}m(v_0 \cos \alpha)^2$. This is accompanied by an increase in potential energy mgh. According to the principle of conservation of energy, the *total* energy is the same at the beginning as at the instant the shell reaches its highest point. That is,

$$\tfrac{1}{2}mv_0^2 = \tfrac{1}{2}mv_0^2 \cos^2 \alpha + mgh \tag{9-39}$$

Solving this equation for h and using the identity $\sin^2 \alpha + \cos^2 \alpha = 1$, we find

$$h = \frac{v_0^2 \sin^2 \alpha}{2g} \tag{9-40}$$

which agrees with the previous result.

It is instructive to compare this method of finding h with that used in Chap. 5. There we first had to find the x and y coordinates as functions of time; then we made deductions from these about the details of the *path* followed by the shell. By using the energy principle, we have been able to avoid these detailed calculations and have related the motion at one point to that at another point without determining what happens in between. This, in fact, is one of the great advantages of the principle of conservation of mechanical energy. Where applicable, it often makes it possible to relate the motions of a system at two different stages without regard for the details of the motion at intermediate points.

The previous example can be simplified even further by using the fact that v^2 can be expressed in terms of the components of velocity as $v^2 = v_x^2 + v_y^2 + v_z^2$, so the kinetic energy is given by

$$\tfrac{1}{2}mv^2 = \tfrac{1}{2}m(v_x^2 + v_y^2 + v_z^2) = \tfrac{1}{2}mv_x^2 + \tfrac{1}{2}mv_y^2 + \tfrac{1}{2}mv_z^2 \tag{9-41}$$

That is, the total kinetic energy is the sum of the kinetic energies associated separately with the x, y, and z components of velocity. In this example, the kinetic energy associated with v_x is constant, since v_x is constant; and v_z is always zero. The kinetic energy corresponding to v_y initially has the value $\tfrac{1}{2}mv_0^2 \sin^2 \alpha$, and at the top of the trajectory it is zero. Correspondingly, the potential energy associated with the vertical motion changes from zero to mgh. There is no change in potential energy associated with the x component of motion, since there is no force in the x direction. Thus we conclude immediately that

$$\tfrac{1}{2}mv_0^2 \sin^2 \alpha = mgh$$

which leads to Eq. (9-40).

Example

A bead of mass m slides without friction on a circular hoop of radius R in a vertical plane, as shown in Fig. 9-9. We start the bead at the top of the

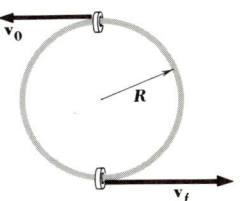

Fig. 9-9 Bead slides without friction on a circular hoop of radius R in a vertical plane. The kinetic energy at the bottom is greater than that at the top, the difference being equal to the change in potential energy as the bead slides from top to bottom.

hoop with initial speed v_0. What is its speed v_f when it reaches the bottom of the hoop?

Solution

This problem is less elementary than the previous ones because the hoop exerts a force on the bead in addition to the force of gravity. If the bead were not sliding on the hoop, it would simply move in a parabolic path like the missile just discussed. But the hoop must always push *radially* on the bead;

there can be no component of force *tangent* to the hoop. If there *were* a tangent force, it would have to be associated with friction, and the bead could not slide freely. So when friction is absent, the contact force between bead and hoop has only a *normal* component. This is a very useful observation; the force which the hoop exerts on the bead is always perpendicular to the velocity of the bead, and hence this force *does no work* on the bead during the motion. Thus we can apply the principle of conservation of energy just as though the hoop were not present at all. That is, the kinetic energy when the bead reaches the bottom is greater than that at the top by an amount $mg(2R)$:

$$\tfrac{1}{2}mv_f^2 = \tfrac{1}{2}mv_0^2 + 2mgR$$

from which v_f can be obtained.

If the hoop can be disregarded in the energy calculation, does this mean that the hoop has no effect on the motion? Of course not. The *direction* of the velocity when the bead reaches the bottom of the hoop is not the same as it would be had the hoop not been present; the forces exerted on the bead by the hoop have produced an acceleration at every point along the motion of the bead although they have done no *work* on it.

Forces which serve to confine the motion to a particular path but which do no work are called *workless forces* or *workless constraints*. Because such forces do no work, there is never any potential energy associated with them, and they may be disregarded in energy calculations.

The force exerted on a charged particle by a magnetic field, $\mathbf{F} = q\mathbf{v} \times \mathbf{B}$, is another example of a workless force. The force is always perpendicular to the velocity of the particle and hence never does any work on the particle. Thus the speed of the particle is constant, provided no other forces act. Detailed proof of these statements is left as a problem.

We have stressed repeatedly that the principle of conservation of mechanical energy is valid only when the work done by the forces on the system is of a *reversible* nature, so that their effect may be represented by means of a potential-energy function. Only then, in fact, is it possible to define a potential-energy function corresponding to the force. There are many cases in which these conditions are not met and in which the principle of conservation of mechanical energy is not valid. The simplest possible example is a body sliding on a rough horizontal surface. We push a box across a rough floor with initial speed v_0; eventually, because of friction, the box comes to rest. During this process, its kinetic energy decreases from its initial value $\tfrac{1}{2}mv_0^2$ to zero. To this extent the situation resembles that in which a moving object attached to a spring was stopped by the stretch of the spring. But here the similarity ends; there is no way to recover the kinetic energy lost by the box sliding on the rough floor. It does not automatically slide back, speeding up as it approaches the original position, as it did when attached to the spring. There is no way to associate potential energy with work done by a force of friction. We can only say the total mechanical energy in this system is not constant but decreases whenever there is motion with friction.

Example

Consider again a box sliding on a rough surface, but now let the surface be inclined, making an angle α with the horizontal. The box is given an initial speed v_0 up the plane. How far does it move before stopping?

Solution

Figure 9-10 shows the situation and a free-body diagram for the box. The force of gravity is shown as the vector sum of a component parallel to the

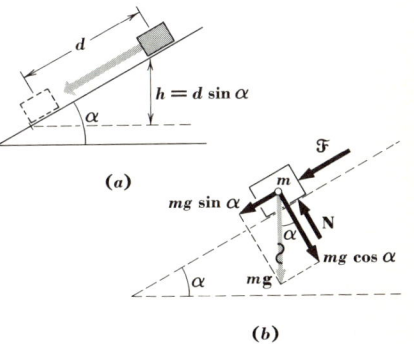

Fig. 9-10 Box slides with friction along an inclined plane. (*a*) Relationship between the change in vertical elevation h and the distance d which the box moves along the plane. (*b*) Free-body diagram for the box; the weight of the box is represented by its components parallel and perpendicular to the plane.

plane and a component perpendicular to the plane; the direction of the force of friction $\mathfrak{F}$ down the plane opposes the motion of the box *up* the plane. Because there is no acceleration perpendicular to the plane,

$$N - mg \cos \alpha = 0$$

Assuming the frictional force $\mathfrak{F}$ is proportional to the normal force N gives

$$\mathfrak{F} = \mu N = \mu mg \cos \alpha \tag{9-42}$$

One way to obtain the distance d the box moves before stopping is to equate the change in kinetic energy of the box, which is $-\frac{1}{2}mv_0^2$, to the total (negative) work done on it by the two forces acting parallel to the plane, i.e., the component of gravity $mg \sin \alpha$ and the force of friction:

$$-\tfrac{1}{2}mv_0^2 = -(\mathfrak{F} + mg \sin \alpha)d \qquad (9\text{-}43)$$

Inserting the expression for $\mathfrak{F}$ given by Eq. (9-42) and solving for d, we find

$$d = \frac{v_0^2}{2g(\mu \cos \alpha + \sin \alpha)} \qquad (9\text{-}44)$$

This is a reasonable result; larger values of v_0 mean greater distance up the plane; greater μ means smaller d.

We can interpret Eq. (9-43) in a different way; $d \sin \alpha$ is just the change in elevation of the block, denoted by h in Fig. (9-10), so we may rewrite Eq. (9-43) as

$$\tfrac{1}{2}mv_0^2 = \mathfrak{F}d + mgh$$

The initial kinetic energy $\frac{1}{2}mv_0^2$ is converted partly into potential energy mgh; the remainder disappears as work done against friction, $\mathfrak{F}d$. The potential energy mgh is recovered as kinetic energy when the box slides back down the plane, but the energy which went into frictional work is lost. When the box slides down the plane, the force of friction reverses its direction; the speed v_f when it returns to the starting point is obtained from

$$mgh = \tfrac{1}{2}mv_f^2 + \mathfrak{F}d$$

This says that the potential energy of the box mgh at the highest point in its travel on the plane is equal to the sum of the energy converted into kinetic energy when it returns to the bottom, $\frac{1}{2}mv_f^2$, and the energy used up as work against friction, $\mathfrak{F}d$. When the box returns to its starting point, its kinetic energy is less by $2\mathfrak{F}d$ than when it left; this much mechanical energy has been *lost* in the process. Hence the principle of conservation of mechanical energy is not applicable here. Nevertheless, the concepts of energy and work are still useful in the analysis of this problem.

9-6 COLLISIONS

Energy considerations play an important part in the analysis of collisions. In the discussion of collisions in Chap. 7, we observed that when two bodies interact with each other but not with their surroundings, i.e., when they are an isolated system, the total momentum is constant. We also introduced a scheme for classifying collisions in terms of the magnitudes of the relative velocities before and after the collision, and we stated without proof that when these are the same, i.e., when the collision is *elastic*, the total kinetic energy of the system is the same before and after the collision. We are now in a position to discuss this matter in more detail.

Suppose two bodies of masses m_1 and m_2 collide. Let their initial velocities be $\mathbf{v}_1$ and $\mathbf{v}_2$ and their final velocities be $\mathbf{v}'_1$ and $\mathbf{v}'_2$. We now prove that if the collision is elastic, the total kinetic energy is the same after the collision as before. The proof involves a modest amount of algebraic acrobatics, but the result justifies the effort. First, momentum conservation gives the relation

$$m_1\mathbf{v}_1 + m_2\mathbf{v}_2 = m_1\mathbf{v}'_1 + m_2\mathbf{v}'_2 \tag{9-45}$$

Second, the condition that the collision be elastic can be stated as follows:

$$(\mathbf{v}_1 - \mathbf{v}_2)^2 = (\mathbf{v}'_1 - \mathbf{v}'_2)^2 \tag{9-46}$$

where the square of a vector denotes the scalar product of the vector with itself, as usual.

Now we square both sides of Eq. (9-45), obtaining

$$m_1^2 v_1^2 + 2m_1 m_2 \mathbf{v}_1 \cdot \mathbf{v}_2 + m_2^2 v_2^2 \\ = m_1^2 v_1'^2 + 2m_1 m_2 \mathbf{v}'_1 \cdot \mathbf{v}'_2 + m_2^2 v_2'^2 \tag{9-47}$$

We multiply Eq. (9-46) by $m_1 m_2$ and expand it:

$$m_1 m_2 v_1^2 - 2m_1 m_2 \mathbf{v}_1 \cdot \mathbf{v}_2 + m_1 m_2 v_2^2 \\ = m_1 m_2 v_1'^2 - 2m_1 m_2 \mathbf{v}'_1 \cdot \mathbf{v}'_2 + m_1 m_2 v_2'^2 \tag{9-48}$$

We add Eqs. (9-47) and (9-48), obtaining

$$m_1(m_1 + m_2)v_1^2 + m_2(m_1 + m_2)v_2^2 \\ = m_1(m_1 + m_2)v_1'^2 + m_2(m_1 + m_2)v_2'^2$$

Finally, we divide by $2(m_1 + m_2)$ to obtain

$$\tfrac{1}{2}m_1 v_1^2 + \tfrac{1}{2}m_2 v_2^2 = \tfrac{1}{2}m_1 v_1'^2 + \tfrac{1}{2}m_2 v_2'^2 \tag{9-49}$$

which shows that in an elastic collision the total kinetic energy is conserved.

The discussion can be extended to collisions which are not perfectly elastic, but that more general case is much easier to handle using the center-of-mass coordinate system, to be introduced in Sec. 11-3. For the present we state without proof that when the coefficient of restitution e is less than unity, the final kinetic energy is less than the initial, and conversely.

In an elastic collision the interaction forces are conservative. If two bodies interact by means of a spring bumper, energy can be stored temporarily as potential energy during the collision, but as the spring returns to its original shape, the potential energy again becomes kinetic energy. But if the bumper is made of chewing gum, the compression is irreversible and corresponds to a frictionlike force; energy expended in deforming the gum is no longer available for reconversion into kinetic energy, and such a collision is not elastic. If the bumper is made of gunpowder, its explosion adds kinetic energy to the system, and the coefficient of restitution is greater than unity.

9-7 CONSERVATIVE FORCE FIELDS

For certain classes of forces, the concept of potential energy provides a convenient way to calculate work for use in energy relations. In particular, in one-dimensional motion, where the force depends only on position and thus may be represented by some function $F(x)$, the corresponding potential energy $V(x)$ can always be found from Eq. (9-30).

Conversely, if $V(x)$ is known, $F(x)$ can always be found. When the body moves a small distance Δx, the work done by the force is $F(x) \Delta x$. But by the definition of potential energy, $V(x)$ must *decrease* by an amount equal to the work done by the force. Therefore, the change in potential energy ΔV is related to the work by

$$\Delta V = - F(x) \Delta x$$

Dividing by Δx and taking the limit as $\Delta x \to 0$ gives

$$F(x) = -\frac{dV}{dx} \tag{9-50}$$

The force at any value of x is the negative of the *slope* of $V(x)$ at the same point.

A hypothetical potential-energy function and the corresponding force are shown in Fig. 9-11. When $V(x)$ has zero slope, F is zero. Points of maximum positive slope for V represent maximum negative values of F, and so on. The minimum points of $V(x)$ correspond to positions of *stable equilibrium*. At these points, $F(x)$ is zero; at neighboring points on both sides, the

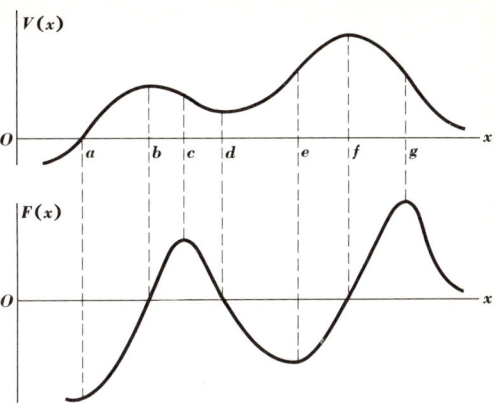

Fig. 9-11 Potential-energy curve and the corresponding force. At points b, d, and f, the slope of $V(x)$ and the force are zero. At points c and g, $V(x)$ has maximum negative slope; the force reaches a maximum positive value at these points. At e, $V(x)$ has maximum positive slope; $F(x)$ has a maximum negative value at this point. Points b and f represent points of unstable equilibrium; point d is a point of stable equilibrium.

force is such as to push the object back to the equilibrium position. Maximum points of $V(x)$, on the other hand, are points of *unstable equilibrium*; a body initially at rest at such a point experiences no force, but when it is displaced in either direction, the direction of the force as given by Eq. (9-50) is such as to push it *farther* away from the equilibrium position. Hence the equilibrium is unstable.

The force $F(x)$ is an example of a *force field*, since it is a force which depends on the position of the object. In the one-dimensional case, it is *always* possible to find a potential-energy function, as discussed above, so that the total mechanical energy (kinetic plus potential) of an object moving under the action of this force is constant. For such a system the principle of conservation of energy is valid, and so this force field is called a *conservative force field*. We have thus shown that in any one-dimensional situation in which the force depends only on the position of the object the force is a conservative force field.

The point x_0, at which we define V to be zero, is completely arbitrary provided we use it consistently through any given problem. In some cases, especially with electric and gravitational forces, it is convenient to take the zero point for potential energy at a point which is very far away from the object producing the force, so that the force, too, is zero at this point.

Consider, for example, the straight-line motion of a particle of mass m under the gravitational attraction of a particle of mass M held stationary at $x = 0$. The force on m is given, according to the law of gravitation, by

$$F(x) = -\frac{GmM}{x^2} \tag{9-51}$$

The minus sign appears because the attractive force is in the direction of *decreasing* x. We define the potential energy of this force to be zero when $x = \infty$; then the potential energy at any other value of x is given by

$$V(x) = -\int_x^\infty \frac{GmM}{x'^2}\,dx' \tag{9-52}$$

which is obtained by substituting Eq. (9-51) into Eq. (9-30).

The infinite limit on the integral deserves a little discussion. What we mean is that we integrate to a very large value of x', say X, and then take the limit of this as X approaches infinity. That is,

$$\int_x^\infty \frac{dx'}{x'^2} = \lim_{X\to\infty} \int_x^X \frac{dx'}{x'^2} = \lim_{X\to\infty} \left(\frac{1}{x} - \frac{1}{X}\right) = \frac{1}{x}$$

The term containing X disappears in the limit. Using this result in Eq. (9-52), we find

$$V(x) = -\frac{GmM}{x} \tag{9-53}$$

In this case the potential energy is a *negative* quantity, corresponding to the fact that when the body moves from a given value of x to the reference value $x = \infty$, the force does a negative quantity of work. Indeed, for *any* attractive force, if the potential energy is taken to be zero at infinity, it is negative for any finite value of x. If the force is *repulsive*, as between like electric charges, then when $V(\infty) = 0$, V is *positive* for finite values of x.

Example
A missile is fired vertically upward from the surface with initial speed v_0. What is its final speed v_f when it is far away from earth, after it has "escaped the earth's gravity," as the newspapers say?

Solution
Let the mass of the missile be m, and idealize the earth to a sphere of mass M and radius R. We further idealize the problem by neglecting gravitational forces due to the sun, moon, and other bodies. Taking the potential energy of the missile to be zero at infinity, its value at the surface of the earth is simply $-GmM/R$. Equating the sum of kinetic and potential energies at launch to the final sum, we find

$$\tfrac{1}{2}mv_0^2 - \frac{GmM}{R} = \tfrac{1}{2}mv_f^2 + 0 \tag{9-54}$$

which can be solved for v_f:

$$v_f = \sqrt{v_0{}^2 - \frac{2GM}{R}} \qquad (9\text{-}55)$$

We note that for v_f to be real, $v_0{}^2$ must be at least as great as $2GM/R$, and thus this represents the minimum speed a missile must have to "escape" the earth's gravitational field. This turns out to be about 7 mi/s. A missile launched with speed greater than this has kinetic energy "left over"; a sufficient fraction of initial kinetic energy is converted into potential energy, and the remaining energy is associated with the final speed v_f.

Forces which hold atoms together in a molecule can be conveniently represented in terms of potential-energy and force graphs. A typical interatomic force in a diatomic molecule is shown in Fig. 9-12, where x is the

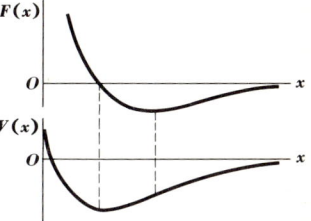

Fig. 9-12 Typical force of interaction between atoms in a diatomic molecule. The corresponding potential energy is shown. The potential energy has been chosen to be zero at infinity.

distance between atoms. We see that there is one value of x for which the force is zero. This is related to the equilibrium separation of the atoms. For values of x greater than this, the force is negative (attractive). For smaller values of x it is repulsive; as x becomes very small, the repulsive force becomes very large. At very large values of x, the force again approaches zero. The corresponding potential-energy function is also shown in Fig. 9-12. The minimum point for V corresponds to the point where the force is zero, a point of stable equilibrium.

For matter as we know it to exist, interatomic forces *must* have the general behavior shown in Fig. 9-12. The force must be attractive in some region of x; otherwise atoms would never be bound together. But at sufficiently small values of x it must become repulsive; otherwise atoms would be drawn closer and closer together, and matter could not have a definite volume. This is a considerable oversimplification of the situation, but the essential idea is clear. Forces between particles (protons and neutrons) in the *nucleus* of an atom are not understood as well as interatomic forces, but they too exhibit behavior of the type shown in Fig. 9-12; this behavior

232 MECHANICAL ENERGY

accounts in part for the observation that all nuclei have roughly the same density.

As we have indicated in Sec. 9-4, the concept of potential energy can be extended to motion in two or three dimensions for some but not all position-dependent forces. For a two-dimensional problem in which the coordinates are x and y and the force given as $F(x,y)$, we may define the potential energy as zero at some point P_0 with coordinates (x_0, y_0). That is, $V(x_0, y_0) = 0$. The potential energy at any other point P with coordinates (x,y) is then given by

$$V(x,y) = \int_{P}^{P_0} \mathbf{F} \cdot d\mathbf{r} \tag{9-56}$$

The difficulty, as we discussed in Sec. 9-4, is that there are many paths leading from P to P_0, as shown in Fig. 9-13. In general there is no guarantee that

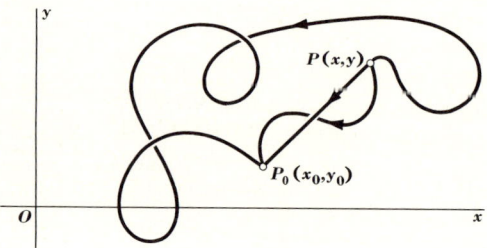

Fig. 9-13 When a particle is displaced from point (x,y) to point (x_0, y_0), it may follow many different paths. In general, the work done by the force on the particle during the displacement depends on the path taken.

the work will be the same for all these paths. Thus Eq. (9-56) is ambiguous and cannot be used in general to define V.

There are however important classes of force fields having the property that the work done between two points is the same for all paths taken between these two points. Such a field is called a *conservative force field*. We have already discussed the uniform gravitational field as an example of such a force field. The work done by gravity on a body moving in such a field is determined entirely by the change in height of the body and is independent of the details of the path taken by the body.

It is not difficult to show, in fact, that *every* gravitational field, whether uniform or not, has this property. For example, the field due to a point mass M at a distance r away from the mass has magnitude GM/r^2 and direction along a line toward the particle. When a body moves in this field, work is done only when the displacement has a radial component, as shown in Fig. 9-14. In a displacement for which r is constant, i.e., a circular arc

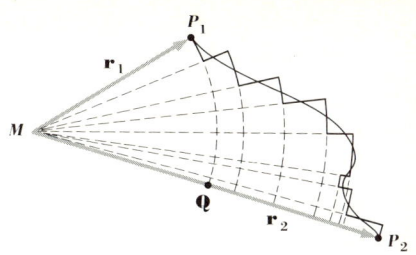

Fig. 9-14 The displacement from P_1 to P_2 can be represented as a series of radial and circular segments. Work is done only during the radial displacements, and the total work is the same as for the straight-line displacement from **Q** to P_2.

centered on the mass M, no work is done. Any path between two points may be represented in many different ways as a series of steps, each of which is either radial or "tangential," as shown in the figure. No work is done in the tangential segments, and the total work done in the radial segments is the same for all paths. Thus such a field is conservative.

Furthermore, any gravitational field may be regarded as a superposition of fields due to point masses, since any mass distribution which creates a field may be regarded as a collection of points. Thus since the field of a point mass is conservative, *every* gravitational field is conservative. A similar argument can be made to show that every electric field produced by charges at rest is a conservative force field. Thus although the conditions for a conservative force field seem rather restrictive, there are a number of important examples of fields that meet these conditions.

Just as in the one-dimensional case, we can obtain a relation between potential energy and force for a conservative force field. If the force has a component F_x in the x direction, then in a displacement Δx (also in the x direction) an amount of work $F_x \Delta x$ is done. Correspondingly, the change in potential energy is

$$\Delta V = -F_x \Delta x$$

Dividing by x and taking the limit as $\Delta x \to 0$,

$$F_x = -\frac{dV}{dx}$$

Similarly, for the y component we find

$$F_y = -\frac{dV}{dy}$$

These derivatives deserve a little more discussion. V is a function of both x and y; we can differentiate it with respect to either while holding the other constant, corresponding to displacement parallel to one coordinate axis or the other. The usual mathematical notation for this situation is

$$F_x = -\frac{\partial V}{\partial x} \qquad F_y = -\frac{\partial V}{\partial y} \qquad (9\text{-}57)$$

These are called *partial derivatives*; the notation $\partial V/\partial x$ is read "partial derivative of V with respect to x." The symbol ∂, another form of Greek delta, emphasizes the fact that we are dealing with a function of more than one variable and its derivative is to be taken with respect to one variable while the others are held constant.

Using unit vectors, we may write

$$\mathbf{F} = -\left(\frac{\partial V}{\partial x}\mathbf{i} + \frac{\partial V}{\partial y}\mathbf{j}\right) \qquad (9\text{-}58)$$

The quantity in parentheses is often called the *gradient* of V and is sometimes abbreviated as

$$\frac{\partial V}{\partial x}\mathbf{i} + \frac{\partial V}{\partial y}\mathbf{j} = \nabla V \qquad (9\text{-}59)$$

in which the inverted delta, read "del," signifies taking the derivative with respect to each coordinate, multiplying by the corresponding unit vector, and adding.

The generalization to a three-dimensional situation is obvious. The force and potential energy are related by

$$\mathbf{F} = -\left(\frac{\partial V}{\partial x}\mathbf{i} + \frac{\partial V}{\partial y}\mathbf{j} + \frac{\partial V}{\partial z}\mathbf{k}\right) \qquad (9\text{-}60)$$

and the corresponding generalization of the definition of the gradient is

$$\nabla V = \frac{\partial V}{\partial x}\mathbf{i} + \frac{\partial V}{\partial y}\mathbf{j} + \frac{\partial V}{\partial z}\mathbf{k} \qquad (9\text{-}61)$$

so that again $\mathbf{F} = -\nabla V$.

Example

A particle is attracted toward the origin of a two-dimensional coordinate system in a plane with a force of magnitude proportional to the distance of the particle from the origin. That is, $\mathbf{F} = -k\mathbf{r}$. Show that this force field is conservative, find a potential energy function, assuming $V = 0$ at the origin, and verify Eqs. (9-57) and (9-58).

Solution
The argument used to show that the gravitational field of a point mass is conservative can also be used in this case. Work is done only in displacements having a radial component, and the work is independent of any angular displacement that may occur at the same time. Thus the potential energy can be a function only of the distance r from the origin. Specifically,

$$V(r) = \int_r^0 F(r')\,dr' = -\int_r^0 kr'^2\,dr' = \tfrac{1}{2}kr^2 \tag{9-62}$$

In terms of cartesian coordinates,

$$V(x,y) = \tfrac{1}{2}k(x^2 + y^2) \tag{9-63}$$

from which Eqs. (9-57) and (9-58) give

$$F_x = -kx \qquad F_y = -ky \qquad \mathbf{F} = -k(x\mathbf{i} + y\mathbf{j}) \tag{9-64}$$

But from the original description of the force we have, in terms of the usual polar coordinate angle θ,

$$F_x = -kr\cos\theta = -k(x^2+y^2)^{1/2}\frac{x}{(x^2+y^2)^{1/2}} = -kx$$

$$F_y = -kr\sin\theta = -k(x^2+y^2)^{1/2}\frac{y}{(x^2+y^2)^{1/2}} = -ky \tag{9-65}$$

which agrees with the above results. A force field of this type, with magnitude depending only on r and with direction always along the radial direction, is called a *central-force field*. On the basis of the above discussion it is easy to show that every central-force field is conservative.

To summarize: In a one-dimensional situation, a force which depends only on the position of a particle on a line is *always* a conservative force field. In two or three dimensions, a force field is conservative only if the work done by the force between any two points is independent of path for every possible path between every possible pair of points. In particular, however, gravitational, electrostatic, and all central-force fields are always conservative.

9-8 NONMECHANICAL ENERGY

We have seen that when only conservative forces are present in a mechanical system, the total mechanical energy of the system is constant. When frictional forces or other nonconservative forces are present, the total mechanical energy is not constant. A frictional force in a dynamic situation always *decreases* the total mechanical energy; in such situations it is observed that *heat* always appears when mechanical energy disappears.

Work may be done on an electric generator to drive it, but this work is not stored as potential energy. Instead, the generator causes an electric current to flow in a circuit. A third example of the disappearance of mechanical energy is a high-energy collision between two protons in which a new particle, a π meson, is produced and the total kinetic energy of the system decreases.

In all these examples, the mechanical energy of the system is not constant; but whenever mechanical energy disappears, something else appears. In the first case it is heat, in another electricity, in the third a new particle. It is reasonable to ask whether it is possible to *generalize* the concept of energy so that the phenomena just mentioned can be interpreted as transformations of mechanical energy into some other form of energy such that the *total* energy is constant.

This is a question which can be answered only by experiment. When we measure the amount of heat produced when a given quantity of mechanical energy disappears in a situation involving friction, we find that the heat produced is always proportional to the loss of mechanical energy; thus we are justified in regarding heat as another form of energy. By defining the quantity of energy corresponding to a given amount of heat, we can generalize the concept of the conservation of energy so that the total energy, including both mechanical energy and heat energy, is conserved. It is of course essential to establish a way of measuring a quantity of heat, so that heat energy is given an operational definition. Some experiments leading to the equivalence of heat and mechanical energy are discussed in detail in Chap. 20. The result is that 4.186 J of mechanical energy becomes 1 calorie of heat.

In the case of the electric generator, we may try to associate energy with electricity in such a way that when mechanical energy disappears, an equal quantity of *electric energy* appears. This, again, turns out to be possible, so we may regard the principle of conservation of energy as valid for the generator, provided we include both the mechanical energy put into the generator and the electric energy taken out. In the production of mesons, we may regard the *mass* of a new particle as equivalent to a certain amount of energy, and it turns out that this equivalence is given simply by Einstein's equation $E = mc^2$, where c is the speed of light. This equivalence is discussed in detail in Chap. 15.

One of the most remarkable facts in all of physical science is that in *every* phenomenon ever observed in which transformation of energy is involved, it is always possible to define an energy of some kind so that the *total* energy of the system under consideration can be accounted for. If the system is isolated, then the total energy in all its various forms is constant. Here, then, is one of the most impressive examples in all of physics of the introduction of a concept (mechanical energy) in a very limited context and the subsequent generalization of this concept to include an extremely large class of physical phenomena.

Problems

9-1 A boy pulls a sled a distance of 5 m along the ground with a rope at an angle of 30° upward from the horizontal; the tension in the rope is 20 N. How much work does he do?

9-2 How much work does a 150-lb man do in climbing a flight of steps 10 ft high?

9-3 A new bucket of paint must be hoisted daily to the third floor of a building. On Monday a workman on the third floor pulls one up; on Tuesday a man on the ground hoists it by pulling down on a rope which passes over a pulley on the third floor. If the pulley is frictionless, how do the quantities of work in the two cases compare?

9-4 A force applied to an object varies with the distance x from the initial position according to the equation $F = kx + cx^3$. The object moves from $x = 0$ to $x = X$. How much work does the force do if it is in the same direction as the displacement?

9-5 A ball is attached to the end of a string the other end of which is tied to the ceiling. The ball is set into motion in a horizontal circular path, as shown in Fig. P9-5. Does the string do work on the ball? Explain.

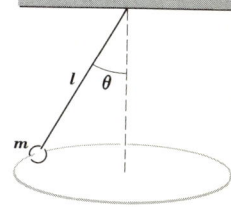

Fig. P9-5

9-6 A block slides on a plane which is stationary. Of the two components (normal and friction) of the contact force *on* the block, one never does work, and the other always does *negative* work. Which is which, and why?

238 MECHANICAL ENERGY

9-7 In the situation of Prob. 7-8, show that the total kinetic energy after the collision is the same as before.

9-8 If a 3,200-lb car is moving at 30 mi/h, what is its kinetic energy? Does it depend on the direction of motion? Explain.

9-9 A 4,000-lb car accelerates uniformly from 50 to 100 ft/s in a distance of 500 ft.
    a Find the acceleration of the car; from this, find the accelerating force.
    b Find the accelerating force using the energy principle, equating the change in kinetic energy to the work done by the force. Compare your result to that of part *a*.

9-10 A projectile is shot with initial speed v_0 at an angle θ upward from the horizontal.
    a What is the initial kinetic energy?
    b What is the kinetic energy at the highest point in the trajectory?
    c How much work is done on the projectile by gravity during flight from the starting point to the highest point?

9-11 A 160-lb man runs up a flight of steps with a vertical rise of 20 ft in 4 s. What is his average power? Express the result in horsepower.

9-12 A 3,200-lb car accelerates from rest to 60 mi/h in 8 s. What average power does the car develop?

9-13 A small car named Seymour weighs 1,000 lb and contains a 17-hp engine. How fast can the car go up a 5° hill?

9-14 An elevator is to lift 1,000 kg of material 50 m vertically in 10 s. What electric power must be fed into the motor driving it if all the electric power is converted into mechanical power?

9-15 Using Eq. (9-26), show that the magnetic field force $\mathbf{F} = q\mathbf{v} \times \mathbf{B}$ never does work on a moving charged particle.

9-16 A certain force field is described by the functions $F_x = x^3 + xy^2$ and $F_y = y^3 + x^2y$. Compute the work done in a displacement from the point (1,1) to the point (2,2) when the path is:
    a A straight line
    b A straight line from (1,1) to (1,2) followed by a straight line from (1,2) to (2,2)

9-17 From what height would a car have to fall freely in order to gain as much kinetic energy as it has when traveling 60 mi/h?

9-18 A simple pendulum is made by tying a string to an ancient Chinese coin with a hole in its center and anchoring the other end of the string to the nearest chandelier. The coin is held so that the string is stretched out horizontally; it is then released with no initial velocity.
    a Find its speed at the lowest point in its motion.
    b Find the tension in the string at this point.

9-19 A ball slides down a smooth, frictionless "loop-the-loop" track as shown in Fig. P9-19. Find the minimum initial height h such that the ball maintains contact with the track all the way around the loop.

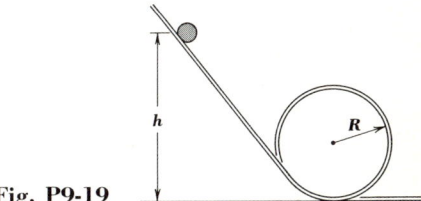

Fig. P9-19

9-20 The pendulum in Prob. 9-18 is given a velocity sufficient for it to move in a vertical circle. Show that the tension in the string at the lowest point in the motion is greater than that at the highest point by six times the weight of the coin.

9-21 In the situation shown in Fig. P9-21, the body is pushed up the plane until the spring is compressed a distance d from its natural length; it is then released from rest. How far down the plane does it travel before stopping if the plane is frictionless?

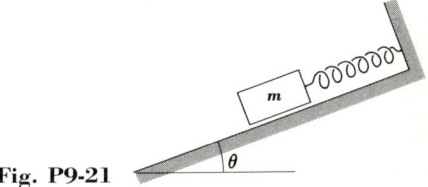

Fig. P9-21

9-22 In Prob. 9-21, if the plane has a coefficient of friction μ, how far down the plane does the body move?

9-23 A cyclotron has an electromagnet which produces a uniform magnetic field B between two circular poles of radius R. If the particles being accelerated move in successively larger circles concentric with the poles, how does the kinetic energy of a particle increase with the radius of its orbit? What is the maximum energy which can be attained in terms of the other quantities? *Hint:* It may be helpful to prove first that the *angular velocity* of the particle is independent of radius.

9-24 The cyclotron discussed in Prob. 9-23 was originally designed to accelerate protons, but now it is to be used to accelerate α particles (containing two neutrons and two protons each).
   a   How is the angular velocity for α particles related to that for protons?
   b   How is the maximum energy for α particles related to that for protons?

9-25 A railroad car of mass m rolls without friction down a hill of height h and length d. At the bottom the track becomes level. The car collides on the level portion with another car of equal mass initially at rest. The two couple together and roll up another hill. How far above the level part of the track do they come to rest?

9-26 A skier starts at the top of a large spherical snowball, 1 mi in radius, left in Antarctica by an extinct race of giants. He skis straight down the side with negligible friction; at what point does he lose contact with the snowball and ski into space; i.e., what angle does a radial line at this point make with the vertical?

9-27 In Prob. 7-3, compare the total kinetic energy of the system before the collision with its value after the collision.

9-28 A body with mass m_1 and initial velocity v collides with body of mass m_2 initially at rest. After the collision the two bodies stick together and thus have the same final velocity. Compute the ratio of final to initial total kinetic energy, and show that it is always less than unity.

9-29 A ball is dropped from a height h_0 and bounces on a smooth horizontal surface. The collision is not perfectly elastic, and the speed just after impact is less than just before impact by a factor equal to the coefficient of restitution ($0 < e < 1$). As a result, after the bounce the ball reaches a height h_1 less than h.
 a Find h_1 in terms of h_0, e, and other relevant quantities.
 b Find the height h_n reached after n bounces.
 c Find the time which elapses before the first bounce and between the first and second.
 d Find the time between the nth and $(n+1)$st bounce. From this find the *total* time for all the bounces. This is finite despite the fact that the number of bounces is infinite. Does this surprise you?

9-30 Find a potential-energy function corresponding to the molecular-force function given in Prob. 8-29. Sketch both the force and the potential energy as functions of r.

9-31 In Prob. 9-30, if the potential-energy function is defined to be zero when the force is zero, what is its value for very large separation r?

9-32 For the force field described in Prob. 9-16, show that the field is conservative by finding a potential-energy function $V(x,y)$ such that $\mathbf{F} = -\nabla V$.

9-33 A particle moves along a straight line under the action of a force which depends only on position and is given by $F(x) = 4x - 3x^2 + 8x^3$, where x is measured in meters and F in newtons.
 a Find the potential energy for this force, assuming $V = 0$ at $x = 0$.
 b Using the result of part a, find the work done when the particle moves from $x = 2$ m to $x = 4$ m.

c Compute the work in part *b* directly by integrating $F(x)$, and compare your result with that of part *b*.

9-34 In the discussion of escape velocity in Sec. 9-7, leading to Eq. (9-55), it was assumed that the missile is launched straight up. Suppose instead that its initial velocity makes an angle α with the vertical. Is the minimum velocity still given by Eq. (9-55), or is it different? *Hint:* Kepler's second law, to be discussed in detail in Sec. 12-1, states that when the position of the body is described in polar coordinates, the quantity $r^2 \, d\theta/dt$ is constant. This can be used to obtain a relation between radial and tangential components of velocity.

9-35 The potential-energy function $V(x,y)$ corresponding to a certain force field in the xy plane is given by $V = k(x^2 + y^2)^2$.
  a Find the x and y components of the force as functions of x and y.
  b Find the magnitude and direction of the force in terms of x and y.
  c Find the magnitude and direction of the force in terms of the polar coordinates (r,θ) of the point.

9-36 Find the force field corresponding to the potential-energy function $V = 4kx^2y^2$.

9-37 For a particle in a plane, show that if the potential energy V depends only on the polar coordinate r, the corresponding force field is a central-force field given by $\mathbf{F} = -\mathbf{n} \, dV/dr$.

9-38 Show that the expression $E = mc^2$ has units of energy.

9-39 The total annual consumption of electric energy in the United States is of the order of 10^{15} watt-hours. If this power were obtained by completely destroying a given quantity of matter, how much matter would have to be consumed?

9-40 For the simple pendulum of Prob. 9-18, find the speed and the tension in the string as functions of the angle of the string.

9-41 The *asteroid belt*, the region of space between the orbits of Mars and Jupiter, contains small bodies (asteroids) of various sizes in orbit around the sun. Consider a spherical asteroid of radius R and the same average density as earth. Suppose a spacecraft lands on such an asteroid and the engine to be used for takeoff refuses to function. If the asteroid is sufficiently small, a man can reach terminal velocity and escape the asteroid's gravitational field simply by jumping. Make a rough estimate of the maximum value of R for which this is possible.

Periodic Motion | 10

Many mechanical systems undergo periodic motion, characterized by repeated cycles. The simplest possible model for periodic motion is the harmonic oscillator. This model is analyzed in detail; then we show how the motions of various other vibrating systems, including diatomic molecules, can be regarded as approximately harmonic motion. The effects of damping forces and additional driving forces are considered briefly.

10-1 THE HARMONIC OSCILLATOR

The motions of many mechanical systems are characterized by repetition. The pendulum of a clock swings back and forth; a struck tuning fork vibrates in a repetitive manner; the internal motions of atoms within molecules are repetitive or vibratory. All these systems perform *cyclic* or periodic motion; the same motion is repeated over and over.

The simplest model for a mechanical system which undergoes periodic motion is the *simple harmonic oscillator*, shown in Fig. 10-1. A body of

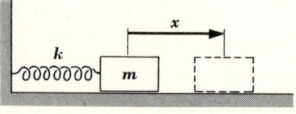

Fig. 10-1 Simple harmonic oscillator. Mass m moves on a frictionless surface in a straight line. Its displacement from its equilibrium position is described by the coordinate x.

mass m moves on a smooth, frictionless, horizontal surface. The body might be, for example, a glider on an air track, as discussed in Chap. 4. It moves in a straight line under the action of a force produced by a spring; one end of the spring is attached to the body, and the other is fixed, as shown in the figure. The displacement of the body from the unstretched position of the spring is denoted by x. We assume that the spring exerts a force directly proportional to x and in the opposite direction, whether it is stretched or compressed from its natural length. The body is given an initial displacement or an initial velocity, or both, to set it in motion.

Before proceeding with a detailed analysis of the motion, we consider some general features. The force is always opposite in direction to the displacement, i.e., in such a direction as to pull the body toward its equilibrium position ($x = 0$). If it is given an initial displacement, however, it does not stop when it returns to the equilibrium position. As it moves toward the point $x = 0$, it gains kinetic energy; thus it overshoots a certain distance and stops on the other side. One might therefore expect that, given an initial displacement, the body will undergo a repetitive or periodic motion. Since the spring action is assumed to be symmetrical, the maximum displacement on one side of equilibrium should be the same in magnitude as that on the other side.

The mechanical system represented by this model is called a *simple harmonic oscillator* (or harmonic oscillator), and its motion is called *simple harmonic motion*. This system is of considerable importance, not only because the situation occurs often in practical situations but also because in a great many situations the force on a body is *approximately* proportional to its displacement from some equilibrium position. In such a case the resulting motion can be represented approximately by the simple model just described. Thus it is useful to try to understand this model as thoroughly as possible.

Our first goal is to obtain a mathematical expression which describes the position of the body as a function of time. If we can find such an expression, we can also find the velocity and acceleration as functions of time and thus have a complete description of the motion.

When the body is displaced a distance x from its equilibrium position, the only unbalanced force acting on it is the spring force $F = -kx$; the minus sign indicates that the force and displacement are always opposite in direction. In Fig. 10-1, we have taken x to be positive when the displacement is to the right of the equilibrium position. Thus when x is positive, F is negative, meaning that the spring pulls to the left. In the figure this corresponds to stretching the spring.

According to Newton's second law, the spring force equals the mass times the acceleration. That is,

$$-kx = m\frac{d^2x}{dt^2} \qquad (10\text{-}1)$$

Both the force and the acceleration are proportional to the negative of the displacement. In fact, the proportionality of the acceleration to $-x$ is equivalent to the force's being proportional to $-x$. Rearranging Eq. (10-1) gives

$$\frac{d^2x}{dt^2} = -\frac{k}{m}x \tag{10-2}$$

To illustrate the sign relations in Eq. (10-2), we note that when the body reaches the extreme negative (left) displacement and starts back, x is negative but the acceleration is positive, corresponding to gaining positive velocity. But after it passes the origin, x becomes positive and the acceleration becomes negative, corresponding to a decreasing positive velocity. After the body reaches the extreme positive (right) displacement, the acceleration is still negative while x is positive, corresponding to a velocity which becomes more and more negative.

Now we know that x is some function of time and that if Newton's laws are to be satisfied, Eq. (10-2) must be obeyed at all times. The equation does not tell us what the function is; it is a *differential equation*, giving a relation between the function and its second derivative. This may not seem very helpful, but in fact it is enough to determine the general form the function must have, because there are very few functions having the property that the second derivative of the function is always equal to a negative constant times the function itself.

It is easy to think of functions that do *not* satisfy this requirement. An example is a power of t, say $x = At^n$, where A and n are constants. The second derivative is $n(n-1)At^{n-2}$. If we try this as a potential solution of Eq. (10-2), we obtain the relation

$$n(n-1)At^{n-2} = -\frac{k}{m}At^n$$

This equation is satisfied when $t = 0$ and when $t^2 = -mn(n-1)/k$, but it clearly cannot possibly be true for *all* values of t because the two sides increase with t at different rates. Thus the function describing the motion cannot have this form. A function that satisfied a differential equation for *all* values of the independent variable (in this case time) is called a *solution* of the differential equation, and we have just shown that $x = At^n$ is *not* a solution of Eq. (10-2)!

Finding the solutions of a differential equation often requires considerable insight and not infrequently a little shrewd guesswork. Fortunately,

245 THE HARMONIC OSCILLATOR [10-1]

it is not difficult to find solutions for Eq. (10-2). We note that $x(t)$ must be a function such that its second derivative is a negative constant times the original function. We immediately think of trigonometric functions; it is known that

$$\frac{d}{dt}(\sin \omega t) = \omega \cos \omega t \qquad \text{and} \qquad \frac{d}{dt}(\cos \omega t) = -\omega \sin \omega t$$

where ω is a constant. Combining these two, we find

$$\frac{d^2}{dt^2}(\sin \omega t) = -\omega^2 \sin \omega t \qquad \text{and} \qquad \frac{d^2}{dt^2}(\cos \omega t) = -\omega^2 \cos \omega t \qquad (10\text{-}3)$$

Both $\sin \omega t$ and $\cos \omega t$ have the property that the second derivative is a negative constant times the original function. The constant, furthermore, is the square of the number ω which multiplies t in the function. Comparing Eqs. (10-3) with Eq. (10-2), we see that they have the same form, provided the constant ω is related to k/m by

$$\omega^2 = \frac{k}{m} \qquad (10\text{-}4)$$

Making this substitution in Eqs. (10-3), we find that two possible functions which satisfy Eq. (10-2), i.e., are *solutions* of this equation, are

$$x = \sin\sqrt{\frac{k}{m}}\,t \qquad \text{and} \qquad x = \cos\sqrt{\frac{k}{m}}\,t \qquad (10\text{-}5)$$

But we can be more general than this; either of Eqs. (10-3) can be multiplied by any constant without destroying its validity. For example,

$$\frac{d^2}{dt^2}(A \sin \omega t) = -\omega^2 (A \sin \omega t) \qquad (10\text{-}6)$$

where A is any constant. Correspondingly, the functions

$$x = A \sin\sqrt{\frac{k}{m}}\,t \qquad (10\text{-}7a)$$

$$x = B \cos\sqrt{\frac{k}{m}}\,t \qquad (10\text{-}7b)$$

satisfy Eq. (10-2) for all values of t and for any values of the constants A and B. Furthermore, the *sum* of these two functions also satisfies the equation, as can be verified by substituting into Eq. (10-2).

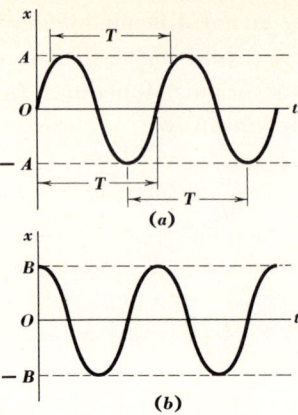

Fig. 10-2 (a) Graph of Eq. (10-7a) showing the position of a simple harmonic oscillator as a function of time. It can be seen that the displacement from equilibrium is never greater in magnitude than A. The period of the motion is the time between corresponding points on successive cycles. All time intervals T are equal. (b) Graph of Eq. (10-7b).

Graphs of Eqs. (10-7) are shown in Fig. 10-2. These are familiar sine curves; they show that the motion is, in fact, of a periodic nature, as originally predicted. Because of the direct dependence on the sine or a related function, the motion is also called *sinusoidal*.

10-2 CHARACTERISTICS OF MOTION

We now consider in more detail the meaning of Eq. (10-7a). Figure 10-2 shows the significance of the constant A: it represents the maximum displacement of the body from its equilibrium position. This is also evident from the equation, since the sine of a quantity is never greater than unity in absolute value. Thus the maximum positive displacement is $x = A$, and the maximum negative displacement is $x = -A$. The maximum displacement from equilibrium is called the *amplitude* of the motion.

The time required for one complete cycle of the motion, labeled T in Fig. 10-2, is called the *period* of the motion. This is most easily determined by considering the times at which the displacement is zero. The sine of an angle is zero when the angle (measured in radians) has the values 0, π, 2π, 3π, ..., or

$$t = 0, \quad \pi\sqrt{\frac{m}{k}}, \quad 2\pi\sqrt{\frac{m}{k}}, \quad 3\pi\sqrt{\frac{m}{k}}, \ldots \tag{10-8}$$

We see from Fig. 10-2 that T is just the time which elapses between the first zero-displacement point and the third. That is,

$$T = 2\pi\sqrt{\frac{m}{k}} \tag{10-9}$$

One might be tempted to say that one cycle of the motion consists of the motion which happens in one-half the interval T, but it is customary to define the period as the time for one *complete* cycle, including both the positive and negative halves of the motion. One can measure T using any two corresponding points in successive cycles. Two other possibilities are shown in Fig. 10-2.

Equation (10-9) is a reasonable result. The larger the mass, the more ponderous we expect the motion to be; thus the time for each cycle should increase with increasing m. Also, for a given value of m, larger values of k (corresponding to a "stiffer" spring) mean more rapid repetition of the cycle, hence smaller values of T.

The period is the time duration of one complete cycle. The reciprocal of the period, the number of cycles which occur per unit time, is called the *frequency*, denoted by f. That is, $f = 1/T$. For the harmonic oscillator,

$$f = \frac{1}{T} = \frac{1}{2\pi}\sqrt{\frac{k}{m}} \tag{10-10}$$

Frequencies are often given in cycles per second, although, strictly speaking, a number of cycles is a pure number without units, so that the units of f are really per second (s^{-1}) in the mks system. One cycle per second is also called 1 *hertz* (Hz). That is,

$$1 \text{ Hz} = 1 \text{ s}^{-1}$$

Especially in electronics, frequencies are often expressed in hertz.

The significance of the constant ω in Eq. (10-4) now emerges; it is directly proportional to the frequency

$$\omega = \sqrt{\frac{k}{m}} = 2\pi f \tag{10-11}$$

The dimensionless quantity ωt in each of Eqs. (10-7) plays the role of an angle, since only its sine or cosine appears. The quantity ω is the time rate of change of ωt; hence it is called the *angular frequency*. Each time the system undergoes a complete cycle, ωt increases by 2π.

Equations (10-9) and (10-10) carry a very important message. They tell us that the period and frequency of motion do not depend on the amplitude A. When we compare two motions with the same mass and spring but different amplitudes, we find that the displacements and velocities at corresponding times differ for the two cases but the *period* is the same for both. It depends only on m and k. For example, the frequency (pitch) of a tuning fork is very nearly independent of its amplitude of vibration.

248   PERIODIC MOTION

How do we know, in a given situation, *which* of Eqs. (10-7), if either, describes the motion and what the values of A and B should be? These depend on the conditions at the beginning of the motion. Clearly, Eq. (10-7a) gives a displacement of zero at time $t = 0$. Thus, if we start the motion by giving the body an initial velocity at its equilibrium position at time $t = 0$, then Eq. (10-7a) is the one to use. If the initial velocity v_0 is known, A can be determined. We differentiate Eq. (10-7a) to find the instantaneous velocity:

$$v = A\omega \cos \omega t \tag{10-12}$$

This equation gives the velocity as a function of time. In particular, at $t = 0$ it must give the initial velocity v_0:

$$v_0 = A\omega \cos 0 = A\omega = A\sqrt{\frac{k}{m}} \tag{10-13}$$

Equation (10-13) then determines the amplitude A if v_0 is known. Note that if v_0 is negative, A is also negative; in this case the motion starts out from the equilibrium position in the negative direction.

Equation (10-7b) describes a situation in which the *maximum* displacement occurs at time $t = 0$. This happens, for example, if we give the body an initial displacement and then release it from rest. In this case B is the initial displacement, which is also the maximum displacement for the motion.

We have just seen one of the important characteristics of solutions of differential equations. It is ordinarily possible to find *many* functions which are solutions of equations such as Eq. (10-2). From this collection of functions, we select the particular one which also satisfies the conditions at the beginning of the motion, called the *initial conditions*.

The solutions of Eq. (10-2) may be written in another form. Consider the function

$$x = C \sin (\omega t + \phi) \tag{10-14}$$

This function is a solution of Eq. (10-2) for *any* values of the constants C and ϕ, as can be verified by substituting Eq. (10-14) into Eq. (10-2). A little thought shows that the motion described by Eq. (10-14) is similar to that shown in Fig. 10-2a except that the zero point on the time scale is shifted. For example, if $\phi = \pi/2$, then at $t = 0$, $x = C \sin (0 + \pi/2) = C$. Thus this motion has a head start of a quarter cycle with respect to the motion described by the same function but with $\phi = 0$. This means that two different motions described by Eq. (10-14) with two different values of ϕ but the same amplitude C differ only in that they are out of step, or out of phase, by an amount which depends on the difference between the two values of ϕ. The constant ϕ is called the *phase angle* or simply the *phase* of the motion.

It is easy to show that Eq. (10-14) is always equivalent to some combination of Eqs. (10-7). That is,

$$A \sin \omega t + B \cos \omega t = C \sin (\omega t + \phi) \qquad (10\text{-}15)$$

provided the constants A, B, C, and ϕ are related as follows:

$$C = (A^2 + B^2)^{1/2} \qquad\qquad A = C \cos \phi \qquad (10\text{-}16a)$$
$$\text{or}$$
$$\phi = \tan^{-1} \frac{B}{A} \qquad\qquad B = C \sin \phi \qquad (10\text{-}16b)$$

This is easily verified by expanding the right side of Eq. (10-15) using the formula for the sine of a sum; the details are left as a problem.

The essential point is that either form of Eq. (10-15) represents the most general solution Eq. (10-2). Every function which is a solution of this equation can be represented in either of these forms, with appropriate values of A and B, or C and ϕ.

Example

A harmonic oscillator is given an initial displacement x_0 and an initial velocity v_0 at time $t = 0$. Describe the subsequent motion in each of the forms of Eq. (10-15).

Solution

We consider first the form $A \sin \omega t + B \cos \omega t$. We know ω if m and k are given; the problem is to find A and B, which depend on x_0 and v_0. At time $t = 0$, we have

$$x_0 = A \sin 0 + B \cos 0 = B$$

Thus the constant B is simply equal to the initial displacement, just as for the case without initial velocity. The velocity at any time t is given by

$$v = \frac{dx}{dt} = A\omega \cos \omega t - B\omega \sin \omega t$$

At time $t = 0$ this becomes

$$v_0 = A\omega \cos 0 - B\omega \sin \phi = A\omega$$

which gives

$$A = \frac{v_0}{\omega} \qquad B = x_0 \qquad (10\text{-}17)$$

Using these results with Eqs. (10-16), we obtain

$$C = \left(x_0^2 + \frac{v_0^2}{\omega^2}\right)^{1/2} \qquad \phi = \tan^{-1}\frac{x_0\omega}{v_0} \qquad (10\text{-}18)$$

If x_0 happens to be zero, $B = 0$, $C = A = v_0/\omega$, and $\phi = 0$, in agreement with Eq. (10-13). If x_0 is different from zero but $v_0 = 0$, we find $A = 0$, $B = x_0$, and $x = x_0 \cos \omega t$. This result is also obtained from the other form, since we have $C = x_0$ and $\phi = \pi/2$, so $x = x_0 \sin(\omega t + \pi/2) = x_0 \cos \omega t$.

It is well to review the procedure for obtaining functions $x(t)$ describing harmonic oscillator motion. Application of $F = ma$ to the idealized model does not lead directly to the function $x(t)$ but to a *differential equation* which the function must satisfy. It is found that because of the presence of arbitrary constants, there are many such functions. The constants must be chosen to fit the specified values of initial position and velocity. There is only one function which satisfies all these requirements, so it must be the correct description of the motion. Obtaining the differential equation (10-2) is thus only the first of several steps. Yet in a physical sense this is the most important step, since it is here that physical principles are used. The remainder of the calculation uses mathematical techniques but no new physical principles. This pattern is very common in physics; often application of a physical principle leads not to a description of a phenomenon but to a differential equation whose *solutions* are the desired description.

10-3 ENERGY RELATIONS

It is illuminating to apply the principle of conservation of energy to the harmonic oscillator. The force exerted by the spring is a conservative force, as discussed in Chap. 9. The corresponding potential energy, also obtained in Chap. 9, is $V = \tfrac{1}{2}kx^2$ if we assume that $V = 0$ when $x = 0$. Denoting the (constant) total energy by H as usual, we have

$$E + V = \tfrac{1}{2}mv^2 + \tfrac{1}{2}kx^2 = H \qquad (10\text{-}19)$$

The total energy H is simply related to the other properties of the motion. Suppose the motion is described by $x = A \sin \omega t$. At either point at which the body reaches its maximum displacement, that is, $x = \pm A$, we have $E = 0$ and $V = \frac{1}{2}kA^2$. Thus

$$H = \frac{1}{2}kA^2 \tag{10-20}$$

Correspondingly, we may find the maximum velocity, which clearly occurs at the point where the potential energy is minimum, namely at $x = 0$. At this point,

$$H = \frac{1}{2}mv_{max}^2 \tag{10-21}$$

Combining this with Eq. (10-20), we find

$$v_{max} = A\sqrt{\frac{k}{m}} = A\omega \tag{10-22}$$

a result which could also have been obtained from Eq. (10-12).

Equation (10-19) also provides an alternative approach to the problem of finding x as a function of t. Combining Eqs. (10-19) and (10-20) and writing dx/dt in place of v gives

$$\frac{1}{2}m\left(\frac{dx}{dt}\right)^2 + \frac{1}{2}kx^2 = \frac{1}{2}kA^2 \tag{10-23}$$

Solving this for dx/dt, we have

$$\frac{dx}{dt} = \sqrt{\frac{k}{m}}\sqrt{A^2 - x^2} = \omega\sqrt{A^2 - x^2} \tag{10-24}$$

This is again a differential equation; it can be solved by a technique known as *separation of variables*. We rearrange the equation so that only x and its differential appear on one side of the equation and only dt appears on the other side. Then we can integrate both sides directly as follows:

$$\int \frac{dx}{\sqrt{A^2 - x^2}} = \omega \int dt \tag{10-25}$$

Performing the indefinite integration and adding a constant of integration ϕ gives

$$\sin^{-1}\frac{x}{A} = \omega t + \phi$$

which can be rewritten as

$$x = A\sin(\omega t + \phi) \tag{10-26}$$

in agreement with our previous result.

252 PERIODIC MOTION

Here, though, one of the deficiencies of the energy approach becomes apparent. Knowing the total energy H of the system is not sufficient to predict all the details of the motion. We can find the amplitude A of the motion from Eq. (10-20), but the phase ϕ cannot be determined unless we know in addition where the body is at some particular time. For a given system, different motions having the same amplitude but different phases have the same total energy.

10-4 SIMPLE PENDULUM

The simple pendulum is a system which, strictly speaking, does not undergo exactly simple harmonic motion. However, as we shall see, its motion is *approximately* simple harmonic if the displacement from equilibrium is sufficiently small.

As a simplified model of the pendulum we consider a point mass m on the lower end of a string of length l the upper end of which is held rigidly fixed in an inertial frame of reference, and the system is placed in a uniform gravitational field **g**. In the equilibrium position, the mass hangs straight down below the point of suspension. When the mass is displaced slightly from this equilibrium position and released, periodic motion results. We now proceed to analyze this motion, assuming that the motion is confined to a vertical plane.

We describe the displacement of the pendulum from its equilibrium position by the angle θ, as in Fig. 10-3. The corresponding arc length,

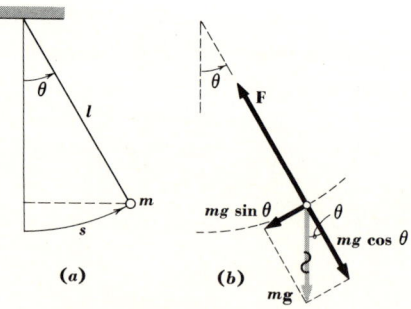

Fig. 10-3 (*a*) Simple pendulum of mass m and length l. The angle θ and the arc length s are shown. (*b*) Free-body diagram showing the forces on mass m. **F** is the force exerted by the string tension; the weight $m\mathbf{g}$ is represented in terms of its components parallel and perpendicular to the string.

denoted by s in the figure, is given by $s = \theta l$, where θ is measured in radians. Also shown in Fig. 10-3 is a free-body diagram for the mass m. The forces acting on m are the tension **F** in the string and the force of gravity $m\mathbf{g}$, which is represented in the figure in terms of components along the string

SIMPLE PENDULUM [10-4]

and perpendicular to it (tangent to the path). The component tangent to the path is $-mg \sin \theta$, negative because it is opposite in direction to the displacement along the path.

We now write an equation for the component of acceleration along the path, d^2s/dt^2. From Newton's second law we have

$$-mg \sin \theta = m \frac{d^2s}{dt^2} \tag{10-27}$$

It is very significant that m can be divided out of this equation. This tells us that the motion does not depend on the value of m; it depends only on the initial conditions, the length l of the pendulum, and the acceleration of gravity. Dividing by m and using the relation $s = \theta l$, we find

$$\frac{d^2s}{dt^2} = -g \sin \frac{s}{l} \tag{10-28}$$

As before, we have obtained a differential equation relating the second time derivative of s to s itself. We now seek a function $s(t)$ which satisfies Eq. (10-28).

We are not so lucky this time. There are no simple functions which satisfy Eq. (10-28); the solutions are a class of functions called *elliptic functions*. But all is not lost; we recall that for a very small angle, the sine of the angle is very nearly equal to the angle itself (measured in radians, as always). For example, if $\theta = 0.1$ (equal to 5.73°), then $\sin \theta = 0.09983$, a difference of less than 0.2 percent. Even for $\theta = 0.5$ (about 29°), $\sin \theta = 0.47943$, a difference of only about 4 percent. Thus, for displacements sufficiently small for s/l to be much smaller than unity, we can replace $\sin (s/l)$ with the quantity s/l. The assumption that the displacement is small then becomes part of the simplified *model* for this problem.

Making this approximation, we find

$$\frac{d^2s}{dt^2} = -\frac{g}{l} s \tag{10-29}$$

Lo and behold! Here is the same equation we obtained for the harmonic oscillator, Eq. (10-2)! The symbols are different but the *form* of the equation is precisely the same. Thus we can immediately write down the solutions of this equation:

$$s = A \sin \sqrt{\frac{g}{l}} t + B \cos \sqrt{\frac{g}{l}} t = C \sin \left(\sqrt{\frac{g}{l}} t + \phi \right) \tag{10-30}$$

We can also find the period and frequency by comparison with the harmonic-oscillator discussion:

$$\omega = \sqrt{\frac{g}{l}} \qquad T = \frac{1}{f} = 2\pi \sqrt{\frac{l}{g}} \qquad (10\text{-}31)$$

Provided that the amplitude of the oscillations is sufficiently small for our approximation to be reasonably good, the simple pendulum behaves like a simple harmonic oscillator.

When the oscillations are made so large that this approximation is no longer accurate, deviations from simple harmonic behavior occur. The restoring force is actually proportional to sin (s/l) rather than s/l itself, and at sufficiently large amplitudes this becomes appreciably *smaller* than the value given by the simplified model. With smaller restoring force, the period *increases* somewhat with increasing amplitude. When the maximum displacement is 90°, the period is about 18 percent greater than that given by Eq. (10-31) for small displacements. This is different from the simple harmonic oscillator, whose period is independent of amplitude.

10-5 MOLECULAR VIBRATIONS

In Sec. 10-1 we remarked that there is a close relation between the harmonic oscillator and the vibrations which take place within molecules. It may not be immediately obvious what the relation is, since a molecule contains at least two different masses in motion, whereas the simple harmonic oscillator has only one. To understand the connection, we construct a simple model of a diatomic molecule. We represent each atom as a point mass; the force of interaction between the atoms is taken as proportional to their displacement from their equilibrium separation. For simplicity, we assume that the atoms move only along the line joining them, so that all motion is confined to one line.

This straight-line motion is not the only kind of motion which may occur, of course. Although the atoms in an isolated molecule experience no forces except the interaction force, the molecule as a whole may move in any direction with uniform velocity. The elimination of this motion from consideration is not a very serious restriction because it corresponds to viewing the system in an inertial frame of reference which moves with the molecule. The molecule may also undergo a *rotational* motion, in which the direction of the line joining the two atoms changes. When this motion is included, the analysis of the system becomes a little more complicated. However, we can discuss the general features of the motion by considering the simple cases in which no rotational motion occurs.

The model to be used is shown in Fig. 10-4. The distances x_1 and x_2

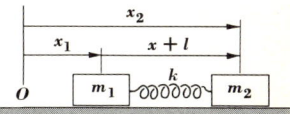

Fig. 10-4 Simple model of a diatomic molecule. The masses of the atoms are m_1 and m_2. They are connected by a spring with natural length l which exerts a force of magnitude proportional to the displacement x from its natural length. The positions of the masses with respect to the origin of an inertial frame of reference are described by the coordinates x_1 and x_2.

are the positions of masses m_1 and m_2 with respect to the origin of an inertial frame of reference. Let the equilibrium length of the spring be l, and let the displacement from this equilibrium length be x. We define x to be positive when the spring is stretched and negative when compressed. Then the total length of the spring is $l + x$, as shown. When $x = 0$, the spring exerts no forces. From the figure, we see that

$$x_2 - x_1 = x + l \tag{10-32}$$

We know that Newton's laws of motion must be formulated in terms of motion with respect to an inertial frame of reference, i.e., using the coordinates x_1 and x_2. But in describing vibrational motion, the significant thing is the *relative* motion of m_1 and m_2, described by the displacement x. Thus the proper way to proceed is to apply Newton's second law to each mass, using the derivatives of x_1 and x_2 to describe the acceleration, and then to try to combine the resulting equations in such a way that only the relative coordinate x appears in the final result. This, as we shall see, is fairly easy to do.

In our idealized model, the spring force has magnitude kx. Furthermore, when x is positive, the spring pulls to the right (the positive direction) on mass m_1 and to the left (the negative direction) on m_2. Thus, the force on mass m_1 is kx, but that on mass m_2 is $-kx$. The equations of motion, from Newton's second law, are

$$kx = m_1 \frac{d^2 x_1}{dt^2} \qquad -kx = m_2 \frac{d^2 x_2}{dt^2} \tag{10-33}$$

When Eq. (10-32) is differentiated twice, we obtain

$$\frac{d^2 x_2}{dt^2} - \frac{d^2 x_1}{dt^2} = \frac{d^2}{dt^2}(x_2 - x_1) = \frac{d^2 x}{dt^2}$$

since l is constant. This result suggests that we combine Eqs. (10-33) to obtain this form. We multiply the first equation by m_2, multiply the second by m_1, and then subtract the first from the second. The result is

$$m_1 m_2 \left(\frac{d^2 x_2}{dt^2} - \frac{d^2 x_1}{dt^2} \right) = -k(m_1 + m_2)x$$

or

$$\frac{d^2 x}{dt^2} = -k \frac{m_1 + m_2}{m_1 m_2} x \qquad (10\text{-}34)$$

When we introduce the abbreviation

$$\mu = \frac{m_1 m_2}{m_1 + m_2} \qquad (10\text{-}35)$$

Eq. (10-34) can be rewritten

$$\frac{d^2 x}{dt^2} = -\frac{k}{\mu} x \qquad (10\text{-}36)$$

We have now accomplished our original goal of obtaining an equation of motion containing only the relative coordinate x, and the simplicity of the final result is gratifying. Equation (10-36) has exactly the same form as the original harmonic-oscillator equation (10-2), which tells us that the *relative* motion of the two masses is just the same as the motion of a simple harmonic oscillator with spring constant k and mass μ [compare Eqs. (10-36) and (10-2)].

The frequency of vibration of the two-body oscillator is exactly the same as the frequency one would obtain by connecting one end of the same spring to a mass μ and holding the other end fixed in an inertial frame of reference. The solutions of the differential equation, which have already been obtained, have the form

$$x = C \sin\left(\sqrt{\frac{k}{\mu}} t + \phi \right) \qquad f = \frac{1}{2\pi} \sqrt{\frac{k}{\mu}} \qquad (10\text{-}37)$$

The quantity μ, which has units of mass, is called the *reduced mass* of the system and is always smaller than either of the two masses m_1 and m_2; the proof of this statement is left as a problem. The relative motion of the masses in the two-body oscillator is completely equivalent to that of a simple harmonic oscillator with the same spring but with the reduced mass μ. In the actual two-body oscillator, of course, both masses have a back-and-forth motion toward and away from each other.

In the model discussed above, the interaction force is proportional to the displacement x from equilibrium; in an actual molecule the interatomic force behaves in a considerably more complicated manner. Figure 10-5 shows

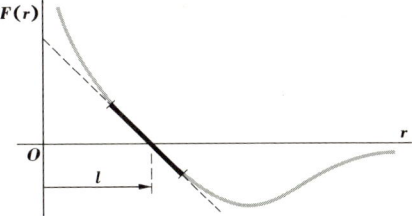

Fig. 10-5 Variation of interatomic force $F(r)$ with the distance r between atoms for a typical diatomic molecule. The curve crosses the r axis at $r = l$; the force is zero at this point. A small section of the curve in the vicinity of this point can be represented approximately as a straight line.

a typical interatomic force as a function of the *total* distance r between the atoms. Following the previous notation, we denote by l the equilibrium distance between atoms at which F is zero; that is, $F(l) = 0$.

If the force *were* proportional to the displacement from equilibrium, the curve in Fig. 10-5 would be a straight line crossing the r axis at $r = l$. This is not the case, but if we consider only small displacements from $r = l$, we can *approximate* a small section of this curve near $r = l$ with a straight line, as shown in the figure. From the slope of this line, we can find an equivalent spring constant to use in the model already discussed. There the force on mass m_2 was $F = -kx$, and the slope of the curve representing F as a function of x was simply $-k$. That is,

$$\frac{dF}{dx} = -k$$

Correspondingly, the slope of the straight line tangent to the curve of Fig. 10-5 at $r = l$ is dF/dr evaluated at that point. So the equivalent force constant for this situation is

$$k = -\left(\frac{dF}{dr}\right)_{r=l} \tag{10-38}$$

in which the right-hand side contains a shorthand notation for the derivative of F with respect to r evaluated at the point $r = l$. To summarize: When the two atoms in a diatomic molecule are displaced slightly from their equilibrium separation l, they vibrate as though they were connected by a spring with unstressed length l and force constant k given by Eq. (10-38). The displacement x from equilibrium and the total separation r are related by $x = r - l$.

Example
A certain diatomic molecule has two atoms with equal mass m. The interatomic force is represented by the function

$$F(r) = -\frac{a}{r^2} + \frac{b}{r^3}$$

where a and b are positive constants. Find the equilibrium distance between atoms and the frequency of small oscillations.

Solution
At the equilibrium distance the force is zero. Setting $F(r) = 0$, we find $r = b/a$. The equivalent force constant is obtained by taking dF/dr and evaluating at $r = b/a$. We find

$$\frac{dF}{dr} = \frac{2a}{r^3} - \frac{3b}{r^4}$$

$$k = -\left(\frac{dF}{dr}\right)_{r=b/a} = \frac{a^4}{b^3}$$

From Eq. (10-35), the reduced mass is $\mu = m/2$. Finally, from Eq. (10-37), the frequency is

$$\omega = \sqrt{\frac{k}{\mu}} = \sqrt{\frac{2a^4}{b^3 m}}$$

Example
An oxygen molecule is observed to vibrate with a frequency of 10^{13} s^{-1}. What is the effective force constant?

Solution
The mass of an oxygen atom is about 2.7×10^{-26} kg, so the reduced mass is about 1.3×10^{-26} kg. Solving the second of Eqs. (10-37) for k, we find

$$k = 4\pi^2 f^2 \mu \simeq 50 \text{ N/m}$$

This is the same order of magnitude as for an ordinary door spring! Of course, the forces and displacements in molecular vibrations are smaller than those of door springs by a factor of the order of 10^{10}.

The energy viewpoint provides additional insight into the vibrations of diatomic molecules. Figure 10-6 shows the potential-energy function corre-

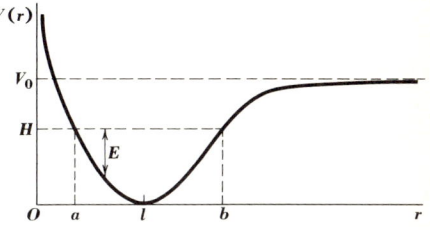

Fig. 10-6 Potential energy V as a function of r corresponding to the force shown in Fig. 10-5. $V(r)$ has a minimum at $r = l$; the potential energy has been defined to be zero at this point. At very large values of r the potential energy approaches V_0. For the total energy H shown, the motion is confined between $r = a$ and $r = b$. The kinetic energy E for any value of r can be found as shown. This potential-energy curve is characteristic of molecules which dissociate in an endothermic reaction and of energy release in nuclear fusion.

sponding to the conservative force shown in Fig. 10-5. The force represented by these graphs has characteristics similar to those of the force shown in Fig. 9-12, but in the present case we have arbitrarily defined the potential energy to be zero at the equilibrium position ($r = l$) rather than at very large r. The value of V at large r is here denoted as V_0. Thus we have $V(l) = 0$ and $V(\infty) = V_0$. The total energy H can be represented on this same graph. Since we are dealing with a conservative system, H is constant and is therefore represented as a straight horizontal line.

Since $H = E + V$ at every point, the *kinetic energy* E of the system for a given r is represented simply by the *difference* between the height of the V curve and that of the H curve at that r. The kinetic energy is maximum at $r = l$, when V is minimum, so the relative *speed* of the atoms is greatest here. At points a and b, where the V curve crosses the H curve, the kinetic energy must be zero. If the system is moving in the direction of increasing r, the speeds of the atoms decrease after passing $r = l$ and become zero at $r = b$. At this point, forces act in the negative r direction to pull the atoms back toward smaller values of r. They continue with increasing speed until the point $r = l$ is reached, when the speeds begin to decrease again. Finally, they stop at $r = a$ and reverse direction once more. The kinetic energy can never be negative, so the system can never be in a region in which V is greater than the total energy H. Thus, for the value of H shown in Fig. 10-6, the motion is confined between the points $r = a$ and $r = b$.

When H is increased, the system moves over a greater range of values of r. If enough energy is added to make H equal to or greater than V_0, r can become very great. In other words, the two atoms in the molecule can move indefinitely far apart. This is what happens when a molecule dissociates. When thermal energy is added to the system, the total energy of molecular vibration increases until some of the molecules acquire vibrational energy equal to or greater than V_0, at which point they dissociate. In order to dissociate, a molecule must gain an amount of energy V_0 above its minimum energy. The substance absorbs heat as it dissociates; this is described in chemical language as an *endothermic*, or heat-absorbing, reaction.

The same picture provides a qualitative understanding of the phenomenon of boiling. In a liquid, the molecules are close enough for strong attractive forces to act between them. As the temperature increases, more and more molecules acquire enough energy to break away from the binding forces and become nearly free, as in the vapor state. A similar phenomenon occurs when a solid melts.

A different potential-energy curve is shown in Fig. 10-7, along with the

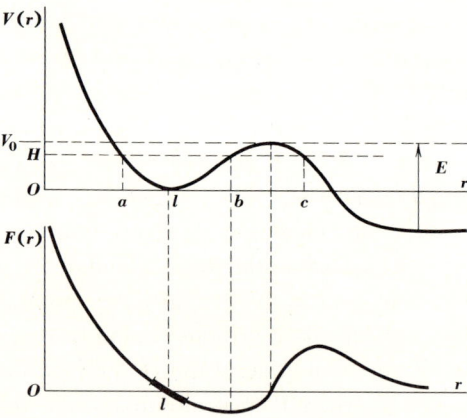

Fig. 10-7 Potential-energy curve in which the value of V at very large values of r is less than at the equilibrium position $r = l$, where it is zero. The corresponding force is shown. Such a potential-energy curve is characteristic of a diatomic molecule which dissociates exothermically and of nuclear fission.

corresponding force. As two atoms approach each other from large values of r, the force is at first repulsive; then at smaller r it becomes attractive; then repulsive again. At $r = l$, there is a possible state of equilibrium. If the total energy is as shown by the horizontal line H in the diagram, the motion may be confined between $r = a$ and $r = b$. If the total energy is increased to V_0, the two atoms are no longer bound and may separate by an indefinitely large distance. But here the kinetic energy E of the system after the atoms are very far apart is *greater* than that which they absorbed in the process of being dissociated. Thus, in this case, the dissociation is one which releases energy, in chemical language an *exothermic*, or energy-producing, reaction.

The same analysis can be applied to *nuclear* reactions. Nuclear fission occurs when a heavy nucleus such as uranium breaks into two smaller pieces, called *fission fragments*, plus a few single neutrons. We can now interpret Fig. 10-7 as representing the interaction between these two fragments. Some energy has to be added to a nucleus initially in order to produce the fission, which requires surmounting the potential-energy barrier between $r = b$ and $r = c$. In an ordinary fission reactor, this is accomplished by the absorption of a neutron by a uranium nucleus. Once the energy has been gained, fission occurs, and the resulting fragments have *greater* kinetic energy than the energy absorbed from the incident particle. Thus the fission process results in the liberation of kinetic energy.

In atomic and nuclear systems it is also possible for the system to go from point a or b to another point c, at which the potential energy is the same, *without* having energy added to get "over the hill." In a mechanical sense, this corresponds to the system having negative kinetic energy during its travel from point b to point c. According to newtonian mechanics this is impossible; understanding this phenomenon requires quantum-mechanical concepts. The central point is that the position is not really well defined, so that it is possible, in a quantum-mechanical sense, for the transition from point b to point c to occur without the addition of extra energy. This effect is sometimes referred to as *tunneling*; it is responsible for the phenomenon of *spontaneous fission* in nuclei and also for the behavior of tunnel diodes.

In nuclear *fusion*, kinetic energy is gained when two simple nuclei come together to form a larger one. In this case, the potential-energy curve must have the general shape shown in Fig. 10-6, so when two nuclei approach each other from great distances, their total kinetic energy in the bound state is greater than in the separated state. Such is the case when two deuterium nuclei (each containing a proton and a neutron) combine to form one helium nucleus, with the release of kinetic energy.

In general, precise calculations of the vibrations which occur under the action of forces such as we have discussed are somewhat complex. The problem becomes very much simpler if the vibrations are small enough for the behavior of the force near equilibrium to be represented as a straight line, as in Figs. 10-5 and 10-7. This corresponds to assuming that $V(r)$ near the equilibrium point is approximately $V = \frac{1}{2}kx^2$ if $V = 0$ when $x = 0$ ($r = l$).

Recalling from Sec. 9-6 that the force is related to the potential energy by $F = -dV/dr$, and combining this with Eq. (10-38), we find

$$k = \left(\frac{d^2V}{dr^2}\right)_{r=l} \tag{10-39}$$

which gives the effective force constant for small oscillations in terms of the potential energy. Correspondingly, the potential energy at points near $r = l$ is expressed as

$$V = \tfrac{1}{2}kx^2 = \frac{1}{2}\left(\frac{d^2V}{dr^2}\right)_{r=l}(r-l)^2 \tag{10-40}$$

The reader familiar with Taylor series expansions will recognize this as the third term in the Taylor expansion of V about the point $r = l$. The first two terms in this expansion are zero, the first because of the way we define the potential energy, and the second because $-dV/dr = F$ and F is zero at $r = l$.

10-6 DAMPED OSCILLATIONS

Up to now, in our discussion of oscillations, we have assumed that the forces are conservative, so the total energy of the oscillating system is constant. The neglect of friction and other nonconservative forces is an idealization leading to a simple model which can be analyzed easily. But we have known all along that this model does not agree completely with observed oscillatory motions. Because of the nonconservative forces which are always present, the amplitude of a pendulum initially set into motion gradually diminishes. A car may bounce when going over a bump, but the oscillations quickly die away after the bump is passed. When a voltmeter is connected to a battery, the needle of the meter may oscillate a few times around the final position, but it eventually comes to rest at a constant value.

In all these cases, an oscillation dies out with time. Thus it is reasonable to try to extend the original model, the harmonic oscillator, to include in an approximate way the nonconservative forces which lead to this dying out of oscillations, usually called *damping* or *damped oscillations*.

In the model illustrated in Fig. 10-1, suppose that instead of being frictionless the surface exerts a frictional force $\mathcal{F}$ on the mass that is constant in magnitude (independent of speed) but in a direction always opposing the motion. In this case, the mass constantly does work against the nonconservative force of friction and thereby *decreases* the total mechanical energy of the system, which eventually approaches zero as the system comes to rest.

It is easy to calculate how the oscillations are damped with time. If at the end point of one oscillation the mass is a distance A away from the equilibrium position, then the next stopping point on the opposite side is at a distance A' somewhat smaller than A, since the potential energy is somewhat smaller. The total distance moved between these points is $A + A'$; the *difference* between the potential energies at the two points is equal to the amount of work done against friction, $A + A'$. That is,

$$\tfrac{1}{2}kA^2 - \tfrac{1}{2}kA'^2 = \mathfrak{F}(A + A') \tag{10-41}$$

which represents the amount of mechanical energy lost to friction in a half cycle.

Equation (10-41) can be solved for A' in terms of the other quantities. The equation is quadratic; choosing the positive root gives

$$A' = A - 2\frac{\mathfrak{F}}{k} \tag{10-42}$$

Successive maximum displacements from equilibrium become smaller by uniform amounts. Thus the motion appears qualitatively similar to that shown in Fig. 10-8. To calculate the position of the mass in detail as a function

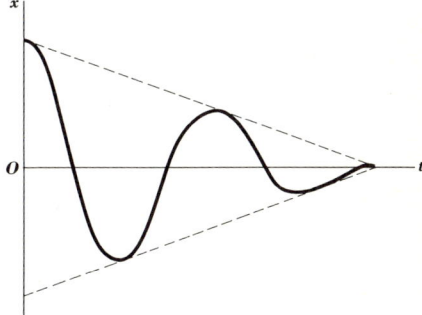

Fig. 10-8 Motion of a simple harmonic oscillator with a frictional damping force which is constant in magnitude. Successive maximum displacements decrease uniformly until the oscillator comes to rest.

of time is rather complicated, but the general nature of the motion is apparent from the preceding discussion. After a number of oscillations, depending on the initial amplitude and the magnitude of the force of friction $\mathfrak{F}$, the oscillator eventually comes to rest, not necessarily at $x = 0$.

The damping force discussed above is characteristic of dry friction between two surfaces. Another kind of damping-force behavior of greater interest in many practical situations is a force whose magnitude is proportional to the *speed* of the mass. This behavior is characteristic of some *viscous* friction, such as that which occurs in a liquid-filled automobile shock absorber. The forces which damp out the oscillatory motion of a voltmeter or a galvanometer are also approximately proportional to the velocity of the coil or needle. When a vibrating molecule loses energy by emitting electromagnetic radiation, the effect on the molecule may be described in terms of a *radiation damping force*, which is approximately proportional to the velocities of the atoms in the vibrating molecule.

Such a damping force can be included in the equation of motion by simply adding a term to Eq. (10-1). The additional damping force is represented as $-R\,dx/dt$, where R is a proportionality constant and the minus sign indicates that the force is opposite in direction to the velocity dx/dt. Then Newton's second law for this situation takes the form

$$-kx - R\frac{dx}{dt} = m\frac{d^2x}{dt^2} \tag{10-43}$$

which differs from Eq. (10-1) in having an extra force term, corresponding to the damping force, on the left-hand side.

The effect of this additional force on the behavior of the system can be explored in more detail. Two approaches are possible, one an approximate analysis using energy considerations, the other using the general solution of Eq. (10-43). The energy approach is simpler, and we consider it first.

Because of the damping force, successive displacements from equilibrium are smaller and smaller. If the maximum displacement (amplitude) for a given cycle is A, how much energy does the system *lose* during that cycle? The instantaneous *rate* of energy loss dH/dt is the rate of doing work against the damping force, which is the magnitude of the force Rv times the velocity v, or Rv^2. This quantity varies during the cycle, but the *average* rate of energy loss is given by

$$\left(\frac{dH}{dt}\right)_{av} = -(Rv^2)_{av} = -R(v^2)_{av} \tag{10-44}$$

To find the average value of v^2, we note that the average kinetic energy $(\tfrac{1}{2}mv^2)_{av}$ and the average potential energy $(\tfrac{1}{2}kx^2)_{av}$ for a harmonic oscillator are equal; each varies sinusoidally with maximum value equal to the total energy H, but the variations are a quarter cycle out of phase. Thus the average value of $\tfrac{1}{2}mv^2$ must equal half the total energy, $H/2$, and we find

$$(v^2)_{av} = \frac{H}{m} \tag{10-45}$$

DAMPED OSCILLATIONS [10-6]

Combining Eqs. (10-44) and (10-45), we obtain

$$\left(\frac{dH}{dt}\right)_{av} = -\frac{R}{m}H \tag{10-46}$$

Now H is not constant but decreases continuously; neither is dH/dt constant, since it is greatest when v is greatest and zero when v is zero. If we ignore these variations and consider how the total energy decreases *on the average*, we can treat Eq. (10-46) as a differential equation for H:

$$\frac{dH}{dt} = -\frac{R}{m}H \tag{10-47}$$

The solution of this equation, which can be verified by substitution, is

$$H = H_0 e^{-(R/m)t} \tag{10-48}$$

where H_0 is the energy at time $t = 0$. Thus the total energy decreases exponentially with time.

We also note that the amplitude A is related to the total energy H by $H = \frac{1}{2}kA^2$. Letting A_0 be the amplitude at time $t = 0$ and substituting in Eq. (10-48), we find

$$A = A_0 e^{-(R/2m)t} \tag{10-49}$$

This shows that with a damping force proportional to velocity, the amplitude decreases *exponentially* with time; this differs from the case of constant friction, in which the amplitude decreased *linearly* with successive numbers of cycles.

Thus an approximate description of the damped oscillation with the viscous damping force, using the decreasing-amplitude concept just developed, is

$$\begin{aligned} x &= A \cos(\omega t + \phi) \\ &= A_0 e^{-(R/2m)t} \cos(\omega t + \phi) \end{aligned} \tag{10-50}$$

A graph of Eq. (10-50) is given in Fig. 10-9. The dashed lines show the product of the constant factor and the exponential function; the motion always oscillates within these two lines, called the *envelope* of the actual curve $x(t)$. The most striking difference between this picture and that in Fig. 10-8 for the constant friction force is that as the amplitude becomes smaller, the oscillations die out more gradually. The exponential envelope curve approaches the t axis asymptotically; the amplitude of oscillation becomes smaller and smaller but never reaches exactly zero because as the amplitude becomes smaller, the velocities also become smaller, so that the damping force has

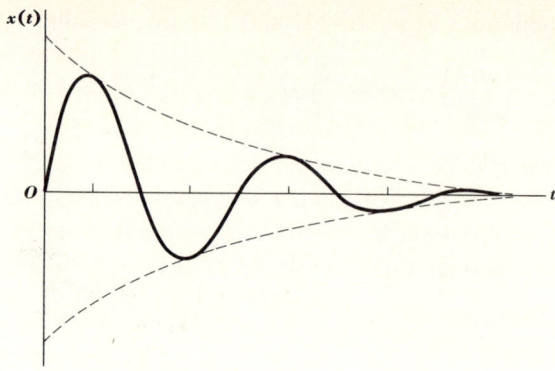

Fig. 10-9 Graph of the function in Eq. (10-50). The exponential envelope, given by the first two factors in this expression, is shown by broken lines. The actual motion, shown by the solid line, oscillates between them. In this example, $\phi = -\pi/2$.

proportionally less effect. This is to be contrasted with the straight-line envelope of Fig. 10-8, corresponding to friction which is independent of amplitude.

Because of the approximations involved in the development of Eq. (10-50), we should not expect it to be a precise description of the motion. That is, this equation is not necessarily an exact solution of the differential equation (10-43). Surprisingly, it turns out that Eq. (10-50) *is* an exact solution, provided we reinterpret the meaning of ω. Instead of using the expression for the undamped oscillator, $\omega = (k/m)^{1/2}$, we must use the generalized expression

$$\omega = \left(\frac{k}{m} - \frac{R^2}{4m^2}\right)^{1/2} \tag{10-51}$$

When this expression for ω is used in Eq. (10-50), it is an *exact* solution of Eq. (10-43); verification of this statement is left as a problem.

It can be seen from Eqs. (10-50) and (10-51) that when the damping force R is small, the exponential function decays very slowly, and the oscillation is nearly that of an undamped oscillator, with ω approximately equal to $(k/m)^{1/2}$. As the damping force becomes larger, the frequency becomes smaller, and the exponential envelope decreases more rapidly. Finally, a point is reached at which the oscillations cease entirely, and the mass returns smoothly to its equilibrium position without oscillating. Then, if the damping force is increased still further, the only additional effect is that the mass returns still more slowly to its equilibrium position. The condition under which the behavior makes the transition from oscillatory to nonoscillatory motion is known as *critical damping*. This occurs when the radical expression $(k/m - R^2/4m^2)^{1/2}$ becomes zero. If the damping force is less than that required for critical damping, the system is said to be *underdamped*; if greater, *overdamped*.

The relation of energy considerations to the differential equation obtained from $F = ma$ can be seen more directly by multiplying Eq. (10-43) by $dx/dt = v$ and rearranging:

$$m\frac{dx}{dt}\frac{d^2x}{dt^2} + kx\frac{dx}{dt} = -R\left(\frac{dx}{dt}\right)^2 \qquad (10\text{-}52)$$

The first term on the left side can be rewritten as

$$mv\frac{dv}{dt} = \frac{d}{dt}(\tfrac{1}{2}mv^2)$$

which is simply the rate of change of kinetic energy. Similarly, the second term on the left side of Eq. (10-52) is the rate of change of potential energy. Thus, Eq. (10-52) can be written

$$\frac{dE}{dt} + \frac{dV}{dt} = \frac{d}{dt}(E + V) = -R\left(\frac{dx}{dt}\right)^2 \qquad (10\text{-}53)$$

The total mechanical energy $E + V$ of the system decreases at a rate given by the right-hand side of Eq. (10-53), which is the force $R\,dx/dt$ times the velocity dx/dt, or the rate at which work is done against the force.

10-7 FORCED OSCILLATIONS

Another phenomenon of interest in periodic motion is the response of a system when a *periodic force* is applied in addition to the restoring and damping forces. One can easily think of situations in which an external periodic force is applied to a system. A child on a swing executes some kind of periodic motion; we can increase the amplitude of this motion by pushing on the swing with a definite frequency. The connecting rods on an automobile engine exert periodic forces on the crankshaft; the crankshaft, being somewhat elastic, is distorted by these forces. It is of interest to know something about the vibratory motion which takes place as a result of these periodic forces. Electrons may be subjected to periodic forces as a result of electromagnetic radiation striking an atom. For some purposes, the electrons may be regarded as elastically bound; in such a case, what motion do they undergo as a result of this periodic force? The answer to this question is the first step in a calculation of the index of refraction of a material for light of a known frequency by use of a microscopic model of the structure of the material.

It is a matter of common experience that the response of a vibrating system to a periodic force depends on the frequency of the force and on its relation to the natural frequency of the system. To build up the amplitude of a child on a swing, we must push with exactly the frequency of the swing; otherwise the oscillations do not increase.

In some cases, the response of an oscillator to a periodic force may be computed fairly simply. We consider here only the simplest possible example. Suppose we have an oscillator similar to that shown in Fig. 10-1 with *no damping force*. Now we apply to the mass an additional force, produced by some external agency, which varies sinusoidally with time according to the equation

$$F(t) = F_0 \sin \omega' t \tag{10-54}$$

where F_0 is a constant representing the maximum magnitude of force and ω' is the angular frequency of the applied force, not necessarily equal to the angular frequency ω of the oscillator itself. With this additional force, $F = ma$ for the oscillator takes the form

$$-kx + F_0 \sin \omega' t = m \frac{d^2 x}{dt^2}$$

It is convenient to rewrite this equation as

$$\frac{d^2 x}{dt^2} + \omega^2 x = \frac{F_0}{m} \sin \omega' t \tag{10-55}$$

where we have used the relationship $\omega^2 = k/m$ to simplify the form.

Here again is a problem in differential equations. What function of time must x be in order to satisfy this equation for all values of t if F_0 and ω' are specified constants? Again, a little physical insight and intuition guide us to a shrewd guess as to a possible solution. Since the driving force is a sinusoidal function of time with frequency ω', we try a solution in which the resulting motion is also sinusoidal with that frequency. That is, we try a solution in the form

$$x(t) = A \sin \omega' t \tag{10-56}$$

To find whether this actually is a solution, we substitute it back into Eq. (10-55). The result is

$$-\omega'^2 A \sin \omega' t + \omega^2 A \sin \omega' t = \frac{F_0}{m} \sin \omega' t$$

Dividing out the common factor, and solving for A, we find

$$A = \frac{F_0/m}{\omega^2 - \omega'^2} \qquad (10\text{-}57)$$

That is, Eq. (10-56) is a solution of the differential equation (10-55) provided we choose for A the value given by Eq. (10-57).

The most interesting part of this result is the amplitude of the motion, given by Eq. (10-57). This turns out to be proportional to the amplitude F_0 of the external force and inversely proportional to the mass, which is to be expected. But A also depends on the frequency ω' of the external force as shown in Fig. 10-10. If ω' is less than ω, A is positive, indicating that the

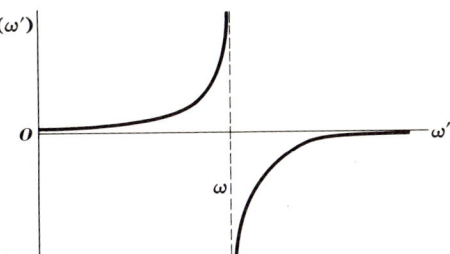

Fig. 10-10 Graph of Eq. (10-57), showing the variation of the amplitude A of a forced oscillation with the frequency ω' of the driving force. When ω' is less than the natural frequency ω, A is positive; when larger, it is negative. When $\omega' = \omega$, A becomes infinite.

force and displacement are in the same direction. When ω' is larger than ω, A becomes negative, indicating that the force and displacement are in opposite directions.

When $\omega = \omega'$, the denominator is zero and A becomes infinite. What does this mean? Mathematically speaking, it means that Eq. (10-56) is not a valid solution of the equation if $\omega' = \omega$. Physically, when a force whose frequency is exactly equal to the natural frequency is applied to the oscillator, this force keeps adding energy to the system by doing work on it. As time goes on, the amplitude of oscillation builds up indefinitely, at least until something breaks.

When damping is included in the foregoing analysis, it is found that the amplitude does not become infinite at any frequency but goes through a maximum value at a particular frequency which is closely related to the natural frequency of oscillation of the system. This phenomenon, in which the amplitude of oscillation reaches a maximum when the frequency of the driving force is close to some natural frequency of oscillation, is referred to as *resonance*.

Physics is full of important examples of resonance. Increasing the amplitude of a child on a swing is one. A tuned circuit in a radio receiver responds strongly only to frequencies very near its resonant frequency; this is what makes it possible to tune in a particular station. If a column of troops marching across a bridge steps with a frequency equal to a natural frequency of oscillation of the bridge, dangerously large vibrations may build up. When an atom or molecule is struck by electromagnetic radiation whose frequency is exactly that of one of the natural vibrations of the atom or molecule, the radiation is absorbed more readily than at other frequencies, a phenomenon referred to as *resonance absorption* of radiation.

Problems

10-1 A harmonic oscillator is observed to move from its equilibrium position to maximum displacement in 1 s. Find its period, frequency, and angular frequency.

10-2 A body with mass 2.0 kg rests on a smooth plane and is attached to a spring as shown in Fig. 10-1. The spring is stretched 0.05 m by a force of 10 N. If the body is displaced 0.5 m from its equilibrium position and released, find the amplitude, frequency, and period of the resulting motion.

10-3 In Prob. 10-2, instead of being displaced 0.05 m from equilibrium, the body is given an initial velocity of 1.0 m/s at its equilibrium position.
 a Find the amplitude, frequency, and period of the resulting motion.
 b Write an equation for the displacement as a function of time.

10-4 A body whose mass is 0.5 kg is hung from a spring, and the spring stretches 0.02 m under this load. The body is now pulled down an additional 0.01 m for this equilibrium position and released.
 a Find the period, frequency, and amplitude of the resulting motion.
 b Using a suitable coordinate system, write an equation describing the position of the body as a function of time.

10-5 Show by direct substitution that both forms of Eq. (10-15) are solutions of Eq. (10-2).

10-6 An automobile with very poor shock absorbers runs over a bump and bounces up and down with a period of about 1 s. If the car weighs 3,200 lb, estimate the spring constants of the springs, assuming that there are four springs with equal constants.

10-7 A body rests on a spring-supported platform, as shown in Fig. P10-7. The body has a mass of 5 kg, and the platform sinks 0.02 m when the body is

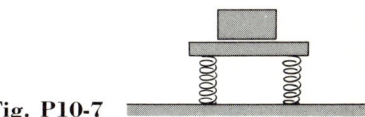

Fig. P10-7

placed on it. The whole system is set into motion by depressing the platform further and then releasing it. What is the maximum amplitude of motion that can occur without the body's losing contact with the platform at any time?

10-8 A simple harmonic oscillator with a frequency $f = 10$ Hz is given an initial displacement of 5 cm and an initial velocity of 10 cm/s at time $t = 0$. Write an equation for the displacement of the oscillator as a function of time.

10-9 Derive the relationships given by Eqs. (10-15) and (10-16) for the various forms of the solutions of the harmonic-oscillator equation.

10-10 Compute the time derivative of Eq. (10-19) to show directly that the total mechanical energy of an undamped harmonic oscillator is constant.

10-11 For displacements on one side of its equilibrium position a certain spring exerts a force $F = -k_1 x$, but for displacements on the other side of equilibrium it exerts $F = -k_2 x$, where k_1 and k_2 are different spring constants. If this spring is attached to a mass m as in Fig. 10-1 and set into motion, show that the period is independent of the amplitude; derive an expression for the period in terms of m, k_1, and k_2.

10-12 In Prob. 10-11, show that the maximum displacements from equilibrium are different on the two sides, and derive an expression for their ratio in terms of the given quantities.

10-13 A certain spring used in an arrangement like that in Fig. 10-1 exerts a force given by $F = -kx - cx^3$, where k and c are positive constants. Qualitatively, how should the period of oscillation of the system depend on the amplitude of the motion?

10-14 If the amplitude of the simple harmonic motion of a particular system is doubled, how does the total mechanical energy change? The period? The maximum velocity?

10-15 A rifle target is mounted as shown in Fig. P10-15. The mass of the cart on which the target is mounted is 2.0 kg, the spring constant is 450 N/m, the mass of the bullet is 5.0 g, and its initial velocity is 200 m/s. Find the period, amplitude, and frequency of the motion which results after the bullet strikes the target.

Fig. P10-15

10-16 At one time the fundamental unit of time (the second) was defined as the time required for a simple pendulum with a length of exactly 1 m to swing from one side to the other. If $g = 9.8$ m/s^2, by how much is this definition in error? For what value of g is it exactly correct?

10-17 Devise a way to estimate roughly the amount by which the period of a simple pendulum changes if the maximum displacement is 30° instead of a very small angle. Does the period increase or decrease? Why?

10-18 Derive an expression for the tension in the string of a simple pendulum as a function of its angle and relevant properties of the system.

10-19 A bead of mass m slides without friction on a wire bent into the shape of a parabola in a vertical plane in a uniform gravitational field. The equation of the parabola is $y = ax^2$. Show that the x coordinate varies sinusoidally with time, and find the frequency of the motion. Is the frequency independent of amplitude?

10-20 In Prob. 10-19, suppose the wire is circular rather than parabolic and is described by the equation $x^2 + (y - R)^2 = R^2$. Show that when x is small, the term in y^2 is small compared to the other term containing y; then the equation becomes approximately $y = x^2/2R$, and the situation reduces to that in the previous problem. Find the frequency of the motion in this approximation, and compare with the result for the simple pendulum derived in the text.

10-21 A particle of mass m moves in the xy plane under the action of a force that attracts it toward the origin with magnitude kr, where r is the usual polar coordinate. That is, $\mathbf{F} = -k\mathbf{r}$.
  a Show that the components of force are $F_x = -kx$ and $F_y = -ky$.
  b Considering the x and y motions separately, show that each is simple harmonic and that both have the same frequency.
  c Suppose the x and y motions have equal amplitudes but a phase difference of $\pi/2$. Show that the motion is circular with radius A and angular velocity ω. What would be the path with a phase difference of $\pi/2$ and unequal amplitudes?

10-22 Show that the reduced mass of a two-body system, defined by Eq. (10-35), is always smaller than either of the masses. What is its value if the two masses are equal?

10-23 The force of interaction between two atoms in a certain diatomic molecule is given by $F = -a/r^2 + b/r^3$, where r is the distance between atoms and a and b are positive constants.
    a Sketch a graph showing the general behavior of F as a function of r.
    b Find the equilibrium separation.
    c Find the effective spring constant for small displacements from equilibrium.
    d Find the period of small oscillations about the equilibrium position.

10-24 For the situation of Prob. 10-23, find a potential-energy function for the system, assuming that $V = 0$ at the equilibrium position. If the system is initially at rest at the equilibrium position, how much energy must be added in order to dissociate the molecule?

10-25 A harmonic oscillator is built as shown in Fig. 10-1 with $m = 0.5$ kg and $k = 200$ N/m. The surface is not perfectly smooth but has $\mu = 0.1$.
    a If the oscillator is displaced 0.1 m from equilibrium and released from rest, how many times does it pass the equilibrium position before it comes to rest permanently?
    b Does it come to rest at the equilibrium position? If not, where?

10-26 Show by direct substitution that Eq. (10-50) is a solution of the differential equation (10-43) for the harmonic oscillator with viscous damping with ω given by Eq. (10-51).

10-27 Show that Eq. (10-50) is a sensible solution of Eq. (10-43) only if $R^2 < 4mk$. This is the condition for underdamped motion.

10-28 For a harmonic oscillator with viscous damping for which $R^2 > 4mk$, corresponding to overdamped motion, show that one solution of Eq. (10-43) is

$$x = A \exp\left[-\left(\frac{R}{2m} + \sqrt{\frac{R^2}{4m^2} - \frac{k}{m}}\right)t\right]$$

10-29 Show that for a harmonic oscillator with viscous damping in the particular case in which $R^2 = 4mk$ (critical damping), the function

$$x = (A + Bt)e^{-Rt/2m}$$

is a solution of Eq. (10-43), where A and B are arbitrary constants.

10-30 Find the instantaneous rate at which the driving force does work on the forced harmonic oscillator whose motion is governed by Eq. 10-55). Show that the average rate of doing work is zero (the average power input to the oscillator is zero) unless $\omega = \omega'$.

PERSPECTIVE III

In Chaps. 7 to 10 a number of new concepts have been introduced, including momentum, energy, and force fields; yet the flavor has been somewhat different from that of earlier chapters. In Chap. 4 we relied heavily on experimental observations as a basis for generalization or induction to the general principles known as Newton's laws of motion. In the more recent chapters the approach has been more deductive; momentum and energy have been defined, and then the properties of these quantities have been derived from the definitions by use of the laws of motion. It is certainly true that in both cases the behavior of physical systems has suggested the usefulness of the definitions we have made, but the actual development of the properties of these quantities has been based on principles already established.

Nevertheless, as we have seen, there are many cases, including almost all collision problems and the motion of a particle in a force field, in which the energy and momentum concepts facilitate the analysis greatly. When a relation is to be derived between the state of motion of a system at one time and that at another time, without concern for the intermediate stages of motion, the energy approach nearly always simplifies the calculations, sometimes to the extent that it yields results which would entail calculations too difficult to be feasible if another approach were used.

Aside from these utilitarian values, however, the principles of momentum and energy have another fundamental significance; they are both readily generalized further to include physical phenomena which are not directly mechanical in nature. We have spoken of the momentum of electromagnetic radiation, the energy associated with heat and with electro-

magnetic fields, and so on. These generalizations lead to very broad conservation principles which have the effect of giving unifying elements to various branches of physics. Thus the very fact that it is possible to generalize these concepts and conservation principles to a wide variety of nonmechanical situations must itself be regarded as a discovery of primary importance. Although the momentum and energy concepts are derived from the laws of motion for mechanical systems, they take on a much more fundamental importance when generalized to include other kinds of systems.

Similar remarks may be made about the discussions of force fields and periodic motion. The force-field concept has roots in electromagnetism, optics, and fundamental-particle interactions, as well as in mechanics, and hence is another thread running through all the fabric of physics and giving it unity. The importance of periodic motion is not by any means limited to understanding vibrations of macroscopic mechanical systems; the concept can also be applied to molecular spectra, specific heats of molecules and solids, and many other areas.

The most obvious deficiency in our study of mechanics thus far is that we have discussed only systems which can be approximated by models treating each body as a particle. In any case where rotational motion of a body is of primary importance, such a model is clearly not adequate, but we lack weapons for dealing with such problems. The next logical step, then, is to generalize the principles developed thus far so that they can be applied directly to systems of particles and extended bodies. This can be done for the most part in a deductive manner, making use of the principles of mechanics and the energy and momentum concepts already available. It is useful to introduce and define two additional concepts, angular momentum and torque.

These generalizations form the subject matter of the next three chapters.

Systems of Interacting Particles 11

Beginning with this chapter, we apply the principles of mechanics to systems containing several particles interacting with each other and to extended rigid bodies. We first define the *center of mass* of a system of particles or of a body. Motion of the center of mass is investigated using the mechanical principles already presented; we then discuss the use of the center of mass as the origin of a moving coordinate system. The kinetic energy of a system of particles is represented by means of motion of the center of mass and motion relative to the center of mass. To compute the kinetic energy of a rotating rigid body, we introduce the concept of moment of inertia. Several examples illustrating the usefulness of these concepts are discussed.

11-1 CENTER OF MASS

In discussing the vibratory motion of a diatomic molecule in Chap. 10, it was mentioned that as the atoms vibrate with respect to each other, the molecule may also move *as a whole*. One can easily think of similar situations in everyday experience. When a body is thrown into the air so that it spins as it moves, the body as a whole is displaced from one position to another while it spins. But just what is meant by moving "as a whole"?

Our study of mechanics thus far has been concerned for the most part with motion of points. In considering the motion of a more complicated system, it is natural to choose one point in the system which in some sense characterizes the motion of the system as a whole. For this purpose, we define a point called the *center of mass* of a system of particles or of a body.

The concept of momentum provides a convenient starting point for defining the center of mass of a system. Our program is to define the center of mass in such a way that the *total* momentum of the system can be expressed as the product of the total mass and the velocity of the point called the center of mass.

Before developing this concept further, we need to introduce some new language. Suppose we have a system including several particles. If there are N particles in all, we may number them from 1 to N. Suppose that each particle has a definite, constant mass; we denote the masses by m_1, $m_2, \ldots, m_N$. Let i be a typical particle number (any integer from 1 to N); then the mass of particle i is m_i. Following the discussion of Sec. 4-3, we define the total mass of the system as the sum of the individual masses and denote it by M. That is,

$$M = m_1 + m_2 + \cdots + m_1 = \sum_{i=1}^{N} m_i \qquad (11\text{-}1)$$

In the second form of this equation we have used the summation notation introduced in Sec. 1-6. Often, in sums of this sort, where there is one term for each particle, we omit the limits $i = 1$ and N on the summation and write simply Σm_i, understanding that the sum always includes one term for each particle.

The position of the first particle can be described with respect to an origin O by the position vector $\mathbf{r}_1$, as shown in Fig. 11-1. Similarly, the

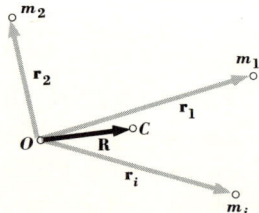

Fig. 11-1 Vector $\mathbf{r}_i$ represents the position of particle i, of mass m_i, with respect to the origin O of an inertial frame of reference. In this system, $\mathbf{R}$ is the position vector of the center of mass, labeled C.

position of particle i is given by the position vector $\mathbf{r}_i$. The velocity and acceleration of particle i are defined in the usual way:

$$\mathbf{v}_i = \frac{d\mathbf{r}_i}{dt} \qquad \mathbf{a}_i = \frac{d\mathbf{v}_i}{dt} = \frac{d^2\mathbf{r}_i}{dt^2} \qquad (11\text{-}2)$$

In this chapter, unless we explicitly indicate otherwise, it is assumed that O is the origin of an inertial frame of reference.

Now we are ready to introduce the concept of center of mass. Just as in Chap. 7, we define the *total momentum* of the system of particles as the vector sum of the momenta of the individual particles. The momentum of particle i is $\mathbf{p}_i = m_i \mathbf{v}_i$. Denoting the total momentum by $\mathbf{P}$, we have

$$\mathbf{P} = \mathbf{p}_i + \mathbf{p}_2 + \cdots + \mathbf{p}_N = \Sigma \mathbf{p}_i = \Sigma m_i \mathbf{v}_i \tag{11-3}$$

Let $\mathbf{V}$ be the velocity of the center of mass, not yet defined. According to the above statement, we want to define the center of mass in such a way that

$$\mathbf{P} = \Sigma \mathbf{p}_i = M\mathbf{V} \tag{11-4}$$

that is, so that the *total* momentum equals the *total* mass M multiplied by the velocity $\mathbf{V}$ of the center of mass.

Equation (11-4) *defines* the velocity of the center of mass. We have

$$\mathbf{V} = \frac{\mathbf{P}}{M} = \frac{\Sigma m_i \mathbf{v}_i}{\Sigma m_i} = \frac{m_1 \mathbf{v}_1 + m_2 \mathbf{v}_2 + \cdots + m_N \mathbf{v}_N}{m_1 + m_2 + \cdots + m_N} \tag{11-5}$$

This does not yet define uniquely the *position* of the center of mass, only its velocity. But Eq. (11-5) should make it obvious how the position should be defined. Let $\mathbf{R}$ be the position vector of the center of mass. We define $\mathbf{R}$ as

$$\mathbf{R} = \frac{\Sigma m_i \mathbf{r}_i}{\Sigma m_i} = \frac{m_1 \mathbf{r}_1 + m_2 \mathbf{r}_2 + \cdots + m_N \mathbf{r}_N}{m_1 + m_2 + \cdots + m_N} \tag{11-6}$$

The time derivative of Eq. (11-6) is Eq. (11-5), since by definition $\mathbf{V} = d\mathbf{R}/dt$ and $\mathbf{v}_i = d\mathbf{r}_i/dt$ for each particle. We therefore adopt Eq. (11-6) as the general definition of the position of the center of mass.

This definition represents an *average* position of the particles in which the position of each particle is given an importance proportional to its mass. The total momentum $\mathbf{P}$ of a system is always given by the product of the total mass M and the velocity $\mathbf{V}$ of the center of mass. Alternatively, we may say that the total momentum of the system is computed as though all the mass were concentrated at the center of mass. Further usefulness of this concept will appear soon.

In introducing the properties of a system of particles, such as the total mass, total momentum, and the position and velocity of the center of mass, we have used capital letters, while lowercase letters have been used for the properties of the individual particles. We shall adhere to this scheme wherever possible.

280 SYSTEMS OF INTERACTING PARTICLES

Equation (11-6) is, of course, a vector equation; it is often convenient to express it in terms of components. In a cartesian coordinate system, the components of the vector $\mathbf{r}_i$ denoting the position of particle i are (x_i, y_i, z_i). Similarly, let (X, Y, Z) be the components of the position vector $\mathbf{R}$ of the center of mass. Then (X, Y, Z) are the coordinates of the center of mass in this coordinate system. Thus Eq. (11-6) is equivalent to the three scalar equations

$$X = \frac{m_1 x_1 + m_2 x_2 + \cdots + m_N x_N}{m_1 + m_2 + \cdots + m_N} = \frac{\Sigma m_i x_i}{\Sigma m_i}$$

$$Y = \frac{\Sigma m_i y_i}{\Sigma m_i} \qquad (11\text{-}7)$$

$$Z = \frac{\Sigma m_i z_i}{\Sigma m_i}$$

Example

Consider two cars connected by a spring which roll on a straight horizontal track. As previously, we idealize the situation by neglecting friction and all other external influences, so that the system of two cars and the spring can be regarded as an isolated system. We also neglect the mass of the spring. This is a one-dimensional situation, and we need only one coordinate for each car. A suitable coordinate system is shown in Fig. 11-2. It is clear that

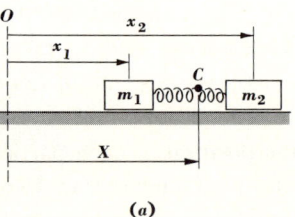

(a)

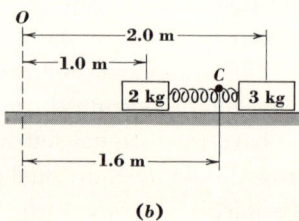

(b)

Fig. 11-2 (a) Idealized model of two cars on a frictionless track connected by a spring. The coordinates x_1 and x_2 describe the positions of the cars with respect to the origin O of an inertial frame of reference. The position of the center of mass, labeled C, is represented by the coordinate X. (b) A numerical example for the case $m_1 = 2$ kg, $m_2 = 3$ kg.

the center of mass, identified as C in the figure, must lie somewhere between the two cars. Its position, denoted by X, is given by

$$X = \frac{m_1 x_1 + m_2 x_2}{m_1 + m_2}$$

If, for example, $m_1 = 2$ kg, $m_2 = 3$ kg, $x_1 = 1.0$ m, and $x_2 = 2.0$ m, then $M = 2$ kg $+ 3$ kg $= 5$ kg, and

$$X = \frac{(2 \text{ kg})(1.0 \text{ m}) + (3 \text{ kg})(2.0 \text{ m})}{5 \text{ kg}} = 1.6 \text{ m}$$

We see that the center of mass is somewhat closer to the 3-kg car than to the 2-kg car, showing that more massive objects have more effect on the position of the center of mass than less massive ones.

Suppose that, at the instant we are considering, car 1 is moving to the right with a speed of 3 m/s and car 2 is moving to the left with a speed of 1 m/s. Then $v_1 = 3$ m/s, and $v_2 = -1$ m/s. The total momentum is then

$$P = (2 \text{ kg})(3 \text{ m/s}) + (3 \text{ kg})(-1 \text{ m/s}) = 3 \text{ kg-m/s}$$

The velocity of the center of mass, according to Eq. (11-5), is

$$V = \frac{(2 \text{ kg})(3 \text{ m/s}) + (3 \text{ kg})(-1 \text{ m/s})}{2 \text{ kg} + 3 \text{ kg}} = 0.6 \text{ m/s}$$

The total mass is 5 kg, so the total momentum, according to Eq. (11-4), is

$$P = MV = (5 \text{ kg})(0.6 \text{ m/s}) = 3 \text{ kg-m/s}$$

which agrees with the previous result.

Example

A simple model of the structure of a water molecule, H_2O, is shown in Fig. 11-3. Each atom is represented as a point, since nearly all the mass of each

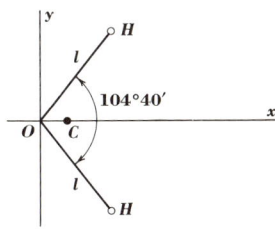

Fig. 11-3 Simple model of a water molecule. The mass of each atom in the molecule is regarded as concentrated at a single point, as shown. The center of mass lies on the x axis; its distance from the origin can be expressed as a multiple of the distance l from the oxygen atom to either hydrogen atom.

atom is concentrated in its nucleus, which is of the order of 10^5 times smaller than the distance between atoms. A convenient location for the coordinate axes is shown in the figure. Where is the center of mass?

Solution
From the symmetry of the situation we see that there is no reason for the center of mass to be either above or below the x axis, so its y coordinate must be zero: $Y = 0$. The x coordinate of each hydrogen atom is $l \cos \frac{1}{2}(104°40')$. That of the oxygen atom, because of the way our coordinate system is defined, is zero. It is convenient to use atomic mass units (amu), in which the mass of a hydrogen atom is 1 amu and that of oxygen 16 amu. The x coordinate of the center of mass, from Eq. (11-7), is

$$X = \frac{(1 \text{ amu})(l \cos 52°20') + (1 \text{ amu})(l \cos 52°20') + (16 \text{ amu})(0)}{1 \text{ amu} + 1 \text{ amu} + 16 \text{ amu}}$$
$$= 0.068 l$$

The center of mass is much closer to the oxygen atom than to either of the hydrogen atoms, as might be expected because of its greater mass. When a water molecule moves, as in water vapor, the total momentum of the molecule is equal to the total mass multiplied by the velocity of the center of mass, whose position we have just found.

Especially in problems with solid bodies, it is often useful to think of matter as a *continuous distribution* of mass, rather than as a collection of particles. Fundamentally, we know that matter is made of particles, but if the dimensions of a body are large compared to the average separation between particles, a model which assumes matter to be continuously distributed is quite precise.

Finding the center of mass of such a distribution requires only a slight extension of the methods already described. We think of the body as consisting of a large number of small elements, which may be described as elements of volume, area, or distance, depending on circumstances. We compute the mass of a typical element; this mass plays the role of m_i in Eqs. (11-6) or (11-7). We then multiply by the appropriate coordinate, sum over all the elements by integrating, and finally divide by the total mass.

Example
A thin uniform rod of mass M lies along the x axis. Its left end is at $x = a$, and its length is L. Where is its center of mass?

Solution

It is obvious from symmetry that the center of mass lies at the center of the rod, a distance $a + L/2$ from the origin. Let us verify this result by the method outlined above. Let dx be an element of length along the rod, and let the mass of this element be dm. Because the mass is uniformly distributed, we have $dm/M = dx/L$, or $dm = M\,dx/L$. The sum $\Sigma m_i x_i$ now becomes the integral

$$\int_a^{a+L} \left(M\frac{dx}{L}\right)x = \frac{M}{L}\left[\tfrac{1}{2}x^2\right]_a^{a+L} = M\left(a + \frac{L}{2}\right)$$

and we find for the position x of the center of mass

$$X = \frac{\Sigma m_i x_i}{M} = \frac{M(a + L/2)}{M} = a + \frac{L}{2}$$

as predicted above.

Example

Find the position of the center of mass of an isosceles right triangle cut out of a uniform sheet of material.

Solution

The triangle is symmetric with respect to a line bisecting the right angle, so the center of mass must lie somewhere along this line, as shown in Fig. 11-4. We consider an elemental strip, a distance x away from the y axis

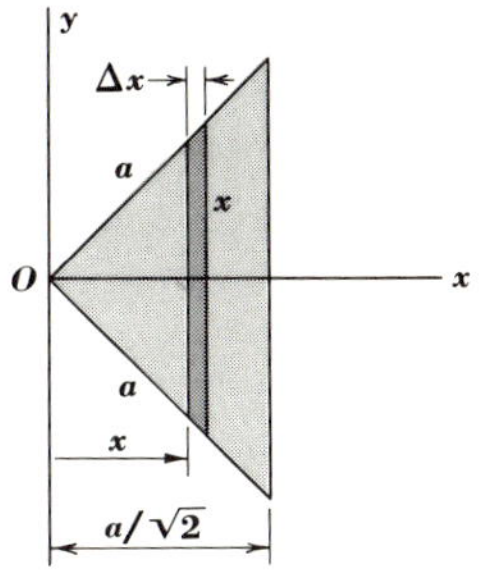

Fig. 11-4 Coordinate system used to calculate the position of the center of mass of an isosceles right triangle. The mass of the strip of width Δx is multiplied by its x coordinate, and all these products are added to find the x coordinate of the center of mass.

and of width Δx. All the mass in this strip is very nearly the same distance from the y axis. Because the smaller triangles in the figure are also isosceles right triangles, the height of this strip is $2x$, so its area is $2x\,\Delta x$. Let the mass per unit area of the sheet be σ; the mass of the strip is

$$m = 2\sigma x \, \Delta x$$

To find the position X of the center of mass, we multiply the mass m by its x distance from the origin, which is simply x, add all the strips by integrating over x, and divide by the total mass. That is,

$$X = \frac{2\sigma}{M} \int_0^{a/\sqrt{2}} x^2 \, dx = \frac{2}{M} \left[\frac{x^3}{3} \right]_0^{a/\sqrt{2}} = \frac{\sigma a^3}{3\sqrt{2}M}$$

The limits of the integral are chosen to include the whole area of the triangle; $a/\sqrt{2}$ is the altitude to the hypotenuse. The total mass can easily be found in terms of the mass per unit area σ, since the total area is simply $\frac{1}{2}a^2$; thus $M = \frac{1}{2}\sigma a^2$, and

$$X = \frac{\sqrt{2}a}{3}$$

which gives the position of the center of mass directly in terms of the dimensions of the triangle. As we should expect, this position does not depend on the total mass if the sheet is uniform in thickness.

Symmetry considerations are often useful in locating the center of mass of an object. For example, it is clear that the center of mass of any body with rotational symmetry about an axis of rotation (such as a wheel, gear, cone, or cylinder) must lie along this axis. As another example, consider the three-bladed airplane propeller in Fig. 11-5. It may not be immediately

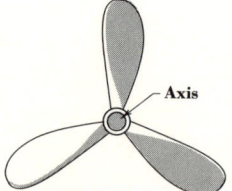

Fig. 11-5 From symmetry considerations it can be concluded that the center of mass of this airplane propeller must lie on its axis of rotation.

obvious that the center of mass lies along the axis of rotation of the propeller. Suppose the center of mass is a point not on the axis. We rotate the propeller by 120°. The position of the center of mass must also rotate 120°. But from symmetry the propeller has exactly the same mass distribution after the rotation as before, and therefore the center of mass must have the same position as initially. This contradiction can be avoided only by assuming that the center of mass lies on the axis, so its position does not change when the body rotates. Just *where* on the axis it lies is another question; locating the exact point may require more detailed calculation. But the calculation has been simplified considerably by the knowledge that the center of mass is on a particular line.

The use of symmetry considerations is of central importance in all branches of physical science. Whenever a physical system possesses symmetry of any sort, analysis of the behavior of this system can almost always be simplified by making use of the symmetry, as in the case of the airplane propeller just discussed.

In defining the center of mass we have always referred to a *system* of particles. In any application of the principles of mechanics to more than one particle, it is absolutely essential to decide at the outset exactly what is included in the system being considered, i.e., which particles or aggregations of matter are part of the system, and which ones are external to the system. The definition of the center of mass of a system, given by Eq. (11-6), has no meaning unless we clearly specify the masses to be included and those to be excluded. The importance of this concept appears in many places; for example, in the next section we distinguish between forces applied by one part of a system to another and forces exerted on a part of the system by an agency outside it. We call these *internal* and *external* forces, respectively; this distinction has no meaning unless we specify carefully what is included in the system.

11-2 MOTION OF THE CENTER OF MASS

The center of mass of a system of particles or of a body has been defined so that the total momentum is equal to the product of the total mass and the velocity of the center of mass. This has the immediate consequence that if the total momentum of a system is constant, the velocity of its center of mass must also be constant provided the total mass is constant.

Under what circumstances is the total momentum of a system constant? In the discussion in Chap. 7, we concluded that if a system is *isolated*, i.e., if it has no external forces acting on it, and if the *internal* forces of interaction between parts of the system obey Newton's third law, then the total momentum of the system is constant. It is also necessary to keep in mind that this statement is correct only when the momenta are computed using velocities measured in an inertial frame of reference and when the *same* inertial system is used for all parts of the system.

Thus *for any isolated system of constant mass the velocity of the center*

286   SYSTEMS OF INTERACTING PARTICLES

of mass, as observed in an inertial frame of reference, is constant. Consider again the two cars connected by a spring (Sec. 11-1). The two cars can oscillate relative to each other, and, in addition, the center of mass can move. But if no external forces act, the velocity of the center of mass is constant, even though the individual velocities of the two cars are not.

In collisions of fundamental particles in which matter is created or destroyed, the total mass of the system may change. An example which has already been mentioned is the production of a π meson in the collision of two protons. If the system is isolated during the collision, the total momentum is constant, and the total energy, including that associated with mass, is conserved. Detailed analysis of this process requires use of the relativistic generalization of momentum introduced in Sec. 15-1.

When a system is *not* isolated, its total momentum is, in general, not constant. We can, however, derive a simple relationship between the rate of change of momentum and the external forces and thus find out something about the acceleration of the center of mass. The rate of change of momentum of each particle in the system equals the total force acting on that particle. That is,

$$\frac{d\mathbf{p}_i}{dt} = \mathbf{F}_i \tag{11-8}$$

where $\mathbf{F}_i$ is understood to mean the *total* force acting on particle i, that is, the vector sum of all the forces on it. The time rate of change of the total momentum of the system is simply the sum of these terms, one term for each particle:

$$\frac{d\mathbf{P}}{dt} = \frac{d}{dt}\left(\sum \mathbf{p}_i\right) = \sum \frac{d\mathbf{p}_i}{dt} = \sum \mathbf{F}_i \tag{11-9}$$

We have used the fact that the derivative of a sum of terms is equal to the sum of the derivatives of the terms. The expression $\sum \mathbf{F}_i$ is the vector sum of *all* the forces acting on *all* the particles.

These forces are of two types: (1) forces exerted by the particles of the system on each other, called *internal* forces, and (2) forces exerted from outside the system, called *external* forces. If particle 1 exerts a force on particle 2, then, according to Newton's third law, particle 2 exerts an equal and opposite force on particle 1 and the vector sum of these two forces is zero. Similarly, *all* the internal forces cancel each other out in pairs. The vector sum of all the internal forces is zero, and all that survives in Eq. (11-9) is the sum of the *external* forces. Thus it is legitimate to rewrite the equation, emphasizing this fact, as follows:

$$\frac{d\mathbf{P}}{dt} = \sum \mathbf{F}_{\text{ext}} \tag{11-10}$$

This important result says that the rate of change of the *total momentum* of the system is determined only by the *external* forces. This is important because the internal forces may be extremely complicated, especially if the system contains a great number of particles, as a solid body does. In computing the total momentum change, we can disregard these internal forces and consider only the forces applied to the system from outside. As a matter of fact, we have been doing this all along. In our initial development of the principles of mechanics, we used a point to represent a body which may have had a finite extension, and we discovered in this case that the acceleration of the body could be predicted from the external forces alone.

When the total mass M of the system is constant, as is usually the case, Eq. (11-10) can be put in an even more useful form. We then write

$$\frac{d\mathbf{P}}{dt} = M\frac{d\mathbf{V}}{dt} = M\mathbf{A} \tag{11-11}$$

where $\mathbf{A}$ is the acceleration of the center of mass: $\mathbf{A} = d\mathbf{V}/dt$. Thus,

$$\sum \mathbf{F}_{\text{ext}} = M\mathbf{A} \tag{11-12}$$

This very useful result shows that the acceleration $\mathbf{A}$ of the center of mass of a system resulting from the application of a given set of external forces is exactly the same as the acceleration of a single particle with the same total mass M acted on by the same forces.

An example will clarify this important result. Suppose a dumbbell is thrown through the air, as shown in Fig. 11-6. Assuming that all forces except

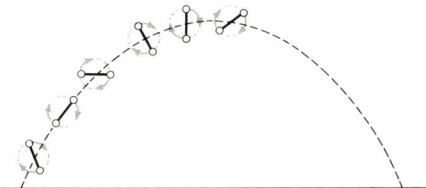

Fig. 11-6 Dumbbell thrown spinning through the air. The center of mass of the dumbbell describes a parabolic trajectory, just as though all the mass were concentrated at that point.

gravity are negligible, the center of mass of the dumbbell describes a parabolic trajectory identical with that of a single particle with the same initial velocity. The masses on the ends of the dumbbell do not, in general, follow this trajectory, since the dumbbell is spinning. But the center of mass always follows this path. It would even continue to follow this path if the dumbbell were to come apart in midair.

288 SYSTEMS OF INTERACTING PARTICLES

An even more spectacular example is the motion of a shell which explodes into fragments in midair. The shell has a parabolic trajectory before it explodes; after the explosion the center of mass of the shell follows the same parabolic trajectory, even if no individual fragment follows this path.

11-3 CENTER-OF-MASS COORDINATE SYSTEM

It is often useful to take the center of mass of a set of particles or a body as the origin of a coordinate system, one reason being that the kinetic energy of such a system can be represented as a sum of two parts, the first depending only on motion *of* the center of mass and the second only on motion of parts of the system *with respect to* the center of mass. The concepts of relative velocity and acceleration used here are just the same as those introduced in Sec. 3-6 in connection with motion described by a moving observer; here the observer is moving with the center of mass.

In Fig. 11-7, O is the origin of an inertial frame of reference, and C

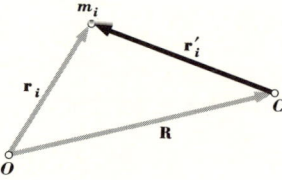

Fig. 11-7 The position of the particle whose mass is m_i may be described by giving its position vector $\mathbf{r}_i$ with respect to the origin O of an inertial frame of reference or by giving its position vector $\mathbf{r}'_i$ with respect to the center of mass C. The position of the center of mass with respect to O is given by the vector $\mathbf{R}$. Equation (11-15) follows directly from this diagram.

is the center of mass of a system of particles or a body. The position of the center of mass is denoted by $\mathbf{R}$ as usual; we denote the position of a typical particle of mass m_i with respect to O as $\mathbf{r}_i$. Let $\mathbf{r}'_i$ be the position vector of particle i with respect to the center of mass, as shown; $\mathbf{r}'_i$ describes its position with respect to the center of mass in precisely the same way that $\mathbf{r}_i$ describes its position with respect to the origin O of the inertial system. Similarly, we define the velocity and acceleration of particle i with respect to the center of mass as

$$\mathbf{v}'_i = \frac{d\mathbf{r}'_i}{dt} \qquad \mathbf{a}'_i = \frac{d\mathbf{v}'}{dt} = \frac{d^2\mathbf{r}'_i}{dt^2} \tag{11-13}$$

in which the primes denote measurement with respect to the center of mass. Figure 11-7 shows that the position vectors of the particles with respect to O and with respect to C are related:

$$\mathbf{r}_i = \mathbf{R} + \mathbf{r}'_i \tag{11-14}$$

This says simply that to get from O to the position of particle i we first go to the center of mass and then from the center of mass to particle i. By differentiating Eq. (11-14), we immediately obtain

$$\mathbf{v}_i = \mathbf{V} + \mathbf{v}'_i \quad \text{and} \quad \mathbf{a}_i = \mathbf{A} + \mathbf{a}'_i \tag{11-15}$$

The coordinate system in which the positions are described relative to C rather than O is called the *center-of-mass coordinate system*. In general, the center-of-mass coordinate system need not be an inertial frame of reference; only when C moves with constant velocity relative to O is it an inertial frame of reference. For example, for an *isolated* system with constant total mass, the total momentum $\mathbf{P}$ and the center-of-mass velocity $\mathbf{V}$ are constant, and the center-of-mass coordinate system *is* an inertial frame of reference. But even in cases where C is *accelerated* with reference to O, it is still a very useful system, as we shall see presently.

Another important characteristic of the center-of-mass coordinate system is that the total momentum with respect to this system, using the velocities $\mathbf{v}'_i$, is always zero. That is,

$$\Sigma m_i \mathbf{v}'_i = \Sigma m_i (\mathbf{v}_i - \mathbf{V}) = \mathbf{0} \tag{11-16}$$

This result can be obtained formally by using Eq. (11-5). It also follows directly from the fact that in *any* coordinate system the total momentum is the total mass multiplied by the velocity of the center of mass; the velocity of the center of mass *with respect to* the center of mass is of course always zero. Thus the center-of-mass coordinate system is also called the *center-of-momentum* coordinate system to emphasize the fact that in this system the total momentum is zero.

Here is an example of the usefulness of the center-of-mass coordinate system. We shall describe the collision of two billiard balls using two coordinate systems, one fixed in the table (assumed to be an inertial frame of reference), the other the center-of-mass coordinate system. Considering two billiard balls of equal mass m, suppose one is initially at rest and the other approaches it with velocity $\mathbf{u}$ as shown in Fig. 11-8a. By use of Eq. (11-5), we find that the center-of-mass velocity is $\mathbf{V} = \mathbf{u}/2$. Thus the initial velocities relative to the center of mass are as shown in Fig. 11-8b. That is, to an observer moving with the center of mass, it appears as though both balls are approaching him with equal speed $u/2$ from opposite directions.

After the collision, the motion of the balls as seen from the center of mass might look as in Fig. 11-8c. Because the total momentum in the

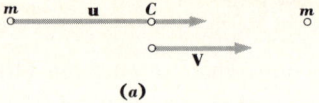

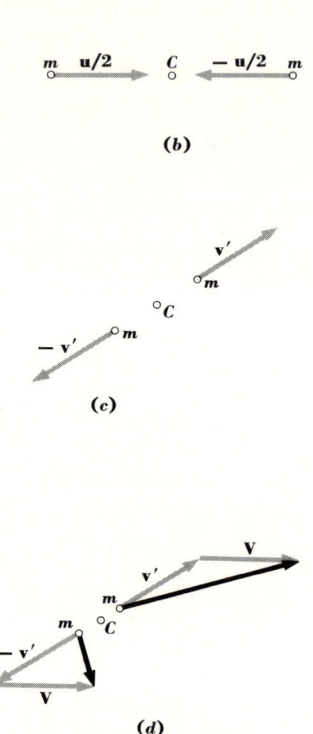

Fig. 11-8 (a) Two billiard balls as viewed from a frame of reference fixed in the billiard table. The left ball moves with velocity **u**; the right one is initially at rest. (b) The same situation viewed from the center-of-mass coordinate system. The velocity **V** of the center of mass is equal to **u**/2. An observer moving with this velocity sees the two balls moving as shown. (c) Motion after the collision, as seen from the center-of-mass system. The magnitudes of the two velocities are equal and are the same as before the collision, but the directions are different. (d) Motion after the collision, as viewed from the "table" coordinate system. The velocity of each ball in this system is obtained by adding the corresponding velocity in the center-of-mass system to the velocity **V** of the center of mass. The heavy lines represent the final velocities of the two balls in the "table" system. By superimposing these diagrams, it can be seen that these two velocities are always perpendicular.

center-of-mass system is zero, the two velocities must be equal in magnitude but opposite in direction, as they were initially. If the total kinetic energy of the system is also the same after the collision as before (very nearly true for good-quality ivory billiard balls), then the *speeds* must be the same as the initial speeds. Thus, **v**′ has the same magnitude as **u**/2 but in general is not in the same direction. To see how the final motion looks from the coordinate system whose origin is O, we add the center-of-mass velocity **V** to the two final center-of-mass velocities, obtaining the picture in Fig. 11-8d. The two final velocities in the "table" coordinate system are shown by heavy lines in Fig. 11-8d.

Now an added dividend appears. Suppose we superimpose the two vector diagrams of Fig. 11-8d so that the vectors **v**′ and −**v**′ coincide. We then see that since **v**′ and **V** have the same magnitude, the two final velocities form the diagonals of a rhombus. It is a fact of plane geometry that the diagonals of a rhombus are always perpendicular, so the final velocities of the two billiard balls, as observed in the table coordinate system, are always perpendicular! This fact is known to every good billiard player, but we have derived it from the principles of dynamics without ever picking up a cue.

In research involving collisions of atomic or nuclear particles, the center-of-mass coordinate system always simplifies calculations, because in this system the total momentum is always zero. Ordinarily, the center-of-mass system has some velocity with respect to an (inertial) coordinate system. Although the particles involved in the above example are not very fundamental, the collision described does indicate the utility of the center-of-mass coordinate system.

11-4 KINETIC ENERGY

The total momentum of a system of particles is determined entirely by the velocity of the center of mass of the system, and not by the motions of the particles relative to the center of mass. This is not true of kinetic energy, but as we shall see, the kinetic energy of a system can always be expressed as the sum of two parts, one determined by the center-of-mass velocity, the other by the velocities relative to the center of mass.

To develop this relationship, we first *define* the total kinetic energy of a system of particles, with respect to the origin O of an inertial frame of reference, as the sum of the individual kinetic energies. That is,

$$E = \Sigma \tfrac{1}{2} m_i v_i^2 \tag{11-17}$$

where the velocities $\mathbf{v}_i$ are measured with respect to O. We now express this in terms of the velocity $\mathbf{V}$ of the center of mass and the velocities $\mathbf{v}'_i$ of the particles relative to the center of mass. To do this, we recall that the square of the magnitude of a vector can be written as the scalar product of the vector with itself. That is, $v_i^2 = \mathbf{v}_i \cdot \mathbf{v}_i$. Using this fact together with the first Eq. (11-15), we find

$$E = \Sigma \tfrac{1}{2} m_i (\mathbf{V} + \mathbf{v}'_i) \cdot (\mathbf{V} + \mathbf{v}'_i) \tag{11-18}$$

We now expand this scalar product, obtaining four terms in all. Some terms have factors common to all terms of the sum; these we take outside the summation sign.

$$\begin{aligned} E &= \Sigma \tfrac{1}{2} m_i (\mathbf{V} \cdot \mathbf{V} + 2\mathbf{V} \cdot \mathbf{v}'_i + \mathbf{v}'_i \cdot \mathbf{v}'_i) \\ &= \tfrac{1}{2} (\Sigma m_i) V^2 + \mathbf{V} \cdot (\Sigma m_i \mathbf{v}'_i) + \Sigma \tfrac{1}{2} m_i v_i'^2 \end{aligned} \tag{11-19}$$

The first and third terms of this expression are readily recognizable. The first is one-half the total mass times the square of the speed of the center of mass, $\frac{1}{2}MV^2$; the last is the sum of the kinetic energies $\frac{1}{2}m_i v_i'^2$ computed using velocities relative to the center of mass. The sum in the middle term represents the total momentum with respect to the center of mass, and, as observed in Sec. 11-3, this is always zero. Thus this term is zero, and Eq. (11-19) becomes simply

$$E = \tfrac{1}{2}MV^2 + \Sigma \tfrac{1}{2}m_i v_i'^2 \qquad (11\text{-}20)$$

This important and useful result shows that the total kinetic energy of a system can be represented as the sum of two parts, the first depending only on motion of the center of mass and the second depending only on motion *relative to* the center of mass.

If all particles in the system have the same velocity as the center of mass, all the $\mathbf{v}_i'$ are zero and the total kinetic energy is given by the first term. But if in addition to the center-of-mass motion there is *internal* motion, in which the particles move relative to the center of mass, then the kinetic energy is larger by an amount given by the second term of Eq. (11-20).

As an example, consider again the exploding shell mentioned in Sec. 11-2. Just before the explosion, all parts of the shell have the same velocity, all the $\mathbf{v}_i'$ are zero, and the total kinetic energy is just $\frac{1}{2}MV^2$. Just after the explosion, when the velocity $\mathbf{V}$ of the center of mass has not changed appreciably, the shell fragments have various velocities $\mathbf{v}_i'$ relative to the center of mass, so the total kinetic energy of the system must be *greater* than before the explosion by an amount equal to the second term in Eq. (11-20). This result is what we should expect, since, when the shell explodes, mechanical energy is added to the system by a chemical reaction taking place within the shell.

This illustrates another important point concerning the kinetic energy of a system. It has been shown that the total momentum of an isolated system is constant; we may ask whether this is true for kinetic energy. As the exploding shell shows, the answer is that it is *not* true in general. The total kinetic energy of the shell fragments, given by Eq. (11-20), is greater just after the explosion than just before; $\frac{1}{2}MV^2$ does not change appreciably, but $\Sigma \frac{1}{2}m v_i'^2$ changes from zero to a quantity different from zero. Thus, in general, the kinetic energy of an isolated system need not be constant. If two particles collide in such a way that the total kinetic energy after the collision *is* the same as before, as in the billiard-ball problem of Sec. 11-3, the collision is called *elastic*. But this is a special case; there are also *semielastic* collisions, in which some kinetic energy is lost, and *inelastic* collisions, in which enough kinetic energy is lost for the two particles to have the same velocity after the collision. In an inelastic collision, the second term of Eq. (11-20) is zero after the collision. In all three cases, however, the term $\frac{1}{2}MV^2$ is constant, provided the system is isolated.

An important example of a system of particles is a body which *rotates* about an axis through its center of mass. If, while the body rotates, the axis is also moving, as in the case of a rolling wheel, then the total kinetic energy can always be represented in terms of a part due to center-of-mass motion, depending only on motion of the axle, and a part due to motion relative to the center of mass, depending only on the rotational motion about the axle. For a rigid body (a body with definite shape), calculation of this rotational kinetic energy is particularly simple. Such calculations are discussed in the following sections.

11-5 RIGID BODIES

A rigid body, as the name implies, is a body having a definite and unchanging shape. Formally, we may define a rigid body as a collection of mass points having the property that the distance between any pair of points is constant.

When a collection of particles can be treated as a rigid body, calculations concerning its motion are enormously simplified. This becomes clear when we realize that a body which can be held in one hand may contain 10^{24} atoms. Each of these requires at least three coordinates to describe its position, making a total of at least 3×10^{24} coordinates. The position of a rigid body, on the other hand, can be described by a much smaller number of coordinates. The position of its center of mass is described by just three coordinates, and it turns out that only three others are required to describe its orientation. Thus the position of a rigid body in space is described completely by six coordinates—quite a saving compared to 3×10^{24}. Correspondingly, the *motion* of a rigid body can be described simply as the combination of a translational motion of the center of mass and a rotation about the center of mass.

In all the applications of dynamics considered thus far, we have assumed that the *rotational* motion of the body under consideration could be ignored, either because there was no rotational motion or because the dimensions of the object were sufficiently small to permit regarding it as a point. But these simplifications are not always possible; sometimes it is essential to take into consideration the rotation of a body.

We consider first the rotation of a rigid body around an axis which is stationary in an inertial frame of reference. We choose the origin of the coordinate system so that it lies on the axis of rotation. The center of mass is not necessarily on the axis. In this situation, the only possible motion of the body is one of rotation about this axis. Just as in Sec. 6-4, we can describe the rotation by means of the vector angular velocity. If that discussion is not fresh in mind, it should be reviewed at this point; if it is, we may proceed immediately.

Since the motion is purely one of rotation, knowledge of the vector angular velocity $\boldsymbol{\omega}$ and the location of the axis of rotation is sufficient to describe the motion completely. It was shown in Chap. 6 that for a point with position vector $\mathbf{r}$ moving in a rigid body which rotates with angular velocity $\boldsymbol{\omega}$, the velocity $\mathbf{v}$ is given by Eq. (6-19): $\mathbf{v} = \boldsymbol{\omega} \times \mathbf{r}$.

The total kinetic energy of the body can be expressed in terms of $\boldsymbol{\omega}$. To do this, we consider first the motion of a typical point in the body, of mass m_i, as shown in Fig. 11-9. The *speed* of this point at any instant is given by

$$v_i = |\boldsymbol{\omega} \times \mathbf{r}_i| = \omega r_i \sin \theta = \omega s_i \tag{11-21}$$

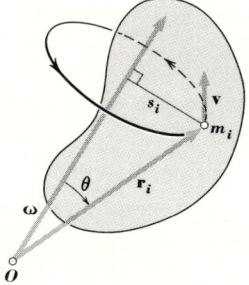

Fig. 11-9 Rigid body rotating with angular velocity ω about a fixed axis passing through the origin O of an inertial frame of reference. A particle with mass m_i at position $\mathbf{r}_i$ moves in a circle of radius s_i as shown. The particle's speed is ωs_i.

where s_i is the perpendicular distance from the point to the axis of rotation, as shown in the figure. Equation (11-21) can also be obtained without using Eq. (6-19) by noting that the point moves in a circle of radius s_i, and thus its speed is ωs_i, the magnitude of the angular velocity multiplied by the radius of this circle, as in Eq. (6-5). Furthermore, since the object is assumed to be a rigid body and the axis stationary, each point moves in a circle whose radius is *constant*. Thus for each point s_i is constant, although of course it is different for different points in the body.

We are now ready to compute the total kinetic energy. Combining Eqs. (11-17) and (11-21) gives

$$E = \Sigma \tfrac{1}{2} m_i (\omega s_i)^2 = \tfrac{1}{2} (\Sigma m_i s_i^2) \omega^2 \qquad (11\text{-}22)$$

where we have taken the common factors out of the summation. The total kinetic energy is proportional to ω^2. The sum in parentheses is a quantity which depends on the shape and configuration of the body. As we have pointed out, all the quantities in this sum are constant, provided the position of the axis does not change. Thus it is sufficient to compute the sum once and for all and to use the same value for every stage of motion.

This sum is given a special name, the *moment of inertia* of the body, and is denoted by I. Specifically, we *define* the moment of inertia of a body about a specified axis as

$$I = \Sigma m_i s_i^2 \qquad (11\text{-}23)$$

where s_i is always the *perpendicular* distance from the particle of mass m_i to the axis of rotation. With this definition, the total kinetic energy becomes

$$E = \tfrac{1}{2} I \omega^2 \qquad (11\text{-}24)$$

This expression is very similar in form to the expression $\tfrac{1}{2} m v^2$ for the kinetic energy of a particle. It contains the factor $\tfrac{1}{2}$, followed by a constant characterizing the inertial properties, followed by the square of a velocity. The individual quantities do not, however, have the same physical units in the two expressions.

11-6 MOMENTS OF INERTIA

Moments of inertia of rigid bodies play a central role in the analysis of rotational motion, and it is useful to examine various techniques for obtaining them. It is essential to keep in mind that the value of I depends on the location of the axis. Thus the moment of inertia of a body does not have the same universal significance as its mass; it is different for different axes in the same body. It is never sufficient to say that the moment of inertia of a body has a certain value; we must always say it has a certain value with respect to a particular specified axis.

Example

A dumbbell-shaped body consists of two point masses m connected by a light rigid rod of length a. This is the simplest possible model of a diatomic molecule. Find the moment of inertia of the body with respect to each of the axes shown in Fig. 11-10.

Fig. 11-10 Dumbbell consisting of two equal masses m connected by a rod of negligible mass and length a. The moment of inertia about the axis drawn as a solid line is $\frac{1}{2}ma^2$, but the moment of inertia about the broken line is ma^2.

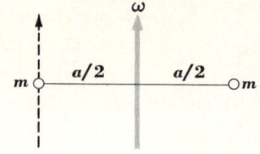

Solution
The moment of inertia about the solid line is

$$I = m\left(\frac{a}{2}\right)^2 + m\left(\frac{a}{2}\right)^2 = \tfrac{1}{2}ma^2$$

Relative to an axis parallel to the first but passing through one of the masses (the broken line) the moment of inertia is

$$I = ma^2 + 0 = ma^2$$

Relative to an axis along the line joining the two masses, the moment of inertia is zero!

Example
Find the moment of inertia about the axle of a bicycle wheel whose whole mass M may be assumed to be concentrated around its rim and whose radius is R (Fig. 11-11).

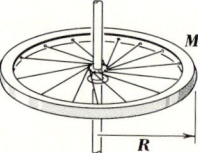

Fig. 11-11 Moment of inertia MR^2 of a bicycle wheel whose mass may be considered as concentrated at a certain radius R.

Solution
The mass is all at the same distance R from the axis. Thus the moment of inertia of such a wheel about an axis through its center perpendicular to its plane is just $I = MR^2$.

Example
Find the moment of inertia of a solid disk of uniform thickness, radius R, and total mass M about an axis through the center perpendicular to the faces of the disk.

Solution

Not all the points of the disk are at the same distance from the axis. It is necessary to divide the disk into small elements in such a way that all parts of a given element are very nearly the same distance from the axis, compute the moment of inertia of each element, and add all these to obtain the total moment of inertia. In this case, we may divide the disk into a series of thin concentric rings. Let the inner radius of a typical ring be r_i, and let its width be Δr_i, as shown in Fig. 11-12b.

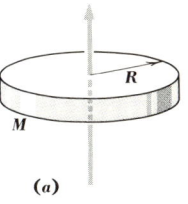

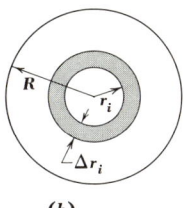

Fig. 11-12 Calculation of the moment of inertia of a uniform disk of radius R. A thin ring of inner radius r_i and width Δr_i is shown. The moment of inertia of this ring is calculated first; the total moment of inertia is obtained by summing the contributions of all such rings.

The moment of inertia of this ring is then $m_i r_i^2$, where m_i is the mass of the ring. To find the mass m_i, we observe that if the disk has uniform thickness, the ratio of m_i to the total mass M is equal to the ratio of the area the ring cuts out of the face to the total area, which is πR^2. The ring area is approximately equal to the circumference $2\pi r_i$ times the width Δr_i. Thus we have

$$\frac{m_i}{M} = \frac{2\pi r_i \, \Delta r_i}{\pi R^2} \quad \text{or} \quad m_i = \frac{2M}{R^2} r_i \, \Delta r_i$$

The total moment of inertia is then approximately

$$I = \sum m_i r_i^2 = \frac{2M}{R^2} \sum r_i^3 \, \Delta r_i \tag{11-25}$$

Finally, in the limit as the number of rings becomes very large and the width of each very small, the sum in Eq. (11-25) becomes an integral. The limits

298 SYSTEMS OF INTERACTING PARTICLES

on r must be $r = 0$ and $r = R$ to include the entire disk. Thus

$$\sum r_i^3 \, \Delta r_i \to \int_0^R r^3 \, dr = \tfrac{1}{4} R^4$$

and we finally obtain

$$I = \frac{2M}{R^2} \frac{1}{4} R^4 = \tfrac{1}{2} MR^2 \tag{11-26}$$

As might be expected, the moment of inertia of a uniform disk about an axis through its center is less than in the preceding example, since most of the mass is at a radius less than R; it turns out to be just half as great as for the thin-rim problem.

Example
Find the moment of inertia of a thin rod of length L and mass M about an axis perpendicular to the rod through one end, as shown in Fig. 11-13.

Fig. 11-13 Moment of inertia of a thin rod about an axis at one end perpendicular to its length. First the moment of inertia of a small element of length Δx_i, a distance x_i from the axis, is calculated. Then the moments of inertia of all such elements are summed to obtain the total moment of inertia.

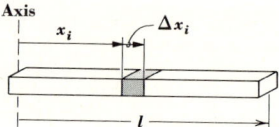

Solution
We may take as an element of mass the mass in the interval Δx_i shown in the figure. The mass m of this element is obtained from the proportion

$$\frac{m_i}{M} = \frac{\Delta x_i}{L} \quad \text{or} \quad m = \frac{M \, \Delta x_i}{L}$$

The moment of inertia is

$$I = \sum m_i x_i^2 = \frac{M}{L} \sum x_i^2 \, \Delta x_i$$

$$= \frac{M}{L} \int_0^L x^2 \, dx = \tfrac{1}{3} ML^2$$

MOMENTS OF INERTIA [11-6]

Calculation of moments of inertia is often facilitated by use of a theorem called the *parallel-axis theorem*, which is stated as follows. Let I_C be the moment of inertia of a body of mass M about an axis through the center of mass and I_0 be its moment of inertia about a second axis parallel to the first but displaced from the center of mass a distance R. The two moments of inertia are related by the equation

$$I_0 = I_C + MR^2 \tag{11-27}$$

That is, the moment of inertia about any axis is always *greater* than a parallel axis through the center of mass, by an amount that depends only on the total mass and the distance between the two axes.

To prove this important theorem, we take the z axis of a coordinate system to be the axis for I_0, as in Fig. 11-14. Let a typical mass element

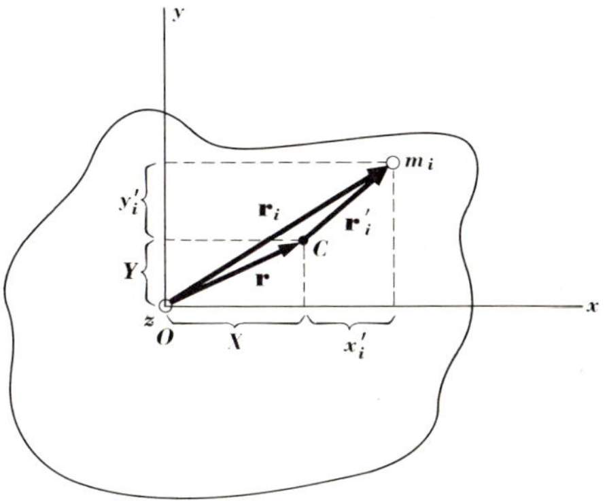

Fig. 11-14 The relation between the position of m_i with respect to O and its position with respect to C is $\mathbf{r}_i = \mathbf{R} + \mathbf{r}'_i$. This is used to derive a relation between the moments of inertia about axes through O and C.

be m_i, and let (x_i, y_i, z_i) be its coordinates in this system. The coordinates (X, Y, Z) of the center of mass C are then given by Eqs. (11-7). The distance s_i of the mass element m_i from the z axis is given by

$$s_i^2 = x_i^2 + y_i^2$$

Thus the moment of inertia about the z axis I_0 is given by

$$I_0 = \Sigma m_i s_i^2 = \Sigma m_i (x_i^2 + y_i^2) \tag{11-28}$$

The coordinates (x_i, y_i, z_i) can also be expressed in terms of the coordinates (x'_i, y'_i, z'_i) relative to the center of mass and the coordinates (X, Y, Z) of the center of mass:

$$x_i = x'_i + X \qquad y_i = y'_i + Y \tag{11-29}$$

which are the component equations corresponding to Eq. (11-14). Furthermore, the amount of inertia I_C about an axis parallel to the z axis but passing through the center of mass C is given by

$$I_0 = \Sigma m_i(x_i'^2 + y_i'^2) \tag{11-30}$$

Combining Eqs. (11-28) and (11-29), we obtain

$$I_0 = \Sigma m_i[(x_i' + X)^2 + (y_i' + Y)^2]$$

Expanding the products and rearranging gives

$$I_0 = \Sigma m_i(x_i'^2 + 2x_i'X + X^2 + y_i'^2 + 2y_i'Y + Y^2)$$
$$= \Sigma m_i(x_i'^2 + y_i'^2) + 2X\Sigma m_i x_i' + 2Y\Sigma m_i y_i' + (\Sigma m_i)(X^2 + Y^2)$$

Now $\Sigma m_i x_i' = \Sigma m_i y_i' = 0$ by definition of the center of mass; $\Sigma m_i = M$, and $X^2 + Y^2$ is the square of the perpendicular distance R between the axes. Thus we obtain $I_0 = I_C + MR^2$, as stated above.

Example
Find the moment of inertia of a solid disk about an axis perpendicular to the faces passing through the edge.

Solution
The moment of inertia about the center of mass, as obtained above, is $I_C = \frac{1}{2}MR^2$. The distance between the two axes is R, so we find

$$I_0 = \frac{1}{2}MR^2 + MR^2 = \frac{3}{2}MR^2$$

Another theorem concerning moments of inertia, less general than the parallel-axis theorem but still useful, is the *perpendicular-axis theorem*. This theorem is applicable only to a thin flat sheet, which we may take to lie in the xy plane of a coordinate system. Let I_x and I_y be the moments of inertia about the x and y axes, respectively. The theorem then states

$$I_z = I_x + I_y \tag{11-31}$$

Proof of this theorem, which is much simpler than for the parallel-axis theorem, is left as a problem.

Example
Find the moment of inertia of a thin uniform disk of mass M and radius R about a diameter.

Solution
Let the disk lie in the xy plane with its center at the origin. Then from previous calculations $I_z = \tfrac{1}{2}MR^2$. We are asked to find I_x and I_y, which from symmetry we know are equal. The perpendicular-axis theorem gives

$$\tfrac{1}{2}MR^2 = I_x + I_y = 2I_x$$

and

$$I_x = I_y = \frac{1}{4}MR^2$$

Figure 11-15 gives the moments of inertia of a few familiar shapes. In calculating a moment of inertia, dimensions measured *parallel* to the axis

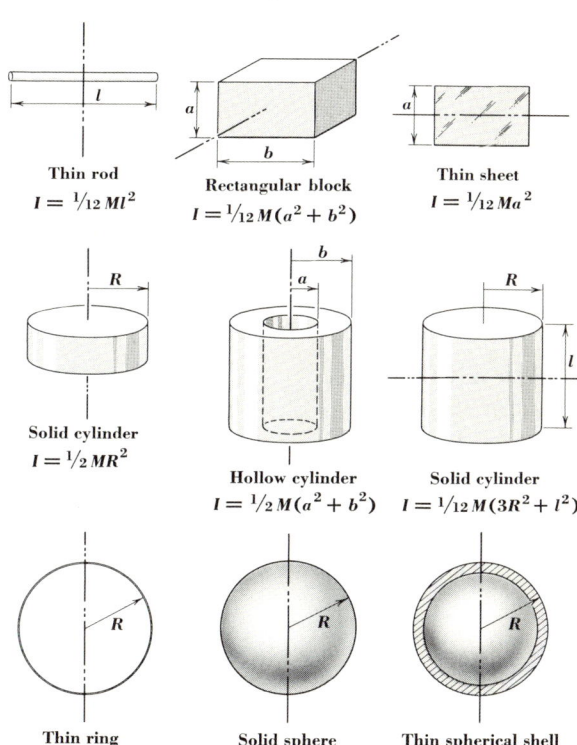

Fig. 11-15 Moments of inertia of a few familiar shapes. In all the solid figures, the density is assumed to be uniform.

Thin rod
$I = \tfrac{1}{12}Ml^2$

Rectangular block
$I = \tfrac{1}{12}M(a^2 + b^2)$

Thin sheet
$I = \tfrac{1}{12}Ma^2$

Solid cylinder
$I = \tfrac{1}{2}MR^2$

Hollow cylinder
$I = \tfrac{1}{2}M(a^2 + b^2)$

Solid cylinder
$I = \tfrac{1}{12}M(3R^2 + l^2)$

Thin ring
$I = \tfrac{1}{2}MR^2$

Solid sphere
$I = \tfrac{2}{5}MR^2$

Thin spherical shell
$I = \tfrac{2}{3}MR^2$

of rotation are of no consequence, since only the distance perpendicular to the axis appears in the definition (Fig. 11-16). For example, in the calculation of the moment of inertia of a uniform disk about its axis, the *thickness* of the disk was immaterial; the result is also correct for a long homogeneous cylinder of the same diameter and total mass. Similarly, the calculation for a thin rod is also valid for a thin rectangular sheet whose axis of rotation lies along one edge. The dimension l in this case is the length of a side of the sheet perpendicular to the axis.

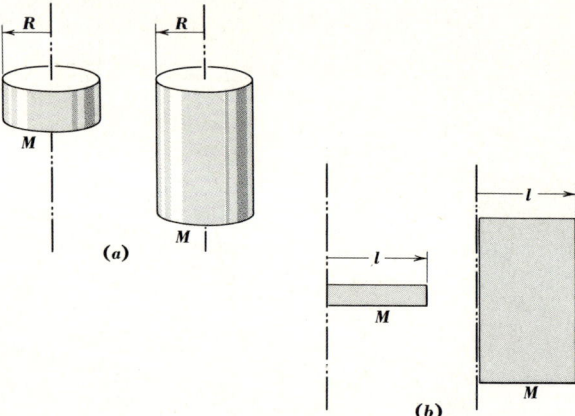

Fig. 11-16 (*a*) Moment of inertia of a homogeneous solid cylinder about its axis is independent of the cylinder's length. The two cylinders shown have equal moments of inertia if their total masses are equal. (*b*) The moment of inertia of a thin rectangular sheet about an axis along one edge is independent of the dimension parallel to the axis. The two sheets shown have equal moments of inertia if their masses are equal.

In addition to kinetic-energy considerations, moments of inertia are useful in other aspects of the dynamics of rotational motion. It will be seen in Chaps. 12 and 13 that a rotating body undergoes an *angular acceleration* when acted upon by *torques*, or *moments*, and it will be shown that the torque necessary to produce a given angular acceleration about a fixed axis is equal to the moment of inertia of the body about this axis multiplied by the angular acceleration. This result, analogous to Newton's second law for a single particle, $\Sigma \mathbf{F} = m\mathbf{a}$, is not a new fundamental principle but is *derived* from Newton's laws by defining and analyzing new quantities.

11-7 EXAMPLES

To conclude this chapter, we consider a few examples of the use of the new concepts, especially those relating to rotation of a rigid body.

Example

In the situation shown in Fig. 11-17, a heavy flywheel in the shape of a uniform disk with mass $M = 32$ kg and radius $R = \frac{1}{2}$ m rotates on a shaft

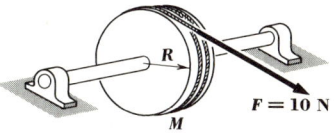

Fig. 11-17 Flywheel in the form of a uniform disk mounted on bearings which may be considered frictionless. A rope wound several times around the rim of the flywheel is pulled with a constant force.

with bearings which may be considered frictionless. A rope wrapped around its rim several times is pulled with a constant force $F = 10$ N. How fast is the flywheel turning after the rope has been pulled a distance of 5 m?

Solution

First, the force does an amount of work on the flywheel equal to

$$W = (10 \text{ N})(5 \text{ m}) = 50 \text{ J}$$

Since there are no frictional or other nonconservative forces, this work must appear as kinetic energy of the flywheel. The moment of inertia of the flywheel is given by

$$I = \tfrac{1}{2}MR^2 = \tfrac{1}{2}(32 \text{ kg})(\tfrac{1}{2} \text{ m})^2 = 4 \text{ kg-m}^2$$

and its kinetic energy, which we know is 50 J at the end of the process, is

$$E = \tfrac{1}{2}I\omega^2 = \tfrac{1}{2}(4 \text{ kg-m}^2)\omega^2 = 50 \text{ J}$$

Solving this for ω, we find

$$\omega = 5 \text{ s}^{-1}$$

The speed of the rope is always equal to the speed of a point on the rim of the flywheel; at the end of the process, this is

$$v = R\omega = (\tfrac{1}{2} \text{ m})(5 \text{ s}^{-1}) = \tfrac{5}{2} \text{ m/s}$$

Thus, using the energy principle, we have been able to find the final velocities of the various components of the system. The energy principle does not, however, enable us to find directly the *accelerations* at various times during the motion. To do this, we must use methods to be presented in the following chapters.

Example

Suppose we make a Yo-Yo out of a uniform disk of mass M and radius R, wind a string around it several times, tie the other end to the ceiling, and release it from rest, with the string initially vertical. The situation is shown in Fig. 11-18. How fast is the Yo-Yo traveling after it has dropped a distance h?

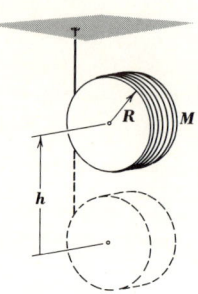

Fig. 11-18 Yo-Yo made from a uniform disk and a string. The top of the string is tied to the ceiling, and the Yo-Yo is released from rest. As it drops, the string unwinds, so the angular velocity of rotation about the center of mass is proportional to the velocity of the center of mass.

Solution

The body loses an amount of potential energy Mgh; this must appear as kinetic energy after it has descended the distance h, assuming that there are no frictional or dissipative forces. But unlike the previous example, the body is not rotating about a fixed axis; there are both motion of the center of mass and rotation about the center of mass. We may use Eq. (11-20) to find the total kinetic energy. The kinetic energy associated with center-of-mass motion is simply $\frac{1}{2}MV^2$. The part associated with motion relative to the center of mass can be computed simply by finding the moment of inertia about an axis through the center of mass and multiplying it by the square of the angular velocity of the rotation about that axis. Thus, equating the loss of potential energy to the total gain in kinetic energy gives

$$Mgh = \tfrac{1}{2}MV^2 + \tfrac{1}{2}I\omega^2 \tag{11-32}$$

To simplify this, we recall that the moment of inertia in question is $I = \frac{1}{2}MR^2$. Also, since the center of mass drops at a rate equal to the rate at which string unwinds, it must be true that $V = R\omega$. Using these equations in the second right-hand term of Eq. (11-32), we have

$$Mgh = \tfrac{1}{2}MV^2 + \tfrac{1}{2}(\tfrac{1}{2}MR^2)\left(\frac{V}{R}\right)^2 = \tfrac{3}{4}MV^2$$

Finally,

$$V = \sqrt{\tfrac{4}{3}gh}$$

We note that M drops out of the final result. This is to be expected; the inertial properties are all proportional to M, but the force producing the acceleration, namely gravity, is also proportional to M, so the mass cancels out of the equations of motion. We also note that V here is *smaller* than the speed acquired by a body in free fall from a height h, which is $(2gh)^{1/2}$. This, too, is to be expected.

For a little comic relief, we consider an *incorrect* solution of the previous problem. Suppose we use a coordinate system which moves with the Yo-Yo. In this system, the center of mass is at rest. With respect to the center-of-mass coordinate system, the total kinetic energy is just the second term in the above energy expressions, namely

$$E = \tfrac{1}{2}I\omega^2 = \tfrac{1}{2}(\tfrac{1}{2}MR^2)\left(\frac{V}{R}\right)^2 = \tfrac{1}{4}MV^2$$

Equating this to the loss of potential energy as before, we find

$$Mgh = \tfrac{1}{4}MV^2 \quad \text{and} \quad V = \sqrt{4gh}$$

This is *larger* than the speed of free fall from the same height and is a quite erroneous result.

What went wrong? The essential point, and an extremely important one to keep in mind, is that the principle of conservation of energy, which relates the potential and kinetic energies, was derived using Newton's laws of motion, which are valid only in an *inertial frame of reference*. But we have computed the kinetic energy in a noninertial frame of reference, namely the center-of-mass coordinate system, which is accelerated with respect to an inertial frame of reference. The principle of conservation of energy may not be used directly in this coordinate system.

In general, when using any kind of energy principle, one must compute the kinetic energy from the motion of the system with respect to an inertial frame of reference. If a noninertial frame of reference is used, the results will be erroneous unless appropriate corrections are made. It may appear at first glance that a noninertial frame of reference is being used in Eq. (11-20). It is true that Eq. (11-20) contains $\mathbf{v}'_i$, the velocities with respect to a point which may be accelerated; we recall, however, that this equation was derived from Eq. (11-17), which contains the velocities in an inertial frame of reference. Thus, although Eq. (11-20) uses the center-of-mass system, it still gives the kinetic energy as observed in an inertial frame of reference whose origin is O and *not* as observed in the center-of-mass system.

Problems

11-1 Find the position of the center of mass of the system consisting of the earth and the moon, regarding each as a point mass. The data in Appendix C will be useful. How far is the center of mass from the surface of earth?

11-2 Find the position of the center of mass of the system consisting of the sun and the earth, regarding each as a point mass. Is the center of mass inside or outside the sun?

11-3 Find the position of the center of mass of a hydrogen atom, consisting of an electron and a proton 1,837 times as massive as the electron, if each particle can be regarded as a point.

11-4 Is it possible for a system to have *no mass* at the center of mass? If so, give an example. If not, explain.

11-5 The distance between nuclei of a molecule of carbon monoxide, CO, is about 1.13×10^{-10} m. Where is the center of mass?

11-6 Two particles whose masses are m_1 and m_2 are separated by a distance r. Derive general expressions for the distance from the center of mass to each particle.

11-7 Find the position of the center of mass of a system consisting of four particles of equal mass located at points whose coordinates are (1 cm, 0, 0), (0, 1 cm, 0), (0, 0, 1 cm), and (1 cm, 1 cm, 1 cm).

11-8 Find the position of the center of mass of a thin rod of total mass M which has a right-angle bend at its midpoint, forming two perpendicular legs, each of length d.

11-9 Two cars are traveling in the same direction on a highway. One has a mass of 1,000 kg and is traveling 10 m/s. Ahead of it a distance of 30 m is another car, whose mass is 2,000 kg and whose speed is 20 m/s.
   a Find the position of the center of mass of the system.
   b Compute the total momentum directly from the mass and velocity of each car.
   c Compute the velocity of the center of mass.
   d Compute the total momentum from the total mass and the velocity of the center of mass. Compare result with that of part *b*.

11-10 Consider a two-body oscillator such as that discussed in Sec. 10-5. Write an equation describing the motion of each mass relative to the center of mass if the relative motion of the two masses is given by Eq. (10-34).

11-11 A ball of mass $3m$ rolls on a smooth flat table toward another ball of mass m with constant velocity v.

a When the two balls are a distance d apart, find the position and velocity of the center of mass.

b Find the total momentum from the velocity of the center of mass. Compare this with the result obtained directly from the velocities of the two balls.

11-12 A foolish canoeist stands up in a canoe and walks from a point 3 ft from the bow end to a point 3 ft from the stern. The total length of the canoe is 16 ft, and its weight is 75 lb. The canoeist's weight is 150 lb. If the canoe is initially at rest, how far does it move during this process?

11-13 A wagon with a flat bottom contains a heavy steel ball. The wagon is pulled with a certain constant force. Consider two possible situations: The ball may be free to roll, or it may be blocked so it cannot roll.

a In which case is the acceleration of the wagon greater? Explain.

b In which case is the acceleration of the center of mass greater? Explain.

11-14 Consider the Atwood's machine of Prob. 5-24 as a system consisting of two masses, the string and the pulley having negligible mass.

a Compute the acceleration of the center of mass directly from the accelerations of the two masses.

b Compute the acceleration of the center of mass by finding the total force on the system and its total mass. Compare with the result of part *a*.

11-15 A body of mass m_1 collides with a body of mass m_2. Using the center-of-mass coordinate system, show that if the total kinetic energy is the same after the collision as before, then the *relative velocity* of the two bodies has the same magnitude before and after the collision.

11-16 A body of mass m and initial velocity **v** collides elastically with a body of mass $2m$ initially at rest.

a Describe the initial motion as viewed from the center-of-mass coordinate system. Draw a vector diagram.

b If the velocity of the first particle after the collision is perpendicular to its original velocity, as viewed from the center-of-mass system, find the directions of the two final velocities in the "laboratory" system.

11-17 A particle of mass m_1 and initial kinetic energy E collides elastically with a particle of mass m_2 initially at rest. What is the maximum energy the first particle can lose during this collision? *Hint:* Use the center-of-mass coordinate system. The maximum energy loss for m_1 occurs during a head-on collision in which the final velocities are along the same line as the initial velocities.

11-18 In a nuclear-fission reactor, neutrons produced by fission of a uranium nucleus must be slowed down so that they can be absorbed by other nuclei and produce more fissions. This slowing down is accomplished by elastic collisions with nuclei in the *moderator* region of the reactor. If it is desired to slow the neutrons down with as few collisions as possible, what elements should be used for the material of the moderator? Why?

11-19 In problems involving two particles, it is often convenient to describe their positions by means of the position vector $\mathbf{R}$ of the center of mass and a vector $\mathbf{r} = \mathbf{r}_2 - \mathbf{r}_1$ which gives the position of particle 2 relative to particle 1.

a Show how each of the position vectors $\mathbf{r}_1$ and $\mathbf{r}_2$ can be expressed in terms of $\mathbf{R}$ and $\mathbf{r}$.

b How can the total momentum be expressed in terms of $\mathbf{R}$, $\mathbf{r}$, and their time derivatives?

c Show that the total kinetic energy is given by $E = \tfrac{1}{2}MV^2 + \tfrac{1}{2}\mu v^2$, where $\mathbf{v} = d\mathbf{r}/dt$ and μ is the reduced mass of the system as defined by Eq. (10-35).

11-20 For the situation described in Prob. 11-9:

a Compute the total kinetic energy directly from the speeds of the cars.

b Compute the speed of each car relative to the center of mass.

c Using the results of part *b*, those of Prob. 11-9, and Eq. (11-20), find the total kinetic energy. Compare the result with that of part *a*.

11-21 It is desired to raise an atom from its lowest energy state to a higher energy state by adding a quantity of energy E by bombardment by an electron. The mass of the atom is M, and that of the electron m. What minimum kinetic energy must the electron have? *Hint:* The answer is *not* E. The electron cannot give all its energy to the atom because the energy associated with center-of-mass motion is constant. Only that part of the total energy which is associated with motion *relative to* the center of mass can be lost.

11-22 Find the moment of inertia of a thin uniform rod of length l and total mass m about an axis through the center perpendicular to the length of the rod. The problem can be solved two ways, by integrating directly or by using the parallel-axis theorem with the result of the example in Sec. 11-6. Check that both methods yield the result given in Fig. 11-15.

11-23 A dumbbell rotates about an axis through its center of mass perpendicular to the dumbbell axis, as shown in Fig. 11-10. Each mass is 0.1 kg, and the distance between the masses is 0.2 m. The angular velocity is 10 rad/s.

a Find the speed of each mass. From this, compute the total kinetic energy.

b Compute the moment of inertia of the system; from this and the angular velocity, find the total kinetic energy. Compare this result with that of part *a*.

11-24 Consider a body with rotational symmetry whose maximum radius is R and whose total mass is M. Is it ever possible for the moment of inertia to be larger than MR^2? Explain.

11-25 Derive an expression for the moment of inertia of a washer with inner and outer radii a and b, respectively, and mass M, about an axis through the center, perpendicular to the plane of the washer.

PROBLEMS

11-26 The flywheel in a certain automobile engine is made in the form of a uniform disk 0.2 m in radius and 0.02 m thick. Its total mass is 10 kg. Find the kinetic energy of rotational motion when the engine is running at 5,000 r/min. Is this large or small compared with the total kinetic energy of a car traveling 60 mi/h?

11-27 Why is the word "approximately" necessary in the sentence preceding Eq. (11-25)?

11-28 Find the moment of inertia of a thin circular ring of mass M and radius R about:
 a A diameter
 b A line tangent to the ring
 c A line perpendicular to the plane of the ring at its edge

11-29 For the rectangular block in Fig. 11-15, find the moment of inertia about an axis parallel to the axis in the figure but lying along an edge perpendicular to the face (a,b). This can be obtained by direct integration or by using the result in Fig. 11-15 with the parallel-axis theorem. Use both methods, and check to see that your results agree.

11-30 Make a rough estimate of the moment of inertia of your body:
 a About a vertical axis from head to toe through your center of mass
 b About a horizontal axis from front to back through your center of mass

11-31 Derive the expression for the moment of inertia of a spherical shell given in Fig. 11-15 by slicing the shell perpendicular to the axis into ring-shaped elements. Be careful to calculate the mass of an element correctly.

11-32 Derive the expression for the moment of inertia of a solid sphere given in Fig. 11-15. This may be done by slicing the sphere into disks perpendicular to the axis or by thinking of it in terms of concentric spherical shells. In either case, the mass of the appropriate element is most easily obtained in terms of the (constant) density of the material.

11-33 In the situation shown in Fig. P11-33, mass m_1 is attached to a string wrapped several times around the pulley, which is a uniform disk free to turn about the axis shown.

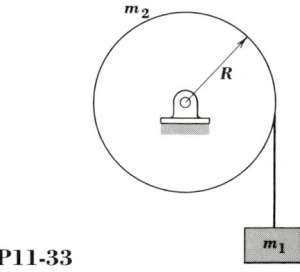

Fig. P11-33

a If the system starts from rest, and if the string unwinds without slipping, find the speed of m_1 and the angular velocity of the pulley after m_1 has dropped a distance h.

b Compute numerical values for part *a* if $m_1 = 2$ kg, $m_2 = 10$ kg, $R = 0.1$ m, and $h = 0.5$ m.

11-34 A Yo-Yo is made from two uniform disks of radius b connected by a light axle whose radius is a, as shown in Fig. P11-34. The total mass is M. A

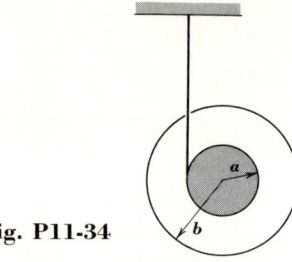

Fig. P11-34

string is tied to the axle, wound around it several times, and held vertically; the Yo-Yo is then released from rest.

a If the thickness of the string may be neglected, find the speed of the axle after the Yo-Yo has dropped a distance h.

b Compute a numerical value for part *a* if $a = 0.4$ cm, $b = 4.0$ cm, $M = 50$ kg, and $H = 50$ cm.

11-35 For the Atwood's machine in Fig. P11-35, assume that m_1 is larger than m_2. The pulley is a uniform disk of mass M, and the string does not slip. Find

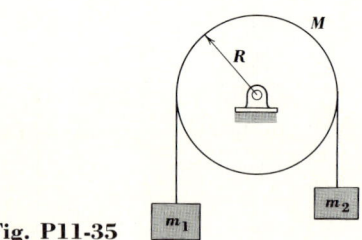

Fig. P11-35

the speeds of the masses after m_1 has dropped a distance h if the system starts from rest.

11-36 A lawn roller is made in the form of a thin cylindrical shell with thin spokes, as shown in Fig. P11-36. It is pulled horizontally a distance d by a constant force F.

Fig. P11-36

- *a* If the roller starts from rest, find its final speed.
- *b* If the initial speed is v, find the final speed.
- *c* Compute a numerical value for part *a* if $M = 50$ kg, $R = 0.4$ m, $F = 100$ N, and $d = 5$ m.

Angular Momentum | 12

The concept of angular momentum provides the key to detailed analysis of the rotational motion of rigid bodies. The concept is introduced first with respect to motion of a point mass. It is shown that Kepler's second law is equivalent to conservation of angular momentum under a central force, and this principle is applied to a problem of scattering of atomic particles. Then the angular momentum of a system of particles is defined, and its simple relation to the torques due to the external forces on the system is developed. It is shown that for a body rotating about a fixed axis which is an axis of rotational symmetry, the angular momentum can be expressed simply in terms of the moment of inertia and angular velocity. These developments form the dynamical principles which relate force and rotational motion.

12-1 PLANETS AND SATELLITES

Although the main task of this chapter is to develop principles applicable to the rotational motion of rigid bodies, we first introduce the concept of angular momentum for a body which can be represented as a single particle. An important application of this principle occurs in the study of the motion of planets in the solar system, which, in fact, provided the seed from which the concept of angular momentum grew. The discussion of angular momentum of a particle also leads naturally to an interesting problem concerning the interaction of atomic particles.

In Sec. 8-3 we discussed empirical laws, discovered by Kepler, dealing with the motions of the planets. In particular, we showed that Kepler's *third* law, relating the period of a planet to the radius of its orbit, is a direct consequence of Newton's laws, together with the law of gravitation. This relationship, of course, was not known to Kepler, whose work preceded that of Newton.

We now consider how another of Kepler's laws is related to Newton's laws. Kepler's *second* law states that the radius line from the sun to any planet sweeps out area at a constant rate. In Fig. 12-1, if the time required

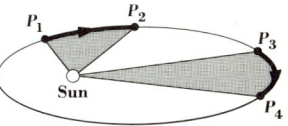

Fig. 12-1 According to Kepler's second law, if the times required to traverse the two segments P_1 to P_2 and P_3 to P_4 are equal, then the shaded areas are equal. The rate at which area is swept out, called the *sector velocity*, is constant in such a case.

for the planet to travel from point P_1 to P_2 is the same as the time required to go from P_3 to P_4, then the two shaded areas are equal. The rate at which the radius line sweeps out area is sometimes called the *sector velocity*; according to Kepler, the sector velocity of each planet is constant, although it is different for different planets.

The sector velocity can easily be expressed in terms of the polar coordinates of the planet. Suppose the sun is located at the origin of a polar coordinate system and the planet in question is located at point P, with coordinates (r,θ), as in Fig. 12-2. When a point moves from P_1 to P_2 in

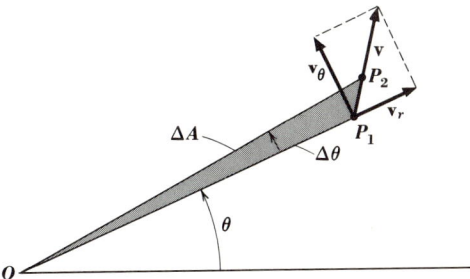

Fig. 12-2 Instantaneous sector velocity can be related to the polar coordinates and their derivatives by computing the element of area swept out in a small time Δt and taking the limit as $\Delta t \to 0$.

a time interval Δt, how much *area* is swept out? The area ΔA of the sector shown in the figure is approximately $\Delta A = \frac{1}{2} r^2 \Delta \theta$. Thus the average rate at which area is swept out is approximately

$$\frac{\Delta A}{\Delta t} \simeq \frac{1}{2} r^2 \frac{\Delta \theta}{\Delta t}$$

In the limit as $\Delta t \to 0$, the instantaneous sector velocity is

$$\frac{dA}{dt} = \tfrac{1}{2} r^2 \frac{d\theta}{dt} \tag{12-1}$$

The statement that the sector velocity of a planet is constant is equivalent to saying that the quantity $r^2 \, d\theta/dt$ is constant.

The sector velocity can easily be related to the velocity of the planet. We recall from Sec. 6-5 that the components of velocity in polar coordinates are

$$v_r = \frac{dr}{dt} \qquad v_\theta = r \frac{d\theta}{dt}$$

Thus the sector velocity can be expressed as

$$\frac{dA}{dt} = \tfrac{1}{2} r v_\theta \tag{12-2}$$

Because we want to relate Kepler's second law to principles of dynamics, it is reasonable to express this law in terms of the momentum **p** of the planet. Clearly, if rv_θ is constant, then so is rp_θ, where $p_\theta = mv_\theta$ is the component of momentum in the θ direction. This is the first of many appearances of the quantity rp_θ; anticipating the future, we give it a special name, *angular momentum*. Denoting this new quantity by L, we make the following definition: The angular momentum L of a point of mass m moving in a plane (described by a polar coordinate system) with respect to point O is defined as

$$L = rp_\theta = rmv_\theta = mr^2 \frac{d\theta}{dt} \tag{12-3}$$

Using this new term, we may restate Kepler's second law as follows: *The angular momentum of a planet moving around the sun is constant.*

The angular momentum of a particle can be expressed in several other ways. Referring to Fig. 12-3, in which **p** is represented by means of its polar components p_r and p_θ, we see that the component p_θ of momentum can be expressed in terms of the *magnitude p* of the momentum vector and the angle ϕ between the directions of **p** and the position vector **r**: $p_\theta = p \sin \phi$. Thus we obtain

$$L = rp_\theta = rp \sin \phi \tag{12-4}$$

This equation can now be reinterpreted as follows. In Fig. 12-3, the quantity

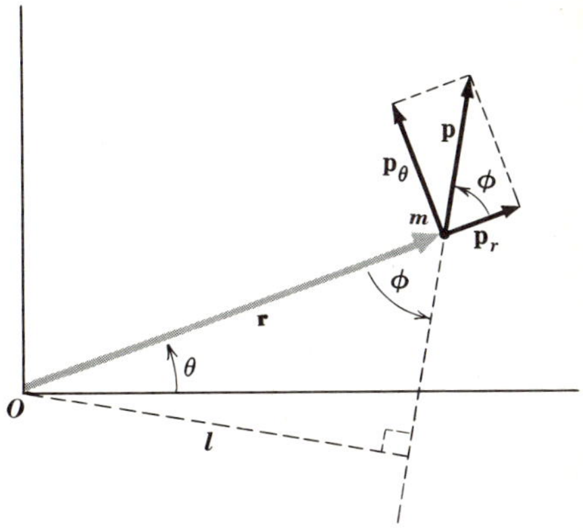

Fig. 12-3 Components p_r and p_θ of the vector **p**. The moment arm l is obtained by extending the line along which **p** lies and drawing a line perpendicular to it from the origin. The equality of the various forms of Eq. (12-5) can be seen from this figure.

$r \sin \phi$ is equal to the perpendicular distance l between the line along which p lies and the origin. Collecting all the equivalent expressions for L, we have

$$L = mr^2 \frac{d\theta}{dt} = rmv_\theta = rp_\theta = rp \sin \phi = lp \tag{12-5}$$

The perpendicular distance l between the line of **p** and the origin is called the *moment arm* or *lever arm* of the vector **p**. Perhaps a more familiar use of this concept is the *moment*, or *torque*, of a force, which can be defined as the product of the magnitude of the force and the perpendicular distance from its line of action to the origin. Because of this similarity, angular momentum is occasionally called the *moment of momentum*; in analogy with the *moment of a force*, it is the product of the momentum and its associated *moment arm* from the origin. The moment of a force is discussed in detail in Sec. 12-2. In the definition of angular momentum, it is essential to specify the origin; if two different origins are used, the results are in general different.

The various expressions in Eq. (12-5) show that the angular momentum L may be positive or negative. If $d\theta/dt$ is positive, corresponding to motion in which v_θ and p_θ are in the same direction as the unit vector **l**, L is positive; in such cases the angle ϕ, as defined in Fig. 12-3, is positive. If the reverse is true, L is negative. In the case of planetary motion, a positive value of L corresponds to a counterclockwise sense of rotation, a negative value to a clockwise sense.

12-2 DYNAMICS OF PLANETARY MOTION

Kepler's second law, the constancy of angular momentum of a planet in its orbit around the sun, was discovered as an empirical law, but like the third

316 ANGULAR MOMENTUM

law, discussed in Sec. 8-3, it can also be *derived* from Newton's laws. In discussing this relationship, we first review the formulation of $F = ma$ in polar coordinates. It was found in Sec. 6-5 that the polar components of acceleration for a point are

$$a_r = \frac{d^2r}{dt^2} - r\left(\frac{d\theta}{dt}\right)^2$$

$$a_\theta = r\frac{d^2\theta}{dt^2} + 2\frac{dr}{dt}\frac{d\theta}{dt}$$

When the total force **F** on a mass m is also represented by means of its polar components F_r and F_θ, the equations of motion are

$$F_r = m\left[\frac{d^2r}{dt^2} - r\left(\frac{d\theta}{dt}\right)^2\right] \tag{12-6a}$$

$$F_\theta = m\left(r\frac{d^2\theta}{dt^2} + 2\frac{dr}{dt}\frac{d\theta}{dt}\right) \tag{12-6b}$$

If Eq. (12-6b) is now multiplied by r, the quantity in parentheses is equal to the time derivative of the quantity $r^2\, d\theta/dt$. So Eq. (12-6b) can be rewritten

$$rF_\theta = m\left(r^2\frac{d^2\theta}{dt^2} + 2r\frac{dr}{dt}\frac{d\theta}{dt}\right) = \frac{d}{dt}\left(mr^2\frac{d\theta}{dt}\right) \tag{12-7}$$

Comparing this with Eq. (12-3), we see that the right side is simply the time rate of change of angular momentum. That is,

$$rF_\theta = \frac{dL}{dt} \tag{12-8}$$

We are now on the trail of something quite interesting. The rate of change of angular momentum is equal to the product of the force component F_θ and the distance r from the origin to the point at which the force is applied.

Just as the angular momentum was expressed in several equivalent forms in Eq. (12-5), the quantity rF_θ may be expressed in various ways. Referring to Fig. 12-4, which is completely analogous to Fig. 12-3, we see that

$$rF_\theta = rF\sin\phi = lF \tag{12-9}$$

where ϕ is the angle between r and F, and l is the perpendicular distance from O to the line of action of the force, called the *moment arm* or *lever arm* of the force.

The quantity rF_θ is called the *moment* or *torque* of the force **F** about

317 DYNAMICS OF PLANETARY MOTION [12-2]

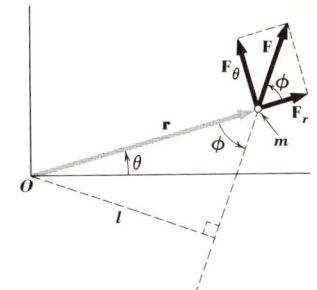

Fig. 12-4 Angle θ between the vectors **r** and **F** and the lever arm of the force. From this figure the equality of the various forms of Eq. (12-9) can be verified.

the point O. It is positive if the force tends to produce a counterclockwise rotation about O, negative if clockwise. Denoting the torque by τ, we may write Eq. (12-8) simply as

$$\tau = \frac{dL}{dt} \tag{12-10}$$

which may be regarded as a rotational analog of $\mathbf{F} = d\mathbf{p}/dt$.

What does all this have to do with Kepler's second law? Simply this: We have shown that this law is equivalent to the statement that the angular momentum of a planet is constant, so that $dL/dt = 0$. Since r is in general not zero, this implies that F_θ is *always* zero. That is, Kepler's second law is obeyed in any situation in which the force is always *radial* (directed along the position vector **r**).

This should not be surprising. In planetary motion the force is just the gravitational attraction of the sun on the planet, and of course it *is* radial. But in Newton's day this was not nearly so obvious; in fact, Newton used Kepler's second law, among other things, as a basis for concluding that the gravitational attraction between a planet and the sun *must* be radial and that the associated force field is a central-force field.

Thus, starting from the opposite end of the problem, we have arrived at the same conclusion as in Sec. 8-4, where we started with the assumption that the force field was central and deduced that the angular momentum must be constant. But now we have an added dividend; if the force field is *not* central, then the force has a θ component as well as an r component. In this case, the angular momentum changes at a rate equal to the torque $\tau = rF_\theta = lF$ due to this force.

Example

An earth satellite moves in an elliptical orbit; its closest distance (perigee) from the center of the earth is 5,000 mi, and its farthest distance (apogee) is 15,000 mi (both distances measured from the center of the earth). If its velocity at perigee is 2,000 mi/h, what is its velocity at apogee?

Solution

The earth's gravitational attraction is a central force, just as in the case of motion of planets around the sun, so the angular momentum of the satellite is constant. At apogee and perigee, $v_r = 0$ (why?) and $v = v_\theta$. Since rv_θ is constant, we have

$$(5,000 \text{ mi})(20,000 \text{ mi/h}) = (15,000 \text{ mi})(v_{ap})$$

and $v_{ap} = 7,000$ mi/h. Note that it is essential to measure distances from the *center* of the earth, not its surface. Why?

12-3 RUTHERFORD SCATTERING

We now consider a problem involving a collision between atomic nuclei of a particular type known as *Rutherford scattering*. This problem is important for a variety of reasons. It provides an illustration of the use of angular momentum, along with other concepts, to simplify the analysis of such problems, and it illustrates how scattering experiments can provide valuable information on the structure of matter. In this case the experiments provided the first concrete evidence for the nuclear model of the atom.

To appreciate the significance of Rutherford scattering, we consider the state of knowledge of atomic structure in 1910. It was known then that atoms are made of electrically charged particles, that the negatively charged particles are electrons whose mass accounts for only about 1/4,000 of the total mass of an atom, and that the overall sizes of atoms are of the order of 10^{-8} to 10^{-7} cm. What was *not* known was the *distribution* of the *positive* charge, which presumably had associated with it most of the mass of the atom. It was commonly supposed to be distributed uniformly throughout the volume of the atom, as shown in Fig. 12-5a, but there was no concrete evidence that this was actually the case. Another hypothesis was that of a concentration of mass and positive charge in a much smaller region at the center of the atom, as in Fig. 12-5b; this later came to be called the *nuclear* model of the atom.

Which, if either, of these models is correct? To investigate this question, Rutherford proposed to fire charged particles at a target and observe how the particles were deflected, or *scattered*, by collisions with the atoms. If the positive charge is uniformly distributed, the scattering should be similar

Fig. 12-5 Two different pictures of the internal structure of an atom. In (a) the positive charge and mass are assumed to be distributed more or less uniformly throughout the volume; in (b) the positive charge and most of the mass are concentrated in a very small nucleus. The small black dots in both figures represent electrons. The concentrated electric charge in (b) permits much larger electric forces than in (a), and α particles are correspondingly deflected more strongly.

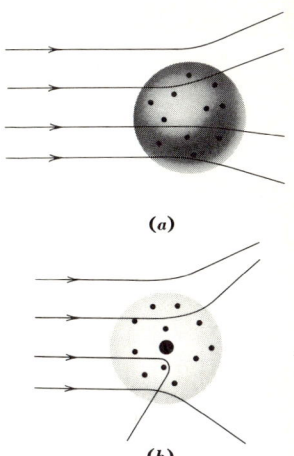

to that shown in Fig. 12-5a, while if the nuclear model is correct, the electric force close to the nucleus should be much larger than in the other model, and a few particles approaching close to the nucleus should be deflected through much larger angles, as in Fig. 12-5b. This procedure has been compared to locating a brick inside a bale of hay by firing a machine gun into the bale and observing how the bullets are scattered. Rutherford's target was a very thin gold foil, and the "missiles" were α particles, which are helium nuclei emitted by naturally radioactive substances.

To make any quantitative comparison of the α scattering predicted by either model with that actually observed, it is necessary to calculate in detail just what each model predicts. Concentrating on the nuclear model, we now calculate the scattering of an α particle on the assumption that the nucleus of a gold atom may be represented by a point. We consider only the interaction of the α particle and the nucleus. The electrons, being much less massive than the α particle, cannot deflect it appreciably, just as a truck can drive through a hailstorm without being deflected appreciably. Also, we regard the gold nucleus as stationary because it is much more massive than the α particle. The interaction force between the two is assumed to be their electrical repulsion, a force of magnitude

$$F = \frac{k}{r^2} \tag{12-11}$$

where r is the distance between the two particles and k is a constant proportional to the product of the electric charges.

Placing the gold nucleus at the origin of a polar coordinate system, we state the simplified problem as follows: If a particle of mass m approaches with an initial velocity v_0 and experiences a repulsive force directed away from the origin with magnitude $F = k/r^2$, through how large an angle is it deflected?

A typical trajectory is shown in Fig. 12-6. When the α particle is very

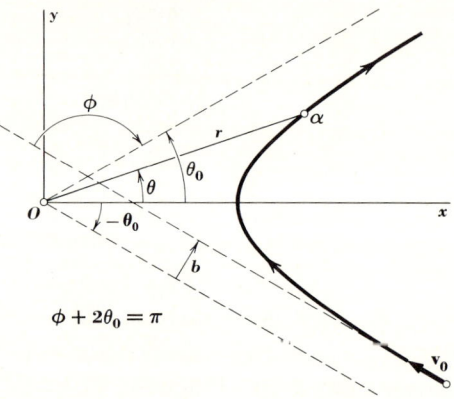

Fig. 12-6 Coordinate system and a typical trajectory for Rutherford scattering. As the α particle approaches from a great distance, θ initially has the value $-\theta_0$ shown; when the α is again a great distance away, it has the value $+\theta_0$. The axes have been chosen so the trajectory is symmetric with respect to the x axis. The impact parameter b and the scattering angle ϕ are shown.

far away from O, where the force is essentially zero, it moves in a straight line with constant velocity $\mathbf{v}_0$. In general, its direction is not exactly toward the gold nucleus. The distance b in the figure represents the minimum distance by which the α particle would miss the gold nucleus if the repulsive force were not present and the α could continue in a straight line; this distance is called the *impact parameter* for the collision. The angle ϕ in the figure represents the total change in the direction of motion of the α as a result of the collision; this is called the *scattering angle*. Our goal is to find a relation between the scattering angle ϕ and the impact parameter b. Once this relationship is known, we can compare it with experimental data to see whether the scattering of particles by gold nuclei actually behaves according to the simple nuclear model of atomic structure.

Because the interaction force between the two particles is a central force, the angular momentum of the α particle is constant. The initial angular momentum is the moment arm b times the initial momentum mv_0, or bmv_0, according to the last form of Eq. (12-5). Thus we have the relation

$$L = mr^2 \frac{d\theta}{dt} = bmv_0 = \text{constant} \tag{12-12}$$

321 RUTHERFORD SCATTERING [12-3]

We now use this equation to represent the force in terms of the angular velocity $d\theta/dt$. Solving Eq. (12-12) for $1/r^2$ and dividing through by m, we have

$$\frac{1}{r^2} = \frac{1}{bv_0}\frac{d\theta}{dt} \tag{12-13}$$

Combining this with Eq. (12-11), we obtain

$$F = \frac{k}{bv_0}\frac{d\theta}{dt} \tag{12-14}$$

We can now use Eq. (12-14) to obtain a relation between the rate of change of momentum and $d\theta/dt$ and thus between the *total* change of momentum and the *total* change of θ, which is related to the scattering angle ϕ. In particular, for the x components, we have

$$F_x = F\cos\theta = \frac{k}{bv_0}\frac{d\theta}{dt}\cos\theta \tag{12-15}$$

Equating F_x to dp_x/dt and integrating over the time during which the motion occurs gives

$$\int \frac{dp_x}{dt}\,dt = \frac{k}{bv_0}\int \cos\theta\,\frac{d\theta}{dt}\,dt \tag{12-16}$$

The left side of Eq. (12-16) is the total change in p_x. Let the initial and final values of θ be $-\theta_0$ and θ_0, respectively, as shown in Fig. 12-6; then the total change in p_x is $2mv_0\cos\theta_0$. Also, the integral on the right side of Eq. (12-16) can be converted into an integral on θ, with the limits $-\theta_0$ and θ_0. Thus Eq. (12-16) becomes

$$2mv_0\cos\theta_0 = \frac{k}{bv_0}\int_{-\theta_0}^{\theta_0}\cos\theta\,d\theta = \frac{2k}{bv_0}\sin\theta_0$$

or

$$b = \frac{k}{mv_0^2}\tan\theta_0 \tag{12-17}$$

The angle θ_0 is related to the scattering angle ϕ by $2\theta_0 + \phi = \pi$, as may be seen from Fig. 12-6. Using this relationship, we have

$$\tan\theta_0 = \tan\tfrac{1}{2}(\pi - \phi) = \cot\tfrac{1}{2}\phi$$

Finally, Eq. (12-17) becomes

$$b = \frac{k}{mv_0^2} \cot \tfrac{1}{2} \phi \qquad (12\text{-}18)$$

Thus we have accomplished the objective of finding a relationship between the impact parameter b and the scattering angle ϕ.

The general form of Eq. (12-18) is to be expected. When b is very large, the cotangent of the scattering angle ϕ is very large, and the angle itself is very small. As b becomes smaller, the scattering angle approaches π (180°), corresponding to a head-on collision in which the particle is scattered exactly backward. Equation (12-18) also shows that the relation between b and ϕ depends directly on the initial α-particle *energy*, $\tfrac{1}{2}mv_0^2$. Figure 12-7 shows a graph of ϕ versus b for gold nuclei and α particles

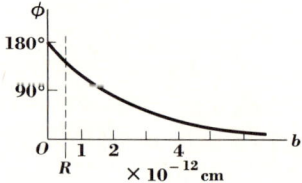

Fig. 12-7 Relation between scattering angle and impact parameter for scattering of 6-MeV α particles by gold nuclei. The value R (about 0.7×10^{-12} cm) is the approximate radius of the gold nucleus as determined by more recent experiments with the scattering of high-energy electrons.

with energies of 6 MeV, a typical value for the particles emitted by natural radioactive elements. For comparison, the approximate radius of the gold nucleus, as determined from more recent experiments, is also shown.

In practice, it is not possible to measure the impact parameter b for a single collision. Instead, a *beam* containing a very large number of α particles is directed at the gold target. The various particles undergo collisions with various impact parameters, and one can calculate what fraction should fall in any given range of values of b. Thus it is possible to calculate the number of particles expected to be scattered out of the beam in any given direction. We observe experimentally how many particles are actually scattered in various directions and compare this with the result of the calculation.

These pioneering experiments in the investigation of atomic structure were first performed in 1910 by Geiger and Marsden at Manchester University in England, under the direction of Rutherford. Their results indicated that the positive charge in an atom is concentrated in a region no larger than 10^{-12} cm in diameter, smaller by 10^4 than the overall diameter of the atom. This led Rutherford to the nuclear model, in which the atom is pictured as made up of a very small, dense, positively charged nucleus containing most of the mass of the atom, surrounded by a swarm of negatively charged, relatively light electrons.

12-4 ANGULAR MOMENTUM OF A RIGID BODY

We are now ready to extend the principle of angular momentum to the motion of a rigid body. We discuss first the case of a body rotating about an axis which is stationary in an inertial frame of reference. In Sec. 12-5 a more general definition of angular momentum is given; this provides the necessary foundation for the general analysis in Chap. 13 of rigid-body motion in which the axis is not stationary.

We begin just as in Sec. 11-5 by regarding a rigid body as composed of a large number of mass elements m_i. We *define* the total angular momentum of the body as the sum of the angular momenta of all these elements. It was pointed out in Sec. 11-5 that when a rigid body rotates about a fixed axis, each point in the body moves in a circle whose plane is perpendicular to the axis. Following the notation of that section, we denote the distance of mass element m_i from the axis as s_i. When the body rotates with angular velocity ω, the speed of the particle is $v_i = \omega s_i$, and so the magnitude of its momentum is $p_i = m_i s_i \omega$. The distance s_i corresponds to the moment arm of Eq. (12-5); hence the angular momentum L_i of this mass element is

$$L_i = m_i s_i^2 \omega$$

and the total angular momentum L, by definition, is

$$L = \Sigma L_i = \Sigma m_i s_i^2 \omega = I\omega \tag{12-19}$$

In this equation we recognize the *moment of inertia I* of the body, defined by Eq. (11-23). Equation (12-19) has the same form as the definition of momentum, $\mathbf{p} = m\mathbf{v}$; a quantity characterizing the inertial properties of the rotating body is multiplied by a kind of velocity.

The rate of change of angular momentum of a rigid body is related directly to the forces acting on it. It was shown in Sec. 12-2 that for a single particle the rate of change of angular momentum with respect to a specified origin is equal to the *torque* with respect to that origin, due to the force on the particle. Since this is true for every particle in a rigid body, a similar relationship holds for the *total* angular momentum L and the *total* torque:

$$\frac{dL}{dt} = \Sigma \tau_i \tag{12-20}$$

where $\Sigma \tau_i$ is understood to mean the sum of all the torques on all the particles.

The quantity dL/dt can be related directly to the motion of the body with the aid of Eq. (12-19). As pointed out in Sec. 11-5, the moment of inertia is constant provided the axis has a fixed position in the body; therefore

$$\frac{dL}{dt} = I \frac{d\omega}{dt} \qquad (12\text{-}21)$$

The quantity $d\omega/dt$ is sometimes further abbreviated $d\omega/dt = \alpha$, where α is called the *angular acceleration*. In terms of this quantity, $dL/dt = I\alpha$; Eq. (12-20) then becomes

$$\Sigma \tau_i = I\alpha \qquad (12\text{-}22)$$

As it stands, Eq. (12-22) is not a very useful result because the sum of the torques includes the torques due to all the internal forces exerted by various parts of the rigid body on each other. There may be an extremely large number of such forces, and it is impossible in practice to calculate this sum. Just as in Sec. 11-2, however, we can show that the torques due to the internal forces add to zero, so that only the torques due to the *external* forces exerted on the body by an outside agency need be considered. To show this, we consider the two particles shown in Fig. 12-8 and the mutual-

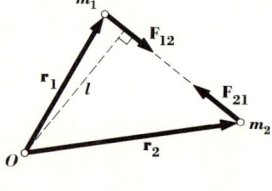

Fig. 12-8 The two forces which particles M_1 and M_2 exert on each other are equal in magnitude and opposite in direction and act along the line joining the particles; the total torque on the system due to these two forces is zero. The sum of the torques due to all internal forces in the system is therefore zero.

interaction forces of these particles. In the figure, F_{12} is the force exerted on particle 1 by particle 2, and F_{21} that on particle 2 due to particle 1. Since these forces obey Newton's third law, they are equal in magnitude but opposite in direction, as shown in the figure. The forces are also directed along the line joining the two particles, as shown, so the perpendicular distance between the line of action of *either* force and the origin O is l; that is, their lever arms have equal lengths. Thus the torques corresponding to these forces have the same magnitude but opposite signs, and they add to zero. In like manner, the torques corresponding to *all* the internal forces cancel in pairs, so that the only terms which survive in the sum in Eq. (12-22) are those corresponding to the *external* forces. To emphasize this point, we write

325 ANGULAR MOMENTUM OF A RIGID BODY [12-4]

$$\Sigma\tau_{\text{ext}} = \frac{dL}{dt} = I\alpha \tag{12-23}$$

where the notation explicitly reminds us that the total rate of change of angular momentum is equal to the sum of the torques that are due to the *external* forces on the body. In particular, when the sum of external torques is zero, the total angular momentum is constant.

Example
Consider again the first example of Sec. 11-7, shown again in Fig. 12-9. Find the acceleration a of the rope and the angular acceleration α of the flywheel.

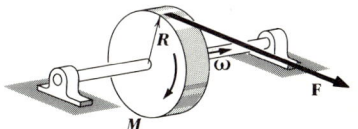

Fig. 12-9 Flywheel rotates about a fixed axis. A force on a rope wrapped around it as shown causes an angular acceleration. Both the angular velocity and the torque are directed along the axis.

Solution
The moment of inertia of the flywheel is $I = \frac{1}{2}MR^2$, and its angular momentum, when rotating with angular velocity ω, is

$$L = \tfrac{1}{2}MR^2\omega \tag{12-24}$$

The torque due to F with respect to the axis of rotation is simply FR. Using this with Eq. (12-23) gives

$$FR = \frac{dL}{dt} = \tfrac{1}{2}MR^2 \frac{d\omega}{dt} = \tfrac{1}{2}MR^2\alpha \tag{12-25}$$

from which we find

$$\alpha = \frac{d\omega}{dt} = \frac{2F}{MR} \quad \text{and} \quad a = R\alpha = \frac{2F}{M} \tag{12-26}$$

Equations (12-26) reveal several things. When the force is constant, the angular acceleration is also constant. Furthermore, as we might expect, the rate of change of angular velocity is inversely proportional to both the total mass and the radius of the flywheel.

If the flywheel starts with some initial angular velocity ω_0 at time $t = 0$, then its angular velocity at any later time can be obtained by simply integrating Eq. (12-26) to obtain

$$\omega = \omega_0 + \alpha t = \omega_0 + \frac{2F}{MR} t \qquad (12\text{-}27)$$

Correspondingly, if the angular *position* at time $t = 0$ is given by the angle θ_0, then by integrating again, one can find the position θ at any later time:

$$\theta = \theta_0 + \omega_0 t + \tfrac{1}{2}\alpha t^2 = \theta_0 + \omega_0 t + \frac{F}{MR} t^2 \qquad (12\text{-}28)$$

This procedure is completely analogous to that used in Chap. 3 to develop relationships among the position, velocity, and acceleration of an object moving with constant acceleration along a straight line. Of course, Eqs. (12-27) and (12-28) take this particularly simple form only because the rate of change of angular velocity is constant, which in turn results from the fact that F is constant. If the forces and their associated torques vary with time, the integration of the corresponding equations becomes more involved.

Example

Suppose two flywheels whose moments of inertia are I_1 and I_2, respectively, are mounted on frictionless bearings on the same axle. The first is set into motion with angular velocity ω_0. Then, without exerting any torque on either flywheel, we push them together so that they rub against each other and finally attain a common final angular velocity ω_f. What is ω_f?

Solution

The essential step in solving this problem is to observe that the total angular momentum of the two flywheels, considered together as a system, is constant, since no external torques are exerted on them. Equating the total angular momentum at the beginning and the end of the process, we find

$$I_1 \omega_0 = I_1 \omega_f + I_2 \omega_f = (I_1 + I_2)\omega_f \qquad (12\text{-}29)$$

from which ω_f can be found in terms of known quantities. One may ask whether the total *kinetic energy* of the system is the same at the end of the process as at the beginning. This question is left as a problem.

Example
Suppose a physics lecturer sits on a stool equipped with frictionless ball bearings and holds two weights straight out to his sides. He is set into rotation, and then he pulls the weights in close to his chest. When he does this, his angular velocity increases noticeably. Why?

Solution
Again, because of the frictionless bearings, there are no torques along the axis of rotation, and the angular momentum is constant. But the lecturer decreases the total moment of inertia of the system when he pulls the weights in, and his angular velocity must increase correspondingly in order to keep the total angular momentum constant.

Example
Consider the Atwood's machine in Fig. 12-10. Find the accelerations of the masses and the tensions in the strings, assuming that the string does not slip

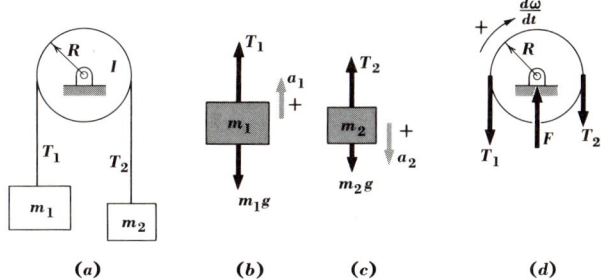

Fig. 12-10 Atwood's machine with a pulley whose moment of inertia I is not negligible. A free-body diagram for each of the three parts of the system is shown. Positive directions for the accelerations are indicated.

on the pulley and that the pulley has frictionless bearings but a moment of inertia which cannot be neglected.

Solution
Shown in Fig. 12-10 is a free-body diagram for each part of the system. Since the pulley has an angular acceleration, it is clear that T_1 and T_2 cannot be equal; otherwise there would be no net torque on the pulley. Choosing

a positive direction for each of the coordinates as shown, and noting that the accelerations are related by

$$a_1 = a_2 = R\frac{d\omega}{dt} = a$$

we write the following equations of motion:

$$T_1 - m_1 g = m_1 a$$
$$m_2 g - T_2 = m_2 a$$
$$T_2 R - T_1 R = I\frac{a}{R}$$

These are three simultaneous equations which can be solved for the quantities a, T_1, and T_2. Completion of the solution is left as a problem.

The choice of positive directions is arbitrary, provided that whatever choice is made is used consistently. The reader is invited to solve this problem using a different choice. For example, the positive direction could be taken to be upward for both masses and counterclockwise as the positive sense of rotation. In this case the relation of accelerations is $-a_1 = a_2 = R\, d\omega/dt$, and the equations will look different, but the final results must be equivalent.

As this last example shows, the use of free-body diagrams to identify forces and torques on various parts of a system is just as useful in problems involving rotational motion as in those considered earlier in this text. The reader is urged to begin the solution of any problem involving forces and accelerations, or torques and angular accelerations, or both, by drawing a free-body diagram showing clearly all forces acting on each part of the system under consideration.

12-5 VECTOR ANGULAR MOMENTUM

We now generalize somewhat the relationship between the angular momentum of a system and the forces acting on this system. According to Eq. (12-4), the angular momentum L of a particle at position $\mathbf{r}$ with momentum $\mathbf{p}$ can be represented as $rp \sin \phi$, where ϕ is the angle between $\mathbf{r}$ and $\mathbf{p}$. But this is equal to the magnitude of the vector product $\mathbf{r} \times \mathbf{p}$. That is,

$$L = |\mathbf{r} \times \mathbf{p}| \tag{12-30}$$

This fact suggests defining a *vector angular momentum* $\mathbf{L}$ with respect to the origin O as follows:

VECTOR ANGULAR MOMENTUM [12-5]

$$\mathbf{L} = \mathbf{r} \times \mathbf{p} \tag{12-31}$$

As an example, the angular momentum of a planet moving in a circular orbit around the sun is a *vector* whose magnitude is rmv, where r is the radius of the orbit and v is the speed of the planet, and whose direction is perpendicular to the plane of the orbit. In addition, in this particular case the direction of $\mathbf{L}$ is the same as that of the vector angular velocity $\boldsymbol{\omega}$ of the planet, defined in Sec. 6-4. Thus the vector angular momentum defined by Eq. (12-31) agrees in magnitude with the previous definition, and its direction has a familiar significance, at least in this particular case.

We now investigate the rate of change of this vector angular momentum by calculating the time derivative of Eq. (12-31):

$$\frac{d\mathbf{L}}{dt} = \frac{d}{dt}(\mathbf{r} \times \mathbf{p}) = \frac{d\mathbf{r}}{dt} \times \mathbf{p} + \mathbf{r} \times \frac{d\mathbf{p}}{dt} \tag{12-32}$$

Now $d\mathbf{r}/dt$ is simply the velocity of the particle, which has the same direction as its momentum. Thus the first term on the right side of Eq. (12-32) is a product of two vectors in the same direction and is therefore zero. Furthermore, $d\mathbf{p}/dt$ is equal to the vector sum of the forces acting on the particle. Denoting this total force by $\mathbf{F}$, we have

$$\frac{d\mathbf{L}}{dt} = \mathbf{r} \times \mathbf{F} \tag{12-33}$$

The vector product $\mathbf{r} \times \mathbf{F}$ is closely related to the torque due to the force, defined by Eq. (12-9); its magnitude is simply $rF \sin \phi$, and it has been shown that this is equal to Fl in Fig. 12-4. That is, the magnitude of $\mathbf{r} \times \mathbf{F}$ is the magnitude of the force multiplied by the perpendicular distance l between the line of action of the force and the origin of coordinates O. This distance has been referred to previously as the lever arm, or moment arm, of the force.

It is thus natural to generalize the definition of *moment*, or *torque*, of a force as follows: When a force $\mathbf{F}$ is applied to a point whose position vector with respect to O is $\mathbf{r}$, the moment, or torque, $\boldsymbol{\tau}$ of this force with respect to O is a vector given by

$$\boldsymbol{\tau} = \mathbf{r} \times \mathbf{F} \tag{12-34}$$

The magnitude of $\boldsymbol{\tau}$, as we have seen, represents the tendency of force to produce a rotation about an axis through O. Moreover, the *direction* of $\boldsymbol{\tau}$ is perpendicular to the plane of $\mathbf{F}$ and $\mathbf{r}$, and this is the direction of the axis about which a body would tend to rotate if only the force $\mathbf{F}$ were present.

Combining Eqs. (12-33) and (12-34), we obtain the vector equation

$$\tau = \frac{d\mathbf{L}}{dt} \tag{12-35}$$

This result may be regarded as a generalization of Eq. (12-10); both are the rotational analogs of $\mathbf{F} = d\mathbf{p}/dt$.

The concepts of angular momentum and torque may be generalized immediately to a system of particles or a rigid body. We define the *total* angular momentum $\mathbf{L}$ of the system as the vector sum of the angular momenta $\mathbf{L}_i$ of the individual particles:

$$\mathbf{L} = \Sigma \mathbf{L}_i = \Sigma \mathbf{r}_i \times \mathbf{p}_i \tag{12-36}$$

where $\mathbf{r}_i$ is the position of particle i and $\mathbf{p}_i$ is its momentum. As in Eq. (12-32), we take the time derivative of Eq. (12-36):

$$\frac{d\mathbf{L}}{dt} = \Sigma \frac{d\mathbf{r}_i}{dt} \times \mathbf{p}_i + \Sigma \mathbf{r}_i \times \frac{d\mathbf{p}_i}{dt} \tag{12-37}$$

Because the velocity of each particle is parallel to its momentum, every term in the first sum in Eq. (12-37) is zero. Furthermore, in the second sum we replace $d\mathbf{p}_i/dt$ with $\mathbf{F}_i$, which represents the total force on particle i, to obtain

$$\frac{d\mathbf{L}}{dt} = \Sigma \mathbf{r}_i \times \mathbf{F}_i = \Sigma \tau_i$$

Now we use exactly the same argument as in Sec. 12-4 to show that all torques corresponding to *internal* forces cancel out in pairs, so the final result is

$$\frac{d\mathbf{L}}{dt} = \Sigma \tau_{\text{ext}} \tag{12-38}$$

This is a general result for any system. If the system happens to be an *isolated* system, the total angular momentum is constant. When the system is a rigid body, the angular momentum can be expressed in terms of the moment of inertia and angular velocity, just as in Sec. 12-4. We consider here only problems in which the body rotates about an axis of symmetry, such as the axis of a circular cylinder or the axle of a wheel. In such cases, because of the symmetry, the total angular momentum is always a vector in the direction of the axis. It cannot have a component perpendicular to this axis, for if it did, the body would have to be lopsided in order to determine a particular direction for the perpendicular component. For a body rotating about an axis of symmetry, then, we have simply

$$\mathbf{L} = (\Sigma m_i s_i^2)\omega = I\omega$$

This result has the same form as Eq. (12-19), but it is a *vector* equation; both **L** and **ω** lie along the axis of rotation, in a direction given by the right-hand rule.

Several additional remarks are in order. First, the angular momentum of an individual mass element need not have the same direction as the axis, but because of the assumed symmetry the *total* angular momentum must lie along the axis. The symmetry is thus an essential condition; in Chap. 13 we consider a case of rotation about an axis which is not an axis of symmetry, and it turns out that the angular momentum does not have the same direction as the angular velocity! Second, we have *not* assumed that **ω** is constant. It may change in magnitude when the rotational motion changes speed, and it may also change in direction. In Chap. 13 we consider the motion of a gyroscope whose axis changes direction with time.

In conclusion, we remark that in developing the concept of angular momentum and its applications to rigid-body motion we have proceeded in a deductive rather than an inductive manner. The principles discussed here result from new definitions combined with principles already established rather than from new generalizations from experimental observation. Thus they do not have quite the same flavor of discovery as some earlier sections of the book, and some defined quantities may seem to have been pulled out of thin air. But it must be remembered that the development of new physical principles always proceeds by a *combination* of inductive and deductive methods, frequently involving considerable guesswork and many theoretical developments which turn out to be unfruitful. In developing the basic laws of mechanics in Chap. 4, we concentrated primarily on the *inductive* approach, beginning with particular observations and using these as a basis for the formulation of general principles. In this chapter, as well as in Chap. 13, we concentrate on a *deductive* approach, using principles already established and applying them to new situations.

Problems

12-1 What is the minimum number of coordinates needed to describe completely the position of a rigid body in space? Describe a coordinate system which might be used.

12-2 How many quantities are needed to describe completely the motion of a rigid body at one instant of time? Give one set of quantities which might be used for this description.

12-3 A satellite moves around the earth in an elliptical orbit. Its closest approach to the surface of the earth is about 400 mi and its farthest distance about 1,600 mi. Its speed when closest to the earth is about 18,000 mi/h.
 a What is the sector velocity?
 b What is the satellite's speed when farthest from the earth?

12-4 A particle moves with constant speed v along a straight line whose perpendicular distance from a fixed point is d. Show that the sector velocity of the particle with respect to this point is constant, and find its value. Is this result consistent with angular-momentum considerations?

12-5 An asteroid moves in an elliptical path around the sun. The lengths of the major and minor axes are $2a$ and $2b$, respectively. The sun is on the major axis. If the asteroid's speed when it crosses the major axis at its point of closest approach to the sun is v_0, how much time is required for a complete trip around the orbit? *Hint:* The area of the ellipse is πab, and the sun is at one focus, at a distance $(a^2 - b^2)^{1/2}$ from the center.

12-6 Derive an expression for the *speed* of a particle which undergoes Rutherford scattering, as discussed in Sec. 12-4, at its position of closest approach, in terms of v_0 and other relevant quantities.

12-7 For Rutherford scattering, derive an expression for the distance of closest approach in terms of the impact parameter and other relevant quantities. *Hint:* The principles of conservation of energy and angular momentum provide two relations for the distance of closest approach and the speed at the point of closest approach.

12-8 A frictionless puck of mass m moves with speed v_0 on a flat table in a circular path with radius r_0 under the action of a string which is tied to the puck, passed through a hole in the table, and held. Now the string is pulled in smoothly, until the distance from puck to hole decreases to $r_0/2$, and again held, so the puck now moves in a circle of radius $r_0/2$.
 a Find the speed of the puck in the final circular path.
 b Is the kinetic energy the same at the end as at the beginning of this process? If not, account quantitatively for the difference.

12-9 A frictionless puck moves on a horizontal table under the action of a string which winds around a vertical peg, as shown in Fig. P12-9.
 a Is the angular momentum constant? Explain.
 b Is the kinetic energy constant? Explain.

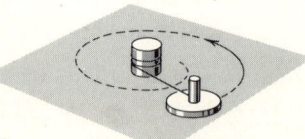

Fig. P12-9

12-10 A particle of mass m is attached to a spring whose natural length is l and whose spring constant is k. The other end of the spring is fixed on a smooth, frictionless horizontal surface on which the particle slides. The particle is pulled until it is a distance $2l$ from the fixed end of the spring and given an initial velocity v_0 in a direction perpendicular to the length of the spring. Using the energy and angular-momentum principles, find the radial and tangential components of the velocity of the particle after it returns to a distance l from the fixed point.

12-11 Two flywheels having moments of inertia I_1 and I_2 are free to rotate on the same shaft with independent angular velocities. The first is given an initial angular velocity, while the second is initially at rest. The two flywheels are then pushed together by a force along the axis so that they attain a common final angular velocity.
  a What is the final angular velocity?
  b Compare the initial and final kinetic energies, and explain the result.

12-12 A certain satellite is in a circular orbit around the earth. By what minimum factor would its speed have to increase for the satellite to escape from the earth's gravity?

12-13 In Sec. 9-7 an expression was obtained for the escape velocity needed for a missile to escape from the earth's gravitational field. If instead of being fired straight up (away from the center of the earth) the missile is fired at an angle α to the vertical, would the minimum speed for escape be the same or different? Explain.

12-14 A certain planet moves about its sun in an elliptical orbit, under the attractive inverse-square gravitational attraction. The ratio of apogee (maximum) radius r_a to perigee (minimum) radius r_p is $r_a/r_p = 3$.
  a Find the ratio of speed at apogee to that at perigee.
  b Find the ratio of potential energy at apogee to that at perigee.
  c Find the ratio of kinetic energy at apogee to that at perigee.
  d Find the ratio of kinetic to potential energy at perigee.

12-15 A hockey puck sliding across a smooth (frictionless) horizontal ice surface with initial speed v_0 collides with a second puck initially at rest. One of the players has carelessly left a bit of chewing gum on the second puck, so the two pucks stick together after the collision. At the moment of collision the line of centers of the two pucks makes an angle of 30° with the initial velocity.
  a Find the velocity of the center of mass after the collision.
  b Find the angular velocity of the system after the collision.
  c Find the final total kinetic energy of the system. Is mechanical energy conserved? Explain.

12-16 A rifle target is mounted on a horizontal turntable which can turn without friction about a vertical axis. The turntable is a uniform disk of radius R and

mass M_1; the target, small compared to R, is at the edge and has mass M_2. A rifle bullet of mass m and speed v_0 is fired so it strikes the target in a direction tangent to the disk and is embedded in the target. Before impact the target and turntable are at rest.
a Find the final angular velocity of the system.
b Is mechanical energy conserved? Explain.

12-17 A door of mass M and width d swings on frictionless hinges. With the door initially at rest, a gob of mud of total mass m is thrown at the door, striking it at the center and sticking. Find the final angular velocity:
a If the mud comes in perpendicular to the door
b If the mud arrives at a horizontal angle of 45° to the plane of the door

12-18 An old-fashioned phonograph turntable in the form of a uniform disk of mass 1.0 kg and radius 0.1 m turns freely with angular velocity 8 rad/s (about 78 r/min). A careless guest drops a gob of cheese dip at a point halfway between the center and outside edge. The mass of cheese dip is 0.01 kg.
a What is the final angular velocity?
b Is mechanical energy conserved? Explain.
c How do you get rid of unwanted guests like this?

12-19 The outer rim of a phonograph turntable is baited with cheese along half its circumference. A mouse runs around the edge of the turntable, starting from rest where the cheese begins. After he has traversed half the circumference, eating cheese along the way, he stops. The turntable is a uniform disk of mass M and radius R which is free to rotate, and the mass of the mouse is m.
a Through what angle does the turntable turn during this process?
b What is the angle if $R = 0.1$ m, $M = 1.0$ kg, and $m = 0.01$ kg?
c Is mechanical energy conserved during this process? Explain.

12-20 A circular toy train track is mounted on a large wheel which is free to turn about a vertical axis. The mass of the wheel is M, and its radius R; the mass of the train is m. Starting with the entire system at rest, the power is turned on, and the train attains a constant speed v *with respect to the moving track*. Find the angular velocity of the wheel.

12-21 A dumbbell-shaped object is made from two masses m_1 and m_2 connected by a light rod of length d. It slides without friction on a smooth horizontal surface. The dumbbell is set into rotational motion in such a way that the center of mass is stationary. Show that the angular momentum with respect to the stationary point can be expressed as $\mu d^2 \omega$, where μ is the reduced mass and ω is the angular velocity of the object. Is this result still correct if the rod is replaced by a spring so that d is not constant? Explain.

12-22 A bullet with mass $m = 5$ g and speed $v = 200$ m/s is fired perpendicularly into the center of a door 2.5 m high and 1.0 m wide whose mass is 20 kg.

The door was initially at rest, but after the bullet lodges in it, it turns on its hinges with angular velocity ω. Find ω.

12-23 A physics lecturer sits on a ball-bearing stool which rotates without friction. He holds a 2-kg weight at arm's length in each hand. He is given an initial angular velocity of 2 rad/s. Then he pulls the weights in close to his chest.
    a Describe a simplified model which can be used to represent this system.
    b Using the model of part *a*, find the final angular velocity.

12-24 Consider the first example in Sec. 11-7.
    a Find the acceleration of the rope and the angular acceleration of the flywheel.
    b How much time is required to pull the rope 5 m if the system starts from rest?

12-25 In the example shown in Fig. 11-17, suppose that instead of being wrapped around the flywheel the rope is wound around the shaft, whose radius is 0.02 m.
    a Find the acceleration of the rope and the angular acceleration of the flywheel.
    b How much time is required to pull the rope 5 m (starting from rest)?
    c What is the angular velocity of the flywheel after the rope has been pulled 5 m?

12-26 In the system shown in Fig. P12-26, assume that the pulley is a uniform disk.
    a Find the acceleration of *m* and the angular acceleration of the pulley.
    b Find the tension in the string.

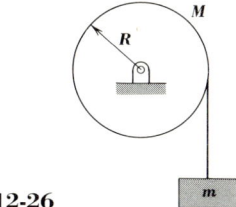

Fig. P12-26

12-27 In the system shown in Fig. P12-27, the wheel is a uniform disk of radius R_2, but the rope is wound around a light hub on the wheel with a smaller radius R_1. Find the acceleration and the tension in the rope.

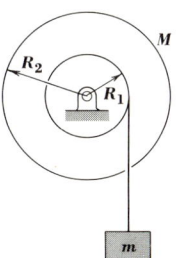

Fig. P12-27

12-28 Complete the solution of the Atwood's machine problem discussed in Sec. 12-4. Check the reasonableness of the solutions in the following particular cases:
   a $m_1 = m_2$
   b $I = 0$
   c $m_1 = 0$

12-29 For the system shown in Fig. P12-29:
   a Is the tension the same in the two parts of the string? Explain.

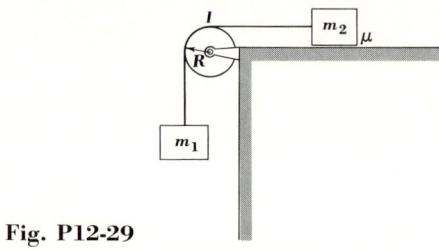

Fig. P12-29

   b If $\mu = 0$, find the acceleration and string tensions.
   c If $\mu = 0$, show that the acceleration of m_1 is the same as it would be if the pulley were massless and m_2 were increased to $m_2 + I/R^2$. Would the tension in the horizontal part of the string be the same in this case? Explain.
   d If $\mu \neq 0$, find the acceleration and string tensions.

12-30 A rotating shaft used for power transmission is driven by a belt running on a pulley on the shaft. Show that the rate of transfer of energy, i.e., the power transmitted, is given by $P = \tau \omega$, where τ is the torque exerted on the shaft and ω is its angular velocity. Is this expression correct if τ and ω are not constant? Explain.

12-31 A bicycle is turned upside down so that it stands on its seat and handlebars. The pedals are turned by hand with forces of magnitude F as shown in Fig. P12-31. The pedal sprocket has n_1 teeth, the rear-wheel sprocket n_2. Find

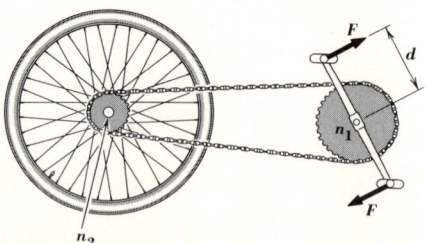

Fig. P12-31

the angular acceleration of the rear wheel. Is this the same as or different from the acceleration which would result if the rear wheel were attached directly to the pedal sprocket? Explain.

Dynamics of a Rigid Body | 13

All possible motions of a rigid body can be described in terms of motion of the center of mass of the body and rotational motion about the center of mass; the general dynamical principles of rigid-body motion are formulated in these terms. The torque exerted by gravity on a body in a uniform gravitational field is shown to be simply related to the center of mass. A number of examples of rigid-body dynamics are discussed in detail, including situations in which the direction of the axis changes. Some generalizations of the principles of rigid-body dynamics are given, and it is shown that in general the angular momentum of a rigid body need not be in the direction of its angular velocity. Finally, equilibrium of a rigid body is discussed.

13-1 ROTATION ABOUT THE CENTER OF MASS

Our discussion of rigid-body motion in Chap. 12 was restricted to cases where the motion consists only of a rotation about an axis which is stationary in an inertial frame of reference. There are of course many possible rigid-body motions which are not readily described in these terms, such as a rolling wheel, a spinning football thrown through the air, or a spinning satellite in orbit around the earth.

In each of these cases, however, we can describe the motion completely at any instant by specifying the velocity **V** of the center of mass and an angular velocity ω which describes the instantaneous rotational motion of the body about an axis passing through the center of mass. In fact, a little thought should convince the reader that *every possible* motion of a rigid body can be described as a simultaneous translational motion of the center of mass

338 DYNAMICS OF A RIGID BODY

and rotational motion about an axis through the center of mass. One can construct a formal proof of this statement, but we may take it to be self-evident on an intuitive basis.

For a complete discussion of rigid-body motion, we need general principles relating the rates of change of center-of-mass velocity **V** and angular velocity **ω** to the forces acting on the body. Part of this problem has already been solved. In Chap. 11, we discovered that the acceleration **A** of the center of mass of a system depends only on the external forces acting on the system:

$$\Sigma \mathbf{F}_{\text{ext}} = M\mathbf{A} \tag{13-1}$$

Regardless of any rotational motion the body may have, the acceleration of its center of mass is *always* determined entirely by the external forces according to Eq. (13-1).

One might surmise that the rotational aspects of the motion might be governed by Eq. (12-38), developed for rotations about a stationary axis. The problem here is that in general the center of mass may be accelerated and hence cannot be used as the origin of an inertial frame of reference. Thus rotations about an axis through the center of mass are in general *not* rotations about an axis stationary in any inertial frame.

Fortunately, one can show that Eq. (12-38) can still be used, provided we reinterpret it so that **L** is the angular momentum defined in terms of positions and velocities with respect to the center of mass and **τ** is the torque with respect to the center of mass. This may be somewhat surprising, since, as just pointed out, the center of mass is in general not stationary in any inertial frame. The statement is nevertheless correct, and we shall now proceed to establish its validity. The proof requires several steps, but it is worth the effort, both as an interesting exercise in vector algebra and for the importance of the final result.

We begin with definitions of angular momentum and torque with respect to the center of mass. Just as we defined the total angular momentum **L** of a system relative to an origin O as $\mathbf{L} = \Sigma \mathbf{r}_i \times \mathbf{p}_i$, we now define a total angular momentum **L'** of a system relative to the center of mass C as

$$\mathbf{L}' = \Sigma \mathbf{r}'_i \times \mathbf{p}'_i \tag{13-2}$$

The notation in Eq. (13-2) is that introduced in Chap. 11. All primed quantities are defined with reference to the center of mass C; $\mathbf{r}'_i$ is the position vector of a mass element with respect to C rather than the origin O of an inertial frame of reference. The momentum $\mathbf{p}'_i$ is computed from the velocity $\mathbf{v}'_i$ relative to C. That is,

$$\mathbf{p}'_i = m_i \frac{d\mathbf{r}'_i}{dt} = m_i \mathbf{v}'_i \qquad (13\text{-}3)$$

Similarly, we define the torque $\boldsymbol{\tau}'$ with respect to C of a force $\mathbf{F}_i$ as

$$\boldsymbol{\tau}'_i = \mathbf{r}'_i \times \mathbf{F}_i \qquad (13\text{-}4)$$

where $\mathbf{r}'_i$ is the position vector of the point of application of the force with respect to C.

Our goal is to prove the statement

$$\frac{d\mathbf{L}'}{dt} = \sum \boldsymbol{\tau}'_{\text{ext}} \qquad (13\text{-}5)$$

In developing this equation, we proceed just as in Sec. 12-5. Taking the time derivative of Eq. (13-2), we have

$$\frac{d\mathbf{L}'}{dt} = \sum \frac{d\mathbf{r}'_i}{dt} \times \mathbf{p}'_i + \sum \mathbf{r}'_i \times \frac{d\mathbf{p}'_i}{dt} \qquad (13\text{-}6)$$

Every term in the first sum on the right is zero since, according to Eq. (13-3), $\mathbf{p}'_i$ and $d\mathbf{r}'_i/dt$ are in the same direction and their vector product is zero. In the second term, we must resist the temptation to say immediately $d\mathbf{p}'_i/dt = \mathbf{F}_i$. This would not be correct, since we are measuring velocity and momentum in a noninertial frame of reference. We may, however, find a simple relationship between $d\mathbf{p}'_i/dt$ and the forces applied to the system by relating $\mathbf{p}'_i$ to the momentum $\mathbf{p}_i$ in an *inertial* frame of reference. First, by definition of the center-of-mass system, $\mathbf{v}'_i = \mathbf{v}_i - \mathbf{V}$, where $\mathbf{v}_i$ is the velocity in an inertial frame of reference and $\mathbf{V}$ is the velocity of the center of mass in the same inertial frame. From this,

$$\mathbf{p}'_i = m_i \mathbf{v}'_i = m_i \mathbf{v}_i - m_i \mathbf{V} = \mathbf{p}_i - m_i \mathbf{V}$$

Using this result in Eq. (13-6), we find

$$\begin{aligned}\frac{d\mathbf{L}'}{dt} &= \sum \mathbf{r}'_i \times \frac{d\mathbf{p}'_i}{dt} = \sum \mathbf{r}'_i \times \left(\frac{d\mathbf{p}_i}{dt} - m_i \frac{d\mathbf{V}}{dt}\right) \\ &= \sum \mathbf{r}'_i \times \frac{d\mathbf{p}_i}{dt} - \left(\sum m_i \mathbf{r}'_i\right) \times \frac{d\mathbf{V}}{dt}\end{aligned} \qquad (13\text{-}7)$$

Now $\sum m_i \mathbf{r}'_i$ is always zero, by definition. If this is not obvious, the reader may rewrite the sum as $\sum m_i(\mathbf{r}_i - \mathbf{R})$ and use the definition of the center of mass, Eq. (11-6), to verify that the sum is always zero. Thus the second term on the right side of Eq. (13-7) vanishes. Moreover, in the first term we may now write $d\mathbf{p}_i/dt = \mathbf{F}_i$, where $\mathbf{F}_i$ is the total force acting on mass element i, since $\mathbf{p}_i$ is measured in an inertial frame of reference. Therefore Eq. (13-7) becomes

$$\frac{d\mathbf{L}'}{dt} = \sum \mathbf{r}'_i \times \mathbf{F}_i = \Sigma \boldsymbol{\tau}'_i \qquad (13\text{-}8)$$

We now use the arguments of Chap. 12 to show that the torques due to all the *internal* forces cancel out in pairs, so that Eq. (13-8) really contains only the vector sum of the torques due to the *external* forces. Thus we obtain the promised result:

$$\frac{d\mathbf{L}'}{dt} = \sum \boldsymbol{\tau}'_{\text{ext}} \qquad (13\text{-}9)$$

Thus we can compute angular momentum and torque with respect to either an inertial origin or the center of mass, and in either case total external torque equals rate of change of angular momentum. Of course, we must not mix the two systems, and we must be sure that our choice of origin is one of these two. If another point is chosen, the relationship is in general not correct. Often it is possible to analyze a rigid-body problem using either of these two approaches, and we shall see some examples in Sec. 13-3.

When a body rotates about an axis of symmetry which passes through the center of mass C, the angular momentum is proportional to the angular velocity, just as in Sec. 12-4. In that case we have simply

$$\mathbf{L}' = I'\boldsymbol{\omega} \qquad (13\text{-}10)$$

where $\mathbf{L}'$ is the angular momentum with respect to an axis through the center of mass coincident with the axis of symmetry and I' is the moment of inertia computed with respect to this axis. If the *direction* of the axis is constant as the center of mass moves, the vector nature of angular momentum and torque can be handled by simply designating one sense of rotation as positive and the opposite as negative.

To summarize: We have now obtained a set of principles which can be applied to the motion of any rigid body which undergoes translational and rotational motion simultaneously. These principles are

$$\sum \mathbf{F}_{\text{ext}} = M\mathbf{A} \qquad \sum \boldsymbol{\tau}'_{\text{ext}} = \frac{d\mathbf{L}'}{dt} \qquad (13\text{-}11)$$

The first of these equations relates the motion of the center of mass to the external forces; the second relates the rotational motion about the center of mass to the torques produced by these forces. In Sec. 13-3 we shall discuss several problems using these relations, but first we need one additional principle.

13-2 GRAVITATIONAL TORQUES

Frequently, in rigid-body problems, one of the forces acting on the body is its *weight*. To compute the acceleration of the center of mass, we need know only the *total force* exerted by gravity on the body. But to determine the rate of change of angular momentum, it is necessary to know the *torque* associated with this force. Because the force of gravity is distributed over the various elements of mass in the body, it is not immediately obvious how to compute the torque. It turns out that we always obtain the correct torque if we treat the entire weight as acting at the center of mass, provided the gravitational field **g** is uniform. We now develop this useful result.

Consider an element of mass m_i located at position $\mathbf{r}_i$ with respect to an origin O, as shown in Fig. 13-1. The force of gravity on this element

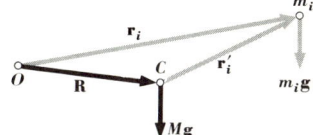

Fig. 13-1 Force of gravity on mass element m_i is $m_i \mathbf{g}$. If the gravitational field is uniform, the total torque due to forces of gravity on all mass elements of the body is the same as though the total weight $M\mathbf{g}$ were concentrated at the center of mass C.

is $m_i \mathbf{g}$, and the corresponding torque with respect to O is $\boldsymbol{\tau}_i = \mathbf{r}_i \times m_i \mathbf{g}$. The total torque due to gravity on all the elements is

$$\boldsymbol{\tau} = \Sigma \mathbf{r}_i \times m_i \mathbf{g} \tag{13-12}$$

If **g** is uniform, it can be factored out:

$$\boldsymbol{\tau} = (\Sigma m_i \mathbf{r}_i) \times \mathbf{g} \tag{13-13}$$

The sum in Eq. (13-13) is reminiscent of the definition of the center of mass,

$$\mathbf{R} = \frac{\Sigma m_i \mathbf{r}_i}{M} \tag{13-14}$$

Combining Eqs. (13-13) and (13-14), we find

$$\boldsymbol{\tau} = \mathbf{R} \times M\mathbf{g} \tag{13-15}$$

342 DYNAMICS OF A RIGID BODY

The quantity $M\mathbf{g}$ is the total weight of the body. Equation (13-15) tells us that the total torque exerted by gravity on a body is exactly the same as that due to a single force equal to the total weight of the body applied at its center of mass.

This is a very important and useful result. It shows that in any calculation of rigid-body motion, the torque associated with gravity can be computed simply by regarding the total weight of the object as being applied at the position of the center of mass. An immediate corollary of this statement is that when one calculates moments with respect to the center of mass, the moment of the force of gravity with respect to the center of mass is *always* zero. Thus, when Eq. (13-9) is used as the basis for the calculations, the torque due to gravity never enters into the calculation.

13-3 EXAMPLES

We now discuss several examples of rigid-body problems. In each example we use one or more free-body diagrams to identify the forces on each body. We then write equations based on Eqs. (13-11) for the translational and rotational motions and perform whatever additional operations are needed to obtain the required results.

Example

A concrete lawn roller is made in the form of a solid cylinder of radius R and mass M, as shown in Fig. 13-2a. A constant force F is applied to the handle. Assuming the roller rolls without slipping, find the acceleration of its center of mass and the frictional force needed to prevent slipping.

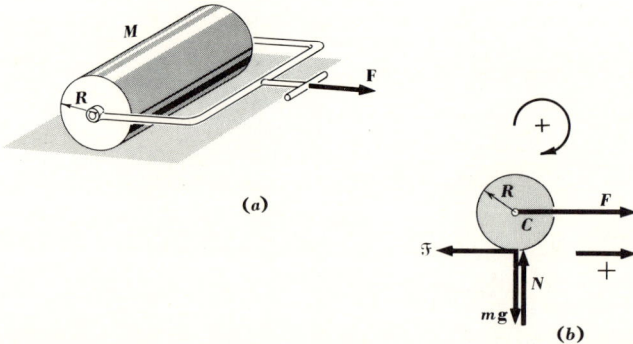

Fig. 13-2 (*a*) Constant horizontal force **F** applied to the handle of a lawn roller which is a homogeneous solid cylinder of radius R and total mass M. The cylinder is assumed to roll without slipping on a horizontal surface. (*b*) Free-body diagram for this body.

Solution

The free-body diagram for the roller is shown in Fig. 13-2b. The force of friction $\mathcal{F}$ acts in the direction opposite that of the force F. If there were no friction, the roller would simply slide without turning; for it to turn in a clockwise sense, the force of friction, which makes it turn, must be in the direction shown. We take the positive direction for forces and acceleration to the right, and we take the positive sense of rotation as clockwise.

We now apply the laws of dynamics given by Eqs. (13-11) to this situation. First, the center of mass has no vertical acceleration, so the sum of the vertical forces must be zero. That is, $N = Mg$. Furthermore, the horizontal acceleration A of the center of mass is determined by

$$F - \mathcal{F} = MA \tag{13-16}$$

So far, both $\mathcal{F}$ and A are unknown. Next we consider the rotational motion. The moment of inertia of the roller is $\frac{1}{2}MR^2$; the magnitude of the angular momentum is $L = \frac{1}{2}MR^2\omega$, and its rate of change is

$$\frac{dL}{dt} = \frac{1}{2}MR^2\alpha$$

where $\alpha = d\omega/dt$. This is equal to the sum of the *torques* due to the external forces with respect to C. With the exception of $\mathcal{F}$, all the forces have lines of action which pass through the axis, and they therefore have no moments with respect to C. The line of action of $\mathcal{F}$ is a distance R from C, so its torque is $\tau = \mathcal{F}R$. Thus the equation for the rotational motion of the roller is

$$\mathcal{F}R = \frac{1}{2}MR^2\alpha \quad \text{or} \quad \mathcal{F} = \frac{1}{2}MR\alpha \tag{13-17}$$

This additional equation introduces an additional unknown, the angular acceleration α of the roller. However, the acceleration of the center of mass A and the angular acceleration α are not independent but are *proportional* because of the assumption that the roller rolls without slipping. Specifically, when the roller turns through an angle θ, it rolls a distance $R\theta$. Correspondingly, when its angular velocity is ω, the velocity V of its center of mass is $V = R\omega$. The time derivative of this equation shows that A and α are related by

$$A = R\alpha \tag{13-18}$$

Equations (13-16) to (13-18) provide three simultaneous equations for $\mathcal{F}$, A, and α; we can solve them to obtain these quantities. First, we combine

344 DYNAMICS OF A RIGID BODY

Eqs. (13-17) and (13-18) to obtain

$$\mathcal{F} = \tfrac{1}{2}MA$$

Next, we add this to Eq. (13-16), which eliminates $\mathcal{F}$:

$$F = \tfrac{3}{2}MA$$

from which we finally find

$$A = \frac{2}{3}\frac{F}{M} \qquad (13\text{-}19)$$

That is, the acceleration of the center of mass is just two-thirds what it would be if the roller could slide without friction instead of rolling. The corresponding force of friction required to prevent slipping is

$$\mathcal{F} = \tfrac{1}{2}MA = \tfrac{1}{2}M\frac{2}{3}\frac{F}{M} = \frac{F}{3} \qquad (13\text{-}20)$$

so the minimum coefficient of friction to prevent slipping is

$$\mu = \frac{\mathcal{F}}{N} = \frac{F}{3Mg} \qquad (13\text{-}21)$$

Example

As an interesting variation of the preceding problem, suppose that instead of pulling on the handle as in Fig. 13-2, we wrap a rope around the roller several times, as in Fig. 13-3, and pull on it with the same horizontal force. Again we are asked to find A and $\mathcal{F}$.

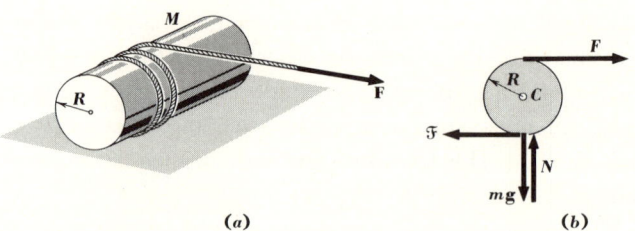

Fig. 13-3 Roller of Fig. 13-2 pulled by a rope wrapped around it. Again a constant horizontal force **F** is applied to the rope.

Solution
Figure 13-3b shows a free-body diagram for this situation. The difference between this and the previous problem is that the force F is applied at the top of the roller rather than at its axle. It may not be entirely clear which direction the force of friction has in this case, but we can assume a direction, as in Fig. 13-3, and solve for its magnitude. If the magnitude turns out to be negative, the actual direction of $\mathcal{F}$ is opposite that assumed.

The equation for the rate of change of angular momentum this time is

$$RF + R\mathcal{F} = \tfrac{1}{2}MR^2\alpha \tag{13-22}$$

We note that torques due to F and $\mathcal{F}$ have the same sign since both correspond to the positive sense of rotation. Dividing Eq. (13-22) by R and adding it to Eq. (13-16), as previously, and using Eq. (13-18), we obtain

$$2F = \tfrac{3}{2}MA$$

from which

$$A = \frac{4}{3}\frac{F}{M}$$

This somewhat surprising result shows that the acceleration of the center of mass is somewhat *greater* than it would be if the roller accelerated without any rotational motion. Returning to Eq. (13-16), for the force of friction we find

$$\mathcal{F} = -\tfrac{1}{3}F$$

The negative sign indicates that the force of friction actually acts in the *forward* direction, rather than backward as assumed. This may not seem to make sense, but further consideration shows that this is just what happens to the back wheels of an automobile when it accelerates. In that case a torque is applied directly by the axles rather than by a force on top of the wheels. The resulting forces of friction on the bottom of the wheels are in the forward direction and are, in fact, the only forces which produce forward acceleration of the center of mass of the automobile. Thus it is reasonable that the force of friction should be in the forward direction.

Example
Consider again the Yo-Yo problem discussed in Sec. 11-7. As previously, we consider a uniform homogeneous disk of radius R and total mass M.

A string is wound around the edge of the disk and tied to the ceiling as in Fig. 13-4, which also shows a free-body diagram for the situation. Find

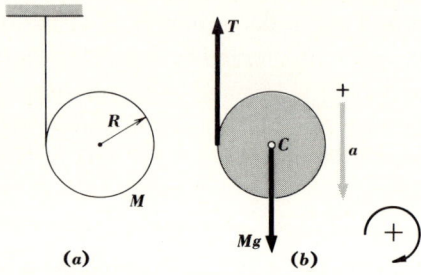

Fig. 13-4 Yo-Yo made from a string wound around a uniform disk. The radius of the disk is R, its total mass is M, and its moment of inertia is $I = \frac{1}{2}MR^2$. The top end of the string is tied to the ceiling, and the Yo-Yo is released from rest with the string initially vertical.

the acceleration of the center of mass of the disk and the tension T in the string.

Solution

As noted in Fig. 13-4b, we choose the downward direction and the clockwise sense of rotation as positive.

The equations for the translational and rotational motions are

$$Mg - T = MA$$
$$RT = \tfrac{1}{2}MR^2\alpha = \tfrac{1}{2}MRA$$

We have used the relationship $A = R\alpha$, which expresses the fact that the string unwinds at a rate proportional to the angular velocity as the disk falls. Dividing the second of these equations by R and substituting the resulting expression for T into the first, we find immediately that

$$Mg - \tfrac{1}{2}MA = MA$$
$$A = 2g/3 \qquad T = \tfrac{1}{2}Mg$$

The acceleration of the center of mass of the disk is two-thirds the acceleration of free fall, and the tension in the string is one-third the weight of the disk, both reasonable results. The role of T in this problem is analogous to that of $\mathcal{F}$ in the previous one.

If the Yo-Yo starts from rest, the speed v of the center of mass after it has dropped a height h is $v = (2Ah)^{1/2}$, since the acceleration is constant. Using the value $A = 2g/3$ just obtained, we find

$$v = \sqrt{\tfrac{4}{3}gh}$$

which agrees with the result obtained in Sec. 11-7 by the energy approach.

Example

A window washer's scaffold is made from a uniform plank of length L and total mass M, and it is supported by a vertical rope at each end, as in Fig.

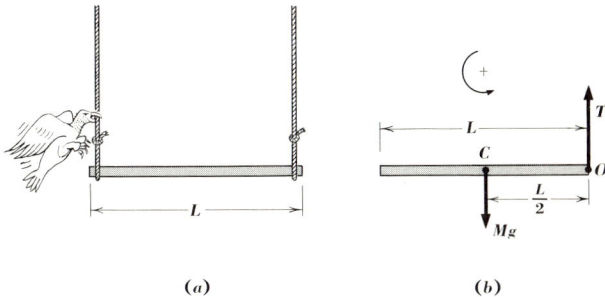

(a)   (b)

Fig. 13-5 (a) A window washer's scaffold. One rope has just been bitten in two by a carrion crow. (b) Free-body diagram for the scaffold.

13-5a. A passing carrion crow suddenly bites one of the ropes in two. Find the tension in the remaining rope and the acceleration of the free end of the plank immediately after the rope is cut.

Solution

Figure 13-5b gives the appropriate free-body diagram. We may describe the motion of the plank either as a rotation about the fixed end or as a translational motion of the center of mass together with a rotation about the center of mass. Taking the first point of view, we observe that the total torque about O is $MgL/2$ and the moment of inertia about O is $ML^2/3$. Thus the angular acceleration is initially

$$\alpha = \frac{\tau}{I} = \frac{MgL/2}{ML^2/3} = \frac{3g}{2L}$$

The initial acceleration a of the free end is given by

$$a = \alpha L = \frac{3g}{2}$$

The acceleration A of the center of mass is just half this:

$$A = \frac{\alpha L}{2} = \frac{3g}{4}$$

The net vertical force equals the mass times the acceleration of the center of mass, and we obtain

$$Mg - T = MA = \frac{3Mg}{4} \quad \text{and} \quad T = \frac{Mg}{4}$$

For comparison, we now start over with the second point of view. For the center-of-mass motion we have again

$$Mg - T = MA$$

Considering rotational motion about the center of mass, the moment of inertia is $I = ML^2/12$, the total torque $TL/2$. The center-of-mass acceleration A is related to the angular acceleration α by $A = \alpha L/2$. (Why?) Thus the equation for rotational motion is

$$\frac{LT}{2} = \frac{ML^2}{12}\alpha = \frac{ML^2}{12}\frac{2A}{L}$$

from which

$$T = \tfrac{1}{3}MA$$

Combining this with the center-of-mass equation, we again obtain

$$T = \frac{Mg}{4} \quad A = \frac{3g}{4}$$

These results agree with those obtained by the first method. This is not astounding, but it may help emphasize the equivalence of the two approaches. It is interesting to note that the free end of the plank has an acceleration *greater* than g and that when one rope is cut, the tension in the other immediately *decreases* from the value $Mg/2$ it has before the rope is cut to just half that value.

13-4 THE GYROSCOPE

Angular momentum is a vector quantity, and the quantity $d\mathbf{L}/dt$ may represent a change in either the magnitude or the direction of $\mathbf{L}$, or both. In the examples considered thus far, the direction of the axis of rotation has not changed; changes in angular momentum have involved only changes of magnitude, and thus we have been able to treat them as scalar quantities. We now consider an example of rigid-body motion in which the direction of the axis of rotation changes and the vector nature of angular momentum must be taken into consideration explicitly.

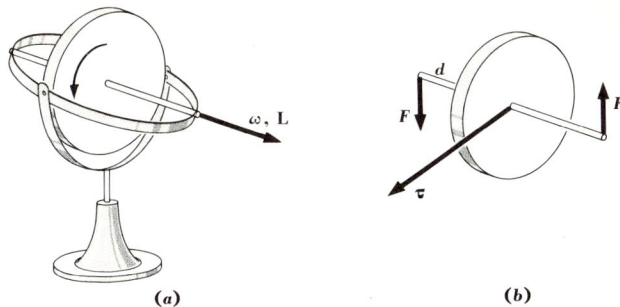

Fig. 13-6 Gyroscope in a double-gimbal mount which permits changes in direction of the axis. If vertical forces are applied to the ends as shown, the axis precesses in a horizontal plane, always moving in a direction perpendicular to that of the forces applied to its ends.

The arrangement shown in Fig. 13-6 is a scheme for mounting a flywheel so that it is free to rotate about its axis of symmetry and the axis can be oriented in any desired direction. The flywheel axle turns in the horizontal ring, which in turn is pivoted about the semicircular ring. Finally, the semicircular ring is free to turn about a vertical axis; we assume all bearings to be frictionless. This arrangement, called a *double-gimbal* mount, permits the axis of rotation of the flywheel to be oriented in any direction. We assume that the construction is symmetric; all the axes of rotation intersect at the center of mass of the flywheel, which therefore never moves.

If the flywheel is given an initial angular velocity about its axis, and if no external torques are exerted on it thereafter, it continues to rotate about this axis with constant angular velocity and angular momentum. There is no tendency for the direction of the axis of rotation to change; for this to happen, the direction of the angular momentum would have to change, which would require an external torque.

Now suppose that two external forces, both vertical and equal in magnitude, are applied to the ends of the axle, as shown in Fig. 13-6b. These forces clearly produce a *torque* on the system. Taking the origin of coordinates at the center of mass, we see that the torque vector is perpendicular to both the axle and the vertical direction and hence is in a horizontal direction as shown. Furthermore, the magnitude of each torque is simply Fd, so the magnitude of the *total* torque is $\tau = 2Fd$.

Because of this torque, the angular momentum must change. Discussion of this change is facilitated by looking down on the flywheel from the top, as in Fig. 13-7. At one instant the angular momentum **L** is as shown; the force at the end of the axle labeled *a* is *into* the paper and that at *b* *out* of the paper. The torque vector is perpendicular to **L**. Using the relation

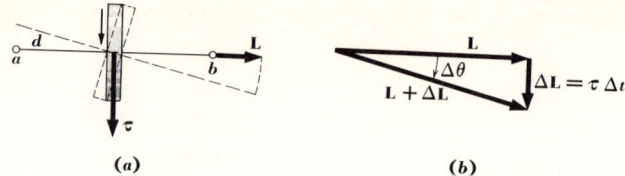

Fig. 13-7 (*a*) Top view of precessing gyroscope, showing directions of angular momentum and torque, both in the horizontal plane. The dotted lines show the position a short time later. (*b*) Angular-momentum vector diagram, showing the change in angular momentum during time Δt and its relation to torque $\boldsymbol{\tau}$. Both **L** and Δ**L** lie in the horizontal plane; Δ**L** is perpendicular to **L**. As the axis precesses, both **L** and $\boldsymbol{\tau}$ turn but remain perpendicular.

$\boldsymbol{\tau} = d\mathbf{L}/dt$, we see that in a small time interval Δt the angular momentum **L** changes by an amount Δ**L** given approximately by

$$\Delta \mathbf{L} = \boldsymbol{\tau}\, \Delta t \tag{13-23}$$

This change is illustrated in Fig. 13-7*b*. We note that Δ**L** is perpendicular to the original direction of **L**; at the end of the interval Δt, the total angular momentum has changed its direction by $\Delta\theta$, as shown, but has the same *magnitude* as at the beginning.

We can now see qualitatively the motion which results from these forces. The angular momentum and therefore the axis of rotation change in *direction*. Because the torque corresponding to the forces is always in the horizontal plane, the change of angular momentum in any time interval is also in a horizontal plane; hence the axis of rotation always remains in this plane. Furthermore, as the axis rotates in a horizontal plane, the direction of the torque $\boldsymbol{\tau}$ also changes so as to remain perpendicular to **L**. Since the instantaneous rate of change of angular momentum is always perpendicular to **L** itself, the *magnitude* of **L** does not change, only its direction. The situation is analogous to that of a particle moving in a circle with uniform speed; there the acceleration is always perpendicular to the velocity, which changes only in direction not in magnitude. If the forces are constant, then the magnitudes of $\boldsymbol{\tau}$ and $d\mathbf{L}/dt$ are also constant, and the axle turns in a horizontal plane at a uniform rate.

The relation of the motion of the axis to the direction of the forces may seem somewhat surprising. One might intuitively expect that these forces would produce a *tilting* of the axis rather than a rotation in a horizontal plane. This, of course, is what would happen if the flywheel had no initial angular velocity; then the torque would simply produce an angular momentum in the direction of $\boldsymbol{\tau}$. But when the flywheel is spinning, as we have assumed, the torque does not simply act to *produce* angular momentum, starting from zero; it acts to *change* the direction of an angular momentum already present, by adding a component in a perpendicular direction. Thus the axle does not move the way our intuition might suggest.

Changing the direction of the axis of rotation under the action of external torques is called *precession* of the rotating body. This precession is described quantitatively by the angle through which the axis turns per unit time, the *angular velocity of precession* of the system. This angular velocity is related simply to the other characteristics of the motion. We denote the instantaneous precession angular velocity by $\Omega = d\theta/dt$. From Fig. 13-7b it can be seen that when Δt is sufficiently small,

$$\Delta\theta = \frac{\Delta L}{L} = \frac{\tau \Delta t}{L} \tag{13-24}$$

where all the symbols in Eq. (13-24) refer to the *magnitudes* of the corresponding vector quantities. Dividing by Δt and taking the limit as $\Delta t \to 0$, we find

$$\Omega = \frac{d\theta}{dt} = \frac{\tau}{L} \tag{13-25}$$

The magnitude of τ is $\tau = 2Fd$, as already pointed out. The magnitude of L is simply $L = I\omega$, where I is the moment of inertia of the flywheel about its axis and ω is the magnitude of the angular velocity of rotation about this axis, sometimes called the *spin angular velocity*. Therefore, Eq. (13-25) becomes

$$\Omega = \frac{2Fd}{I\omega} \tag{13-26}$$

As expected, the precession angular velocity Ω is proportional to F and d, which determine the torque. It is interesting that Ω is *inversely* proportional to ω; the faster the flywheel spins, the more slowly it precesses under a given set of forces.

We have not been completely honest in the foregoing analysis. Since the direction of the axis of rotation of the flywheel turns, there is an additional component of angular velocity in the vertical direction and also a component of angular momentum in the vertical direction. But in this example these components are constant and are not involved in the relation $\Sigma\boldsymbol{\tau} = d\mathbf{L}/dt$.

Figure 13-8 shows a somewhat more spectacular example of the same

Fig. 13-8 Gyroscope supported at one end of its axis. If it precesses in a horizontal plane, the vertical force F exerted on the axis is equal in magnitude to the weight Mg of the flywheel. If the spin angular velocity is as shown, the precession is such that the right end of the axis moves into the paper. The horizontal force F_c exerted on the end of the axis by the pivot is necessary to produce the centripetal acceleration of the center of mass.

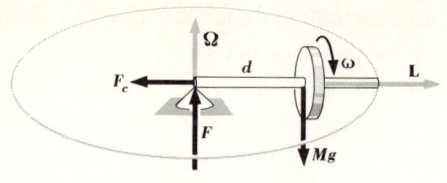

kind of system, an idealized model of the familiar toy gyroscope supported at one end of its axis. The mount at the fixed end of the axis in Fig. 13-8 may be pictured as a ball-and-socket arrangement which permits the axis to change direction but keeps the end at a fixed point. A possible motion of such a toy gyroscope is one in which the axis moves around the pivot point in a horizontal plane with constant (vertical) angular velocity of precession. The situation is a little different from that considered above because in this case the center of mass is not stationary.

Following the general program of Sec. 13-1, we consider separately the motion of the center of mass and the rotational motion about the center of mass. In particular, when the center of mass of the flywheel moves in a horizontal plane, it has no vertical component of acceleration; hence the sum of the vertical forces must be zero, and the upward force **F** exerted on the pivot must be equal in magnitude to the weight of the flywheel. This force therefore exerts a torque *with respect to the center of mass* whose magnitude is Mgd. The angular momentum of the flywheel is again $I\omega$, where I is the moment of inertia about the axis and ω the spin angular velocity.

Just as previously, the rate of change of **L'** is always perpendicular to **L'** itself, where both quantities are measured with respect to the center of mass. The relation between precession angular velocity and torque is precisely the same as previously, and we may use Eq. (13-25) directly. We find

$$\Omega = \frac{Mgd}{I\omega} \tag{13-27}$$

In the particular case in which all the mass is concentrated around the rim at a radius R, we have $I = MR^2$, and M drops out of the final result:

$$\Omega = \frac{Mgd}{MR^2\omega} = \frac{gd}{R^2\omega}$$

If, instead, the mass is a uniform disk as in Fig. 13-8, then $I = \frac{1}{2}MR^2$; again M drops out, and

$$\Omega = \frac{2gd}{R^2\omega}$$

In any case, I is proportional to M, so Ω is independent of the total mass; it does, however, depend on the *distribution* of mass, as shown by the above examples. Again Ω is inversely proportional to the spin angular velocity ω; it is also proportional to the gravitational field g, as should be expected, since without gravity there would be no torque to cause precession.

We are not quite finished with this example. The center of mass of the flywheel moves in a circle of radius d with angular velocity Ω. Thus it is accelerated toward the fixed point about which the axle is pivoted; the magnitude of this acceleration is $a = \Omega^2 d$. There must be a horizontal force F_c acting on the end of the axle to produce this centripetal acceleration; its magnitude is

$$F_c = M\Omega^2 d \tag{13-28}$$

where Ω is given by Eq. (13-27). So, in addition to the vertical force F acting on the end of the axle, there must be a horizontal force F_c pulling along the axle toward the pivot point.

Precession is not by any means the only possible motion for a gyroscope. A precessional motion may have superimposed on it another kind of motion, called *nutation*, in which the direction of the axis of rotation actually *oscillates* periodically. A possible combination of precession and nutation for the system of Fig. 13-8 is shown in Fig. 13-9. A detailed analysis of this more general kind of motion is fairly involved, and we shall not discuss it here.

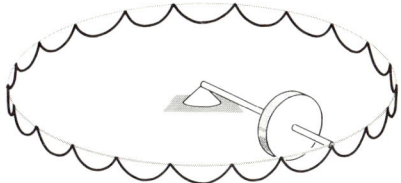

Fig. 13-9 Possible combination of precession and nutation for a gyroscope. Superimposed on the smooth precession motion is a "bobbing" of the axis; the end of the axis describes the path shown.

13-5 SPIN AND ORBITAL ANGULAR MOMENTA

We have seen that the principles of rigid-body motion can be formulated in terms of angular momentum and torque with respect to either the origin of an inertial frame of reference or the center of mass. There is a simple relationship between the angular momentum **L** with respect to a fixed origin O and the angular momentum **L'** with respect to the center of mass C.

As an example of the usefulness of such a relation, suppose we want to compute the total angular momentum of the earth with respect to the sun, considering the sun as the origin O of an inertial frame of reference. The earth's motion is described most easily as an orbital motion of its center of mass along with a "spin" about an axis through the center of mass. If we were dealing with a point mass in orbit, the angular momentum would be simply $\mathbf{L} = \mathbf{R} \times \mathbf{P} = \mathbf{R} \times M\mathbf{V}$, with **R** and **V** measured in the sun's frame of reference; this is termed *orbital angular momentum*. Associated with the spin about the earth's axis is angular momentum **L'** with respect to its center of mass, called *spin angular momentum*. The theorem we are about to prove states that the total angular momentum with respect to the sun is always equal to the sum of these two contributions; that is,

$$\mathbf{L} = \mathbf{L}' + \mathbf{R} \times \mathbf{P}$$

The derivation of this result requires a little vector algebra, but it is healthful exercise and the final result is useful!

Treating the moving body as a collection of mass elements as before, we write the angular momentum L with respect to O:

$$\mathbf{L} = \Sigma \mathbf{r}_i \times \mathbf{p}_i \qquad (13\text{-}29)$$

where $\mathbf{r}_i$ and $\mathbf{p}_i$ are both measured relative to the origin O of the inertial frame of reference. We express these quantities in terms of the position $\mathbf{r}'_i$ and momentum $\mathbf{p}'_i$ relative to the center of mass, just as previously:

$$\mathbf{r}_i = \mathbf{r}'_i + \mathbf{R}$$
$$\mathbf{p}_i = \mathbf{p}'_i + m_i \mathbf{V}$$

Substituting these into Eq. (13-29) and expanding the vector product, we have

$$\begin{aligned}\mathbf{L} &= \Sigma(\mathbf{r}'_i + \mathbf{R}) \times (\mathbf{p}'_i + m_i\mathbf{V}) \\ &= \Sigma \mathbf{r}'_i \times \mathbf{p}'_i + \Sigma \mathbf{r}'_i \times m_i \mathbf{V} + \Sigma \mathbf{R} \times \mathbf{p}'_i + \Sigma \mathbf{R} \times m_i \mathbf{V} \\ &= \Sigma \mathbf{r}'_i \times \mathbf{p}'_i + (\Sigma m_i \mathbf{r}'_i) \times \mathbf{V} + \mathbf{R} \times \Sigma \mathbf{p}'_i + \mathbf{R} \times M\mathbf{V}\end{aligned} \qquad (13\text{-}30)$$

The quantity in parentheses in the second term of Eq. (13-30) vanishes because of the definition of the center of mass; for the same reason, $\Sigma \mathbf{p}'_i = \mathbf{0}$, and the third term vanishes. The first term is recognized as the angular momentum **L'** with respect to the center of mass, defined by Eq. (13-2). We have called this the *spin angular momentum*. In the last term the quantity $M\mathbf{V}$ is simply the total momentum $\mathbf{P}$ of the system. Putting all this information together, we finally find

$$\mathbf{L} = \mathbf{L}' + \mathbf{R} \times \mathbf{P} \tag{13-31}$$

$\mathbf{R} \times \mathbf{P}$ is just the angular momentum the system would have if all the mass were concentrated at the center of mass and moved with the velocity $\mathbf{V}$ of the center of mass. We have called this the *orbital angular momentum*.

Thus Eq. (13-31) states that the total angular momentum $\mathbf{L}$ of any system with respect to O can be computed by adding its angular momentum $\mathbf{L}'$ with respect to its center of mass to the angular momentum $\mathbf{R} \times \mathbf{P}$ corresponding to motion of the center of mass with respect to O. To find the total angular momentum of the earth with respect to the sun, we compute its spin angular momentum $\mathbf{L}'$ corresponding to rotation about its axis and add this to the orbital angular momentum $\mathbf{R} \times \mathbf{P}$ corresponding to motion of its center of mass in its orbit around the sun. The *total* angular momentum is the vector sum of the orbital and spin angular momenta.

A similar technique is useful in the analysis of the angular momentum of an electron in an atom with respect to its nucleus. The electron possesses an intrinsic angular momentum which may be pictured roughly as due to rotation or spin about an axis through its center of mass. In addition, it has orbital angular momentum due to its motion around the nucleus. The total angular momentum is the vector sum of the orbital and spin contributions.

Equation (13-31) provides a convenient method of attack in situations in which a rigid body rotates about a fixed axis which does *not* pass through its center of mass. Consider a body of total mass M rotating about an axis fixed in an inertial frame of reference and passing through the origin O of the frame of reference (Fig. 13-10). Let the center of mass be a distance

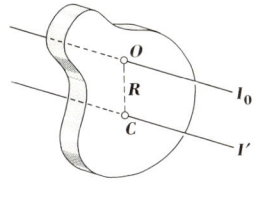

Fig. 13-10 Two axes of rotation, one stationary and passing through the origin O of an inertial frame of reference, the other parallel to it but passing through the center of mass. The moments of inertia I_0 and I' about these axes are related by
$I_0 = I' + MR^2$.

R from this axis. We may describe the motion of the body simply as a rotation ω about the axis through O; the component of angular momentum along the axis is $L = I_0\omega$, where I_0 is the moment of inertia about the fixed axis through O.

An alternative description of the motion regards it as a rotation about a parallel axis through the center of mass C combined with a motion of the center of mass. In this description, the angular velocity of rotation is again ω, although the axis is different; in addition the center of mass moves with a speed $V = \omega R$ with respect to O. The total angular momentum with respect to O may be represented according to Eq. (13-31) as the sum of two contributions, which in this case have the same direction. The angular momentum with respect to the center of mass is $L' = I'\omega$, where I' is the moment of inertia about the axis through the center of mass. The magnitude of the momentum P associated with the center-of-mass motion is $MV = M\omega R$, so the magnitude of the angular momentum due to motion of the center of mass is MR^2. Putting these together, we find that the total angular momentum with respect to O is

$$L = I_0 \omega = I'\omega + MR^2\omega$$
$$= (I' + MR^2)\omega \qquad (13\text{-}32)$$

Equation (13-32) represents the equality of the angular momentum computed by two different methods, one using the fixed axis directly, the other using the axis through the center of mass together with motion of the center of mass. Since these two expressions must be equal, it must be true that

$$I_0 = I' + MR^2 \qquad (13\text{-}33)$$

The reader will recognize this as the *parallel-axis theorem*, developed by a different method in Sec. 11-6. It states that the moment of inertia I_0 about any axis a distance R away from the center of mass is *greater* than that about a parallel axis through the center of mass by an amount MR^2.

Example

Suppose a uniform circular disk is pivoted at a point on its edge so that it can swing freely in its plane, as shown in Fig. 13-11a. Find the moment of inertia about the axis through O, and analyze the pendulum motion which occurs when the body is displaced from its equilibrium position.

Solution

If the radius of the disk is R and its total mass M, the moment of inertia about the axis shown through the center of mass C is $I' = \tfrac{1}{2}MR^2$. Correspondingly, using Eq. (13-33), we find that the moment of inertia about the

Fig. 13-11 (a) Illustration of the parallel-axis theorem. The moment of inertia of a uniform disk about the axis through C is $\tfrac{1}{2}MR^2$; about the axis through O, $\tfrac{3}{2}MR^2$. (b) The disk suspended from point O, acting as a pendulum. The displacement from equilibrium is described by the angle θ. The moment arm of the weight Mg with respect to O is $l = R \sin \theta$.

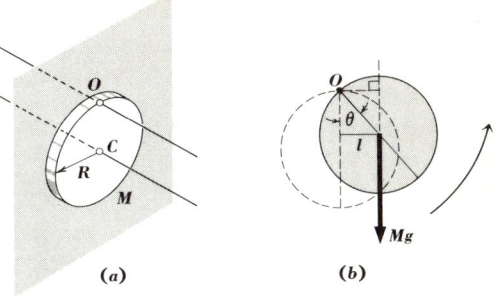

axis through point O in the figure is

$$I_0 = \tfrac{1}{2}MR^2 + MR^2 = \tfrac{3}{2}MR^2$$

If the disk is displaced slightly from its equilibrium position as shown in Fig. 13-11b, it acts as a pendulum. Labeling its angular displacement from equilibrium as θ, we see that the magnitude of the angular momentum with respect to O is given by

$$L = I_0 \omega = \tfrac{3}{2}MR^2 \frac{d\theta}{dt}$$

The torque produced by the weight of the disk, regarded as acting at its center of mass, is equal in magnitude to $\tau = MgR \sin \theta$. Equating the torque to the rate of change of angular momentum, keeping in mind that the torque always acts in a sense opposite to the displacement, we find

$$-MgR \sin \theta = \tfrac{3}{2}MR^2 \frac{d^2\theta}{dt^2}$$

This equation, not too surprisingly, has the same form as the equation of motion of the simple pendulum discussed in Sec. 10-4. As in Sec. 10-4, we may approximate $\sin \theta$ with θ if the oscillations are sufficiently small; rearranging the resulting equation, we obtain

$$\frac{d^2\theta}{dt^2} + \frac{2g}{3R} \theta = 0$$

Comparing this equation with Eqs. (10-2) and (10-9), we see immediately that the period and angular frequency are

$$T = 2\pi \sqrt{\frac{3R}{2g}} \qquad \omega = \sqrt{\frac{2g}{3R}}$$

358 DYNAMICS OF A RIGID BODY

In general, if the moment of inertia about the fixed axis is I_0, and if the center of mass is a distance d from the axis, the equation of motion is

$$\frac{d^2\theta}{dt^2} + \frac{Mgd}{I_0}\theta = 0$$

and the corresponding period and frequency are

$$T = 2\pi\sqrt{\frac{I_0}{Mgd}} \qquad \omega = \sqrt{\frac{Mgd}{I_0}}$$

Proof of these statements is left as a problem.

13-6 GENERALIZATIONS

Here is an example of rigid-body motion which introduces some new features. Consider the situation in Fig. 13-12. A dumbbell-shaped arrangement with a mass m on each end rotates about an axis through the center of mass not

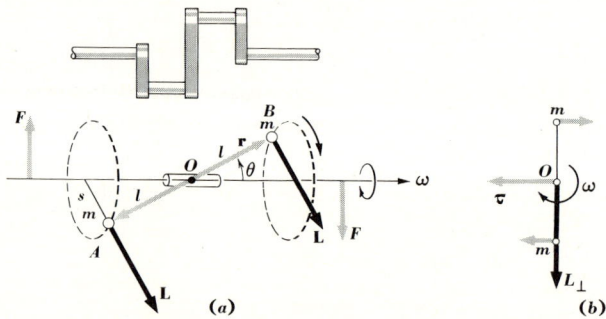

Fig. 13-12 (a) Crankshaft rotating with constant angular velocity about an axis making an angle θ with its own axis. The angular momentum of each mass is shown by the heavy black arrows. The total angular momentum is not in the direction of the axis of rotation but rotates around this axis. At the instant shown, forces must be applied to the axle as shown to produce this change in **L**. (b) End view of the situation, showing the instantaneous directions of the torque.

perpendicular to the dumbbell axis. This arrangement may be regarded as a very simple model representing an unbalanced crankshaft in a gasoline engine or an unbalanced automobile wheel; its behavior illustrates the problem of dynamic balancing of engines and wheels.

We begin by computing the angular momentum directly from the definition $\mathbf{L} = \Sigma \mathbf{r}_i \times \mathbf{p}_i$. At the instant the body lies in the plane of the figure, the velocity of mass A is directed into the page with magnitude $v = s\omega = l \sin \theta \, \omega$. Correspondingly, the magnitude of its momentum is $p = ml \sin \theta \, \omega$. According to the right-hand rule, the vector product $\mathbf{L} = \mathbf{r} \times \mathbf{p}$ is in the direction shown; since the two vectors are perpendicular, the magnitude of $\mathbf{L}$ for this particle is

$$L = rp = ml^2 \sin \theta \, \omega \tag{13-34}$$

At this instant, $\mathbf{L}$ lies in the plane of the figure, as shown. The position vector of mass B is the negative of that of A, and so is its velocity; thus its angular momentum is in the *same* direction as that of A. The total angular momentum is the sum of the two vectors indicated by heavy lines in Fig. 13-12a. *The total angular momentum is not in the direction of the axis of rotation!* As the body rotates, the total angular-momentum vector rotates with it, lying always in the plane of the body axis and the axis of rotation. This varying angular momentum may be represented in terms of a component $L_\parallel$ along the axis, which is constant if ω is constant, and a component $L_\perp$, which is constant in magnitude but which rotates with the body.

Thus if the angular velocity of the body is constant, its angular momentum is *not* constant. The component perpendicular to the axis changes in direction as the body rotates. For this to happen, *torques* must be applied to the system. How do these come about? If one holds the ends of the shaft in one's hands, the body tends to wobble as it rotates; to prevent this wobbling, it is necessary to exert forces on the ends; these give rise to the necessary torque. When the shaft is held by bearings, they supply the torque.

To compute the magnitude of the torque required, we note that the component of L perpendicular to the axis has magnitude

$$L_\perp = L \cos \theta = 2ml^2 \sin \theta \cos \theta \, \omega \tag{13-35}$$

We also note that when $\theta = 90°$ and the two point masses move in the same plane, the perpendicular component of angular momentum vanishes, as we should expect. Otherwise, the perpendicular component rotates with the body as shown in Fig. 13-12b, which is a view looking in the direction of the axis. Proceeding just as in the previous section, we see that

$$\left| \frac{d\mathbf{L}_\perp}{dt} \right| = \omega L_\perp = 2ml^2 \sin \theta \cos \theta \, \omega^2 \tag{13-36}$$

and the torque must be in the direction shown in Fig. 13-12b. This torque could be supplied, for example, by forces on the ends of the axis as in Fig. 13-12a. The details of this analysis are left as a problem.

The fact that it is possible for the angular momentum of a rigid body to have a direction different from that of the angular velocity may come as something of a shock. This shows why, in computing the angular momentum of a body directly from its angular velocity, we were always careful to specify that the axis of rotation must be an axis of symmetry. In the rotating-dumbbell problem, the axis of rotation is *not* an axis of symmetry, and the angular velocity and angular momentum turn out to have different directions. Even in this case, it is possible to prove that the component of angular momentum parallel to the axis is given by $L_\parallel = I\omega$, where I is the moment of inertia. For example, in the present problem,

$$L_\parallel = L \sin \theta = 2ml^2 \sin^2 \theta \, \omega \tag{13-37}$$

The moment of inertia is

$$I = \Sigma m_i s_i^2 = 2m(l \sin \theta)^2 = 2ml^2 \sin^2 \theta \tag{13-38}$$

Comparison of Eqs. (13-37) and (13-38) shows that $L_\parallel = I\omega$ even though **L** also has a component perpendicular to **ω**.

The discovery that **L** and **ω** are not in the same direction unless the axis is an axis of symmetry indicates the need for a more general formulation of rigid-body dynamics to include the possibility of rotation about *any* axis. Here we can only hint at the direction of this development. First, it can be shown that for *any* body, even if it has no discernible symmetry at all, there always exist three special axes, mutually perpendicular, such that in a rotation about any one of these, **L** and **ω** *do* have the same direction. These are called *principal axes of inertia*. For an object with rotational symmetry, the axis of symmetry is always one of the principal axes of inertia. The moments of inertia about the principal axes are called the *principal moments of inertia*.

The three principal axes may be used as the coordinate axes of a cartesian coordinate system. Denoting the three principal moments of inertia by I_x, I_y, and I_z, one can show that each component of **L** is proportional to the corresponding component of ω:

$$L_x = I_x \omega_x \qquad L_y = I_y \omega_y \qquad L_z = I_z \omega_z \tag{13-39}$$

which may be summarized in the single vector equation

$$\mathbf{L} = I_x \omega_x \mathbf{i} + I_y \omega_y \mathbf{j} + I_z \omega_z \mathbf{k} \tag{13-40}$$

In general, the vectors **L** and **ω** have different directions since there is not a direct proportionality between them but a different relation for each component.

Because the formulation of angular momentum is simplified considerably by using coordinate axes which coincide with the principal axes of inertia, this choice is always made in solving practical problems. There is a difficulty inherent in the use of principal axes, however. Since the body moves, a set of axes which are principal axes at one instant may *not* be principal axes the next instant. One way to avoid this problem is to use a set of axes attached to the body and rotating with it. Such a rotating coordinate system, of course, is in general not an inertial frame of reference, and appropriate modifications must be made in the equations of motion. Nevertheless, the advantages of having a set of coordinate axes which are *permanently* principal axes of inertia are so great that they overshadow the small disadvantage of having to formulate the principles in a noninertial frame of reference.

13-7 EQUILIBRIUM OF A RIGID BODY

The principle of angular momentum, in a somewhat degenerate form, can be used to analyze situations in which a rigid body is in a state of *equilibrium*, i.e., in a state of rest or uniform translational motion (without rotation) in an inertial frame of reference. For such a state, the angular momentum is clearly either zero or constant, and in either case the vector sum of the *torques* on the system is zero. The vector sum of the external forces must also be zero in order that the acceleration of the center of mass be zero.

Thus the two fundamental conditions for equilibrium of a rigid body are

$$\Sigma \mathbf{F} = 0 \quad \text{and} \quad \Sigma \boldsymbol{\tau} = 0 \tag{13-41}$$

In the second of Eqs. (13-41) we have not specified an origin for computing torques. If the body is in equilibrium, *any* point in the body may be used as the origin of an inertial frame of reference, so the vector sum of the torques must be zero *no matter what point is chosen as an origin*.

In equilibrium problems, as in dynamics problems, it is necessary to specify clearly the extent and boundaries of the system to which Eqs. (13-41) are to be applied. For this reason, it is always advisable to draw a free-body diagram showing all the forces acting on the system under consideration, and no others. The following examples illustrate this procedure.

Example

Consider the boom of a crane, as shown in Fig. 13-13a. The bottom end is pivoted, and the boom is held up by a horizontal cable attached at its center. An object of weight W is to be lifted; we are asked to find the tension

362 DYNAMICS OF A RIGID BODY

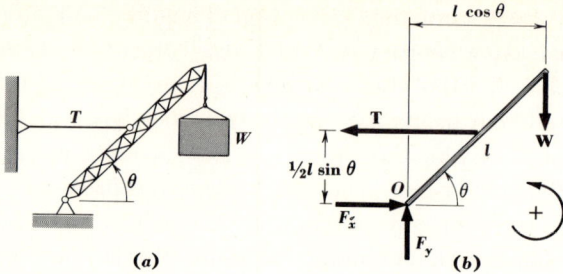

Fig. 13-13 (*a*) The boom of the crane, which lifts a body of weight W, is supported at the bottom by a pivot and is held in position by a horizontal cable at its midpoint, with tension T. (*b*) Free-body diagram for the problem. The moment arms of the forces with respect to O are shown.

T in the cable and the forces exerted on the bottom of the boom by the supporting pivot.

Solution

Figure 13-13*b* is a free-body diagram for the problem. The force **F** at the pivot O is represented by its x and y components. The condition $\Sigma \mathbf{F} = 0$ shows immediately that $F_y = W$ and $T = F_x$. The latter relation is not immediately helpful, because both quantities are unknown; to determine them, we must use the condition that the sum of the *torques* is zero. All torques are vectors perpendicular to the plane of the problem; we may call those tending to produce counterclockwise rotations (vectors directed out of the paper) *positive* and those in the opposite direction *negative*. To compute the magnitudes of the torques, we multiply the magnitude of each force by its perpendicular lever arm as obtained from Fig. 13-13*b*.

As already pointed out, any point whatsoever may be used to compute torques. The problem is simplified by choosing the pivot point; the forces acting at this point produce no torques about it, so the torque equation contains only two terms. The torque due to W with respect to O is $-l \cos \theta \, W$, and that due to T is $+\tfrac{1}{2} l \sin \theta \, T$. Requiring the sum of the torques with respect to O to be zero leads to the equation

$$\tfrac{1}{2} \sin \theta \, T - l \cos \theta \, W = 0$$

from which we find

$$T = 2W \cot \theta = F_x$$

The magnitude and direction of **F** can be computed from F_x and F_y. It should be noted that the direction of **F** does *not* lie along the length of the boom.

Example

An Iowa farmboy with an analytic turn of mind has to tie a barn door so that the wind will not blow it shut (Fig. 13-14a). Assuming that the force of the wind may be represented as a constant force **F** acting at the center

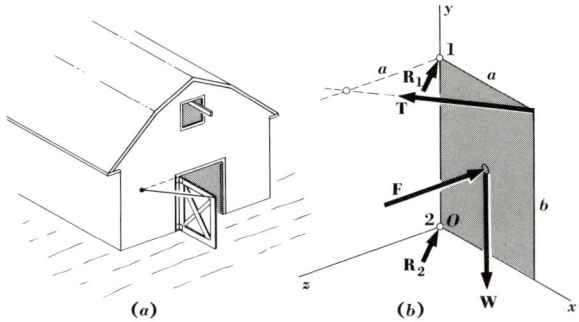

Fig. 13-14 (a) Barn door held open by a rope, with the wind blowing. (b) Free-body diagram for the door. The force **F** exerted by the wind is represented as acting at the center of the door. The construction of the door is assumed to be symmetric, so the center of mass is also at the center.

of the door and perpendicular to it, find the tension T in the rope and the three components of force on each hinge. For simplicity we assume that each hinge is at a corner of the door.

Solution

Figure 13-14b shows a free-body diagram for the problem. The forces can be represented in terms of their components as follows:

$$\mathbf{F} = -F\mathbf{k} \qquad \mathbf{R}_1 = R_{1x}\mathbf{i} + R_{1y}\mathbf{j} + R_{1z}\mathbf{k}$$
$$\mathbf{W} = -W\mathbf{j} \qquad \mathbf{R}_2 = R_{2x}\mathbf{i} + R_{2y}\mathbf{j} + R_{2z}\mathbf{k}$$
$$\mathbf{T} = \frac{T}{\sqrt{2}}(\mathbf{k} - \mathbf{i})$$

The position vectors of the points of application of the forces are

$$\mathbf{r}_W = \mathbf{r}_F = \tfrac{1}{2}a\mathbf{i} + \tfrac{1}{2}b\mathbf{j} \qquad \mathbf{r}_{R_1} = b\mathbf{j}$$
$$\mathbf{r}_T = a\mathbf{i} + b\mathbf{j} \qquad \mathbf{r}_{R_2} = 0$$

The torques due to these forces with respect to O can now be computed in a straightforward manner. For example, the torque due to **F** is

$$\tau_F = \mathbf{r}_F \times \mathbf{F} = (\tfrac{1}{2}a\mathbf{i} + \tfrac{1}{2}b\mathbf{j}) \times (-F\mathbf{k})$$
$$= \tfrac{1}{2}aF\mathbf{j} - \tfrac{1}{2}bF\mathbf{i}$$

The torques due to the other forces can be computed in the same manner. Having made these computations, we impose the requirements of Eqs. (13-41). These are both vector equations; the sum of the components of force in each of the three coordinate directions is zero, as is the sum of the components of torque in each direction. Thus we obtain six simultaneous scalar equations. This seems strange, since we have seven unknown quantities, the three components each of $\mathbf{R}_1$ and $\mathbf{R}_2$ and the tension T. We have more unknowns than equations.

The difficulty arises from the fact that the door is pivoted about the y axis; there is no way to determine how much vertical force is exerted at each hinge, since these two forces act along the same line. Therefore we cannot determine R_{1y} and R_{2y} separately but only the sum $R_{1y} + R_{2y}$, which should be regarded as a single unknown. Completion of the solution is left as a problem.

Choice of the point O for the computation of torques simplifies the torque equations considerably, because all components of $\mathbf{R}_2$ are automatically eliminated by this procedure. If some other point had been chosen, the moment equations would have contained all seven unknowns and would have been much more complicated to solve.

Problems

13-1 A stick of length d rotates with constant angular velocity about an axis through one end perpendicular to its length. The axis is fixed in an inertial frame of reference. Discuss how the motion can be represented in terms of:
   a Rotation about a fixed axis
   b Motion of the center of mass and rotation about an axis through the center of mass

13-2 If the vector sum of the external forces on a system of particles is zero, show that $\mathbf{F}_i = d\mathbf{p}'_i/dt$, where $\mathbf{F}_i$ is the vector sum of the forces on particle i and $\mathbf{p}'_i$ is its momentum with respect to the center of mass.

13-3 An electric dipole consists of two electric charges of equal magnitude but opposite sign separated by a certain distance. If the charges are $+q$ and $-q$

and the position of $+q$ relative to $-q$ is given by the vector **d**, it is customary to define the *electric dipole moment* of the system as $\mathbf{M} = q\mathbf{d}$. Show that if this system is placed in a uniform electric field **E**, the torque $\boldsymbol{\tau}$ exerted on the system by the field is given by $\boldsymbol{\tau} = \mathbf{M} \times \mathbf{E}$ and that in this case the torque is the same no matter what point is used as the origin of coordinates for computing torques.

13-4 A trapdoor in a ceiling is hinged on one side and supported on the other by a rope, as shown in Fig. P13-4.
 a Find the tension in the rope.
 b If the rope is cut, find the instantaneous acceleration of the point at which the rope was attached.

Fig. P13-4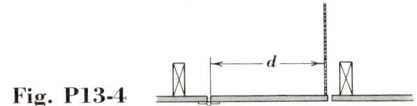

13-5 A pencil is held vertically with its point on a desk and then allowed to fall over. Assuming that the point does not slip, find the speed of the eraser just as it strikes the desk. Compare this with the speed that would result from free fall of a horizontal pencil from a height equal to the length of the pencil.

13-6 In Prob. 13-5 suppose the desk top is perfectly smooth, so the point can slide freely.
 a Find the instantaneous acceleration of the center of mass when the pencil is at an angle of 45° to the desk top.
 b Find the speed of the eraser just as it strikes the desk. Compare with the result of Prob. 13-5.

13-7 The circular bottom of a birdcage is held in place by two clamps on opposite sides. As the bird is sitting near one clamp, it suddenly releases and the bottom drops out, pivoting around the other clamp. Find the initial acceleration of the cage bottom and that of the bird. In particular, does the bottom drop away from the bird or do the two remain in contact?

13-8 For the Yo-Yo described in Prob. 11-34, find the tension in the string and the acceleration of the center of mass. Solve symbolically first; then insert the numerical values given in Prob. 11-34*b*.

13-9 A Yo-Yo made from two uniform disks of radius b connected by an axle of radius a is placed on edge on a horizontal table; the string, wound several times around the axle, is pulled as shown in Fig. P13-9. The coefficient of friction with the floor is large enough to prevent slipping.

366 DYNAMICS OF A RIGID BODY

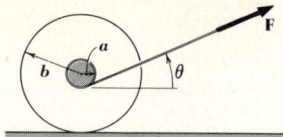

Fig. P13-9

a Show that if the string is vertical, the Yo-Yo rolls one way but if it is horizontal, it rolls the other way.

b Show that there is a critical value of the angle θ such that if θ is greater than this critical angle, the Yo-Yo rolls one way; if less, the other way. Find the value of the critical angle. Does it depend on the moment of inertia of the Yo-Yo? Explain.

c Describe the motion which takes place if the string is pulled at exactly the critical angle.

13-10 A uniform disk rests on a smooth frictionless horizontal surface. A string wrapped around its edge several times is pulled with a constant force, as shown in Fig. P13-10. If the disk is initially at rest, describe its motion in detail.

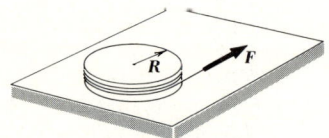

Fig. P13-10

13-11 A uniform solid cylinder rests on a horizontal surface with its axis horizontal. A rope wrapped around it several times is pulled vertically upward with a constant force F.

a Find the acceleration of the center of mass of the cylinder.

b What is the maximum force which can be exerted for which the cylinder rolls without slipping if the coefficient of friction is μ?

c Is it possible to exert a force so large that the cylinder is lifted off the plane? If so, what force is needed?

13-12 A uniform solid sphere of radius R rolls without slipping down a plane inclined at an angle θ to the horizontal.

a Find the acceleration of the center of the sphere. Compare this with the acceleration it would have if it could slide without friction.

b What minimum coefficient of friction is needed to prevent slipping?

13-13 A solid cylinder, a cylindrical shell, and a solid sphere, all with the same radius, roll down an inclined plane. Which reaches the bottom first? Explain.

13-14 A round object with mass M, radius R, and moment of inertia $I = kMR^2$, where k is a dimensionless constant between zero and unity, rolls without slipping down a plane inclined at an angle θ to the horizontal.

a Find the acceleration of the center of mass of the object.
b What coefficient of friction is needed to prevent slipping?

13-15 A dumbbell is made from two uniform disks, each of radius *a* and mass *M*, connected by a light rod of smaller radius *b*. The dumbbell is permitted to roll down a board resting on the connecting rod, as shown in Fig. P13-15. If the board is inclined at an angle θ to the horizontal, find the acceleration.

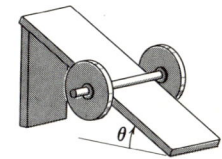

Fig. P13-15

13-16 A uniform solid cylinder is given an angular velocity about its axis and then dropped on a flat horizontal surface. The surface is not frictionless, so the cylinder begins to move as it slips; eventually the motion becomes one of rolling without slipping. Find the final velocity of the center of mass. *Hint:* Neither energy, nor momentum, nor angular momentum is conserved, but the changes in momentum and angular momentum are directly related, since the force of friction is responsible for both.

13-17 Two mountaineers tied to opposite ends of a rope are ascending a sheer vertical cliff. The top man grabs a crumbly handhold. The second man is not adequately anchored, so both men fall. Describe qualitatively the subsequent motion.

13-18 In the situation of Fig. 13-8, suppose that instead of being horizontal, the axle of the gyroscope is inclined at an angle θ upward from the horizontal. Show that the gyroscope precesses with a motion in which θ is constant. Find the precession angular velocity. How does it depend on θ?

13-19 An ancient form of mill for grinding grain is shown in Fig. P13-19. The millstone is a uniform disk with a light axle through its center. One end of the axle is held fixed, so it rolls around in a circle with constant angular velocity. Find the force which the millstone exerts on the flat surface and the forces at the fixed end of the axle.

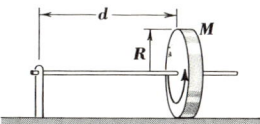

Fig. P13-19

13-20 A proton has an intrinsic angular momentum **S** which may be pictured crudely as due to spinning about an axis. It also has a magnetic moment $\boldsymbol{\mu}$ which is proportional to the angular momentum: $\boldsymbol{\mu} = \beta \mathbf{S}$, where β is a proportionality constant called the *gyromagnetic ratio*. The magnetic moment causes the proton to interact with a magnetic field much as a compass needle does. In particular, in a uniform magnetic field **B**, the proton experiences a torque τ given by $\tau = \mu \times \mathbf{B}$. Show that under the action of this torque the proton's axis precesses around the direction of **B**. Find the precession angular velocity in terms of the other quantities. (The precession frequency can be measured electrically with great precision; this phenomenon can be used for very precise measurements of magnetic fields and magnetic moments by methods called *magnetic resonance*.)

13-21 A bicycle wheel and its supporting fork are mounted as shown in Fig. P13-21. The axle is vertical, and the entire assembly is also free to rotate about a vertical axis at the end of the fork. With the axle stationary, the wheel is given an angular velocity. A device attached to the fork then rotates it 180°, so the entire wheel is turned over. Describe the subsequent motion of the system.

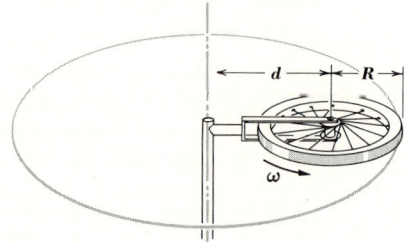

Fig. P13-21

13-22 Using the data in Appendix C, compare the magnitudes of the spin and orbital parts of the earth's angular momentum with respect to the sun.

13-23 Compare the spin and orbital parts of the angular momentum of the moon with respect to the earth. *Hint:* The moon always has the same side turned toward the earth.

13-24 A thin circular ring hangs from a peg on a wall. It is displaced sideways slightly and then released, so it swings back and forth in its own plane.
   a Find the frequency of the resulting small oscillations.
   b How long must a simple pendulum be to have the same frequency?
   c Would the frequency be the same or different if the ring had been displaced perpendicular to the plane? Explain.

13-25 A thin uniform rod hangs vertically, suspended by a pivot at its top end. It is displaced slightly from the vertical and released.
   a Find the frequency of the resulting oscillations.
   b Compare this frequency with that of a simple pendulum having the same length.
   c How long should the rod be in order to have a period $T = 2$ s?

PROBLEMS

13-26 A uniform thin rod of length l is hung vertically, pivoted at a point a distance x from the center. It is displaced slightly from the vertical and then released.
   a   Find the frequency of the resulting oscillations.
   b   For what value of x is the frequency greatest? Smallest?
   c   For what value of x is the frequency the same as that of a simple pendulum of length equal to the length of the rod?

13-27 A body of irregular shape is suspended from each of two points P_1 and P_2, as shown in Fig. P13-27, and for each point the period of small oscillations about the equilibrium position is measured. It is found that by suitable choice

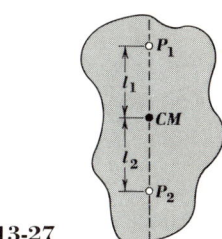

Fig. P13-27

of these points the periods in the two cases can be made equal. Assuming that a straight line from P_1 to P_2 passes through the center of mass, show that the period depends only on the total distance $l_1 + l_2$ between the two points and that $T^2 = 4\pi^2(l_1 + l_2)/g$. (This arrangement is called *Kater's pendulum*, after its inventor; it provides a means of making very precise measurements of g.)

13-28 Verify in detail the last two equations of Sec. 13-5.

13-29 For the system of Fig. 13-12, show in detail that the rate of change of angular momentum is given by Eq. (13-36). If the shaft is supported at each end by a bearing a distance d from the center of mass, find the forces which these bearings must exert.

13-30 A thin rod of length L is made to rotate with constant angular velocity about a fixed axis making an angle of 45° with the rod and passing through its center. This axis is supported by two bearings, each a distance L from the center of the rod.
   a   Find the angular momentum of the rod at one instant.
   b   Find the instantaneous rate of change of angular momentum.
   c   What forces must the bearings exert on the axis?
   d   Is this axis a principal axis of inertia? Explain.

13-31 A uniform disk of mass M and radius R rotates at constant velocity on a shaft passing through its center but making an angle θ with a line perpendicular to the disk faces. The shaft is supported by bearings on either end a distance d from the center of the disk.

    a Find the magnitude of the forces the bearings must exert on the shaft.

    b Obtain a rough estimate of the magnitudes of these forces if $M = 5$ kg, $R = 0.1$ m, $\theta = 1°$ and the disk rotates at 3,600 r/min. Compare with the weight of the disk.

13-32 A large crate containing a refrigerator is to be pulled across a rough floor ($\mu = 0.50$) by a rope attached as shown in Fig. P13-32. The center of mass of the crate is at its geometric center.

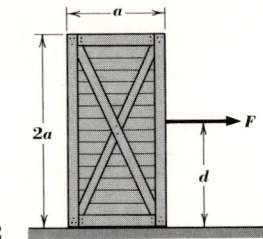

Fig. P13-32

    a How much force is required to move the crate at constant velocity?

    b Show that if the force determined in part *a* is applied too high, the crate will tip over instead of sliding. Find the maximum value of d to avoid tipping.

    c If the force is applied halfway up the crate ($d = a$) but is larger than the above value so that the crate accelerates, find the maximum force that can be applied without tipping the box over.

13-33 The crane shown in Fig. P13-33 has a boom 100 ft long and supports a weight of 5 tons. It is held up by the pinned support at the bottom and the horizontal cable at its midpoint. If the weight of the boom is negligible, find the tension in the horizontal cable and the components of the force on the boom at the bottom end.

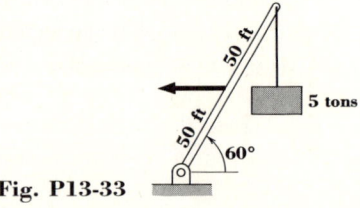

Fig. P13-33

13-34 In Prob. 13-33 suppose the weight of the boom is not negligible but is 2 tons. Find the tension in the horizontal cable and the forces at the bottom support.

13-35 A 20-ft ladder leans against a wall at an angle of 60° to the horizontal, as shown in Fig. P13-35. It weighs 50 lb, and a 150-lb man wants to climb it halfway. The floor and wall are both perfectly smooth. What horizontal force must be applied to the ladder at its base to prevent slipping? What then are the other forces on the ends of the ladder?

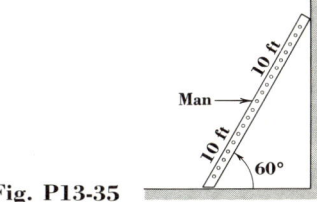

Fig. P13-35

13-36 The frame shown in Fig. P13-36 stands on a perfectly smooth floor. It is kept from collapsing by the rope connecting the midpoints of the two sides. Find the tension in this rope and the force transmitted from one side to the other by the pin joint at the top.

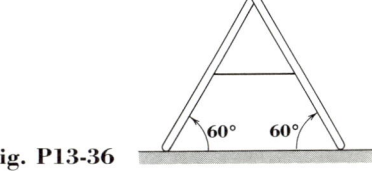

Fig. P13-36

13-37 A neon sign weighing 500 lb is suspended by a pinned joint and a cable, as shown in Fig. P13-37. Find the tension in the cable and the forces at the pinned joint.

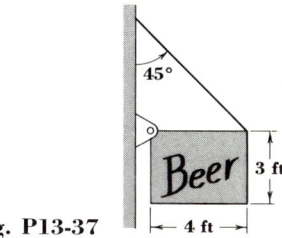

Fig. P13-37

13-38 Complete the solution of the barn-door problem in Sec. 13-7.

PERSPECTIVE IV

The accomplishments of the last three chapters have centered in two areas, the extension of Newton's laws and the momentum and energy principles so they can be applied directly to a system containing several particles, and the development of new principles to permit analysis of the rotational motion of rigid bodies. In both these areas, our procedure has been mostly deductive, starting with Newton's laws for a particle, defining new quantities when necessary, and using the laws of motion to derive relations among them.

In the first area, the motion of the center of mass of any system is simply related to the external forces on the system, even though the internal motion may be very complex. Furthermore, using the center of mass as the origin of a coordinate system often facilitates the analysis of collision problems; this is especially useful if the system is isolated so that the total momentum is conserved, since in the center-of-mass system the total momentum is always zero. Correspondingly, the total kinetic energy of the system can be represented conveniently in terms of motion of the center of mass and internal motion relative to the center of mass.

With respect to rigid-body motion, the concepts of angular momentum and torque provide the key to the analysis of rotational motion. Angular momentum was introduced first in the context of central-force problems for a single particle. In these problems angular momentum is a convenience but not an absolute necessity; it is always possible to use Newton's laws directly, with at most a slight increase in the amount of labor involved. But for a rigid body made up of a very large number of particles, the concept of angular momentum is indispensable. It would be hopeless to try to *describe individually the motion of each of the particles, but it is relatively easy to describe the rotation of a rigid body. As we have seen, the rotational motion is directly related to the external forces on the body.*

We now have all the necessary principles for a complete analysis of general rigid-body motion. Motion of the center of mass is governed by one principle and rotation about the center of mass by another; both are directly related to the external forces. The remaining problems in the analysis of rigid-body motion are mathematical in nature; the physical principles developed here are complete.

In the next two chapters we discuss the modifications of the principles of mechanics which are necessary when an object moves with a speed approaching that of light; these modifications are contained in the special theory of relativity. They do not in any way invalidate the mechanics we have developed thus far, which holds in all cases in which the speeds are much less than the speed of light. Rather, relativistic mechanics is a generalization from newtonian mechanics, applicable without restrictions on speeds of moving objects.

From relativistic mechanics we proceed into an extensive discussion of the mechanical and thermal properties of matter. In this discussion the unifying theme is the continuing effort to relate the macroscopic behavior and properties of matter to its microscopic structure and the behavior of the atoms and molecules of which it is composed. Many principles of mechanics are used in the analysis, and the basic unity of mechanics, thermodynamics, and the behavior of matter becomes apparent.

Relativity | 14

The theory of relativity is based on the principle that all inertial frames of reference are completely equivalent with respect to the formulation of physical laws. This requirement can be satisfied for both the laws of mechanics and those of electromagnetism only when certain modifications are made in the familiar relationships between distances and time intervals measured in two different coordinate systems moving relative to each other. The modified kinematic relations which result necessitate a reformulation and generalization of the laws of mechanics. The resulting principles have greater generality than those based on Newton's laws. The principles of newtonian mechanics emerge as a special case of the more general formulation; newtonian mechanics retains its validity in situations in which all speeds are much smaller than the speed of light.

14-1 INVARIANCE OF PHYSICAL LAWS

In earlier chapters we have stressed repeatedly the importance of inertial frames of reference. Newton's laws of motion are valid only in an inertial frame of reference, but they are valid in *all* inertial frames. Any frame of reference moving with constant velocity relative to an inertial frame is also an inertial frame, and either of these is as good as the other in stating the basic principles of mechanics. To be sure, some frames of reference are more *convenient* than others for specific problems, but there is no difference in principle.

An observer in a moving train who drops an object may see it drop straight down, while to an observer standing beside the track the body appears to follow a parabolic path. To *each* observer, using velocities and accelerations measured in his own frame of reference, Newton's laws are obeyed. For mechanics, at least, there is no evidence for a single coordinate system that is better than all others for describing motion; all inertial frames of reference are equivalent. To put it another way: *The laws of mechanics are the same in every inertial frame of reference.*

Einstein proposed in 1905 that this principle should be extended to all of physics. This is a natural suggestion; when a phenomenon is observed in two different inertial frames, it would be very strange if some of the governing physical laws were the same in both systems while others were different. Obvious though Einstein's proposal may seem, it has very subtle and far-reaching consequences; we shall explore these consequences in the area of mechanics.

We begin with a brief look at wave propagation in matter. This may appear to be extraneous, but the basic ideas are central to our discussion. We consider a source of sound waves that is stationary in a mass of air through which the waves travel out from the source. Experiment shows that to an observer at rest in the frame of reference of the source and the air, the waves travel with a definite speed in all directions. To a moving observer, however, the speed appears different. In Fig. 14-1, observer A is moving away from the source, and the apparent speed in his frame of reference is less than in the stationary frame. If the wave speed relative to the air is v and the speed

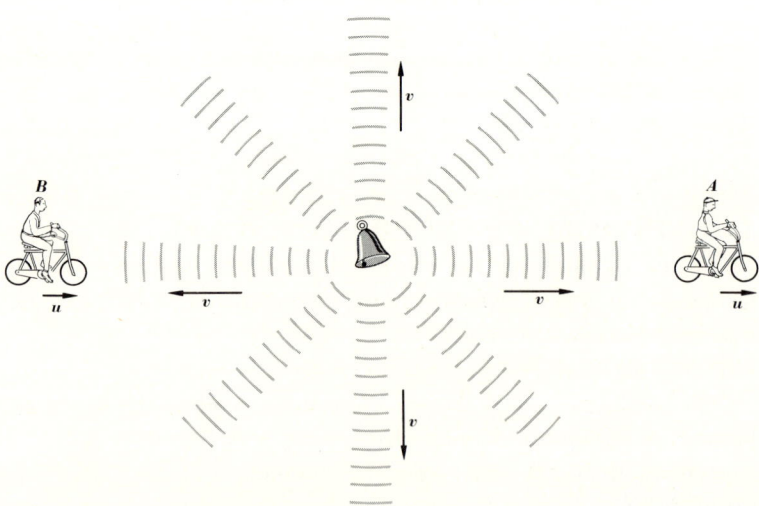

Fig. 14-1 The speed of sound emitted by a moving source is different for observer B moving toward the source and for observer A moving away from the source.

of the observer u, then A measures a wave speed of not v but $v - u$. Similarly, observer B approaching the source measures a speed relative to his frame of reference of $v + u$. This asymmetry does not violate our basic premise of equivalence of inertial frames because the air is at rest in only one frame.

Do these considerations apply also to optical phenomena? In the early nineteenth century it was known with a fair degree of certainty that light is a wave phenomenon. Physicists of that era tended strongly to try to picture everything in mechanical terms, and it was commonly thought that light should have some material medium in which to travel, analogous to air or other matter as a medium for sound waves. This hypothetical medium was given the name *the ether*. Its properties had to be strange indeed; its density would have to be immeasurably small, yet it must be extremely rigid to account for the large value of the speed of light.

If the ether existed, one ought to be able to determine our motion relative to it by measuring the speed of light in various directions, just as with sound waves as discussed above. A great deal of effort was put into experiments to measure this *ether drift*, as it came to be called. The details of these experiments need not concern us; the best-known are those of Michelson and Morley beginning in 1881 and extending over several decades. The results were entirely negative; no evidence for any motion relative to the ether was found, despite the fact that the earth in its orbital motion moves through space in different directions at different times of the year. From these and other experiments it became more and more clear that *there is no ether*. The behavior of light in vacuum cannot be understood on the basis of propagation through a material medium.

In the absence of such a medium, we are forced to conclude that if all frames of reference are to be equivalent for light propagation as well as mechanical phenomena, then the speed of light in vacuum must be the same in all inertial frames of reference. It follows that the speed of light is independent of the motion of its source. Once light is emitted by a source, it "forgets" whatever motion the source may have relative to a given observer and travels with a definite speed independent of this motion.

In recent years this conclusion has been confirmed directly by experiment. In 1962 J. G. Fox and collaborators measured the speed of electromagnetic radiation emitted by decay of π^0 mesons moving at high speeds relative to the observer. Their results confirmed the conclusion that the speed of light is independent of the motion of source or observer.

The speed of light, a universal physical constant, is one of the most precisely known of all physical constants. It is usually denoted by c; to six significant figures its value is

$$c = 2.99793 \times 10^8 \text{ m/s} \qquad (14\text{-}1)$$

The approximate value $c = 3.00 \times 10^8$ m/s is in error by less than 1 part in 1,000 and is often used when greater accuracy is not needed.

To make certain the reader is sufficiently impressed by this conclusion concerning the speed of light, we return to Fig. 14-1. If instead of a bell we have a light source emitting radiation with speed c relative to the source and observer A measures the speed of this radiation in *his* frame of reference, he will obtain the value c despite the fact that he is moving away from the source with speed u. This conclusion is the first of several in this chapter which appear at first not to agree with common sense, but it is correct. We must keep in mind that what we call common sense is intuition growing out of everyday experience, and this experience does not include measuring speeds of the magnitude of the speed of light. Thus intuition may often be misleading in domains far removed from everyday experience.

14-2 GALILEAN TRANSFORMATION

To explore the consequences of the special role of the speed of light in kinematics, we now proceed to a detailed analysis of the relationships between two inertial frames of reference. We consider two coordinate systems, which we call S and S', as shown in Fig. 14-2. Let the x and x' axes coincide,

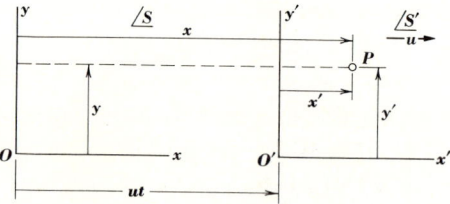

Fig. 14-2 Position of point P described in two different coordinate systems S and S'. S' moves relative to S with constant velocity u along the x and x' axes; the two origins O and O' coincide at time $t = 0$.

but let S' move along this line, relative to S, with constant speed u. Furthermore, we assume for simplicity that the two origins O and O' coincide at time $t = 0$. Then at time t the distance between O and O' is ut.

Any point P can be located by its coordinates (x,y,z) in system S or, alternatively, its coordinates (x',y',z') in S'. The figure shows that these coordinates are related by

$$x' = x - ut \qquad y' = y \qquad z' = z \qquad (14\text{-}2)$$

GALILEAN TRANSFORMATION [14-2]

By taking the second derivative of each of Eqs. (14-2), recalling that u is constant, we see immediately that the components of acceleration of P are the same in the two systems; hence Newton's second law is *invariant* under the transformation of coordinates given by Eqs. (14-1). Such a coordinate transformation is called a *galilean transformation*.

The relationship between the *velocities* of a point P measured in the two systems is of particular interest. For example, suppose a rifle bullet is fired in the $+x$ direction with a speed v as observed in S. What is its speed v' with respect to S'? This question can be answered immediately by taking the time derivative of the first of Eqs. (14-2):

$$\frac{dx'}{dt} = \frac{dx}{dt} - u \quad \text{or} \quad v' = v - u \tag{14-3}$$

which says that the speed v' measured in S' is less than the speed v measured in S, the difference u simply being the speed of the system S' with respect to S, as we might have guessed. Comparing this result with our discussion of the speed of light, we see a clear contradiction. If Eq. (14-3) is correct for the motion of light as well as material bodies, then the speed c' in frame S' must be related to the speed c in S by

$$c' = c - u$$

Einstein's postulates and the absence of an ether require instead that $c' = c$.

We are thus forced to conclude that if the equivalence of all inertial frames holds for both mechanics and electrodynamics, the formulation of newtonian mechanics is in need of modification. For example, perhaps Eqs. (14-2) are correct only in the limit of very small values of u but need to be modified when u approaches the speed of light. There is also direct evidence that at high speeds newtonian mechanics is not precisely correct. One example is an experiment in which the relationship between kinetic energy and speed is examined.[1] Electrons are accelerated in a vacuum tube by an arrangement similar to that used to produce the electron beam in a TV picture tube. They are then sent down a long evacuated pipe, and their speeds measured directly by measuring the time of flight down the tube. Their energies are determined by measuring the electrical potential difference through which they are accelerated. In principle, the energy can also be obtained independently by having the beam collide with a target and measuring the heat developed in the target by the electron bombardment, but this is much more difficult to do with good precision.

---

[1] This experiment is beautifully illustrated in a film *The Ultimate Speed*, by W. Bertozzi, Education Development Center, Newton, Mass., 1962.

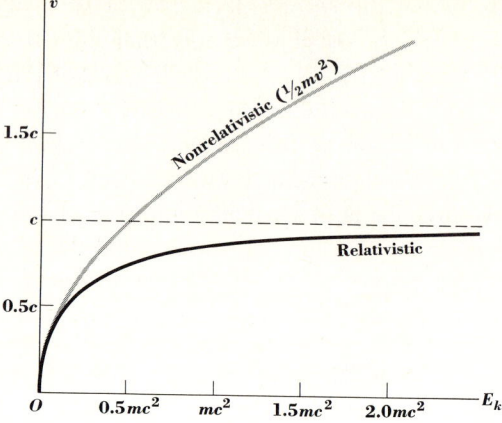

Fig. 14-3 Relation between the speed of a particle and its kinetic energy. The speed never exceeds c, no matter how great the kinetic energy. At speeds small compared to c, the kinetic energy is approximately $\frac{1}{2}mv^2$, but at larger speeds this expression deviates more and more from the relativistically correct equation.

The result of this experiment, shown graphically in Fig. 14-3, is an empirical determination of the functional relationship between speed and kinetic energy. In this graph, speeds are expressed as multiples of c and energies as multiples of the quantity mc^2, which has units of energy. For comparison, the newtonian relation $E = \frac{1}{2}mv^2$ is also plotted.

Two striking features are evident. The empirical result agrees with the newtonian one at small speeds but departs more and more significantly from it at higher speeds. In addition, the speed appears to be tending toward an asymptotic value c, as though the speed might never exceed c no matter how much the energy is increased. We shall return to this experiment in Chap. 15, after developing some needed background material. We glance at it here as an additional piece of evidence for the conclusion that newtonian mechanics needs to be reexamined carefully and modified wherever necessary.

In beginning this examination, we mention several assumptions made above in the derivation of Eq. (14-3). These may seem too obvious to require discussion, yet they lie at the very heart of the difficulties in newtonian mechanics which necessitate the development of the theory of relativity. First, in deriving Eqs. (14-2), we have assumed that when two observers moving relative to each other measure the distance between two points at a given time, they obtain the same result; rulers do not shrink or expand when moving. We have also assumed that the *time scale* is the same for both coordinate systems. This assumption can be stated formally by adding to Eqs. (14-2) a fourth equation $t' = t$. This assumption, in turn, implies that it is possible to observe two clocks moving relative to each other and to determine with certainty that they remain synchronized during their motion.

The validity of these assumptions seems obvious, and they are certainly in accord with everyday observation. Yet this comprises a very small part of the totality of physical phenomena, and we must be prepared to consider the possibility that these seemingly obvious conclusions require modification when applied to situations far removed from everyday experience. Conversely, however, it is also necessary that any modified formulations which describe motion in various coordinate systems must reduce to the familiar ones when applied to familiar situations. It is shown in the next section that both assumptions, the invariance of length and the universal time scale, require modification in order to conform to a more general principle of invariance of physical laws under coordinate transformations.

In the experiments on which newtonian mechanics is based, speeds are always slow compared with that of light; hence it is essential that the modified mechanical principles reduce to the newtonian principles when the speeds are sufficiently small. Newtonian mechanics must retain its validity for the class of phenomena for which it was originally developed.

14-3 RELATIVITY OF TIME

In developing the consequences of Einstein's postulates, it is often helpful to consider imaginary experiments which illustrate particular points. Such experiments, usually called *thought experiments* (or, in German, *Gedankenexperimente*), are very useful in clarifying new physical theories. In discussing thought experiments, it is necessary to be extremely careful lest assumptions incompatible with the basic postulate of relativity sneak in unnoticed, for we are on ground where our intuition may be misleading, and it is essential to proceed with utmost caution.

Frequently thought experiments involve the description of one or more *events* in various frames of reference. A description of an event includes the position and time at which it occurs. An event described in a frame of reference S as (x,y,z,t) may be described in a second frame S' as (x',y',z',t'). One of our goals will be to derive a general transformation which relates these two descriptions; the result, the *Lorentz transformation*, is a generalization of the newtonian relations given by Eqs. (14-2).

We begin by questioning the newtonian assumption that $t = t'$, that is, that there exists a universal time scale for all reference frames. The question is best stated as follows: When two events occur, each observed in two frames of reference S and S', will the time interval between the two events appear the same or different?

In attempting to answer this question qualitatively, we first point out that measuring times and time intervals involves the concept of simultaneity of two events. When we say that a bus leaves the station at noon, we mean that the event of the bus passing the end of the platform is simultaneous with the event of the hour hand of the clock arriving at the number 12. The fundamental difficulty with time measurements is that two events that appear simultaneous in one frame of reference do not in general appear simultaneous in a second frame.

The following thought experiment, devised by Einstein, illustrates this point. Consider a long train moving with uniform velocity, as shown in Fig. 14-4a. Two lightning bolts strike the train, one at each end. Each bolt leaves

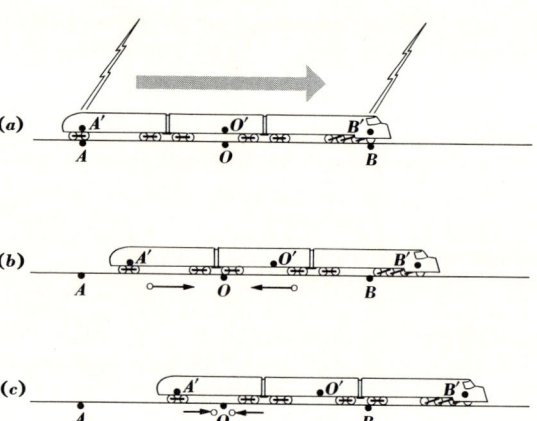

Fig. 14-4 (a) To the stationary observer O, two lightning bolts appear to strike the two ends of the train simultaneously. (b) The light pulse from the front of the train arrives at the position of the moving observer O' sooner than that from the rear; hence O' thinks the front bolt struck first. (c) The two light pulses arrive at O simultaneously, and this observer thinks the two bolts were simultaneous.

a mark on the train and on the ground at the same instant. The points on the ground are labeled A and B in the figure and the corresponding points on the train A' and B'. An observer on the ground is located at O, midway between A and B; another observer is at O', moving with the train and midway between A' and B'. Both these observers use the light signals from the lightning to observe the events.

Suppose the two lightning bolts strike at such times that the two light pulses reach the observer at O simultaneously; he concludes that the two events took place at A and B simultaneously. But the observer at O' is moving with the train, and the light pulse from B' reaches him before the light pulse from A'; he concludes that the event at the front of the train happened *earlier* than that at the rear. Thus the two events appear simultaneous to one observer but not to the other. More generally, whether or not two events which occur at points distant from an observer are simultaneous depends on the state of motion of the observer. Since this is the case, the time interval between two distant events is in general different for two observers in relative motion.

A reader steeped in the newtonian tradition might argue that in this example the lightning bolts really *are* simultaneous and that if the observer at O' could communicate with the distant points without time delay, he would realize this. But the finite speed of information transmission is not the problem. If O' is really midway between A' and B', then in his frame of reference the time for a signal to travel from A' to O' is the same as from B' to O'. Two signals arrive simultaneously at O' only if they were emitted simultaneously at A' and B', and since in this example they *do not* arrive simultaneously at O', O' must conclude that the events at A' and B' were not simultaneous.

Furthermore, we have no basis for saying that O is right and O' is wrong, or the reverse, since according to our basic postulate of relativity, no inertial frame of reference is preferred over any other in the formulation of physical laws. Each observer is correct in his own frame of reference, and we have to conclude that simultaneity is not an absolute concept. Whether or not two events are simultaneous depends on the frame of reference in which the observation is made, and the time interval between two events is in general different for observers in different frames of reference.

To obtain a quantitative relationship between time intervals, we consider another thought experiment. A frame of reference S' moves with velocity u relative to a frame S. An observer O' in S' has a source of light which he directs at a mirror a distance d away, oriented so that the light is reflected back to him as shown in Fig. 14-5. This observer measures the time interval

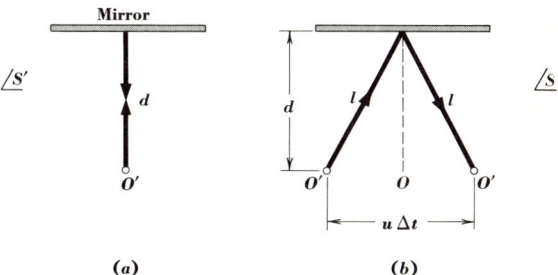

Fig. 14-5 (a) Light pulse emitted at O' and reflected back along the same line, as observed in S'. (b) Path of the same light pulse, as observed in S. The initial and final positions of O' are shown. The light path is longer in S than in S', but the speed of the pulse is the same in both.

$\Delta t'$ required for a light pulse to make the round trip to the mirror and back. The total distance, as measured in S', is $2d$, the speed is c, and the time required is

$$\Delta t' = \frac{2d}{c} \qquad (14\text{-}4)$$

Consider how this experiment looks to an observer at O, with respect to which O' is moving with speed u. Let the time interval observed by O for the round trip be Δt. During this time, the source moves relative to O a distance $u\,\Delta t$, as shown in Fig. 14-5b. The total round-trip distance as seen by O is not $2d$ but

$$2l = 2\sqrt{d^2 + (\tfrac{1}{2}u\,\Delta t)^2} \qquad (14\text{-}5)$$

According to the basic postulate of relativity, the speed of light c is the same with respect to both observers, and so the relation in S analogous to Eq. (14-4) is

$$\Delta t = \frac{2l}{c} = \frac{2}{c}\sqrt{d^2 + (\tfrac{1}{2}u\,\Delta t)^2} \qquad (14\text{-}6)$$

Equations (14-4) and (14-6) can be combined to obtain a relationship between Δt and $\Delta t'$ which does not contain the distance d. We solve Eq. (14-4) for d, substitute the result in Eq. (14-6), and solve for Δt. The result of this algebraic manipulation is

$$\Delta t = \frac{\Delta t'}{\sqrt{1 - u^2/c^2}} \qquad (14\text{-}7)$$

We may generalize this important result: If two events occurring at the same space point in a frame of reference S' (in this case, the departure and arrival of the light signal at O') are observed to be separated in time by an interval $\Delta t'$, then the time interval Δt between these two events as observed in the frame of reference S is *larger* than $\Delta t'$, and the two are related by Eq. (14-7). Thus a clock moving with S' appears to an observer in S to run at a rate which is *slower* than the rate observed in S'.

Example

The π^+ meson (pion) is an unstable particle with mass about 273 times that of the electron. After it is produced in a high-energy collision between nuclear particles, it lives on the average about 2.6×10^{-8} s before decaying into a muon and a neutrino. This time is measured in a frame of reference in which the particle is at rest (the *rest frame* of the particle). If such a particle is created with a speed $u = 0.99c$, what is its lifetime as measured in the laboratory, and how far does it travel during that time?

Solution

We identify S' with the particle's frame and S with that of the laboratory. Thus $\Delta t' = 2.6 \times 10^{-8}$ s. Using Eq. (14-7), we find

$$\Delta t = \frac{\Delta t'}{\sqrt{1 - u^2/c^2}} = \frac{2.6 \times 10^{-8} \text{ s}}{\sqrt{1 - (0.99c)^2/c^2}} = 18.3 \times 10^{-8} \text{ s}$$

This represents a dilation of the time interval of about a factor of 7. The distance d traveled in S during this interval is

$$d = u \, \Delta t = (0.99)(3.0 \times 10^8 \text{ m/s})(18.3 \times 10^{-8} \text{ s}) = 54 \text{ m}$$

Of course, not all the particles live the same time in S'; there is a statistical distribution of decay times. But if all particles have the same velocity, their lifetimes are all dilated by the same factor, so the *average* lifetime in S is increased relative to that in S' by this same factor. This dependence of lifetime on velocity is a well-established experimental fact.

The derivation of Eq. (14-7) is not complete without two additional remarks. First, the observer in S who measures the time interval Δt between the departure and arrival of the light pulse cannot make this measurement with only one clock, because the two events occur at different points in S. If he tries to use just one clock at the point of departure, there arises the question of the time required for the signal to travel to the clock from the point at which the light pulse returns. This difficulty is easily circumvented by using *two* clocks, one where the light pulse departs and one where it returns. Since these two clocks are both stationary in S, it is possible to synchronize them unambiguously. It is of course essential to avoid errors introduced by the finite propagation time for light signals, but this is not difficult. One possible scheme, for two clocks separated by a distance l, is for the first clock to send out a light pulse at a prearranged time, say t_1. When the pulse arrives at the second clock, its operator sets that clock to read $t_1 + l/c$. Alternately, we may place a light source midway between the clocks and send a light pulse. The operators of both clocks then set their clocks at the prearranged time when they receive the signal. Many other schemes are possible.

In any event, there is no difficulty in synchronizing two or any number of clocks unambiguously in a single frame of reference, and it is often useful in thought experiments to consider many synchronized clocks distributed in a frame of reference. Only when a clock is moving relative to a frame of reference do synchronization difficulties arise.

Second, we have assumed the distance d in Fig. 14-5 to be the same in S and S'. This assumption, though correct, is not quite as trivial as it may appear. To justify it in detail we must consider how to measure the length of a ruler stationary in S' but moving in S. Such a measurement requires simultaneous observation of the positions of the two ends. When the ruler is moving perpendicular to its length, as in the measurement of the distance d in Fig. 14-5, there is no problem; we need only station observers along the line described by the midpoint of the ruler. All such observers, whether moving or not, agree on the simultaneity of arrival of the ends of the moving ruler at predetermined points and thus on the length measured in the various frames of reference. When the ruler moves *parallel* to its length, the situation is quite different, as we shall see in the next section.

Because a time interval between two events at the same point in a given frame of reference is more fundamental than one between two events at different space points, we use the term *proper time* to designate a time interval between two events at the *same point* in a given frame of reference. Equation (14-7) shows that when the relative velocity of the two observers is small compared with c, the quantity u^2/c^2 is extremely small; then Δt is very nearly equal to $\Delta t'$. Hence, in the limit as $u \to 0$, this relationship approaches the newtonian $\Delta t = \Delta t'$, which assumes an absolute time scale for all frames of reference.

14-4 RELATIVITY OF LENGTH

Because of the relative nature of time intervals and simultaneity, we anticipate that there may be difficulties in comparing length measurements in different frames of reference, especially when the lengths are measured parallel to the direction of motion. To measure the length of a ruler, we must observe the positions of its two ends simultaneously, but what is simultaneous in one frame may not be in another. Suppose a ruler is at rest in a coordinate system S' and lies parallel to the x' axis. Both ruler and S' move with uniform velocity u with respect to another frame of reference S. If observers in S and S' measure the length of this ruler, how will their results compare?

One way to measure the length in each coordinate system, at least in principle, is to attach a source of light to one end of the ruler and a mirror to the other end. One can then send a pulse of light toward the mirror and measure the amount of time required for it to be reflected back to the source. This thought experiment is illustrated in Fig. 14-6, which shows the process as it appears both in S' and in S. Denoting by l' the length measured in S', we observe that the time $\Delta t'$ required for the light pulse to travel the

Fig. 14-6 (a) Light pulse emitted from a source at one end of a ruler, reflected from a mirror at the opposite end, and returning to the source, as seen in S'. (b) Same light pulse, reviewed from S. The light pulse travels farther than the length of the ruler, since the ruler is moving in S with speed u. The speed of the light pulse in S is the same as in S'.

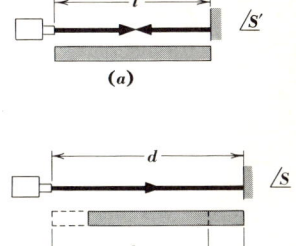

length of the ruler twice is given simply by

$$\Delta t' = \frac{2l'}{c} \tag{14-8}$$

This is a *proper* time interval, since it is measured between two events, the departure and arrival of the light pulse, at the same space point in S'.

From the point of view of S, the light pulse leaving the left end must travel farther than l (the ruler length in S) because the ruler is displaced during its travel. We let the total light path from source to mirror be d and the time of travel from source to mirror be Δt_1. Then, during this time, the mirror moves a distance $u\,\Delta t_1$, so the total light path, as shown in Fig. 14-6, is

$$d = l + u\,\Delta t_1 \tag{14-9}$$

It is also true, of course, that $d = c\,\Delta t_1$, since the light pulse travels with constant speed c. Combining this with Eq. (14-9) to eliminate d, we find

$$\Delta t_1 = \frac{l}{c - u}$$

In exactly the same way, one can show that the time Δt_2 required for the return trip is given by

$$\Delta t_2 = \frac{l}{c + u}$$

Thus the total time $\Delta t = \Delta t_1 + \Delta t_2$ required for the entire trip, as observed in S, is

$$\Delta t = \frac{l}{c - u} + \frac{l}{c + u} = \frac{2l}{c(1 - u^2/c^2)} \tag{14-10}$$

This relationship differs from Eq. (14-8) by the factor $1 - u^2/c^2$ in the denominator.

We also know that the time intervals Δt and $\Delta t'$ are related by Eq. (14-7), since $\Delta t'$ is a proper time interval in S' and Δt is the interval between the same two events as observed in S. Substituting the expressions for the time intervals, Eqs. (14-8) and (14-10), into Eq. (14-7) and rearranging, we obtain

$$l = l' \sqrt{1 - \frac{u^2}{c^2}} \tag{14-11}$$

This important result shows that the length l measured in S, in which the ruler is moving, is *shorter* than the length l' in S', in which it is at rest. That is, a moving ruler appears *contracted* in the direction of its motion. As with time intervals, it is customary to call a length measured in the rest frame of the body a *proper length*. Thus Eq. (14-11) shows that the proper length of a body is greater than the length observed by a moving observer.

Example
In the π^+ meson example in Sec. 14-2, what distance does the pion travel as measured in its rest frame?

Solution
In this case the path d measured in the laboratory is the proper length, and in the pion's rest frame the path length d' appears contracted by the factor given in Eq. (14-11):

$$d' = d \sqrt{1 - \frac{u^2}{c^2}} = (54 \text{ m}) \sqrt{1 - \frac{(0.99c)^2}{c^2}} = 7.7 \text{ m}$$

We note that an observer moving with the pion can calculate his speed relative to the laboratory (frame S) from this information:

$$u = \frac{d'}{\Delta t'} = \frac{7.7 \text{ m}}{2.6 \times 10^{-8} \text{ s}} = 3.0 \times 10^8 \text{ m/s}$$

which agrees with the direct measurement to two significant figures.

As in the case of time dilation, we note that when $v \ll c$, the length-contraction relation, Eq. (14-11), reduces to the newtonian one, $l = l'$.

14-5 THE LORENTZ TRANSFORMATION

We have seen that the galilean coordinate transformation, Eqs. (14-2), relating the position and time of an event in one coordinate system with those in another system, are inconsistent with the basic postulates of relativity. We have examined two specific problems involving time and length transformations and have obtained results that *are* consistent with those postulates, and we are now in a position to derive a relativistically correct coordinate transformation. Called the *Lorentz transformation*, these equations were actually developed by H. A. Lorentz in 1887, 18 years before Einstein first published the theory of relativity; but not until the work of Einstein was their real significance understood.

Consider, as previously, two inertial frames of reference, S and S', the latter moving in the $+x$ direction with constant speed u relative to the former. Furthermore, suppose that at the instant the two origins O and O' coincide, clocks in the two frames of reference located at the respective origins are synchronized, so that when $x = x'$, $t = t' = 0$. Note that this procedure does not conflict with our previous discussion of simultaneity because we are comparing two clocks at the same point in space. We now ask: If a certain event occurs at point (x', y', z') at time t', as observed in S', at what point and at what time does the event occur as observed in S?

First, as already pointed out, there is no difficulty in comparing lengths measured perpendicular to the direction of motion. Hence we can say immediately that $y = y'$ and $z = z'$. A relation between the two x coordinates can also be obtained quite simply. In S, the distance between O and O' at time t is simply ut. The coordinate x' is a proper length in S', so when viewed from S it appears foreshortened by the factor given by Eq. (14-11). Then the total x distance in S from O to the point at which the event takes place is

$$x = ut + x'\sqrt{1 - \frac{u^2}{c^2}} \tag{14-12}$$

Rearranging this expression to obtain an equation for x' in terms of x and t gives

$$x' = \frac{x - ut}{\sqrt{1 - u^2/c^2}} \tag{14-13}$$

which differs from the first of Eqs. (14-1) in having an additional factor in the denominator.

Equation (14-13) is half of the Lorentz transformation. The other half is the equation giving t' in terms of x and t. This is obtained most easily by noting first that since S and S' are completely equivalent, the transformation giving x in terms of x' and t' must have exactly the same *form* as Eq. (14-13) for x' in terms of x and t. The only difference is that the sign of u must be reversed, since the velocity of S relative to S' is the negative of that of S' relative to S. Thus we can write immediately

$$x = \frac{x' + ut'}{\sqrt{1 - u^2/c^2}} \qquad (14\text{-}14)$$

We can now solve this equation for t' as a function of x and x' and then use Eq. (14-13) to eliminate x' in favor of x and t, as follows:

$$\begin{aligned} ut' &= x\sqrt{1 - \frac{u^2}{c^2}} - x' \\ &= x\sqrt{1 - \frac{u^2}{c^2}} - \frac{x - ut}{\sqrt{1 - u^2/c^2}} \end{aligned}$$

Simplifying the right side and dividing by u gives

$$t' = \frac{t - ux/c^2}{\sqrt{1 - u^2/c^2}} \qquad (14\text{-}15)$$

The form of Eq. (14-15) shows the time-dilation factor derived previously and also shows that the time for an event observed in S' depends on both its time and position in S. This reflects the fact that two clocks that are synchronized in S are out of synchronization in S' by an amount proportional to the distance between them in S. This equation is the second half of the Lorentz transformation.

Collecting all the transformation equations, we have

$$\begin{aligned} x' &= \frac{x - ut}{\sqrt{1 - u^2/c^2}} \\ y' &= y \\ z' &= z \\ t' &= \frac{t - xu/c^2}{\sqrt{1 - u^2/c^2}} \end{aligned} \qquad (14\text{-}16)$$

These are the Lorentz transformation. In this form, the equations show explicitly that if the relative velocity u of the two frames of reference is much smaller than c, so that $u/c \ll 1$, the transformation reduces to the galilean

THE LORENTZ TRANSFORMATION [14-5]

transformation, Eqs. (14-2). If u is not small compared to c, the transformations are quite different.

As an example of further useful results which can be derived from the Lorentz transformation equations, we consider the relativistic generalization of Eq. (14-3), relating the velocities of a moving object observed in two different coordinate systems. For simplicity, we consider only the case of an object moving with constant velocity parallel to the x axis in each coordinate system. To apply the Lorentz transformations, we may consider as events the arrival of the moving object at two different points. In S', let the object be at position x'_1 at time t'_1 and at x'_2 at time t'_2. The velocity v' observed in S' is then

$$v' = \frac{x'_2 - x'_1}{t'_2 - t'_1} = \frac{\Delta x'}{\Delta t'} \tag{14-17}$$

We now use Eqs. (14-16) to express the quantities in this equation in terms of quantities measured in S:

$$x'_2 - x'_1 = \frac{x_2 - x_1 - u(t_2 - t_1)}{\sqrt{1 - u^2/c^2}} = \frac{\Delta x - u\,\Delta t}{\sqrt{1 - u^2/c^2}}$$

$$t'_2 - t'_1 = \frac{t_2 - t_1 - (x_2 - x_1)u/c^2}{\sqrt{1 - u^2/c^2}} = \frac{\Delta t - \Delta x\, u/c^2}{\sqrt{1 - u^2/c^2}}$$

$$v' = \frac{\Delta x - u\,\Delta t}{\Delta t - \Delta x\, u/c^2} \tag{14-18}$$

We note that it would *not* be correct to substitute Eqs. (4-7) and (14-11) directly into Eq. (14-17), since these are valid only when $\Delta t'$ and $\Delta l'$ are proper in S'; this is not the case here because both the locations and times of the two events are different. Hence the general coordinate transformation given by Eqs. (14-16) must be used.

The velocity observed in S, of course, is simply $v = \Delta x/\Delta t$. To relate this to v', we divide numerator and denominator of Eq. (14-18) by Δt:

$$v' = \frac{\Delta x/\Delta t - u}{1 - \Delta x/\Delta t\, u/c^2} = \frac{v - u}{1 - vu/c^2} \tag{14-19}$$

This equation can also be rearranged to give v in terms of v'. The details of this calculation are left as an exercise; the result is

$$v = \frac{v' + u}{1 + v'u/c^2} \tag{14-20}$$

which has the same form as Eq. (14-19) except that the terms containing u have signs opposite those in Eq. (14-19), as might be expected.

If both u and v are much smaller than c, the second term in the denominator of Eqs. (14-19) and (14-20) is negligible and these expressions reduce to the newtonian equation (14-3). But if the speeds are comparable to that of light, the results may be quite different. The extreme case occurs when $v' = c$; then no matter what value u has, Eq. (14-20) gives $v = c$, in agreement with the basic postulate of relativity.

Example
Suppose a body moves in S' with a speed $v' = 0.9c$ and S' moves relative to S with a speed $u = 0.9c$. What is the body's speed in S?

Solution
If we used the nonrelativistic velocity relations to find the speed of the object relative to S, we would find $v = 1.8c$. But when we insert the numerical values in Eq. (14-20), we find

$$v = \frac{0.9c + 0.9c}{1 + (0.9c)(0.9c)/c^2} = 0.994c$$

We can also derive a velocity transformation for the more general case of a point moving in the xy ($x'y'$) plane with components of velocity v'_x and v'_y in S'. The x-component relation is still given by Eq. (14-20), since the y coordinate is not involved at all in this equation. The y component of velocity in S is given by

$$v'_y = \frac{y'_2 - y'_1}{t'_2 - t'_1} = \frac{\Delta y'}{\Delta t'}$$

Applying the Lorentz transformation equations (14-16), we obtain

$$v'_y = \frac{(y_2 - y_1)\sqrt{1 - u^2/c^2}}{(t_2 - t_1) - u(x_2 - x_1)/c^2} = \frac{\Delta y \sqrt{1 - u^2/c^2}}{\Delta t - u\,\Delta x/c^2}$$

Again we divide numerator and denominator by Δt and identify $v_x = \Delta x/\Delta t$, $v_y = \Delta y/\Delta t$ to obtain

$$v'_y = \frac{v_y \sqrt{1 - u^2/c^2}}{1 - v_x u/c^2} \tag{14-21}$$

The general velocity transformation can be summarized:

$$v'_x = \frac{v_x - u}{1 - v_x u/c^2} \qquad v'_y = \frac{v_y \sqrt{1 - u^2/c^2}}{1 - v_x u/c^2} \qquad (14\text{-}22)$$

The inverse transformation is obtained easily by interchanging v_x and v'_x, v_y and v'_y, and changing the sign on u:

$$v_x = \frac{v'_x + u}{1 + v'_x u/c^2} \qquad v_y = \frac{v'_y \sqrt{1 - u^2/c^2}}{1 + v'_x u/c^2} \qquad (14\text{-}23)$$

The most striking feature of these equations is that the y component of velocity in one frame depends on *both* the x and y components in the other, quite unlike the situation with the galilean transformation. These relations are very useful when we consider conservation of momentum in various frames of reference in Chap. 15.

Example

A light source at rest at the origin O' of S' emits a ray of light in the $x'y'$ plane at an angle θ' with the x' axis. What is its direction as seen in S?

Solution

The components of velocity of the light ray in S' are

$$v'_x = c \cos \theta' \qquad v'_y = c \sin \theta'$$

We use Eqs. (14-23) to obtain the components of velocity in S:

$$v_x = \frac{c \cos \theta' + u}{1 + cu \cos \theta'/c^2} \qquad v_y = \frac{c \sin \theta' \sqrt{1 - u^2/c^2}}{1 + cu \cos \theta'/c^2}$$

The direction θ in S, that is, the angle the light ray makes with the x axis in S, is given by

$$\tan \theta = \frac{v_y}{v_x} = \frac{c \sin \theta' \sqrt{1 - u^2/c^2}}{c \cos \theta' + u}$$

We note that when $u = 0$, $\tan \theta = \tan \theta'$, as expected. This formula is useful in the analysis of a phenomenon known as *stellar aberration*, in which the true position of a star differs somewhat from the direction a telescope is aimed if there is relative motion of the two.

We have now completed the development of the modified coordinate transformations required by Einstein's principle of the invariance of physical laws. The next step is to explore the modifications of *dynamics,* including the relationships between force, momentum, and energy, which are required for consistency with Einstein's principle. These considerations are the topics of the next chapter.

Problems

14-1 A π^+ meson at rest decays on the average approximately 2.6×10^{-8} s after it is produced. This time is called its *lifetime.*
    a A π^+ meson is moving with a speed of $0.8c$ with respect to an observer in a laboratory. He measures the lifetime of the particle; what result does he obtain?
    b What distance (measured in the laboratory) does the particle travel between production and decay?

14-2 The μ mesons (muons) are unstable particles which decay into electrons and neutrinos after an average lifetime (in the rest frame of the mesons) of 2.3×10^{-6} s. For a meson traveling at a speed of $0.99c$ relative to the laboratory, how far does the meson travel on the average before decaying? Compare your result with a nonrelativistic calculation.

14-3 For the train struck by lightning in Sec. 14-3, construct the corresponding argument to show that if the two lightning bolts appear simultaneous to an observer on the train, they do not appear simultaneous to an observer on the ground. Which one appears to come first?

14-4 Two spaceships, each of which has a length of 100 m in its rest frame, meet in outer space. An observer in the cockpit at the front of one ship observes that a time of 2.0×10^{-6} s elapses while the second moves past him, heading in the opposite direction.
    a How much time (measured in the first ship) elapses while the front of the second ship moves from front to back of the first?
    b What is the relative velocity of the ships?

14-5 A race-car driver at Indianapolis passes a timer holding a stopwatch at a speed of 1.0×10^8 m/s. The driver observes that 1.0×10^{-7} s elapses between the time the watch is started and stopped, but he cannot see its face. What does the stopwatch read?

14-6 The Orient Express moves past a small-town station at a speed of 1.5×10^8 m/s. It has mirrors attached to both ends; the (proper) length of the train is 100 m. After it passes, the stationmaster turns on a light. The light travels to both mirrors and is reflected back to him. How much time elapses between the arrival of the two reflected light beams?

14-7 In Prob. 14-6 describe a method by which the stationmaster can measure the length of the train as it passes. What result does he obtain?

14-8 A Porsche is driving on the autobahn between Munich and Salzburg at a speed of $0.6c$. One of the fuel injectors has a leak, so a drop of gasoline falls to the pavement 100 times each second, as measured by the driver.
    a At what rate do the drops fall as measured by an observer beside the road?
    b What is the distance between adjacent drops, measured in the frame of reference of the road?
    c Why is the accident rate on the autobahn so high?

14-9 Solve Eqs. (14-16) for x and t in terms of x' and t', and show directly that the resulting inverse transformations have the same form as the original transformations except that the sign of u is changed.

14-10 In a frame of reference S, consider two events (x_1, t_1) and (x_2, t_2). Using the notation $\Delta x = x_2 - x_1$, $\Delta t = t_2 - t_1$, show that in a frame S' moving so that the two events occur at the same space point in S', the time interval $\Delta t'$ between the two events is given by

$$\Delta t' = \sqrt{(\Delta t)^2 - \left(\frac{\Delta x}{c}\right)^2}$$

Hence show that if $\Delta x > c\,\Delta t$, there is *no* frame S' in which the two events occur at the same space point. The interval $\Delta t'$ is often called the *proper time* for the two events. Is this term appropriate?

14-11 For the two events in Prob. 14-10, show that if $\Delta x > c\,\Delta t$, there exists a frame of reference S' in which the two events occur *simultaneously*. Find the distance between the two events in S'. Such a distance is often called a *proper length*. Is this term appropriate?

14-12 A light pulse is emitted at the origin O' of frame S' at time $t' = 0$. Its distance x' from the origin after a time t' is given by $x'^2 = c^2 t'^2$. Use the Lorentz transformation equations with this relation to show that in a frame of reference S the pulse motion appears exactly the same; i.e., after a time t its distance x from O is given by $x^2 = c^2 t^2$.

14-13 A source of electromagnetic radiation at rest in S' emits radiation uniformly in all directions. Show that in S the radiation appears to be more concentrated along the x axis and less in directions perpendicular to this axis. The angle transformation derived in the example of Sec. 14-5 may be helpful.

14-14 Consider two successive Lorentz transformations, the first from S to S', moving at speed u_1 relative to S, the second from S' to S'', moving at speed u_2 relative to S'. Write the transformations from S to S', giving x' and t' in terms of x and t, and those from S' to S'', giving x'' and t'' in terms of x' and t'. Combine these to obtain x'' and t'' in terms of x and t, and show that the result is equivalent to a single Lorentz transformation from S to S'' with relative speed u given by

$$u = \frac{u_1 + u_2}{1 + u_1 u_2/c^2}$$

Relativistic Dynamics | 15

The kinematic relations of relativity require corresponding modifications of the principles of dynamics. For the principle of conservation of momentum of an isolated system to hold in all inertial frames, the definition of momentum must be generalized. The generalized definition points the way to the new equation of motion which is the generalization of Newton's second law. The corresponding modification of the definition of kinetic energy leads naturally to consideration of energy associated with mass of a body, and the principles of conservation of mass and energy emerge as two aspects of a single conservation law. The relation between energy and momentum for a massless particle also emerges naturally from the new definitions.

15-1 MOMENTUM

We have seen that the laws of mechanics as formulated by Newton are invariant under the galilean coordinate transformation, Eqs. (14-2), but this transformation is inconsistent with the basic postulates of relativity and must be replaced by the more general Lorentz transformation, Eqs. (14-16). Corresponding modifications are needed in the principles of dynamics if they are to be brought into harmony with relativity theory. For example, we shall see that if conservation of momentum holds in one inertial frame, it *does not* hold in a second inertial frame unless we modify the definition of momentum.

To illustrate, we consider an elastic collision between two bodies of equal mass. We arrange the collision so that in frame S body B has no x component of velocity either before or after the collision, as shown in Fig. 15-1a. We

398 RELATIVISTIC DYNAMICS

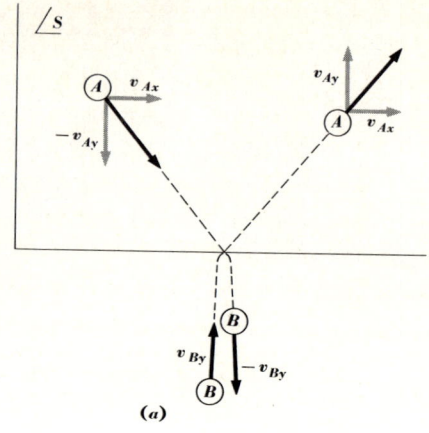

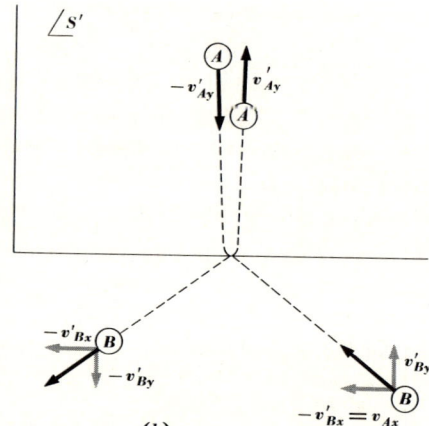

Fig. 15-1 (*a*) Collision of two bodies A and B as observed in S; body B has no x component of velocity in S. (*b*) The same collision viewed in a frame S' moving with speed $u = v_{Ax}$ relative to S. In this frame body A has no x' component of velocity.

label the various velocity components as shown in the figure. We now view this same collision from a second frame S' moving relative to S just fast enough so that in S' body A has no x component of velocity, as shown in Fig. 15-1*b*. The velocity u of S' relative to S is then given simply by $u = v_{Ax}$; we shall make use of this fact below in Eq. (15-1).

Using the newtonian definition of momentum, $\mathbf{p} = m\mathbf{v}$, we consider conservation of the y component of momentum. If momentum is conserved in S, $v_{Ay} = v_{By}$. In S' the magnitudes of the y components of initial velocity, obtained from Eq. (14-22), are

$$v'_{Ay} = \frac{v_{Ay}\sqrt{1 - u^2/c^2}}{1 - v_{Ax}u/c^2} = \frac{v_{Ay}}{\sqrt{1 - u^2/c^2}} \qquad v'_{By} = v_{By}\sqrt{1 - \frac{u^2}{c^2}} \qquad (15\text{-}1)$$

The fact that these have different magnitudes shows that the total y component of momentum in S' is not zero, despite the fact that it *is* zero in S. Furthermore, since the final y velocities in S' are just the negatives of the corresponding initial y velocities, the total y momentum in S' also reverses sign after the collision. That is, in S' the y component of momentum is *not* conserved!

This result contradicts the basic postulate of relativity that the laws of physics, and in particular the law of conservation of momentum, should be equally valid in *all* inertial frames of reference. Thus the *definition* of momentum itself must be in need of modification. As we shall see, the problem is resolved by defining the momentum of a particle not as $\mathbf{p} = m\mathbf{v}$ but as

$$\mathbf{p} = \frac{m\mathbf{v}}{\sqrt{1 - v^2/c^2}} \tag{15-2}$$

In developing Eq. (15-2), we first note that the generalized momentum expression must still be proportional to m to preserve its additive properties and that it must contain $\mathbf{v}$ as a factor in order for $\mathbf{p}$ and $\mathbf{v}$ to have the same direction. Thus it must have the general form

$$\mathbf{p} = m\gamma\mathbf{v} \tag{15-3}$$

where γ is a quantity depending only on the speed of the particle. Because this must reduce to the newtonian expression $\mathbf{p} = m\mathbf{v}$ in the limit of small velocity, we must also require $\gamma = 1$ when $v = 0$. The problem is to find the specific functional dependence of γ on v.

Returning to the collision discussed above, we consider the special case where all the y components of velocity are very small, a sort of "grazing" collision. We make no restriction on the magnitudes of the x components. Since we have dropped the classical definition of momentum, we may no longer assume at the outset that $v_{Ay} = v_{By}$. Instead, we require that the y component of momentum be conserved in both S and S'. Beginning with S and using Eq. (15-3), we write

$$m\gamma v_{Ay} = m v_{By} \tag{15-4}$$

We have placed $\gamma = 1$ for body B because by assumption its speed is very small. Similarly, in S' we have

$$m v'_{Ay} = m\gamma v'_{By} \tag{15-5}$$

where here $\gamma = 1$ for body A. Equations (15-4) and (15-5) express conservation of y momentum in S and S', respectively.

We can also use the velocity-transformation equations to express Eq. (15-5) in terms of components in S. Using Eqs. (15-1) and recalling that $v_{Ax} = u$, we obtain

$$\frac{m v_{Ay}}{\sqrt{1 - u^2/c^2}} = m\gamma v_{By} \sqrt{1 - \frac{u^2}{c^2}} \tag{15-6}$$

This must be consistent with Eq. (15-4) if momentum is to be conserved in both S and S'. One way to test this consistency is to solve Eqs. (15-4) and (15-6) for v_{Ax} and equate the results, as follows:

$$\frac{v_{By}}{\gamma} = \gamma v_{By}\left(1 - \frac{u^2}{c^2}\right)$$

This result could also be obtained by dividing Eq. (15-4) by Eq. (15-6) and rearranging. Solving for γ,

$$\gamma = \frac{1}{\sqrt{1 - u^2/c^2}} \tag{15-7}$$

Now by assumption the y components of velocity are very small, so u is very nearly equal to the speed of body A in S and that of B in S'. Thus we may replace u by v, obtaining finally

$$\mathbf{p} = \frac{m\mathbf{v}}{\sqrt{1 - v^2/c^2}} \tag{15-8}$$

as predicted.

The approximations in the above development should not distract one's attention from the generality of the result. One can consider a more general collision between two bodies of unequal mass with no restrictions on the velocities. Using the generalized definition of momentum and the velocity-transformation equations, one can show by straightforward calculation that if momentum is conserved in S, it is also conserved in S'. Verifying this statement in detail is somewhat involved and is not essential for our purposes.

The new definition of momentum requires a revision of the concept of center of mass. If we try to use the old definition, Eq. (11-6) or (11-7), we find that the total momentum is no longer determined by the motion of the center of mass. Instead, we define the center-of-mass coordinate system as a system in which the total momentum of the system is zero and the center of mass as the origin of this system. This defines the velocity but not the position of this system. Usually in the analysis of high-energy collisions only the velocity of the center of mass is of interest, so this definition is often sufficient.

Example
A particle of mass m is at rest at the origin O of a system S, and a second particle with the same mass moves along the positive x axis with velocity v. Find the velocity of the center of mass.

Solution
We want to find a frame S' moving with velocity u relative to S such that in S' the total momentum is zero. The velocities in S', according to Eq. (14-19), are

$$v'_1 = \frac{v - u}{1 - vu/c^2} \qquad v'_2 = -u$$

For the total momentum in S' to be zero, these must be equal in magnitude:

$$\frac{v - u}{1 - vu/c^2} = u$$

This is a quadratic equation for u; one root is larger than c; the other is

$$u = \frac{1 - (1 - v^2/c^2)^{1/2}}{v/c^2} \tag{15-9}$$

We can check whether this reduces to the nonrelativistic result $u = v/2$ in the limit when $v \ll c$. We expand the radical using the binomial theorem, retaining only the first two terms:

$$\left(1 - \frac{v^2}{c^2}\right)^{1/2} = 1 - \frac{v^2}{2c^2} + \cdots$$

Equation (15-9) becomes

$$u = \frac{v^2/2c^2}{v/c^2} = \frac{v}{2}$$

as expected.

15-2 FORCE AND MOTION

Equation (15-8) can be interpreted in two ways. The point of view presented above is that it is a generalization of the definition of momentum. To this should be added the remark that m is a constant for a given particle and that it represents the inertial properties of the particle exhibited in low-speed (newtonian) collisions. The alternate viewpoint is that although momentum is still mass times velocity, the mass to be used is not just m but a relativistic mass given by

$$m_{\text{rel}} = \frac{m}{\sqrt{1 - v^2/c^2}} \tag{15-10}$$

The choice between these viewpoints is largely a matter of taste; one cannot say that either is right and the other wrong, since they say the same thing in different ways. It turns out that the first is more useful in finding the correct generalization of $\mathbf{F} = m\mathbf{a}$, our next problem. In either case, m is usually called the *rest mass* to distinguish it from the velocity-dependent relativistic mass used in some literature. In the following discussion m is *always* a constant, *not* a velocity-dependent quantity.

One may make two reasonable guesses how Newton's second law should be generalized to bring it into harmony with the principle of relativity. One of these is to retain the form $\mathbf{\Sigma F} = m\mathbf{a}$, using for m the relativistic mass given by Eq. (15-10). The other is to return to Newton's original form, $\mathbf{\Sigma F} = d\mathbf{p}/dt$, with $\mathbf{p}$ given by Eq. (15-8). The two are *not* equivalent; in the first case we have

$$\mathbf{\Sigma F} = \frac{m}{\sqrt{1 - v^2/c^2}} \frac{d\mathbf{v}}{dt}$$

while in the second,

$$\mathbf{\Sigma F} = \frac{d}{dt}\left(\frac{m\mathbf{v}}{\sqrt{1 - v^2/c^2}}\right)$$

Which form agrees with the behavior of the physical world can be decided only by experiment.

This question has been investigated experimentally in a wide variety of situations, especially those concerning the motion of electrically charged particles with speeds comparable to c in electric and magnetic fields. It is found that the correct equation of motion is

$$q(\mathbf{E} + \mathbf{v} \times \mathbf{B}) = \frac{d}{dt}\left(\frac{m\mathbf{v}}{\sqrt{1 - v^2/c^2}}\right) \tag{15-11}$$

which agrees with the second of the two possibilities. The left side of Eq. (15-11) is the Lorentz force expression, introduced in Sec. 8-5. In this equation it is assumed that $\mathbf{E}$ and $\mathbf{B}$ are measured in the same frame of reference as $\mathbf{v}$ and that m and q are constants which characterize the inertial and electrical properties of the particle. Specifically, neither m nor q depends on the speed v of the particle; they are *invariant* under transformations from one inertial frame to another.

Observations with other types of forces, such as gravitational forces, are much more difficult, but all such observations are consistent with the assumption that the generalized law of motion is

$$\sum \mathbf{F} = \frac{d\mathbf{p}}{dt} = \frac{d}{dt} \frac{m\mathbf{v}}{\sqrt{1 - v^2/c^2}} \qquad (15\text{-}12)$$

In the above discussion, the term *force* has been used somewhat loosely and without a careful analysis of the role of the observer in its definition. The reader may raise the question whether if two observers in motion relative to each other measure a force, they obtain the same or different results. We return to this briefly in Sec. 15-4. In the case of electromagnetic forces, this question does not arise provided we use one coordinate system consistently in applying Eq. (15-11), measuring **E, B,** and **v** always with respect to the same inertial frame of reference.

Example

A charged particle with mass m and charge q is traveling at a speed $v = 0.8c$. Find the magnitude of electric field required to give the particle an acceleration a along the original direction of motion. If this same field were applied to a particle at rest, what acceleration would it produce?

Solution

From Eq. (15-11),

$$qE = m \frac{d}{dt} \left(\frac{v}{\sqrt{1 - v^2/c^2}} \right)$$

$$= m \left[\frac{1}{\sqrt{1 - v^2/c^2}} + \frac{v^2}{(1 - v^2/c^2)^{3/2}} \right] \frac{dv}{dt}$$

Simplifying using $dv/dt = a$, we obtain

$$qE = \frac{m}{(1 - v^2/c^2)^{3/2}} a \quad \text{or} \quad a = \frac{qE(1 - v^2/c^2)^{3/2}}{m}$$

If the particle had been initially at rest, the factor $(1 - v^2/c^2)^{3/2} = {}^{27}\!/_{125}$ would not appear, and the acceleration with a given E would be larger by ${}^{125}\!/_{27} \simeq 4.6$.

15-3 WORK AND ENERGY

The modifications of the laws of motion discussed in the previous section imply corresponding modification of the relationship between work and energy. Just as we used Newton's second law in its original form to derive the relationship (for one-dimensional motion)

$$\int_{x_1}^{x_2} F\, dx = \tfrac{1}{2}mv_2^2 - \tfrac{1}{2}mv_1^2$$

between work done on a particle and the change in its kinetic energy, we can now use the modified equation of motion, Eq. (15-12), to derive a relativistic generalization of this relationship. Our program is similar to that of Sec. 9-2. We retain the original definition of work:

$$W = \int_{x_1}^{x_2} F\, dx \tag{15-13}$$

We use Eq. (15-12) to convert this integral into a form containing only the *speed* of the particle, and then we perform the integration. The result is an expression containing the initial and final speeds, from which we can deduce an appropriate generalization of the definition of kinetic energy.

For simplicity, we consider only motion along a straight line. The force may vary during the motion but is assumed to act always along the x axis. Let the particle have velocity v_1 at point x_1 and time t_1 and velocity v_2 at x_2 and t_2. The momentum may be regarded as a function of x, v, or t, since these are functionally related. Using the chain rule and $v = dx/dt$, we successively transform the integration variable in Eq. (15-13) as follows:

$$W = \int_{x_1}^{x_2} F\, dx = \int_{x_1}^{x_2} \frac{dp}{dt} dx = \int_{x_1}^{x_2} \left(\frac{dp}{dv}\frac{dv}{dt}\right) dx$$

$$= \int_{x_1}^{x_2} \frac{dp}{dv}\left(\frac{dv}{dx}\frac{dx}{dt}\right) dx = \int_{x_1}^{x_2} \frac{dp}{dv} v \frac{dv}{dx} dx = \int_{v_1}^{v_2} \frac{dp}{dv} v\, dv \tag{15-14}$$

Since p is given as a function of v by Eq. (15-8), we have

$$W = \int_{v_1}^{v_2} v \frac{d}{dv}\left(\frac{mv}{\sqrt{1 - v^2/c^2}}\right) dv \tag{15-15}$$

All that remains is to evaluate the integral, best done by first integrating by parts. The details need not concern us; the final result is

$$W = \frac{mc^2}{\sqrt{1 - v_2^2/c^2}} - \frac{mc^2}{\sqrt{1 - v_1^2/c^2}} \tag{15-16}$$

Equation (15-16) shows that the effect of the work is to produce a change in the quantity

$$E = \frac{mc^2}{\sqrt{1 - v^2/c^2}} \tag{15-17}$$

and it therefore seems natural to define this as the *energy* of the particle. Equation (15-17), however, is conspicuously different from the classical kinetic-energy expression, $\frac{1}{2}mv^2$; it does not become zero when $v = 0$. Instead, at $v = 0$ it reduces to $E = mc^2$. Thus to obtain the analog of the classical kinetic energy, we must subtract mc^2, defining the kinetic energy E_k as

$$E_k = \frac{mc^2}{\sqrt{1 - v^2/c^2}} - mc^2 \tag{15-18}$$

If this is a correct relativistic generalization of kinetic energy, it must reduce to approximately $\frac{1}{2}mv^2$ when $v \ll c$. We can show that this is so by expanding the binomial in Eq. (15-18), using the binomial theorem:

$$\left(1 - \frac{v^2}{c^2}\right)^{-1/2} = 1 + \frac{1}{2}\frac{v^2}{c^2} + \frac{3}{8}\frac{v^4}{c^4} + \frac{5}{16}\frac{v^6}{c^6} + \cdots$$

Combining this with Eq. (15-18) gives

$$E_k = mc^2\left(1 + \frac{1}{2}\frac{v^2}{c^2} + \frac{3}{8}\frac{v^4}{c^4} + \cdots\right) - mc^2$$

$$= \tfrac{1}{2}mv^2 + \tfrac{3}{8}m\frac{v^4}{c^2} + \cdots$$

The three dots stand for omitted terms. When $v \ll c$, all terms containing v/c or higher powers of this ratio can be neglected, and we have, approximately, $E_k = \frac{1}{2}mv^2$.

The general validity of Eq. (15-18) has been tested directly by experiments making use of high-energy particle accelerators, as described in Sec. 14-2. These experiments directly confirm the relation between speed and kinetic energy given by Eq. (15-18). This relation is illustrated graphically in Fig. 15-2, which also shows, for comparison, the nonrelativistic expression $\frac{1}{2}mv^2$.

The forms of Eqs. (15-17) and (15-18) suggest that in addition to having energy associated with motion, a particle of mass m has an energy mc^2 even when it is not moving. This energy may be thought of as associated with

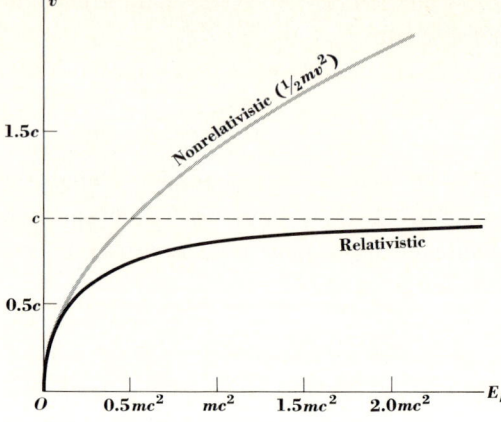

Fig. 15-2 Relation between the speed of a particle and its kinetic energy. The speed never exceeds c, no matter how great the kinetic energy. At speeds small compared to c, the kinetic energy is approximately $\frac{1}{2}mv^2$, but at larger speeds this expression deviates more and more from the relativistically correct equation.

the mass, rather than with the motion. On the basis of this interpretation, Eq. (15-17) represents the *total* energy of a particle, including both its kinetic energy and the energy mc^2 associated with its mass, which we call *rest energy*. This, of course, is not a proof that the concept of rest energy is a physically meaningful one, but it suggests a path for further inquiry.

There is, in fact, direct and dramatic confirmation of the point of view that mc^2 represents an energy associated with mass. The π^0 meson is an unstable particle produced during collisions between nuclear particles; it decays into electromagnetic radiation (gamma rays) with a half-life of about 10^{-16} s. When one measures the total energy of the radiation resulting from the decay of the π^0 meson, using a frame of reference in which the meson is originally at rest, it is found to be exactly equal to mc^2, where m is the mass of the meson. Similar phenomena occur in many other interactions involving fundamental particles. Whenever the total mass of the system changes, there is found to be a corresponding energy change consistent with associating an energy mc^2 with mass m.

These and other experimental confirmations of the energy associated with mass and of the equivalence of mass and energy are of the utmost importance. Although historically the principles of conservation of mass and energy developed more or less independently, we now see that they are directly related. These two principles, in fact, now emerge as two aspects of a more general and inclusive principle, usually called the *principle of conservation of mass and energy*.

Additional insight into the relation of mass and energy can be obtained by deriving an equation relating the total energy (including rest energy) of a particle and its momentum. To do this, we combine Eqs. (15-8) and (15-17) to eliminate v and obtain an expression relating E and p. This is most easily accomplished by the following manipulation: We divide Eq. (15-17) by mc^2 and square the resulting equation; we divide Eq. (15-8) by mc and square the result:

$$\left(\frac{E}{mc^2}\right)^2 = \frac{1}{1 - v^2/c^2} \qquad \left(\frac{p}{mc}\right)^2 = \frac{v^2/c^2}{1 - v^2/c^2} \qquad (15\text{-}19)$$

We then subtract the second of these expressions from the first and rearrange, obtaining

$$E^2 = (mc^2)^2 + (pc)^2 \qquad (15\text{-}20)$$

This result shows again that the energy of a particle depends on its mass and momentum. When expressed in this form, the total energy is not simply the sum of two contributions but a sort of pythagorean sum. Furthermore, Eq. (15-20) also shows directly that a particle which possesses energy also must have momentum even if it has no mass. When $m = 0$, $E = pc$. This is the case, for example, for photons and neutrinos, discussed in Sec. 7-4. Thus the theory of relativity provides the basis for Eq. (7-18), which we have already used in the analysis of collisions between fundamental particles.

15-4 RELATIVITY AND NEWTONIAN MECHANICS

It may now appear that we have succeeded in demolishing the very foundations of newtonian mechanics. In one sense this is true; we have shown that the absolute meanings of length and time in Newton's mechanics are inconsistent with the principle of relativity and that nearly all the other concepts have to be modified in one way or another. Yet, as we have seen, all the relativistic formulations become equivalent to the newtonian ones in situations in which velocities are sufficiently small. In such cases, the phenomena of time dilation, contraction of length, and nonabsoluteness of simultaneity and the various modifications of the dynamical principles are not observed. Thus the theory of relativity should be regarded not as a refutation of classical mechanics but as a generalization of it. Every one of the principles of newtonian mechanics is contained in the relativistic theory as a special case, valid when the velocities are sufficiently small.

The relationship between newtonian mechanics and relativistic mechanics is an example of a general guiding principle of great importance in all scientific thought, the *correspondence principle*. According to this principle, when a new theory is in partial disagreement with an older, established theory, it must give the same results as the old theory in any areas in which the old theory has been found experimentally to be correct. Thus the correspondence principle insists that relativistic mechanics predict the same results as newtonian mechanics in situations where the latter has been found valid; as we have seen in several instances, relativistic mechanics does in fact satisfy this requirement.

On the other side of the balance, there are many physical situations for which newtonian mechanics is clearly inadequate. These include, as we have seen, the interaction of electrically charged particles and electromagnetic fields at speeds approaching that of light and, particularly, the direct conversion of mass into energy, a phenomenon which newtonian mechanics is powerless to explain. But there still remains a large area, including practically all the behavior of macroscopic bodies and familiar mechanical systems, for which newtonian mechanics is perfectly adequate.

In Sec. 15-3, it was mentioned briefly that there are difficulties associated with a relativistically correct treatment of the concept of force. The difficulties arise because it is hard to make an operational definition of force which can be applied when bodies move at very high velocities. It would be difficult indeed to measure with a spring balance a force exerted on an object moving relative to the observer with nine-tenths the speed of light.

In the case of electric and magnetic forces on charged particles, this whole problem can be circumvented. In principle, one can measure the electric and magnetic fields $\mathbf{E}$ and $\mathbf{B}$ in a given frame of reference, using measurements on particles *at rest* in that frame. For a particle moving with velocity v with respect to that same frame, it is found experimentally that the correct equation of motion is Eq. (15-11). Thus in this instance the question of what force acts on the particle is really irrelevant; the important thing is a correct expression of the relationship between the particle's motion and the electric and magnetic fields, measured in a specified coordinate system, say S.

But what happens if we measure the electric and magnetic fields by the same techniques in another coordinate system S' moving with respect to S? For example, suppose a particle moves with constant velocity $\mathbf{v}$ in S, in which uniform electric and magnetic fields exist which are of just the right magnitude and direction so that in this frame of reference $q(\mathbf{E} + \mathbf{v} \times \mathbf{B}) = 0$. Now we look at the situation from a frame of reference S' which moves relative to S with a velocity $\mathbf{v}$ (so that the particle is at rest in S'). In this frame of reference, $\mathbf{v} \times \mathbf{B} = 0$. But the particle has no acceleration in S'; hence there must be no force. Therefore $\mathbf{E}$ must be zero, although it was *not* zero in S.

This example shows that electric and magnetic fields look different to observers in different coordinate systems. Furthermore, it turns out that the components of *each* field **E′** and **B′** as measured in *S′* depend on *both* fields **E** and **B** as measured in *S*. This is the reason we speak of *electromagnetic* fields rather than *electric* and *magnetic* fields, since when one makes a Lorentz transformation from one frame to another, the two fields become mixed. One can derive equations, analogous to Eqs. (14-16), relating the components of the fields in two coordinate systems. Furthermore, by using the Lorentz transformation one can derive all the properties of the interactions between electric currents or moving charges, which are usually expressed in terms of magnetic fields, from purely *electrostatic* considerations. This is an extremely important result and a beautiful example of the unifying effect relativity has had on the various parts of electromagnetic theory and its relation to other areas of physics.

One further question may be raised in connection with the status of the theory of relativity: In stating the basic postulate of relativity, we insisted that all *inertial* frames of reference be equivalent. Why did we restrict this requirement to *inertial* frames of reference? Why not *all possible* frames of reference? This question provides the seed of Einstein's *general theory of relativity* (so called to distinguish it from the *special theory of relativity*, which we have been discussing up to now) and his theory of gravitation, developed in 1915. We cannot discuss these theories in any detail here, but an example will indicate their direction.

Consider the following situation: A man is standing in an elevator; the elevator cables have all broken, and the safety devices have all failed at once, so that the elevator is falling freely with an acceleration relative to the earth of $g = 9.80$ m/s^2 downward. The man observes that he is weightless, but he can interpret this observation in two ways: He may think that the elevator is in fact falling freely, and he with it, or he may think that somehow the force of gravity has suddenly been turned off, so that he is no longer attracted to the earth. If he confines his observations to those which can be made within his own frame of reference, there is no way for him to distinguish experimentally whether he is in an accelerated (noninertial) frame of reference or whether the force of gravity has been turned off.

A more realistic example concerns an astronaut in a space station in a circular orbit around the earth. Objects in the orbiting station appear weightless because both objects and station are constantly accelerated toward the earth under the action of its gravitational attraction. But there is no way to distinguish, by means of experiments performed in the station, between this situation and one in which the station is moving with constant velocity in a region of zero gravitational field. In this case, just as with the falling elevator, there is no way to distinguish between a gravitational field and an accelerated reference system.

The heart of the *general* theory of relativity is the hypothesis that an accelerated reference system and a gravitational field are really two aspects of the same thing; if one cannot distinguish between them experimentally, it makes no sense to distinguish between them at all. This is called the *principle of equivalence*. Einstein postulated further that *any* gravitational field may be represented as equivalent to some modification of the coordinate system. It turns out that if the gravitational field is not uniform, we cannot simply use an accelerated frame of reference, but more drastic steps must be taken; in general, one must use a non-euclidean geometry. Thus if one accepts the principle of equivalence, one is led to the assumption that the space of the physical universe is in general non-euclidean.

The correspondence principle insists that if this new theory is correct, it must reduce to the older theory (in this case, the special theory of relativity) in areas in which the latter has been found valid. In fact, the disagreements between the special and general theories are minute and have been observed in only a small number of experimental situations. The basic assumptions of the general theory of relativity are well established, but some details are still regarded as somewhat speculative in nature. The chief interest in this theory centers around investigation of the structure of the universe and related problems involving astronomical dimensions and masses. It is not now thought to have any relevance to the understanding of atomic or nuclear phenomena or to the mechanics of systems of less than astronomical dimensions.

Problems

15-1 A student proposed to compute the kinetic energy of a particle relativistically by using the expression $\frac{1}{2}mv^2$ with the "relativistic" mass of the particle. Is this correct? Explain.

15-2 Protons emerge from a certain synchrocyclotron with kinetic energy of about $0.5mc^2$. What is the speed of these particles? Compare your result with that obtained from the nonrelativistic relation between mass and kinetic energy.

15-3 What is the speed of a particle whose kinetic energy is equal to its rest energy? What percentage error is made if the nonrelativistic kinetic-energy expression is used?

PROBLEMS

15-4 A proton emerges from a particle accelerator with a speed of $0.75c$ and collides with another proton initially at rest.
    a What is the kinetic energy of the proton?
    b Find the velocity of a moving coordinate system such that in this system the total momentum of the two protons is zero.

15-5 Carry out the integral in Eq. (15-15) to obtain Eq. (15-16).

15-6 Show that Eq. (15-12) can be rewritten as

$$\mathbf{F} = \frac{m\mathbf{a}}{\sqrt{1 - v^2/c^2}} + \frac{m(\mathbf{v} \cdot \mathbf{a})\mathbf{v}}{(1 - v^2/c^2)^{3/2}}$$

15-7 Write Eq. (15-12) in component form to show that for a particle with instantaneous velocity $\mathbf{v}$ along the x axis the components of acceleration are given by

$$F_x = \frac{ma_x}{(1 - u^2/c^2)^{3/2}} \qquad F_y = \frac{ma_y}{(1 - u^2/c^2)^{1/2}}$$

In the early days of relativity theory the coefficients of a_x and a_y were sometimes called the *longitudinal* and *transverse* masses, respectively, reflecting the fact that the relation of $\mathbf{F}$ to $\mathbf{a}$ is different for components parallel to $\mathbf{v}$ and perpendicular to $\mathbf{v}$.

15-8 Show that a charged particle of mass m and charge q moving perpendicular to a uniform magnetic field B at relativistic speed v moves in a circle of radius R given by

$$R = \frac{mv}{qB\sqrt{1 - v^2/c^2}}$$

The results of Prob. 15-7 may be helpful.

15-9 In a hypothetical nuclear-fusion reactor two deuterium nuclei combine, or fuse, to form one helium nucleus. The mass of a deuterium nucleus, expressed in atomic mass units, is 2.0147 amu; that of a helium nucleus is 4.0039 amu (1 amu = 1.66×10^{-24} g).
    a How much energy is released when 1 g of deuterium undergoes fusion? (Neglect the masses of the electrons, which do not participate in the nuclear reaction.)
    b The annual consumption of electric energy in the United States is on the order of 10^{16} watt-hours. How much deuterium must react to produce this much energy?

15-10 According to newtonian mechanics, when a charged particle moves in a magnetic field, it moves with constant speed. Is this result correct in relativistic mechanics? Explain.

15-11 Consider the motion of a charged particle in a magnetic field in the situation of Prob. 15-8.
  a Show that the nonrelativistic formulation predicts that the angular velocity of the particle is independent of its speed and hence of its energy.
  b Using relativistic mechanics and the result of Prob. 15-8, derive a relationship between the angular velocity and the total energy of the particle.
  c In a certain synchrocyclotron the angular velocity of very slow protons is about 30×10^6 r/s. What is the angular velocity when the particles have been accelerated to a kinetic energy equal to $0.5mc^2$?

15-12 A particle with mass m and initial speed v collides with a second particle of mass m, initially at rest, as observed in a frame S. The collision is completely *inelastic*; the two particles move off as one mass M after the collision.
  a Show that v is related to the speed u of the center-of-momentum frame of reference S' in which the total momentum is zero by

$$v = \frac{2u}{1 + u^2/c^2}$$

  b Find the final velocity of the composite particle M in S.
  c Using conservation of momentum in S, show that M is not equal to $2m$ but is given by

$$M = \frac{2m}{\sqrt{1 - u^2/c^2}}$$

In working out this result, the calculation is simplified somewhat by first establishing the relation

$$\frac{1}{\sqrt{1 - v^2/c^2}} = \frac{1 + u^2/c^2}{1 - u^2/c^2}$$

  d Find the kinetic energy in S' before and after the collision, and show that the difference is accounted for by the increased rest mass.

15-13 An electron in a uniform electric field E in the x direction starts from rest at the origin at time $t = 0$.
  a Show that its speed v after time t is given by

$$v = \frac{qEt/m}{\sqrt{1 + (qEt/mc)^2}}$$

  b Show that its position x after time t is given by

$$x = \frac{mc^2}{qE}\left[\sqrt{1 + \left(\frac{qEt}{mc}\right)^2} - 1\right]$$

  c Show that after a very long time v approaches c and x approaches ct.

413 PROBLEMS

15-14 A positron with initial speed v collides with an electron initially at rest. Both particles are annihilated, and two photons of electromagnetic radiation are produced. Find the direction and energy of each photon. Why is it not possible for only one photon to be produced?

15-15 A particle moving along the x axis has momentum p and total energy E as measured in a frame S. Show that in a frame S' moving with speed u relative to S, the momentum p' and energy E' are given by

$$p' = \frac{p - uE/c^2}{\sqrt{1 - u^2/c^2}} \qquad E' = \frac{E - up}{\sqrt{1 - u^2/c^2}}$$

Hence the transformation for the pair of quantities p, E/c^2 is the same as for x, t.

Fluid Mechanics | 16

A fluid is a substance which can flow and which has no definite shape. A fluid at rest is described simply in terms of the pressure and density at various points. Moving fluids exhibit more complex behavior and are usually described using idealized models in which one or more complicating features of the motion, such as compressibility or viscosity, are omitted. Useful analytical results can be obtained from such models, but there are some fluid phenomena, such as turbulent flow, which can be studied analytically only to a limited degree and for which an experimental approach is most productive.

16-1 HYDROSTATICS

The study of fluids (gases and liquids) at rest in static equilibrium is called *hydrostatics*. A force applied to a fluid cannot be applied to a point but must be distributed over a surface, for example, the wall of a container. The force on a fluid at rest always acts perpendicular to the surface at each point. If we were to attempt to exert a component of force parallel or tangent to the surface, layers of fluid would simply slide past each other to eliminate this force. This situation contrasts with that for a solid body, in which a force can be applied at any point and in an arbitrary direction.

Thus a fluid exerts a force on any surface with which it is in contact. This force is most easily described in terms of the force per unit area, called *pressure*. When a fluid exerts a force F perpendicular to an element of area A, the pressure P is defined as

$$P = \frac{F}{A} \tag{16-1}$$

Often the pressure varies from point to point in the fluid, so it is more precise to define pressure for a small area ΔA near a given point, with the corresponding force ΔF, as

$$P = \lim_{\Delta A \to 0} \frac{\Delta F}{\Delta A} \tag{16-2}$$

Pressure exists within a fluid as well as at its surfaces. Figure 16-1 shows an idealized pressure-measuring device. When this device is immersed

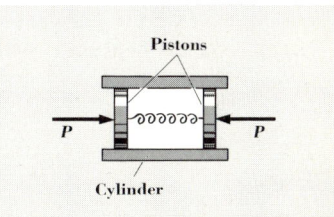

Fig. 16-1 An idealized pressure-measuring device. The pistons slide without friction in the cylinder. The compression of the spring can be measured to find the forces the fluid exerts on the pistons; from these the pressure can be obtained.

in a fluid, the forces on the two frictionless pistons compress the spring and permit a measurement of the force and thus the pressure. At a given point, the pressure is independent of the orientation of the device, assuming it is small enough to permit neglecting the variation of pressure over the dimensions of the device. Thus pressure has no direction; it is a *scalar* quantity. A surface in contact with a fluid at rest at a given point, either as part of the container, an immersed body, or an imaginary surface in the fluid, experiences a force per unit area always perpendicular to the surface, with magnitude independent of the orientation of the surface.

For a fluid in a uniform gravitational field the pressure varies with vertical position and is independent of horizontal position. The independence of horizontal position is easily seen by considering Fig. 16-2a; if the forces on the two ends of the cylindrical element were not equal in magnitude, the element could not be in equilibrium. Similarly, we can consider the vertically oriented element in Fig. 16-2b. The forces on this element are the upward force $P_1 A$ at the bottom, the downward force $P_2 A$ at the top, and the weight of the cylinder. If it has uniform density ρ (mass per unit volume), the mass is the density times the volume Ah and the weight is $W = \rho A g h$. For static equilibrium these forces must add to zero, and we have

$$P_1 - P_2 = \rho g h \tag{16-3}$$

416 FLUID MECHANICS

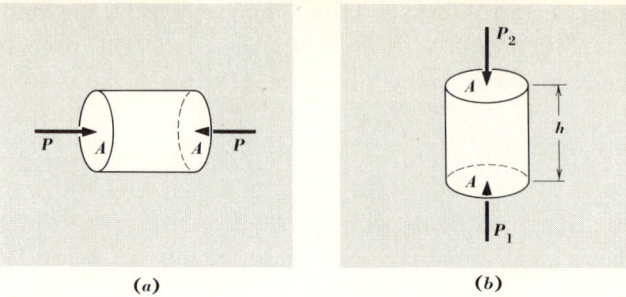

Fig. 16-2 (*a*) Imaginary cylinder in a fluid. For this element of fluid to be in static equilibrium, the forces on the two ends must be equal in magnitude. Hence the pressure in a fluid at rest is independent of horizontal position. (*b*) Imaginary cylinder. For this element to be in static equilibrium, the force $P_1 A$ on the bottom must be greater than that on the top $P_2 A$ by the weight of fluid in the cylinder $\rho A g h$. Equation (16-3) follows.

Example
An ordinary mercury barometer has a column of mercury 76.0 cm high at "normal" temperature and pressure conditions. Find the value of normal atmospheric pressure in newtons per square meter.

Solution
In Fig. 16-3, the pressure at the top surface of the column is zero, so in this case $P_2 = 0$. The pressure at the level of the surface in the reservoir, which is also equal to atmospheric pressure, is given by

$$P = \rho g h$$

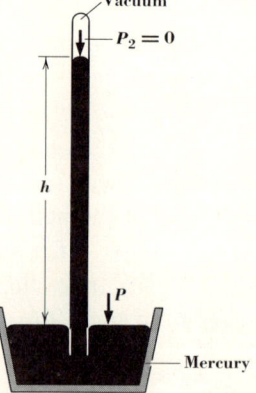

Fig. 16-3 A mercury barometer, as invented by Torricelli about 1640. The space above the mercury in the tube is vacuum, neglecting the very small pressure due to mercury vapor.

The density ρ of mercury is $13.6 \text{ g/cm}^3 = 13.6 \times 10^3 \text{ kg/m}^3$, so the pressure is given by

$$P = (13.6 \times 10^3 \text{ kg/m}^3)(9.80 \text{ m/s}^2)(0.76 \text{ m}) = 1.01 \times 10^5 \text{ N/m}^2$$

One atmosphere, defined to be the pressure of a column of mercury 76 cm high at a temperature of $0°C$, is a commonly used unit of pressure. We have just derived the conversion factor

$$1 \text{ atm} = 1.013 \times 10^5 \text{ N/m}^2$$

In British units, $1 \text{ atm} = 14.7 \text{ lb/in.}^2$.

Example
The pressure at the top of a compressed-air tank 2.0 m high is 10 atm, or about $1.0 \times 10^6 \text{ N/m}^2$. The density of air at this pressure is about 13 kg/m^3. What is the pressure at the bottom of the tank?

Solution
According to Eq. (16-3), the *increase* in pressure from top to bottom is given by

$$\rho g h = (13 \text{ kg/m}^3)(9.8 \text{ m/s}^2)(2.0 \text{ m})$$
$$= 255 \text{ N/m}^2 = 2.52 \times 10^{-3} \text{ atm}$$

Thus the variation in pressure from top to bottom is a very small fraction of the average pressure. This is usually the case with gases, unless the distances are extremely large, and it is common to neglect the variation of pressure with height for gases in laboratory-scale situations.

Example
Find the variation of pressure of the earth's atmosphere with height above the surface, assuming the density of air is proportional to pressure.

Solution
We cannot use Eq. (16-3) directly because the density is not constant; air is compressible, and the density increases with pressure. The assumption of direct proportionality of density and pressure is an idealization based on the

ideal-gas equation of state and the assumption of constant temperature. The ideal-gas model will be discussed in detail in Sec. 18-1. Let y be the distance above earth's surface, $P(y)$ and $\rho(y)$ the pressure and density, respectively, at this height, and P_0 and ρ_0 the pressure and density at $y = 0$. Then the proportionality of P and ρ can be formulated as

$$\frac{P(y)}{\rho(y)} = \frac{P_0}{\rho_0}$$

We now use Eq. (16-3), taking for h an increment of height dy small enough to ensure that P and ρ do not vary appreciably over this interval. We obtain

$$P(y) - P(y + dy) = \rho g \, dy$$

or

$$\frac{dP}{dy} = -\rho g = -\frac{\rho_0 g}{P_0} P$$

This is a differential equation whose solution gives P as a function of y. By noting that P must be a function whose derivative is proportional to P itself, we conclude that the function is an exponential; the reader can verify that the solution is

$$P(y) = P_0 e^{-\rho_0 g y / P_0} \tag{16-4}$$

The constant $P_0/\rho_0 g$ for air at $0°C$ turns out to be about 8×10^3 m, which shows that, according to this model, the pressure at 8,000 m is less than at sea level by a factor of $1/e$, or about 0.37. This is roughly equal to the height of Mt. Everest and the height at which commercial airplanes typically fly; thus we see why pressurized cabins or supplementary oxygen equipment are essential at these altitudes.

Various devices are used to measure pressure. The simplest in concept is the open-tube manometer, shown in Fig. 16-4. One side is open to the atmosphere, and the pressure at the top of this column of liquid is therefore atmospheric pressure P_0. Thus according to the previous discussion, the pressure in the container attached to the other side is $P_0 + \rho g h$. Various mechanical devices have been invented to measure pressure, reporting the result as a pointer position on a circular scale. Such pressure gauges are often calibrated to read zero under normal atmospheric pressure so the dial reading is the *difference* between actual (absolute) pressure and atmospheric pressure, often called *gauge pressure*. When a tire is inflated to 30 lb/in.^{2},

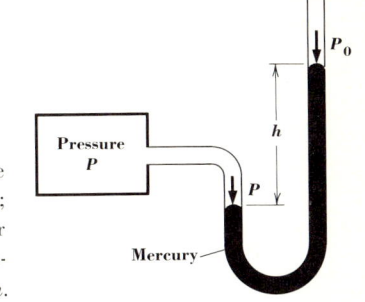

Fig. 16-4 An open-tube manometer. One side is open to atmospheric pressure P_0; the other side is connected to the container in which the pressure P is to be measured. The difference $P - P_0$ is given by $\rho g h$.

the actual pressure in the tire is 30 lb/in.^{2} + 1 atm or about 44.7 lb/in.^{2}.

A familiar principle of hydrostatics is Archimedes' principle, giving the apparent loss of weight of a body immersed in a fluid. We need not recount the tale of Archimedes jumping from his bathtub and running naked through the streets of Athens, yelling "eureka." The reason for his jubilation was the realization that the apparent loss of weight of a body immersed in a fluid is exactly equal to the weight of fluid which is displaced by the body, i.e., the weight of fluid which would occupy the same volume as the body.

Archimedes' principle is easily derived from the hydrostatic principles developed above. A body appears to lose weight when immersed in a fluid because the fluid exerts a net upward force on it. This force depends on the configuration of the surface of the body but not on the details of the matter inside that surface. We can imagine replacing the body by a quantity of the fluid with the same shape as the body. In that case the fluid would be in equilibrium, so the net upward force on the fluid would be just sufficient to balance its weight. We now replace the fluid again by the body of the same shape; the net upward force is the same, equal to the weight of fluid displaced by the body.

Example
A steel ball of mass 10 kg is immersed in water. What upward force must a diver exert to lift it?

Solution
The weight in air is $W = mg = (10 \text{ kg})(9.8 \text{ m/s}^2) = 98$ N. The density of steel is about 7.8×10^3 kg/m^3, or about 7.8 times that of water. Thus the mass of water displaced by the ball is 10 kg/7.8 = 1.28 kg. The weight

of water displaced is $(1.28 \text{ kg})(9.8 \text{ m/s}^2) = 12.6$ N, and this is the upward buoyant force of water. Thus the additional force needed to lift the ball, i.e., its apparent weight, is

$$98 \text{ N} - 12.6 \text{ N} = 85.4 \text{ N}$$

16-2 FLUID FLOW

Describing the motion of a fluid is a more complex undertaking than for motion of a point or a rigid body because a fluid is made up of a large number of particles, no two of which need have the same motion. Two general approaches are usually used. In the first, one tries to describe the motion of characteristic points in the fluid. This approach facilitates pictorial representation of the motion in terms of *lines of flow*, the paths traced out by typical elements in a fluid. Figure 16-5 uses lines of flow to show the motion

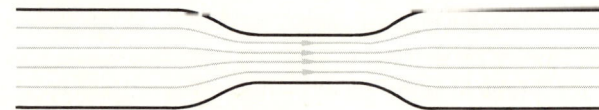

Fig. 16-5 Flow lines representing motion of a fluid in a tube with a constriction. If dye is injected into the fluid (or smoke into a gas), representative flow lines can actually be seen.

of fluid in a pipe with a constriction. At each point on a line of flow, the velocity of the fluid is tangent to the line.

The other approach, usually more convenient for detailed calculations, is to describe the motion by giving the velocity as a function of position for the region of space in which the fluid moves. If a cartesian coordinate system is used, each component of velocity is a function of x,y,z and possibly also of t. Thus the velocity is a *vector field*, as the term was used in Sec. 8-2. In general the pressure and density are different at different points, so they are also functions of x,y,z, and t. They are of course *scalar fields*. When viscous effects must be included, the internal stress in the fluid is described by a tensor field, which we cannot discuss in detail here.

The general problem of fluid flow is so complex that various idealizations are used to simplify the problem. Which of these are appropriate depends on the specific situation, as we shall see. The most important simplifying assumptions are the following:

1 *Incompressible flow.* The fluid is assumed to be incompressible, so that its density is constant. This is clearly not a realistic assumption for gases, but it is quite reasonable for most liquids except in special cases such as wave propagation, where compressibility is an essential feature.

2 Nonviscous flow. Adjacent layers of fluid are assumed to be able to slip past each other without resisting forces. The validity of this assumption depends on the properties of the fluid and the flow rate. It is more likely to be valid for slowly flowing water than for molasses.

3 Nonturbulent flow. The flow pattern is assumed not to contain irregular eddies and whirlpools, or vortices. The presence of these features in the motion is one characteristic of *turbulent flow*. Nonturbulent flow is also called *laminar* flow, referring to the smooth motion of *laminae*, or layers, of the fluid past each other. Turbulent flow cannot be described in these terms.

4 Steady-state flow. At each point in space, the flow velocity, pressure, and density do not vary with time, although they may be different at different points. Transient effects associated with starting or stopping the flow are absent in steady-state flow; wave propagation *cannot* be steady-state flow.

An incompressible, nonviscous fluid is called an *ideal* fluid. The absence of viscosity makes it possible to use energy conservation in the analysis, while viscosity provides a mechanism for *dissipation* of mechanical energy, making energy considerations less straightforward. Our discussion in this chapter will be confined entirely to steady-state flow. In this case, when one element of the fluid traces out a line of flow, any other element lying on this line traces out the same path, and in this case the line is called a *streamline*. We may consider all the streamlines passing through a cross-section area A as in Fig. 16-6. The streamlines passing through the boundary of this area form a tube called a *tube of flow*. For a fluid in a pipe, the walls of the pipe define a tube of flow, but this concept is also useful in the interior of a fluid. By definition of streamlines, no fluid crosses the side surfaces of a tube of flow.

An important example of the usefulness of the concept of the tube of

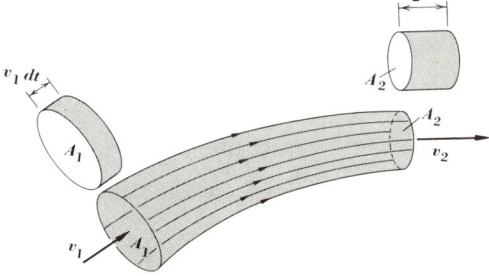

Fig. 16-6 A tube of flow, defined by the flow lines passing through the boundary of an area A_1 perpendicular to the direction of flow. No fluid crosses the boundary surface of such a tube. The elements of fluid passing through areas A_1 and A_2 in time dt are shown; in steady-state flow these must be equal.

flow is the *continuity* relation, an expression of conservation of mass in steady-state flow. To derive this relation we consider a flow tube sufficiently narrow for the density and flow velocity to be considered constant across any cross section of the tube. In any small time interval dt, the mass entering one end of a tube must equal that leaving the other end; otherwise the total mass in the tube would not be constant, and we would not have steady-state flow.

Let the cross-section areas at the two ends be A_1 and A_2, the velocities v_1 and v_2, and the densities ρ_1 and ρ_2, as in Fig. 16-6. In a time dt, the fluid entering moves a distance $v_1\,dt$, and the volume of fluid entering is $A_1 v_1\,dt$. The *mass* of this fluid, given by the density times the volume, is $\rho_1 A_1 v_1\,dt$. Similarly, the mass of fluid leaving the other end is $\rho_2 A_2 v_2\,dt$. From the above conservation-of-mass argument, these must be equal. Dividing out dt, we obtain the continuity equation for a tube of flow:

$$\rho_1 A_1 v_1 = \rho_2 A_2 v_2 \tag{16-5}$$

showing that when the tube of flow becomes narrower, the flow velocity increases, and conversely. Thus in Fig. 16-5 the region where the streamlines are closely spaced corresponds to the region of high flow velocity. For an incompressible fluid, the density of streamlines is proportional to the velocity, but if the fluid is compressible, the relationship is more complex.

Equation (16-5) holds for *all* steady-state flow. For nonsteady flow, of course, the concept of a tube of flow becomes ambiguous. In the special case of incompressible flow, the density is constant, and $\rho_1 = \rho_2$. In that case the continuity equation takes the simpler form

$$A_1 v_1 = A_2 v_2 \tag{16-6}$$

Example

Gaseous CO_2 flows through a tapered pipe from a tank where the pressure is 10 atm and the pipe cross section 0.2 cm^2 to a region at 1 atm, where the cross section is 1.0 cm^2. If the flow velocity as the gas leaves the tank is 50 m/s, what is the flow velocity at the other end of the pipe? Assume the density of the gas is proportional to the pressure.

Solution

From Eq. (16-5) we have

$$v_2 = \frac{\rho_1 A_1}{\rho_2 A_2} v_1$$

Since density is proportional to pressure, $\rho_1/\rho_2 = 10$, and we also have $A_1/A_2 = 0.2 \text{ cm}^2/1.0 \text{ cm}^2 = \frac{1}{5}$. Thus

$$v_2 = (10)(\tfrac{1}{5})(50 \text{ m/s}) = 100 \text{ m/s}$$

16-3 BERNOULLI EQUATION

Equation (16-5) relates the density, velocity, and cross-section area at various points along a tube of flow, but as yet we have no principle relating the *pressures* at various points. There are many practical problems in which knowledge of the variation of pressure with location in the fluid is of primary importance. The Bernoulli equation, to be discussed next, provides the needed relation.

First, we note that when the flow velocity changes, the fluid must be accelerated, requiring a net force on an element of fluid. Thus in general there *must* be pressure changes along a tube of flow, otherwise the net force on an element would be zero and it could not accelerate. The precise relationship between pressure and the kinematic quantities A, v, and ρ was obtained first by Daniel Bernoulli in 1738. As we shall see, the derivation is nothing more than a careful application of energy conservation to fluid in a tube of flow.

For simplicity we consider only the case where the flow is incompressible, irrotational, nonviscous, and steady-state. We consider again the displacement occurring in a time dt in a tube of flow, but this time we examine the work-energy relation. In Fig. 16-7, the total force on the left end of the fluid in the tube is $P_1 A_1$. During time dt this fluid moves a distance $v_1 \, dt$, and the work done at this end on the fluid in the tube is $P_1 A_1 v_1 \, dt$. A similar relation holds at the other end, but the work here is negative because force

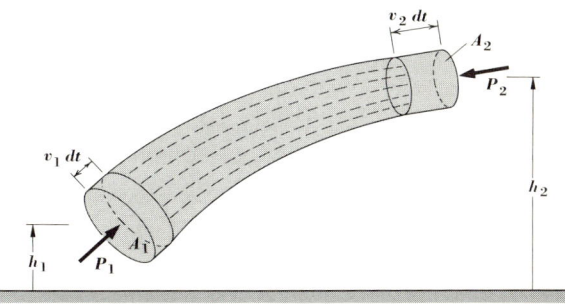

Fig. 16-7 A tube of flow, showing displacements $v_1 \, dt$ and $v_2 \, dt$ at the two ends during the interval. The forces at the two ends are $F_1 = P_1 A_1$ and $F_2 = P_2 A_2$; *these are used to compute the net work done during this interval.*

and displacement are opposite in direction. Thus the net work done on this portion of fluid is

$$W = P_1 A_1 v_1 \, dt - P_2 A_2 v_2 \, dt \tag{16-7}$$

What becomes of this work? Part of it goes to increase the kinetic energy of the fluid. In time dt a mass of fluid $\rho A_1 v_1 \, dt$ enters the tube, as we noted in deriving Eq. (16-5). The kinetic energy of this mass is $\tfrac{1}{2}(\rho A_1 v_1 \, dt)v_1^2$. Similarly, the kinetic energy of the mass leaving the other end in this interval is $\tfrac{1}{2}(\rho A_2 v_2 \, dt)v_2^2$, and the net change in kinetic energy is

$$\Delta E = \tfrac{1}{2}(\rho A_2 v_2 \, dt)v_2^2 - \tfrac{1}{2}(\rho A_1 v_1 \, dt)v_1^2 \tag{16-8}$$

If we ignore the effect of gravity, this change in kinetic energy must account for all the work done on the fluid. When gravity is included, we must add the change in *potential energy*. Since the mass element enters at height h_1 and leaves at height h_2, the change in potential energy is given by

$$\Delta V = (\rho A_2 v_2 \, dt)gh_2 - (\rho A_1 v_1 \, dt)gh_1 \tag{16-9}$$

Now all that remains is to put the pieces together. We have assumed nonviscous flow, so there is no mechanism for energy dissipation and we are dealing with a conservative system. The appropriate energy relation is thus

$$W = \Delta E + \Delta V \tag{16-10}$$

Combining this with Eqs. (16-7) to (16-9) and dividing out dt, we obtain

$$P_1 A_1 v_1 - P_2 A_2 v_2 = \tfrac{1}{2}(\rho A_2 v_2)v_2^2 - \tfrac{1}{2}(\rho A_1 v_1)v_1^2 \\ + (\rho A_2 v_2)gh_2 - (\rho A_1 v_1)gh_1 \tag{16-11}$$

This looks a bit messy, but we can simplify it using the continuity equation for incompressible flow, Eq. (16-6), to divide out the common factor Av. The final result is

$$P_1 - P_2 = \tfrac{1}{2}\rho(v_2^2 - v_1^2) + \rho g(h_2 - h_1) \tag{16-12}$$

An alternate form is

$$P_1 + \tfrac{1}{2}\rho v_1^2 + \rho g h_1 = P_2 + \tfrac{1}{2}\rho v_2^2 + \rho g h_2 \tag{16-13}$$

which shows that the quantity $P + \tfrac{1}{2}\rho v^2 + \rho g h$ is the same at all points in a tube of flow.

When the fluid is at rest, $v_1 = v_2 = 0$, and Eq. (16-12) reduces to

$$P_1 - P_2 = \rho g(h_2 - h_1)$$

in agreement with Eq. (16-3).

It might be thought that Eq. (16-12) could be generalized to include compressible flow by simply using ρ_1 and ρ_2 wherever the density appears in the derivation. The generalization is actually a little more complex than this because for a compressible fluid there is additional potential energy associated with the compression, and this has to be added to Eq. (16-10). To calculate this additional energy requires knowledge of the pressure-density relation for the specific material under study. Exploring this generalization would take us beyond our present scope.

Example
Water enters a house through a pipe 2.0 cm in inside diameter, at an absolute pressure of 4×10^5 N/m² (about 50 lb/in.² gauge). The pipe leading to the second-floor bathroom 5 m above is 1.0 cm in diameter. When the flow velocity at the inlet pipe is 4 m/s, find the flow velocity and pressure in the bathroom.

Solution
The flow velocity is obtained from the continuity equation, Eq. (16-6):

$$v_2 = \frac{A_1}{A_2} v_1 = \frac{(2.0 \text{ cm})^2}{(1.0 \text{ cm})^2} (4 \text{ m/s}) = 16 \text{ m/s}$$

The pressure is now obtained from Bernoulli's equation, Eq. (16-12):

$$\begin{aligned} P_2 &= P_1 - \tfrac{1}{2}\rho(v_2^2 - v_1^2) - \rho g(h_2 - h_1) \\ &= 4 \times 10^5 \text{ N/m}^2 - \tfrac{1}{2}(1.0 \times 10^3 \text{ kg/m}^3)(256 \text{ m}^2/\text{s}^2 - 16 \text{ m}^2/\text{s}^2) \\ &\qquad - (1.0 \times 10^3 \text{ kg/m}^3)(9.8 \text{ m/s}^2)(5 \text{ m}) \\ &= 4 \times 10^5 \text{ N/m}^2 - 1.2 \times 10^5 \text{ N/m}^2 - 0.5 \times 10^5 \text{ N/m}^2 \\ &= 2.3 \times 10^5 \text{ N/m}^2 \simeq 19 \text{ lb/in.}^2 \text{ gauge} \end{aligned}$$

We note that when the water is turned off, the second term on the right vanishes and the pressure rises to 3.5×10^5 N/m².

16-4 APPLICATIONS OF BERNOULLI'S EQUATION

In this section we consider several applications of Bernoulli's relation. The first is the venturi tube, a scheme for measuring the flow velocity in an incompressible liquid. The setup is shown in Fig. 16-8. We orient the tube horizontally so that gravity can be neglected, and the Bernoulli equation becomes simply

$$P_1 + \tfrac{1}{2}\rho v_1^2 = P_2 + \tfrac{1}{2}\rho v_2^2 \tag{16-14}$$

426 FLUID MECHANICS

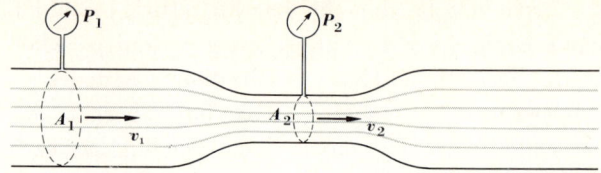

Fig. 16-8 A venturi tube, used to measure the flow velocity v_1 in a pipe by measuring the pressure drop at a constriction where the area decreases from A_1 to A_2. The gauges shown may be replaced by an open-tube manometer to measure the difference $P_1 - P_2$.

Since the pressures can be measured, the only unknown quantities are the two velocities. Furthermore, these are related by the continuity equation, which for an incompressible liquid is simply $A_1 v_1 = A_2 v_2$. We use this to eliminate v_2 from Eq. (16-14) and then solve the resulting equation for v_1. The details of this manipulation are left as a problem; the result is

$$v_1 = A_2 \sqrt{\frac{2(P_1 - P_2)}{\rho(A_1{}^2 - A_2{}^2)}} \qquad (16\text{-}15)$$

Since this result depends only on the difference $P_1 - P_2$, it is often convenient to use a simple U-tube manometer to measure this pressure difference. The areas and the fluid density are constant for any given setup, so Eq. (16-15) can be expressed as $v_1 = C(P_1 - P_2)^{1/2}$, where C is a constant determined by the geometry and the fluid. Of course the actual volume flow can also be computed; this is simply $A_1 v_1$.

Our second example, similar in principle to the venturi tube, involves the rate of flow of a liquid out of a container, as in Fig. 16-9. The space

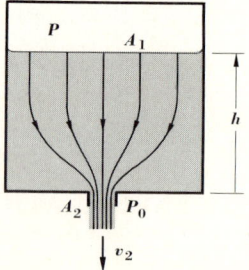

Fig. 16-9 The velocity v_2 of fluid leaving a tank in which the liquid height is h and the pressure above the liquid P can be calculated from Bernoulli's equation. The resulting formula is called *Torricelli's theorem*.

above the liquid contains air at pressure P, and the opening at the bottom is open to the air at pressure P_0. Considering the entire volume of liquid as a tube of flow, we apply Bernoulli's equation to obtain

$$P - P_0 = \tfrac{1}{2}\rho(v_2^2 - v_1^2) - \rho g h \tag{16-16}$$

where v_1 and v_2 are the velocities at the top and bottom, respectively.

A special case of practical interest is that in which the top of the tank is also open to the atmosphere; in this case $P = P_0$, and Bernoulli's equation reduces to

$$v_2^2 - v_1^2 = 2gh \tag{16-17}$$

If in addition the cross-section area A_1 of the tank is much larger than that A_2 of the outlet, then $v_1 \ll v_2$ and we may approximate $v_1 = 0$. In this case

$$v_2 = \sqrt{2gh} \tag{16-18}$$

That is, the speed with which the liquid leaves the orifice is the same as though it had fallen freely from height h! This simple result was discovered by Torricelli, a contemporary of Galileo and the inventor of the mercury barometer.

A variation of this problem leads to a discussion of the thrust of a rocket engine. Suppose the tank is the interior of the rocket, the fluid is a mixture of gases from combustion in the rocket, and the pressure P is large enough for the gravitational term in Bernoulli's equation to be neglected. If again we assume $v_1 \ll v_2$, we obtain for the exhaust velocity of the gases

$$P - P_0 = \tfrac{1}{2}\rho v_2^2 \tag{16-19}$$

This result can be used to calculate the *thrust* of the rocket. The volume of gas leaving per unit time is $A_2 v_2$, the mass leaving per unit time is $\rho A_2 v_2$, and the momentum change per unit time, equal in magnitude to the thrust, is

$$\frac{dp}{dt} = F = (\rho A_2 v_2) v_2 = \rho A_2 v_2^2 \tag{16-20}$$

with v_2 given by Eq. (16-19). Combining Eqs. (16-19) and (16-20), we obtain

$$F = 2A_2(P - P_0) \tag{16-21}$$

Thus the thrust is independent of the density and depends only on the area of the orifice and the pressure. This treatment, however, should be taken with a grain of salt; in using Bernoulli's equation, we have assumed the gas to be incompressible, which is hardly realistic; in addition the flow may be turbulent. These and other factors severely limit the validity of Eq. (16-21).

Bernoulli's equation adds at least qualitative insight into a number of familiar phenomena, including the lift of airplane wings, the curve-ball pitch in baseball, and the flight of a golf ball. In each case the essential point is that along a tube of flow regions of highest velocity correspond to regions of lowest pressure, and conversely. Airfoils are always designed with greater curvature on the upper surface than the lower, resulting in an increased flow velocity over the upper surface. Thus the pressure on the top surface is less than on the bottom surface, and the result is a net upward force.

The curving path of a baseball results from spin about a vertical axis. In Fig. 16-10a, the flow velocity (as seen in a frame of reference moving

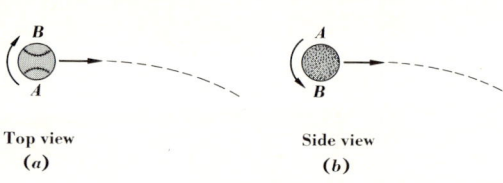

Fig. 16-10 For a spinning ball moving through air, the spin motion adds to the flow velocity at points A and decreases it at points B, and the pressure is decreased at points A and increased at points B. The resulting forces are responsible for the curve of a baseball and the lift of a golf ball.

with the ball) is greater at point A and smaller at point B than the average flow velocity. Thus the pressure at A is less than at B, and the resulting net sideways force makes the ball curve as shown. A similar effect occurs with golf balls, which always have backspin from impact with the slanted club face, as in Fig. 16-10b. The resulting aerodynamic effect is a lift force, which keeps the ball in the air longer than would be possible without air. In this connection, the dimples on a golf ball play an essential role. An undimpled ball has a much shorter trajectory than a dimpled one given the same impact. One manufacturer claims that polygonal dimples are better than round ones!

16-5 VISCOSITY

The assumption that layers of a fluid can slip past each other without resistance is reasonably realistic for some fluids, especially with small flow velocities, but with other fluids, such as molasses or a batch of fudge being beaten, it is not at all realistic. The resistance which all real fluids exhibit, to a greater or lesser degree, to relative motion of adjacent layers, is called *viscosity*. It may be thought of as a kind of internal friction in a fluid.

It is found experimentally that when two surfaces of area A separated by a distance d have a relative velocity v, as shown in Fig. 16-11, the force

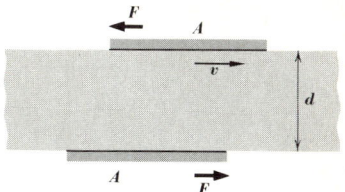

Fig. 16-11 Two parallel surfaces of area A separated by a distance d have a relative velocity v. The forces exerted on the surfaces by the fluid are in the directions shown. It is found that the force is proportional to A and to v and varies inversely with d.

resisting this *shearing* motion is described approximately by

$$F = \eta \frac{vA}{d} \tag{16-22}$$

where η is a constant characteristic of the fluid, called the *coefficient of viscosity* of the fluid or simply its *viscosity*. The ratio v/d is the rate of change of velocity with distance, transverse to the direction of flow, sometimes called the *velocity gradient*. To maintain the motion of the surfaces in Fig. 16-11 requires forces proportional to the *area*, which determines the amount of fluid being sheared, and to the transverse rate of change of velocity.

From Eq. (16-22), viscosity has units of (force)(time)/distance². The unit in most common use is the *poise* (P). Based on cgs rather than mks units, the poise is defined as

$$1 \text{ P} = 1 \text{ dyn·s/cm}^2 = 10^{-1} \text{ N·s/m}^2 \tag{16-23}$$

Viscosities of a few common fluids are given in Table 16-1. As the table shows, viscosities are strongly temperature-dependent; in most cases the viscosity of gases increases with temperature, while that of liquids usually decreases with temperature. An important consideration in the design of oils for engine lubrication is to reduce this temperature dependence as much as possible.

An important application of Eq. (16-22) occurs in the study of fluid flow in pipes. Bernoulli's equation, which holds only for nonviscous flow, predicts that in a horizontal pipe of uniform cross section the pressure is uniform along the pipe. It is a matter of common experience, however, that the pressure always drops along the direction of flow; this is an effect of the viscosity of the fluid. The layers of fluid near the wall of the pipe cling to it and have a smaller flow velocity than those near the center; just at the wall of the pipe the velocity is zero. One may visualize the motion as like a number of telescoping tubes, with the central one moving most rapidly and the outer ones successively more slowly. We may apply Eq. (16-22) to the problem, obtaining an expression for the velocity variation across the pipe, and hence

430 FLUID MECHANICS

Table 16-1 Viscosities of Common Fluids

Material	Temperature, °C	Viscosity, P
Water	0	1.79×10^{-2}
	20	1.00×10^{-2}
	40	0.65×10^{-2}
Ethanol	0	1.77×10^{-2}
	20	1.20×10^{-2}
	40	0.83×10^{-2}
Ethylene glycol	20	0.19
	40	0.091
	60	0.049
Glycerin	0	121
	20	14.9
	30	6.29
Mercury	0	1.68×10^{-2}
	50	1.40×10^{-2}
	100	1.24×10^{-2}
Air	0	1.71×10^{-4}
	40	1.90×10^{-4}
Hydrogen	0	0.83×10^{-4}
	300	1.38×10^{-4}

find the relationship between the total flow rate and the pressure drop per unit length. This latter relationship is called *Poiseuille's law*, after its discoverer.

We consider a cylindrical pipe of inner radius R. Within the fluid, we consider a cylinder of fluid of radius r and length l, as shown in Fig. 16-12. We use Eq. (16-22) to find the viscous force in this element. The appropriate area is $2\pi r l$, the side surface area of the element, and we obtain

$$F = -2\pi r l \eta \frac{dv}{dr} \tag{16-24}$$

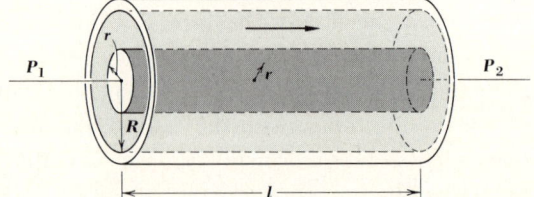

Fig. 16-12 A fluid flows in a cylindrical pipe of radius R. We analyze the viscous force on an element of fluid of radius r and length l in order to derive an expression for the pressure difference between the ends.

The negative sign is needed because v *decreases* with increasing r. For this element of fluid to maintain steady motion there must be a driving force to counteract the resisting viscous force, and this requires a variation in pressure with length. The forward force on the left end is the pressure P_1 at that end times the cross-section area πr^2 of the element. Similarly, there is a backward force on the right end; the net force resulting from this pressure difference is

$$F = (P_1 - P_2)\pi r^2 \tag{16-25}$$

We now equate this to the viscous force, obtaining

$$-2\pi r l \eta \frac{dv}{dr} = (P_1 - P_2)\pi r^2 \tag{16-26}$$

This is a differential equation for v as a function of r. To solve it we separate variables and integrate:

$$dv = -\frac{P_1 - P_2}{2l\eta} r\, dr \qquad v = -\frac{P_1 - P_2}{4l\eta} r^2 + C \tag{16-27}$$

where C is an integration constant. To evaluate C we recall that $v = 0$ at $r = R$; thus

$$0 = -\frac{P_1 - P_2}{4l\eta} R^2 + C$$

Finally, we obtain

$$v = \frac{P_1 - P_2}{4l\eta}(R^2 - r^2) \tag{16-28}$$

which gives the variation of velocity with radius within the pipe. At $r = 0$ the velocity has its maximum value; at $R = r$ it is zero.

To compute the total volume of flow per unit time, we consider the flow in a cylindrical shell of inner radius r and outer radius $r + dr$; the cross-section area of this shell is $2\pi r\, dr$, and the flow through it is $v(2\pi r\, dr)$, with v given by Eq. (16-28). To find the total flow dV/dt we add up the contribution of all such shells by integrating on r from $r = 0$ to $r = R$:

$$\frac{dV}{dt} = \int_0^R 2\pi v r\, dr = \int_0^R \frac{2\pi(P_1 - P_2)}{4l\eta} r(R^2 - r^2)\, dr \tag{16-29}$$

Evaluation of this integral is left as a problem; the final result is

$$\frac{dV}{dt} = \frac{R^4(P_1 - P_2)}{8\eta l} \tag{16-30}$$

This result, called Poiseuille's law, shows that the rate of flow is proportional to the fourth power of the radius and to the pressure gradient $(P_1 - P_2)/l$, or rate of change of pressure with length along the pipe. It is also inversely proportional to the viscosity of the fluid.

Equation (16-30) provides a practical way to *measure* the viscosity of a fluid by measuring the flow rate in a tube under measured pressure conditions. Of course, this relation is valid only when the flow is laminar (nonturbulent). At sufficiently high velocities, turbulent flow sets in, and the flow rate is always *less* than that predicted by Eq. (16-30).

16-6 TURBULENCE

When the velocity of fluid flow reaches a certain critical value, depending on the properties of the fluid, the assumption of laminar, or nonturbulent, flow is no longer valid, and the flow pattern becomes much more complex. Vortices, or eddies, appear, the pattern changes continuously, and there is no longer any such thing as steady-state flow. This condition is called *turbulent flow*; nearly all fluids exhibit turbulent flow at sufficiently high velocities.

Turbulent flow is an extremely complex phenomenon, not all aspects of which are well understood, and we cannot discuss it in detail. The onset of turbulent flow depends on the properties of the fluid in an interesting way, best expressed using a characteristic number called the *Reynolds number* for the particular problem at hand. For fluid flow in a pipe of diameter D with average velocity v_{av} the Reynolds number R is defined as

$$R = \frac{\rho v_{av} D}{\eta} \tag{16-31}$$

where ρ is the density and η the viscosity, as usual. The Reynolds number is a *dimensionless number*; for a given situation its value is the same in all unit systems. Verification of this statement is left as a problem. The significance of the Reynolds number was first discovered experimentally, but now it also has some theoretical foundation.

It is found experimentally that laminar, nonturbulent flow occurs whenever R is less than about 2,000. When R is greater than about 3,000, the flow is nearly always turbulent, and in the region between 2,000 and 3,000 the flow is often unstable, changing from one type of flow to the other and depending critically on any irregularities in the path of the fluid. For water at room temperature in a 1-cm-diameter pipe, turbulent flow occurs above about 0.2 m/s, while for air the critical speed is of the order of 4.0 m/s. Verification of these numbers is left as an exercise.

Turbulent flow in pipes is characterized by greater pressure drops for a given flow velocity than predicted by the Poiseuille formula. Usually the flow is still laminar in a region close to the wall of the pipe, called the *boundary layer*, but in the interior it is irregular, and the velocity and pressure fluctuate with time. For sufficiently large values of R the drag forces on the fluid may actually decrease with increasing flow velocity; this phenomenon is not yet well understood.

The Reynolds number concept is important in the experimental investigation of fluid flow, for example, testing airplane structures in wind tunnels. It is almost always desirable to test a scaled-down model of the real structure rather than building a full-sized model. The Reynolds number tells us how the velocity should be scaled in the reduced-size model situation. We choose the velocity so that the Reynolds number for the model is the same as for the actual situation; the flow characteristics will then be the same for both.

Problems

16-1 An automobile is immersed in water 5 m deep. If the area of a door is 1.0 m², find the total force the water exerts on the door.

16-2 A hydraulically operated elevator has a cylinder and piston 20 cm in diameter. What must the oil pressure be if the elevator and its load have a total mass of 2,000 kg? Express your result in newtons per square meter and in atmospheres.

16-3 A hot-water heating system in a house has an expansion tank in the attic, open to the atmosphere. If the water surface is 10 m above the pressure gauge on the furnace in the basement, what is the gauge pressure at that point? The absolute pressure?

16-4 Air pressure in pipe organs is usually measured in inches of water. How many inches of water is 1 atm?

16-5 A dam in an aqueduct of rectangular cross section is 2.5 m wide and 2.0 m high. The water on one side comes up to the top of the dam, and there is no water on the other side. Find the total force of the water on the dam.

16-6 A fisherman in the north woods proposed to use a slab of ice 0.5 m thick as a raft to ferry his car across a lake.
  a If the mass of the car is 2,000 kg, what minimum area must the slab have if it is to float with the top surface barely out of water?
  b How does he get his car back across the lake in the spring, after the ice melts?

16-7 The density of air at standard temperature and pressure is about 1.29 kg/m^3, that of helium 0.178 kg/m^3, and that of hydrogen 0.090 kg/m^3. If a balloon is filled with helium, what must its volume be in order for the buoyant force of air to support a load of 200 kg? If the balloon is spherical, what is its radius? Would the radius be appreciably smaller if hydrogen were used instead of helium?

16-8 A block of steel (density $7.8 \times 10^3 \text{ kg/m}^3$) floats in mercury (density $13.6 \times 10^3 \text{ kg/m}^3$). What fraction of the volume of the block is beneath the surface?

16-9 Sailors claim that only one-ninth of an iceberg is above the surface of the water in which it floats. From this number, calculate the ratio of the density of ice to that of water. Is the distinction between fresh water and seawater (salt water) relevant? Explain.

16-10 If the temperature of the atmosphere is uniform, at what altitude is the pressure half as great as at sea level?

16-11 A fire hose has a diameter of 10 cm, and the diameter of the orifice of the nozzle is 1.25 cm. If the nozzle is pointed straight up, the stream reaches a height of 20 m. Find the flow velocity in the hose and the volume of water flowing per unit time, neglecting the effect of air resistance on the stream.

16-12 A pipe of diameter 2.0 cm carries water at 10 m/s. It is connected to a pipe of diameter 1.0 cm. Find the velocity in the smaller pipe and the volume flow rate.

16-13 In a television commercial, holes are punched in a can of antifreeze, and the fluid squirts out in a parabolic path, hitting the table at a distance x from the base of the can. Find x as a function of height h of the hole, assuming the flow is slow enough for the fluid level in the can to be approximately constant.

16-14 Show that when a body is immersed in a fluid whose density is k times as great as that of the fluid, its apparent weight on the fluid is $W(k-1)/k$. What error does this effect introduce in weighing meat at the meat market? If the meat were weighed in a vacuum, would the housewife get more or less meat for her money?

16-15 A venturi flowmeter has a pipe diameter of 0.20 m and a throat diameter of 0.10 m. The pressure in the pipe is $1.0 \times 10^5 \text{ N/m}^2$, and that in the throat is 0.8 N/m^2. Find the flow velocity and the volume rate of flow in the pipe.

PROBLEMS

16-16 Derive Eq. (16-15) from the preceding discussion.

16-17 A water line runs parallel to a hilly street, becoming smaller as it approaches the end of the street. At one point the flow velocity is 1.0 m/s, the gauge pressure 2.0 atm, and the radius of the pipe 10 cm. At another point 50 m vertically below the first the radius is 5.0 cm. Find the gauge pressure at that point, the flow velocity, and the volume flow rate.

16-18 A glider on an air track is supported by a layer of air 0.1 mm thick. If the total area of the glider adjacent to the track is 25 cm^2 and it is moving at 20 cm/s, find the damping force due to air viscosity.

16-19 Complete the derivation of the Poiseuille equation by evaluating Eq. (16-29).

16-20 By examining the data in Table 16-1 or a handbook, try to establish an empirical equation between viscosity and temperature for gases. For example, is η proportional to the absolute temperature T, or to T^2, or something similar?

16-21 Water at 20°C flows in a pipe of radius 0.01 m, with a speed at the center of 0.05 m/s. Find the pressure drop per unit length of pipe due to viscosity.

16-22 Verify that the Reynolds number, Eq. (16-31), is dimensionless.

16-23 For a water pipe of radius 0.01 m, what is the maximum flow velocity without turbulent flow?

16-24 Show that in laminar flow of a viscous fluid in a cylindrical pipe, the total volume flow is the same as though the flow velocity were uniform over the cross section with a value half the actual velocity on the axis.

16-25 Water at 20°C flows through a pipe of diameter 0.3 cm with a speed of 50 cm/s. Find the Reynolds number, and predict the nature of the flow.

Properties of Matter | 17

With this chapter we begin a detailed study of the properties and behavior of matter in its various states and especially of the relationship between these properties and microscopic structure. We first review briefly several qualitative ideas concerning the structure of matter. Next we discuss the concepts of temperature and heat and procedures for measuring them—mostly from a macroscopic viewpoint; the microscopic basis is discussed qualitatively, but a quantitative microscopic treatment of temperature and heat is postponed until later chapters.

17-1 STRUCTURE OF MATTER

There is a vast body of evidence to indicate that matter in its familiar forms is composed of fundamental species which we call *elements*. Ninety-two elements are known to exist in nature, and several more have been produced artificially. Two or more elements can combine to form new substances, called *compounds*. The physical characteristics of a compound are quite different from those of the elements which it contains.

When elements combine to form compounds, they always do so in definite proportions by mass. This and many other observations establish that each element consists of identical basic particles called *atoms*. An atom is the smallest unit of an element which can exist. The fact that elements combine in definite proportions to form compounds is understood on the basis that several atoms combine in a definite way to form a basic particle of the compound, called a *molecule*. A molecule therefore is the smallest particle of a compound which can exist. A molecule may also be formed of two or more atoms of a single element; examples are O_2, N_2, and H_2.

Atoms are not indivisible, as was once thought. The internal structure of atoms has been investigated in considerable detail by a wide variety of methods. One of these, Rutherford scattering, has been discussed in Sec. 12-3. Many similar experiments have been performed more recently; some of these use protons, electrons, or neutrons, instead of α particles, as the "bullets."

Such experiments have shown that an atom always has at its center a *nucleus*, with a positive electric charge, whose mass includes almost all the mass of the atom. This is surrounded by negatively charged *electrons*, with masses much smaller than that of the nucleus, in sufficient number to balance the electric charge of the nucleus. This swarm of electrons extends, in most atoms, to a distance of the order of 10^{-10} m away from the nucleus. The nucleus itself is very much smaller than this; nuclear radii range from 1.2×10^{-15} m for the smallest to about 8×10^{-15} m for the largest. The nucleus has about 2,000 to 5,000 times as much mass as the total mass of all the electrons, depending on the particular element in question.

The nucleus of an atom can be broken up into even more fundamental constituents, called *protons* and *neutrons*. These particles are about 1,840 times as massive as electrons. The neutron, discovered in 1932, has no electric charge; the charge of the proton is equal in magnitude to that of the electron but opposite in sign. In high-energy collisions involving protons and neutrons, other kinds of particles may be *created*; these are always unstable, and they decay to the familiar particles.

Ordinary chemical reactions involve changes in the arrangements of atoms in molecules and do not alter the *internal* structure of atoms, except in the configuration of the outermost electrons. When an atom is broken into its constituent electrons, neutrons, and protons, it loses its identity as an atom of a definite chemical element. Similarly, when an atomic nucleus is taken apart, it loses its identity as belonging to a specific element.

A molecule of a compound behaves as an indivisible unit so long as the compound is not chemically altered. Thus, in speaking of the structure of a gas, we often refer to the molecules of the gas as particles, despite the fact that they have a complex internal structure. Provided no chemical reaction occurs, the molecules may be treated as indivisible bodies, and their internal structure is significant only in considerations of the internal motion which may occur. The energy associated with internal motion has an important effect on specific heats and other properties, however. Similarly, as long as no nuclear reactions take place to alter the internal structure of a nucleus, this structure is irrelevant, and the nucleus may be regarded as a rigid body.

Several of the concepts of newtonian mechanics are directly applicable to the behavior of atoms and molecules; the momentum and energy principles

438 PROPERTIES OF MATTER I

are especially useful. The forces which atoms exert on each other and which are responsible for binding atoms together in a molecule are electromagnetic in nature. These have been discussed briefly in Sec. 5-7. The forces binding nuclear particles to form nuclei which are stable despite the electrical repulsions between protons are of quite a different nature. They fall off more rapidly with distance than electric forces, but at sufficiently small distances they are much stronger. These are called *nuclear forces*. Gravitational forces are very much weaker than either electromagnetic or nuclear forces, where these exist, and are of no importance in atomic, molecular, or nuclear structure.

It is useful to classify materials as being solid, liquid, or gas. A solid has a definite shape; liquids and gases, both of which are *fluids*, take the shape of their container.

The gaseous state is the easiest one to understand. In a gaseous element or compound, the atoms or molecules are, on the average, quite far apart compared to their dimensions and are free to move more or less independently of each other. This accounts for the fact that gases are quite compressible, since the amount of empty space between molecules is much greater than the volume of the molecules themselves. Since the molecules are not rigidly held in position, a gas has no definite shape but takes the shape of its container and always fills it completely.

In a solid, the atoms are separated by distances comparable to their size. As a result of this close proximity, the atoms exert strong forces on each other; these tend to keep them in more or less definite positions relative to each other. Thus a solid always has a definite shape and resists any attempt to change it.

Solids may be *amorphous* or *crystalline*. An amorphous solid is one whose atoms or molecules are not arranged in any definite pattern; in a crystalline solid the molecules have an orderly, well-defined arrangement. A typical structure for an amorphous solid, such as glass, is shown schematically in Fig. 17-1. In this figure, the atomic nuclei are represented by

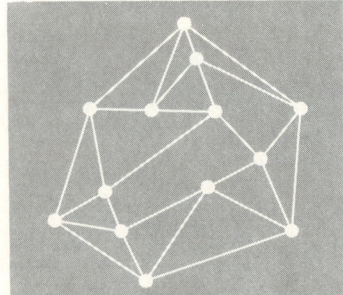

Fig. 17-1 Schematic diagram showing arrangement of molecules in an amorphous solid. The circles represent molecules, the lines the intermolecular forces. Intermolecular distances are expanded for clarity.

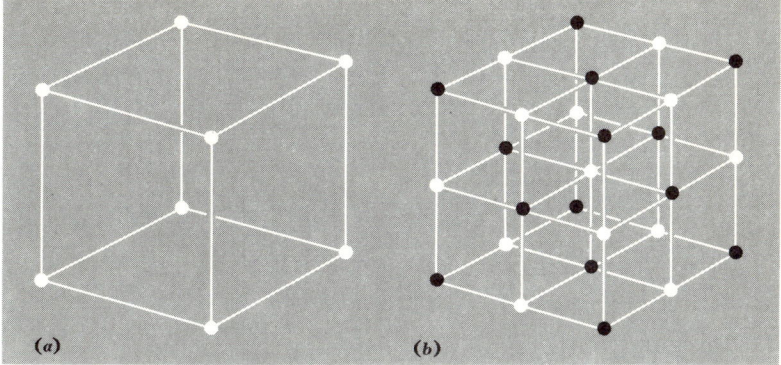

Fig. 17-2 Cubic crystal lattices. (*a*) Crystal of an element; the atoms are located at the corners of the cubic cells. (*b*) Crystal of sodium chloride; the sodium and chlorine atoms are placed alternately in the lattice. Circles represent the positions of the nuclei.

circles; the lines represent interatomic forces. Figure 17-2 shows two very simple crystalline structures. Both of these are called *cubic* structures, because the atoms are located at the corners of cubes. In a sodium chloride crystal, the sodium and chlorine atoms occupy alternate positions in the lattice.

There are many other types of crystal structure, most of them more complicated than the cubic. Two additional types are shown in Fig. 17-3. All crystal structures are characterized by recurring patterns which extend over a considerable distance in the crystal. This is sometimes referred to as *long-range order*, the term *order* denoting a systematic arrangement. Because of this order and the more or less rigid binding of atoms, a crystal is very similar in principle to a *molecule* containing a very large number of atoms.

The structure of liquids is less well understood than that of either gases

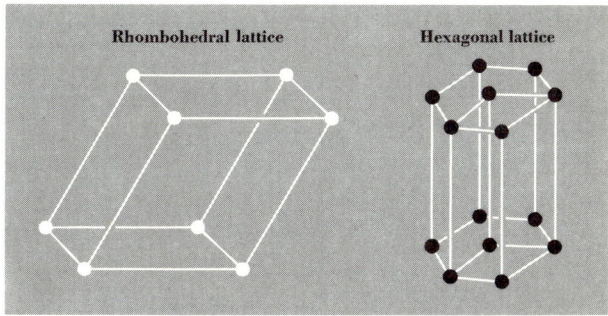

Fig. 17-3 Two examples of noncubic crystal structures.

or solids. As with solids, the molecules are relatively close together, with separations of the same order of magnitude as their dimensions, and there are strong cohesive forces between the molecules. But the forces are not sufficient to keep the molecules in definite positions relative to each other. They are comparatively free to move, so a liquid does not have a definite shape. On the average, each molecule has other molecules arranged around it in a somewhat orderly manner, so there is *short-range order* in a liquid. There is not, however, the long-range order found in crystalline solids. In this sense, liquids resemble amorphous solids in their structure. In fact, in many respects amorphous solids such as glass are more similar to very viscous liquids than to crystalline solids. They undergo fluidlike deformations over long periods of time, and they usually do not have definite melting temperatures, instead softening progressively with increasing temperature.

The mechanical behavior of solids, liquids, and gases can thus be understood, crudely at least, on the basis of the amount of freedom of motion the molecules possess. In a solid, the molecules are more or less rigidly bound to each other; in a liquid, there is more freedom of motion; in a gas, they are almost entirely free. There is also a fourth state of matter, the *plasma* state, in which the atoms are all dissociated from each other and may have some or all of their electrons removed. An atom which has either more or less electrons than the number needed for electrical neutrality is called an *ionized* atom or simply an *ion*. Matter in the plasma state, then, consists of free electrons and atoms from which various numbers of electrons have been detached. Such a state can exist stably only at extremely high temperatures. This is the state of matter which is involved in thermonuclear reactions and which must be produced in order to make controlled thermonuclear reactions possible. It is also the state of matter occurring in the interior of stars.

Microscopic structure serves as a basis for understanding many different physical properties of matter. Because molecules of a gas can move nearly independently of each other, the gaseous state is the easiest one for which to make a detailed analysis of a microscopic model. It is more difficult to make quantitative calculations for solids, and more difficult still for liquids. For these, we often have to be content with qualitative understanding of physical properties.

17-2 MOLECULAR MASS

Most of the mass of an atom is concentrated in its nucleus, which always contains a definite number of protons and neutrons. All protons have the same mass, as do all neutrons, and the two masses are very nearly equal. Thus it is not surprising that the masses of many different kinds of atoms are approximately whole-number multiples of the mass of the lightest atom, hydrogen, whose nucleus is a single proton.

This fact has led to the introduction of a new unit of mass, the *atomic mass unit* (amu). The original idea was to define this unit so that the mass of one neutral hydrogen atom would be exactly 1 amu. For various reasons it has been found more convenient to define this unit instead as one-twelfth the mass of a neutral carbon atom (whose nucleus contains six protons and six neutrons), so the mass of a carbon atom is exactly 12 amu; this is now the internationally accepted definition. With this definition, the atomic mass of hydrogen is about 1.008 amu. The mass of a molecule can be expressed in the same units. The mass of a molecule of water, H_2O, is about 18 amu. The mass of one molecule of a compound, measured in atomic mass units, is called the *molecular mass* of the compound. Similarly, the mass in atomic mass units of one atom of an element is the *atomic mass* of the element. Atomic and molecular masses are sometimes called atomic weight and molecular weight. This is an incorrect use of the term *weight*; the correct term is mass. There is a tendency toward universal adoption of the terms *atomic mass* and *molecular mass*, and we adhere to this usage throughout this book.

The conversion factor from grams to atomic mass units is denoted by N_0. Specifically, 1 g = N_0 amu. The numerical value of N_0 has been found to be

$$N_0 = 6.025 \times 10^{23} \text{ amu/g}$$

Especially in chemical calculations, it is often convenient to measure a quantity of a substance by the number of molecules it contains rather than by its mass. The number of N_0 provides a convenient unit for this purpose; a quantity of any substance containing N_0 molecules is called *one mole* (mol) of the substance. That is,

$$1 \text{ mol} = 6.025 \times 10^{23} \text{ molecules}$$

The mass of 1 mol of any substance can be computed from the value of N_0 and the atomic or molecular mass of the substance. Carbon, for example, has a molecular mass of 12 amu; 1 mol of carbon contains 6.025×10^{23} atoms with a total mass of $12 \times 6.025 \times 10^{23}$ amu, or simply 12 g. More generally, the mass of 1 mol of *any* substance in grams has the same numerical value as the atomic or molecular mass in atomic mass units. Thus 1 mol of water has a mass of 18 g, and so on.

The number N_0 is called *Avogadro's number*. Its value has been measured experimentally by a number of different methods. One of these makes use of a measurement of the mass of a partially ionized atom, obtained by observing its deflection while it moves in electric or magnetic fields. This measurement gives the ratio of electric charge to mass for the atom; the charge is always a whole-number multiple of the charge of the electron. The electron charge was measured directly in 1909 by the famous Millikan oil-drop experiment, in which one observes the motion of very small electrically charged oil drops under the action of an electric field and the viscous force of air resistance.

Atomic and molecular masses can also be expressed in terms of moles. For example, the atomic mass of carbon is

$$12.000 \text{ amu/atom} = 12.000 \text{ g/mol} = \frac{12.000 \text{ g}}{6.025 \times 10^{23} \text{ atoms}}$$

Thus the mass of one carbon atom is 1.992×10^{-23} g.

17-3 TEMPERATURE

One of the most familiar properties of matter is *temperature*. We associate temperature with ideas of "hot" and "cold" based on our sense of touch; roughly speaking, the temperature of a body is a quantitative measure of its hotness or coldness. More generally, many physical properties of matter change when the temperature changes. The length of a steel bar increases measurably when it is heated. The pressure of gas in a closed container increases with heat; if too large a fire is built in a steam-boiler furnace, the boiler may explode. The electrical conductivities of materials change with temperature. When materials are very hot, they emit visible light of a color which depends on the temperature; everyone knows that "white heat" is hotter than "red heat."

Microscopically, temperature is associated with energy of molecular motion. In any of the various states of matter, the atoms and molecules are constantly in motion; even in solids, atoms vibrate back and forth around their equilibrium positions. It has been found that higher temperatures correspond to increased molecular motion and increased energy associated with this motion. The molecules in matter at a given temperature do not all have the *same* energy, but temperature is directly related to the *average* energy of molecular motion. It will be shown in Chap. 18 that the average kinetic energy of the molecules of a gas under certain conditions is directly proportional to the absolute temperature. For other forms of matter the relationship is less simple. In any case, temperature is intrinsically an *average* property; to say that any individual molecule has a certain temperature has no meaning.

Although the sense of touch gives a qualitative idea of temperature, it does not provide a means of measuring temperature quantitatively. To establish a numerical scale for temperature, we may use any physical property which depends on temperature. Any instrument for establishing such a scale is called a *thermometer*. The common type of thermometer makes use of the volume expansion of a fluid with temperature. A glass bulb containing mercury or alcohol is attached to a tube with a very small bore, as shown in Fig. 17-4. A change in the height of fluid in the tube corresponds to a change in temperature.

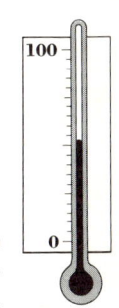

Fig. 17-4 Ordinary liquid thermometer in cross section. The tube diameters are exaggerated. The liquid is usually mercury or alcohol.

To establish a numerical temperature scale, we assign numbers to a series of states corresponding to different degrees of hotness or coldness. The way in which this assignment is made is, for the present, quite arbitrary. One widely used scheme is to assign the number zero to the temperature at which ice melts at standard atmospheric pressure and the number 100 to the boiling temperature of water at the same pressure. A temperature between these two points can then be described by a number between zero and 100. This assignment forms the basis of the Celsius temperature scale.[1]

To calibrate the liquid thermometer described above, we can observe the liquid levels when the thermometer is in contact with melting ice and with boiling water, label these 0°C and 100°C, respectively, and divide the region between them into 100 subintervals. It is customary to designate a numerical temperature as a certain number of *degrees*. Thus we refer to the

---

[1] This scale was formerly called the *centigrade* scale, because it has 100 units between the melting and boiling points of water. It is now officially called the Celsius scale, after its inventor, A. C. Celsius, a Swedish astronomer (1701–1744).

temperature of melting ice on the Celsius scale as 0°C, the temperature of boiling water as 100°C, and the temperature of a hot summer day as 33°C.

Another temperature scale, used in everyday life in the United States, is the *Fahrenheit* scale. In this scale, the melting point of ice is 32°F, and the boiling point of water 212°F. There are 180° between these two points, compared with 100° on the Celsius scale. Thus a temperature difference of one Fahrenheit degree (1 F°) is $100/180 = 5/9$ as great as one Celsius degree (1 C°). Using this fact, it is easy to convert from one temperature scale to the other. For example, to find the Celsius temperature corresponding to 68°F, we note that 68°F is $(68 - 32)$F°, or 36 F°, above the melting point. This corresponds to $36° \times 5/9$, or 20°, above the freezing point on the Celsius scale, which is 0°C. Thus the temperature is 20°C. One can derive a general formula for converting from Fahrenheit to Celsius and back, but it is just as easy to go through the reasoning we have outlined. This procedure also avoids the danger of remembering the formulas incorrectly.

Suppose now that we build two thermometers, using two different liquids. If they are calibrated at the melting and boiling points of water, they automatically agree at 0 and 100°C. Do they also agree at all intermediate temperatures? There is no reason to think that they should. In fact, when precise observations are made on intermediate temperatures, discrepancies between the two measurements are observed, indicating that the two materials have different expansion characteristics.

In order to have a well-defined temperature scale, we must specify not only the temperatures of two reference points, but also the means of interpolating between these reference points. For practical use, the ordinary glass thermometer with mercury or alcohol as the fluid often suffices. In some cases, especially in measuring very low temperatures, it is preferable to define a temperature scale based on the changing electrical resistance of a material. In any event, it is important to recognize that any such temperature scale depends on the physical properties of the material used; there is no reason to expect two scales using different materials or different physical properties to agree exactly at all temperatures.

So far, all we have really done in defining temperature is to make the somewhat circular statement that "temperature is what one measures with a thermometer." After we acquire some additional background concerning temperature and heat, we shall be able to give a considerably more fundamental definition of temperature and to establish a temperature scale which is independent of the properties of any particular material. Such a scale is discussed in Chap. 21.

17-4 GAS THERMOMETER AND ABSOLUTE TEMPERATURE

A thermometer widely used for calibration purposes and other careful measurements is the *constant-volume gas thermometer*. It can be made very precise and reproducible, yet it is relatively simple in construction. The constant-volume gas thermometer makes use of the fact that the pressure which a constant volume of a gas exerts on its container varies with temperature. Furthermore, if the temperature is measured with an ordinary mercury thermometer such as we have described, it is found that the pressure is very nearly a *linear* function of temperature. This result is embodied in the law of Charles, which states that when the volume of a gas is kept constant, its pressure increases in proportion to its temperature. Thus it is reasonable to use this phenomenon to *define* a temperature scale.

The construction of a typical constant-volume gas thermometer is shown schematically in Fig. 17-5. The pressure of the gas in the bulb, which is

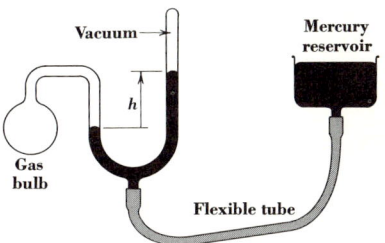

Fig. 17-5 Constant-volume gas thermometer. The gas bulb is placed in thermal contact with the body whose temperature is to be measured. The position of the mercury reservoir is adjusted to keep the level on the left side of the manometer constant so that the volume is constant. The pressure is proportional to the height h.

used as the temperature-sensing element, is proportional to the difference in height h of mercury between the two sides of the tube. The mercury reservoir is included so that for any temperature the height of the left-hand column of the manometer can be adjusted to a fixed level in order to keep the volume of gas constant.

One obvious advantage of a gas thermometer is that it can be used to measure very low temperatures. Mercury freezes at a temperature of $-39°C$ and ethanol at $-130°C$, so these thermometers are useless below these temperatures. A hydrogen-gas thermometer, on the other hand, is useful down to temperatures of at least $-240°C$. Hydrogen liquefies at a temperature of about $-253°C$ at atmospheric pressure.

To establish a scale, we calibrate this thermometer at two temperatures, say 0 and $100°C$, and then use a straight-line relationship between pressure and temperature for other temperatures. Figure 17-6 shows a temperature scale for a particular gas thermometer. A significant feature of Fig. 17-6 is that when we extrapolate the line until it crosses the T axis, we find a

Fig. 17-6 Temperature scale defined by the pressure P in a constant-volume gas thermometer. The absolute zero temperature, corresponding to extrapolation of the line to zero pressure, is shown.

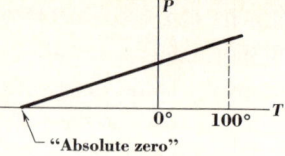

temperature at which the gas would exert *no pressure at all* if the proportionality of pressure and temperature were to hold. For the line shown, this temperature is about $-273°C$.

Every gas thermometer has an extrapolated temperature at which the pressure would go to zero if liquefaction or some other phenomenon did not occur. One might expect that this "zero-pressure" temperature would be different for different gases. It turns out, however, that it is very nearly the same for a wide variety of gases; the extrapolated temperature for zero pressure is always $-273.2°C$, independent of the choice of gas. Furthermore, temperature scales defined with many different gases turn out to agree closely over a wide temperature range. Thus we have a temperature scale which is the same for a large class of materials and which is not dependent on one particular gas. In the remainder of this chapter we use the gas thermometer as a standard.

The temperature $T = -273.2°C$, at which all the gas thermometers predict zero pressure, is called the *absolute zero* of temperature. We now define a new temperature scale which has its zero point at the absolute zero by simply adding 273.2° to every Celsius temperature. This provides a more convenient scale to use with the gas thermometer, since on this scale the pressure exerted by the gas is directly proportional to the temperature. The scale resulting from this shift is called the *Kelvin* temperature scale. On the Kelvin scale 1° represents the same interval as on the Celsius scale, but the zero point is shifted. Thus $0°C = 273.2$ K. In general, the relation between the two temperatures is given by

$$T(K) = T(°C) + 273.2° \tag{17-1}$$

In using the Kelvin scale, it is increasingly common practice to drop the degree sign and refer to a temperature as, say, 50 K, read "fifty kelvins."

A temperature scale whose zero point coincides with absolute zero is called an *absolute* temperature scale. Microscopically, absolute zero corresponds to a state in which the molecules have the minimum possible energy of motion. For this and other reasons, absolute temperature scales are of fundamental importance in the relationships between mechanical energy and heat, developed in detail in the following chapters.

In defining temperature and temperature scales, we have made an assumption which must be discussed explicitly as preparation for the next section. The thermometers discussed made use of changes in some physical property of a material in the thermometer as a result of changes in the temperature *of the thermometer.* Thus, what a thermometer really measures is *its own temperature.* But what good is a thermometer if it can measure only its own temperature? How can we measure the temperatures of other bodies?

A consideration of how thermometers are actually used shows the way out of this predicament. To measure the temperature of a bathtub full of warm water, we immerse the thermometer bulb in the water, wait until the thermometer reaches its final value, and then read it. In doing this, we assume that the temperature of the thermometer is the same as that of the water. We say that the thermometer and the water have reached a state of *thermal equilibrium,* a state in which their temperatures no longer change as a result of their contact with each other.

How do we know that two bodies in thermal equilibrium have the same temperature? This question must be investigated experimentally. Suppose we put the same thermometer in contact with each of two bodies and find that the equilibrium temperature of the thermometer is the same for both. If the temperatures are *not* all equal under these conditions, then there is no reason to expect that the two bodies have the same temperature. If we then place them in thermal contact with each other, their temperatures may change, and these changes should be detectable with the same thermometer. Many experiments with this general type of situation have shown that in fact at thermal equilibrium the temperatures *are* the same. It is important to realize, however, that this is an experimental result and is *not* one which follows from the definition of temperature given above.

The basic principles of thermodynamics are often referred to as the first, second, and third laws of thermodynamics. The principle that *at thermal equilibrium the temperatures of two objects in contact are the same* is basic to all of these, and so it is often called the *zeroth law of thermodynamics.* For systems having a microscopic structure simple enough to permit detailed analysis (such as ideal gases), this principle can be derived from a microscopic model, but its more general validity must be regarded as an empirical result.

17-5 QUANTITY OF HEAT

We are now ready to study in more detail the changes which occur when two bodies having different temperatures are put into thermal contact with each other and eventually reach thermal equilibrium at a common final temperature. The process of attaining thermal equilibrium in such a situation may be described as a transfer of *heat* from one object to the other. That is, it is possible to account for the temperature changes by assuming that the amount of heat *lost* by one object equals the amount of heat *gained* by the other object. A precise quantitative formulation of this statement is the essence of the field of *calorimetry*, or the measurement of heat.

It is not necessary at this point to specify precisely what heat *is*. In the early days of thermodynamics (in the early nineteenth century) there was considerable confusion concerning the physical basis of heat. Some scientists called it *caloric fluid* and regarded it as a material fluid which flowed from one object to another. We now know that the physical basis of heat is *energy*; a transfer of heat from one object to another corresponds to a transfer of mechanical energy associated with the motions of individual atoms and molecules. Heat is microscopic mechanical energy in transit from one body to another.

What determines *how much* heat must be added to a body to produce a given temperature increase? Clearly, this depends both on the amount of temperature change and on the quantity of matter contained in the body. Denoting the quantity of heat added to a body by Q, we may say

$$Q = mc\,\Delta T \qquad (17\text{-}2)$$

where m is the mass of the object, ΔT the temperature change, and c a proportionality factor. In general c is different for different materials, but it depends only on the *kind* of material and not on the *quantity*. It is usually called the *specific heat* of the material; it is the quantity of heat which must be added to unit mass to increase its temperature by $1°$.

One commonly used unit of heat quantity is the *calorie*, which is defined as the amount of heat which must be added to 1 g of pure water to increase its temperature $1\,C°$, from 14.5 to $15.5°C$. The specific heats of a few familiar substances are listed in Table 17-1.

In Eq. (17-2), when the temperature of the object *decreases*, the temperature change $\Delta T = T_2 - T_1$ is a negative quantity, and therefore Q is negative. In such a case heat is taken out of the object rather than added. A positive value of Q means that heat is added, a negative value that heat is taken out.

Example

Ten copper pennies, with total mass of about 30 g, are immersed in boiling water. They are then taken out and dropped in a small glass of water at

Table 17-1 Specific Heats

Substance	c, cal/g C°
Aluminum:	
20°C	0.21
−240°C	0.009
Brass	0.094
Copper	0.093
Ethanol	0.58
Glass (typical value)	0.12
Ice	0.48
Lead:	
20°C	0.031
−259°C	0.0073
Mercury	0.033
Steel	0.11
Water	1.00
Wood (typical value)	0.5

room temperature. If the glass contains 100 g of water at 20°C before the pennies are dropped in, find the final temperature.

Solution
While the pennies are in thermal equilibrium with the boiling water, their temperature is 100°C. Let the final temperature of the cooler water and pennies be T. Then the amount of heat gained by the water is, using Eq. (17-2),

$$(100 \text{ g})(1 \text{ cal/g C°})(T - 20°)$$

and the amount of heat lost by the pennies is

$$(30 \text{ g})(0.092 \text{ cal/g C°})(100° - T)$$

Since the heat lost by the pennies is to equal the heat gained by the water, these two quantities must be equal. Setting them equal, we obtain an algebraic equation for T; the solution of this equation yields $T = 22.1$°C.

An equivalent procedure is to consider the quantity of heat added to each component of the system, as calculated from Eq. (17-2), and then to equate the *algebraic* sum of these quantities to zero, recognizing that some are positive and some negative, because some temperature changes are positive and some are negative. Using this approach to the present problem, we find ΔT (water) $= T - 20$°C, ΔT (pennies) $= T - 100$°C, and

$$(100 \text{ g})(1 \text{ cal/g C}°)(T - 20°\text{C}) + (30 \text{ g})(0.092 \text{ cal/g C}°)(T - 100°\text{C}) = 0$$

Again $T = 22.1°\text{C}$.

When a material undergoes a change of state, as when ice melts or water boils, heat is added without change of temperature. It has been found that a definite quantity of heat per unit mass is associated with such a transformation. For example, at ordinary pressures 80 cal of heat is required to change 1 g of ice at $0°\text{C}$ to liquid water at $0°\text{C}$. This heat does not change the temperature but only the *state* of the material. It is called the *latent heat of fusion*; i.e., the latent heat of fusion of water is 80 cal/g. Similarly, to change liquid water at $100°\text{C}$ and atmospheric pressure to steam at the same pressure requires 540 cal for each gram of water; the *latent heat of vaporization* of water is 540 cal/g. These transformations are particular examples of a class of phenomena called *phase transformations*, which are discussed in more detail in Chap. 19.

The calorie is not the only unit of heat in common use. Sometimes it is more convenient to use a unit called a *kilocalorie*, equal to 1000 cal. The usual unit of food energy, ordinarily called a calorie, is actually a kilocalorie. Another unit, frequently used in engineering work, is the *British thermal unit* (Btu), the amount of heat necessary to produce a temperature increase of $1 \text{ F}°$ in 1 lb of water.

Thus far we have treated specific heat as a constant for any given material. Careful measurements show, however, that it changes somewhat with temperature. For example, to heat 1 g of water from 14.5 to $15.5°\text{C}$ requires exactly 1 cal, but to heat the same amount from 90 to $91°\text{C}$ requires 1.0044 cal. Thus Eq. (17-2) is exactly correct only when $\Delta T \rightarrow 0$, and a more precise statement of this relationship is

$$\frac{dQ}{dT} = mc(T) \tag{17-3}$$

where we show explicitly that c may be a function of T. If c is really a constant independent of temperature, then Eq. (17-2) is exactly correct; otherwise, the *total* heat which must be added to mass m to produce a change from temperature T_1 to temperature T_2 is given by

$$Q = \int_{T_1}^{T_2} dQ = m \int_{T_1}^{T_2} c(T) \, dT \tag{17-4}$$

That is, the total amount of heat which must be added is the sum (obtained by integrating dQ) of the quantities of heat dQ, given by Eq. (17-3), necessary to produce all the temperature changes dT which collectively make up the interval T_1 to T_2.

Example
At very low temperature, the specific heats of some materials are found to be directly proportional to the absolute temperature. That is, $c = c_0(T/T_0)$, where c_0 is a constant having the same units as specific heat and T_0 is a constant reference temperature. Find the amount of heat which must be added to a mass m of material to change its temperature from T_0 to $2T_0$.

Solution
The amount of heat dQ necessary to produce a temperature change dT is

$$dQ = mc\,dT = mc_0 \frac{T}{T_0} dT$$

The total amount of heat which must be added for the given change in temperature is

$$Q = \int_{T_0}^{2T_0} \frac{mc_0}{T_0} T\,dT = \tfrac{3}{2} mc_0 T_0$$

Does this result have the correct units?

To be precise, we must formulate the definition of specific heat in terms of an *infinitesimal* process involving a very small quantity of heat and a very small temperature change, as in Eq. (17-3). In many practical problems, however, the variation of c with temperature may be neglected; then Eq. (17-2) can be used.

17-6 HEAT TRANSFER
We have discussed the transfer of heat from one body to another, leading to a state of thermal equilibrium, but nothing has been said about the *rate* at which heat is transferred. This question is of considerable practical importance. If we are trying to get a pint of ice cream home from a store on a hot summer day without its melting, we must delay the attainment of thermal equilibrium as long as possible, for when the ice cream reaches

thermal equilibrium with its surroundings, it is no longer frozen. This delay is often accomplished by using multiple-layer bags, which reduce the rate of heat flow from the outside air into the ice cream and thus slow the melting process.

The process of heat transfer which is easiest to describe quantitatively is that of *conduction* of heat. Whenever two parts of a body are at different temperatures, a spontaneous flow of heat takes place from the region of higher temperature to that of lower temperature. This process is known as conduction of heat. A qualitative understanding of the microscopic basis of conduction of heat can be gained by recalling that heat is microscopic mechanical energy of individual atoms and molecules. When the molecules in one region have, on the average, more kinetic energy than those in a neighboring region, they transfer energy to their neighbors in collisions with them. This transfer of energy takes place, on the average, from a region of higher temperature (and greater molecular motion) to one of lower temperature. In addition, in some materials, particularly metals, some of the electrons are more or less free to move throughout the material, rather than being bound tightly to individual atoms. These electrons can carry energy readily from one region to another. In metals this is usually the most important mechanism for heat conduction; it is the reason that materials which are good conductors of electricity are generally also good conductors of heat.

As a simple experiment to investigate the nature of heat transfer by conduction, consider a bar of uniform cross-section area A and length L, as shown in Fig. 17-7. One end of this bar is kept at temperature T_1, the

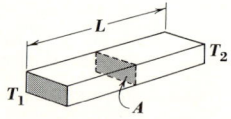

Fig. 17-7 Conduction of heat along the length of a bar of uniform cross-section area A whose ends are kept at different temperatures T_1 and T_2. The sides of the bar are thermally insulated.

other end at T_2, and the sides are insulated. We can investigate experimentally how the rate of transfer of heat from one end of the bar to the other depends on L, A, and the temperatures. For example, we can immerse one end in boiling water and the other end in ice water, measure the amount of ice which melts in a given time, and compute the amount of heat which must have been added as latent heat of fusion.

The results of these experiments are in agreement with common-sense notions. Of two different bars with different cross-section areas, the one with the larger cross section transfers more heat per unit time. The *rate* of transfer of heat, which we denote by q, is directly proportional to the cross-section area. It also depends on the temperature difference $T_2 - T_1$ between the two ends and on the length L of the bar. More specifically, the rate of heat transfer turns out to be proportional to the temperature change *per unit of length*, $(T_2 - T_1)/L$. Finally, it depends on the material of which the bar is made, so we introduce a proportionality factor k for the material.

The results of all these experiments can be embodied in a simple formula:

$$q = kA \frac{T_2 - T_1}{L} \tag{17-5}$$

The factor k is called the *thermal conductivity* of the material. The units of k in the mks system are calories per second per meter per Celsius degree. The thermal conductivities of several common materials are given in Table 17-2.

Example

The door of a freezer is 0.75 m wide, 1.00 m high, and 0.03 m thick and is insulated with plastic foam. The inside temperature is 0°C and the outside temperature 20°C. During a power failure, how much ice inside the freezer melts per hour because of the heat which enters by conduction through the door?

Solution

According to Eq. (17-5), the rate of heat flow into the refrigerator is given by

$$q = \frac{(1 \times 10^{-2} \text{ cal/m s C°})(0.75 \text{ m})(1.00 \text{ m})(20°\text{C})}{0.03 \text{ m}}$$

$$= 5 \text{ cal/s}$$

We have used the value of k for plastic foam given in Table 17-2. In 1 h, which contains 3,600 s, the total heat flow is 18,000 cal. The quantity of heat necessary to melt 1 g of ice is 80 cal, so the amount of ice melted per hour is

$$\frac{18,000 \text{ cal/h}}{80 \text{ cal/g}} = 225 \text{ g/h}$$

Table 17-2	Thermal Conductivity
Substance	k, cal/s m C°
Aluminum	48
Copper	92
Cork	0.010
Glass (typical value)	0.14
Gold	70
Ice	0.53
Lead	8
Plastic foam	0.01
Silver	101
Steel	11
Wood (typical value)	0.02

This is about 0.5 lb/h or 12 lb/d. There are also heat losses through the other sides, of course, but if the freezer is filled with frozen food and the door is not opened, the food will stay frozen for a day or two during a power failure.

As Eq. (17-5) shows, the rate of heat flow is determined by the temperature change per unit length, $(T_2 - T_1)/L$. This observation gives us a clue to generalizing this formulation to situations with a more complicated geometry. The rate of change of temperature with position is called the *temperature gradient*. Considering a point in the bar in Fig. 17-7, we denote by x its distance from one end of the bar; then T is a function of x, and the temperature gradient is simply dT/dx. Thus we can generalize Eq. (17-5) to the following:

$$q = -kA \frac{dT}{dx} \tag{17-6}$$

The minus sign denotes that the heat flow is toward the direction in which the temperature *decreases*.

How does the temperature vary from point to point along the bar in Fig. 17-7? In general, the temperature distribution depends on the end temperatures and on how long the ends have been kept at these temperatures. It is found that after a sufficiently long time the temperature at each point reaches a final value, and the temperature distribution is then independent of time. This condition is called the *steady state*; we note that it is *not* the same as thermal equilibrium, in which all parts of the bar must have the *same* temperature. We now proceed to find the steady-state temperature distribution.

The rate of heat flow into one side of any section of the bar must equal the rate of heat flow out the other side. If this were not true, the temperature of this section would increase or decrease, and we are considering the case in which this does not happen. The two heat flows are equal only if the *temperature gradient* is the same at both sides. Thus dT/dx is the same for every point in the bar; since dT/dx is constant, the temperature increases uniformly from one end of the bar to the other.

Example

Consider the radial flow of heat out of a steam pipe (Fig. 17-8). The inner surface of the cylindrical pipe is kept at the temperature T_1 of the steam

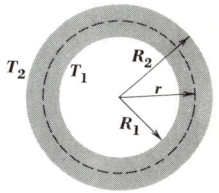

Fig. 17-8 Cross section of a steam pipe. The difference in temperature between inner and outer surfaces leads to a flow of heat out of the pipe.

in the system, and the outside temperature T_2 is that of the surrounding air. How much heat flows through the pipe per unit time from inside to outside, and how can we find the temperature at any point within the pipe?

Solution

Consider an imaginary cylindrical surface coaxial with the surfaces of the pipe, with radius r intermediate between R_1 and R_2. The temperature T is some function of r, $T(r)$, which we want to determine. The heat flow through the imaginary surface with radius r can be obtained from Eq. (17-6). Since the heat flow is radial, and since the temperature depends on r, we should replace dT/dx in this equation with dT/dr. The relevant area is the total area of this imaginary surface, which is its circumference multiplied by the length L of the pipe; that is, $A = 2\pi r L$. The rate of heat flow through the imaginary surface is then given by

$$q = -2\pi r L k \frac{dT}{dr} \tag{17-7}$$

If we assume steady-state conditions as before, this heat flow must be the same for every value of r. Otherwise, heat would have to accumulate somewhere within the pipe. Thus, q is a constant.

Rearranging Eq. (17-7), we obtain

$$\frac{dT}{dr} = -\frac{q}{2\pi Lk}\frac{1}{r} \tag{17-8}$$

This is a differential equation for the function $T(r)$. To solve it, we must find a function whose derivative with respect to r satisfies it. The most general possibility is

$$T(r) = -\frac{q}{2\pi Lk}\ln r + C \tag{17-9}$$

where C is an integration constant whose value must be determined. At $r = R_1$ the temperature is T_1, and at R_2 it is T_2. For these values, Eq. (17-9) gives

$$T_1 = -\frac{q}{2\pi Lk}\ln R_1 + C$$

$$T_2 = -\frac{q}{2\pi Lk}\ln R_2 + C$$

These two equations are sufficient to determine the constants q and C, since all other quantities are known. The results are

$$q = -\frac{2\pi kL(T_2 - T_1)}{\ln(R_2/R_1)}$$
$$C = T_1 - \frac{(T_2 - T_1)\ln R_1}{\ln(R_2/R_1)} \tag{17-10}$$

Substituting these expressions into Eq. (17-9), we find for the temperature

$$T(r) = T_1 + (T_2 - T_1)\frac{\ln(r/R_1)}{\ln(R_2/R_1)} \tag{17-11}$$

The derivative of this function is proportional to $1/r$, as required by Eq. (17-8). A graph of T as a function of r is shown in Fig. 17-9.

The first of Eqs. (17-10) shows that q is proportional to the length of pipe L and to $T_2 - T_1$, as we expect; its dependence on R_1 and R_2 is more complicated.

Fig. 17-9 Variation of temperature with radius within the wall of a pipe with inner and outer temperatures T_1 and T_2. The temperature changes more rapidly at small values of r than at larger values; the slope is proportional to $1/r$, as shown by Eq. (17-8).

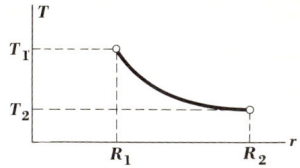

There are two important mechanisms of heat transfer in addition to conduction; one of these is *convection*. In convection, heat is transferred by mass motion of the material. In a hot-water heating system, for example, heat is transferred from the furnace to the room being heated by flow of hot water from the furnace into the radiator, then by conduction from the water to the radiator and to the room. It is much more difficult to make quantitative statements about heat transfer by convection than about conduction, and we shall not attempt to do so here. In situations in which matter is free to move, as when fluids of any kind are involved, convection may be an important effect. Within a solid, convection does not occur.

A third mechanism of heat transfer is *radiation*. In radiation, thermal energy is transferred through emission and absorption of electromagnetic radiation by matter. For this reason radiation involves a somewhat generalized concept of heat transfer. The rate of heat loss by radiation from a hot body whose absolute temperature is T is proportional to $T^4 - T_0^4$, where T_0 is the absolute temperature of the surroundings. The result of this very strong dependence on temperature is that at room temperature radiation is usually negligible as a mode of heat transfer, but at high temperatures, such as in the filament of an incandescent light bulb, it is usually far more important than either of the other mechanisms discussed. Since thorough analysis of energy transfer by electromagnetic radiation necessitates some background in electromagnetic theory and in elementary quantum theory, we shall not pursue the subject further here.

It should be emphasized that a difference in temperature between two bodies is necessary for heat transfer. When a body is in thermal equilibrium with its surroundings, the body and the surroundings have the same temperature; in this case there is no net transfer of heat between the two. Conversely, in steady-state heat flow, the system is *not* in thermal equilibrium; different points have different temperatures, but heat transfer occurs in such a way that the temperature at each point is constant in time.

17-7 HEAT AND ENERGY

We have seen that the temperature of a body is directly related to its average molecular mechanical energy and that heat may be regarded as microscopic mechanical energy in transit from one body to another. It is also evident from the discussion of this chapter, however, that many calculations concerning temperature, heat, and heat transfer can be made *without* any fundamental understanding of the nature of heat. In the early nineteenth century, when the science of thermodynamics was in its infancy, many scientists spoke of heat as "caloric fluid" and endowed it with mechanical properties such as mass and viscosity. This is now known to be an erroneous concept, but the point is that in purely thermal calculations it is not necessary to inquire into the fundamental nature of heat. Any theory which correctly describes and correlates phenomena associated with temperature and heat should be regarded as complete, and even the caloric-fluid theory satisfies this requirement.

Serious difficulties arise, however, when we try to apply these restricted theories to the relations between heat and other physical phenomena which are not directly thermal in nature. For such cases a more general theory is essential. As an example, consider the familiar observation that whenever mechanical energy disappears, as in a situation involving friction, heat is produced. The work of Joule and Mayer in the mid-nineteenth century showed that there is a definite proportionality between the two; whenever 4.186 J of mechanical energy disappears, 1 cal of thermal energy appears. This observation did not clarify completely the physical basis of heat, and a controversy concerning its fundamental nature raged for some time after this proportionality was discovered. Joule's experiments and their consequences are discussed in detail in Sec. 20-1.

It is now known with a high degree of certainty that what we call *heat* is really energy associated with motion and positions of the molecules in a material. The conversion of mechanical energy into heat or thermal energy is really a conversion of *macroscopic* energy into *microscopic* energy. These facts make possible a much more fundamental understanding of thermal phenomena than would otherwise be possible. For example, one can understand the behavior of a gas by applying the rules of mechanics (Newton's laws) to the individual molecules which constitute the gas. By computing the energies of these molecules and their dependence on temperature, one can actually *calculate*, from these same laws of mechanics, the specific heats of gases. Calculations of this sort are discussed in Chap. 18.

Problems

17-1 What is the mass of an atom of oxygen in kilograms?

17-2 By means of high-energy electron-scattering experiments it has been determined that the radius of the nucleus of a lead atom (atomic mass 206 amu) is about 7×10^{-15} m. From this, calculate the density of the nuclear matter. How does this compare with the macroscopic density of lead, which is about 11.0 g/cm^3, or 11.0×10^3 kg/m^3?

17-3 Experiments with high-energy electron scattering indicate that the radius of an atomic nucleus is given approximate by $r = 1.2 \times 10^{-15}$ m $\times A^{1/3}$, where A is the total number of particles (neutrons and protons) in the nucleus. Assuming that neutron and proton have equal masses, show that if this equation is valid, the density of nuclear matter is independent of the size of the nucleus. Also give an estimate of the size of a single nucleon (proton or neutron).

17-4 From the density of lead, make a rough estimate of the diameter of an atom of lead.

17-5 Atomic masses occurring in nature range from 1 to 238 amu; densities of elements in the solid state range from about 0.1 g/cm^3 for the lightest element to about 20 g/cm^3 for some of the heaviest. What do these facts imply regarding the dependence of the sizes of atoms on their atomic masses? That is, is diameter proportional to atomic mass, or to its square root, or to some other function? Or is it nearly independent of atomic mass?

17-6 The density of liquid oxygen at atmospheric pressure and $-184°$C is 1.14 g/cm^3. What is the approximate volume of an oxygen molecule?

17-7 The density of gaseous oxygen at atmospheric pressure and 0°C is 1.42×10^{-3} g/cm^3. From this and the data of Prob. 17-6, find the fraction of the total volume of the gas that is occupied by the molecules themselves and the fraction that is empty space.

17-8 Sodium chloride (ordinary salt) crystallizes in the cubic lattice shown in Fig. 17-2. By x-ray diffraction experiments it has been found that the lattice spacing (edge length of a cube) is 5.63×10^{-10} m. From this, compute the density of a crystal of salt.

17-9 The atomic mass of manganese is about 55 amu, and its density in the simple cubic crystal form is 7.42 g/cm^3. What is the spacing of atoms in the crystal lattice?

17-10 What Celsius temperature is equal to 86°F?

17-11 What is the temperature of absolute zero on the Fahrenheit scale?

17-12 A constant-volume gas thermometer is calibrated by observing its pressure at the freezing and boiling points of pure water at atmospheric pressure. The ratio is found to be $P_{100}/P_0 = 1.35$. Find the temperature of absolute zero on the Celsius scale according to these observations.

17-13 A thermometer commonly used for precise temperature measurements makes use of the change of electrical resistance of a circuit component with temperature. It has been found that the dependence of resistance on temperature is described fairly well by the equation $R = R_0(1 + AT)$, where T is measured in Celsius degrees by a gas thermometer and A is a constant. A certain temperature-sensing element is found to have a resistance of 100.0 ohms (Ω) at the freezing point of water and 163.5 Ω at the melting point of lead (327°C).
    a Find the constants A and R_0.
    b What is the resistance at the boiling temperature of water?

17-14 How much heat must be added to 0.25 kg of water to increase its temperature from 10 to 20°C?

17-15 How much heat must be added to 0.10 kg of ice at -10°C to convert it to liquid water at $+10$°C?

17-16 A copper container of mass 0.2 kg contains a mixture of 0.5 kg of water and 0.5 kg of ice in equilibrium. If 1.0 kg of water at the boiling temperature is added, what is the final temperature?

17-17 In Prob. 17-16, how much boiling water would be required to melt the ice and leave the mixture at the freezing temperature?

17-18 Is ethanol a more or less effective cooling liquid than water for use in automobile engines? Explain.

17-19 An unidentified piece of matter of mass 50 g is immersed in boiling water. It is then removed and dropped in a copper container of mass 100 g containing 200 g of water at 20.0°C. After this system attains thermal equilibrium, the final temperature is found to be 22.0°C. What is the specific heat of the unknown material?

17-20 An aluminum container of mass 100 g contains 200 g of ice at -20°C. Heat is added to the system at the rate of 100 cal/s. Plot a graph showing the temperature of the system as a function of time.

17-21 The specific heat of a certain substance at very low temperature is found to depend on the absolute temperature T according to the equation $c = AT^3$, where A is a constant. (Some metals exhibit this behavior at temperatures below 15 K.)
    a If the units are calories, kilograms, and kelvins, what units must the constant A have?
    b How much heat must be added to raise the temperature of a mass m of the material from T_1 to T_2?

17-22 A uniform copper rod 1.0 m in length and 5.0 cm² in cross-section area has its sides insulated to prevent heat loss. One end is immersed in boiling water, the other in a mixture of water and ice. At what rate is heat transferred along the rod?

17-23 In Prob. 17-22, how much ice melts per second?

17-24 Two metallic rods of uniform and equal cross-section areas and equal lengths with thermal conductivities k_1 and k_2 are welded together end to end. The free ends are kept at temperatures T_1 and T_2. If there is no heat transfer at the sides of the rods, and if steady-state conditions prevail, derive an expression for the temperature of the point at which the two rods are joined.

17-25 An aluminum pan containing water is placed on a stove; after the water comes to a boil, it evaporates at the rate of 0.12 kg/min. If the area of the bottom of the pan is 200 cm² and its thickness is 2.0 mm, what is the temperature of the bottom surface adjacent to the fire?

17-26 One end of a uniform rod is heated to a constant temperature of 200°C, and the other end is immersed in ice water at 0°C. The sides are not insulated, and there is heat transfer with the air in the room, which is at 20°C. Sketch a graph showing qualitatively how the temperature of the rod varies from one end to the other. On the same graph, show how the temperature distribution would look if there were no heat transfer at the sides.

17-27 A young lady's apartment is on the top floor of a house, just under the attic. The ceiling, which has an area of 50 m², is insulated with a layer of cork 5 cm thick. On a hot summer day the temperature of the attic reaches 40°C, while the temperature of the apartment is kept at 20°C by an air conditioner. At what rate does heat enter the apartment?

17-28 Air conditioners are sometimes rated in terms of the number of tons of ice which can be frozen from water at 0°C in a 24-h period. In Prob. 17-27, what rating must the air conditioner have?

17-29 A cubical oven 0.5 m on a side is insulated with a material whose thermal conductivity is comparable to that of cork. The outside temperature is 20°C. At what rate must heat be supplied to keep the inside temperature at 300°C if the insulation is 5 cm thick?

17-30 Two steam pipes are insulated with an asbestos material. One pipe is twice as large as the other in outside diameter, but both have the same quantity of insulating material per unit length. Which pipe loses heat more rapidly? Explain.

17-31 Consider a spherical shell of inner and outer radii R_1 and R_2, respectively. If the inside and outside temperatures are T_1 and T_2, respectively, and if the thermal conductivity is k, find the rate of heat flow through the shell. Also find the temperature as a function of radius for a point inside the material of the shell.

17-32 A tapered rod has a circular cross section but a diameter that increases uniformly, being twice as large at one end as at the other. The small end is kept at temperature T_1, the large end at T_2. The length of the rod is L. The taper may be assumed gradual enough for the temperature to be uniform across a plane cross section, as with a rod of uniform cross section.
    a Sketch a graph showing the temperature distribution along the rod, assuming the sides are thermally insulated.
    b Derive an equation which gives the temperature as a function of position along the rod.

17-33 In Prob. 17-32, suppose the rod tapers gradually so that the cross-section area, rather than the diameter, changes uniformly with distance along the rod; obtain the required graph and equation.

17-34 Some materials, especially metals and stone, feel cold to the touch. Others, such as fabrics, feel warm. How are these characteristics related to the specific heat and the thermal conductivity of these materials?

17-35 The electromagnet in a cyclotron has coils cooled by circulating oil. The oil in turn is cooled by water in a heat exchanger consisting of a number of tubes, immersed in water, through which the oil circulates. Suppose the tubes are made of copper with 2.0 cm inside diameter and 0.2 cm wall thickness. The temperature of the oil is $150°C$, and that of the water is $50°C$. What is the rate of heat flow through a 1-m length of tube?

17-36 In Prob. 17-35, is it realistic to assume that the oil temperature is the same everywhere within the tube? The water temperature? Discuss.

17-37 A lake has a layer of ice 0.5 m thick on its surface. The temperature of the water under the ice is $0°C$, while the temperature of the top surface in contact with air is $-10°C$. At what rate does additional ice form if the latent heat of fusion is all conducted through the ice to the air surface?

17-38 Suppose only a given amount of insulating material per unit length of steam pipe is available. Can the heat loss be reduced by using two pipes rather than one if the total cross-section area of the pipes is the same in each case? Explain.

17-39 In a bar of uniform cross section whose sides are thermally insulated, the temperature distribution was found to be as shown in Fig. P17-39. Is this

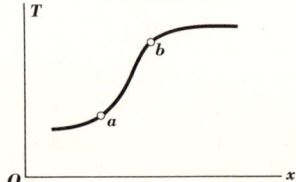

Fig. P17-39

a steady-state condition? Explain. Is the temperature increasing or decreasing with time at point *a*? At point *b*?

17-40 Two identical bodies, one with initial temperature T_1, the other T_2, are connected thermally by a thin uniform rod. The system is thermally insulated from its surroundings.
   a   Sketch graphs showing qualitatively how the temperature of each body varies with time. Explain the reasons for the shapes of the curves.
   b   Derive equations giving the temperatures of the bodies as functions of time. *Hint:* The time rate of change of temperature of each object is proportional to the rate of heat flow along the rod.

17-41 Derive Eqs. (17-10) from the equations preceding them.

Kinetic Theory of Gases | 18

With this chapter we begin a detailed study of some of the relationships between the macroscopic behavior of matter and its microscopic structure. The kinetic theory of gases is one of the simplest examples of the fundamental understanding of the bulk behavior of matter which can be gained from analysis of the motions of individual molecules.

18-1 IDEAL-GAS LAW

The ideal-gas law is a relation concerning the pressure, volume, and temperature of a gas. It is based on a variety of experiments on the behavior of gases. One involves the relation of pressure to volume; to compress a gas one must increase the pressure. Robert Boyle, a contemporary of Newton, found that when the temperature is held constant, the pressure varies *inversely* with the volume. When the pressure P doubles, the volume V decreases to one-half the original value, and so on. That is, at constant temperature,

$$PV = \text{constant}$$

A second experiment concerns the relation between pressure and temperature when the volume is constant. Pressure *increases* with temperature; the air pressure in automobile tires increases as they heat up during high-speed driving. Experiment shows that when we use the absolute temperature scale introduced in Chap. 17, pressure is directly proportional to temperature T. That is,

$$P = \text{constant} \times T$$

465 IDEAL-GAS LAW [18-1]

This principle was formulated first by Gay-Lussac and Charles.

In stating these experimental results, we have tacitly assumed that the pressure and temperature are the same everywhere in the gas. This implies that the gas is in mechanical and thermal equilibrium and that the weight of gas is small enough for any variation in pressure or density between top and bottom of the container due to gravity to be negligible. One can, of course, find exceptions; the most obvious one is the earth's atmosphere, in which the pressure decreases markedly with increasing distance above the surface of the earth, owing to the weight of the air itself. In laboratory situations the variations in pressure due to gravity are usually negligible.

The two results described above can be summarized by a single equation:

$$PV = \text{constant} \times T \tag{18-1}$$

If we double the amount of gas, keeping the temperature and pressure constant, the volume doubles; thus the constant in Eq. (18-1) is directly proportional to the quantity of gas present. It is reasonable to express it in terms of the number of moles n of gas and another constant R:

$$PV = nRT \tag{18-2}$$

We recall that, according to the definition of the mole in Sec. 17-2, n is directly proportional to the number of molecules of gas.

It might be expected that every gas should have a different value of R, characteristic of that particular gas. Instead, experiment shows that R has the same value for *all* gases. Thus the behavior of all gases is described by Eq. (18-2), with the same numerical constant R. Furthermore, the *quantity* of gas enters the relation only in the total number of molecules of gas; the *kind* of molecules does not enter the relation at all. This equation is known as the *ideal-gas equation*. The fact that the behavior of a large number of gaseous substances can be described by the same simple equation suggests strongly that there should be a corresponding simplicity in the *structure* of gases. We shall see soon that this is in fact the case.

The constant R is called the *universal gas constant*. Its value must, of course, be determined by experiment, just as the ideal-gas equation itself is a generalization from experimental observations. Under conditions referred to as *standard temperature and pressure*, that is, 0°C ($T = 273.2$ K) and normal atmospheric pressure, the volume of 1 mol of any gas is found to be about 22.4 liters (l). Using this fact in Eq. (18-2) gives

$$R = \frac{PV}{nT} = \frac{1 \text{ atm}}{1 \text{ mol}} \frac{22.4 \text{ l}}{273 \text{ K}} = 0.0821 \text{ l-atm/mol K} \tag{18-3}$$

Example

A tank whose volume is 40,000 cm³ contains oxygen under a pressure of 4 atm at a temperature of 30°C. How many moles of oxygen does the tank contain, and what is the total mass of oxygen?

Solution

First we convert the data to a consistent set of units in order to use the value of the gas constant given above. The volume is 40 l, the temperature 303.2 K. We therefore find

$$n = \frac{PV}{RT}$$

$$= \frac{(4 \text{ atm})(40 \text{ l})}{(0.0821 \text{ l-atm/mol K})(303.2 \text{ K})} = 6.42 \text{ mol}$$

The molecular mass of oxygen is 32 amu, which means that 1 mol of oxygen has a mass of 32 g. The total mass is

$$(32 \text{ g/mol})(6.42 \text{ mol}) = 206 \text{ g}$$

For chemical calculations the system of units used above is often the most convenient one. However, it is often more convenient to use a system of units in which the pressure is expressed directly in mechanical units, force per unit area. In the mks system the unit of pressure is the newton per square meter, and the unit of volume is the cubic meter. The units of the product PV are newton-meters. Thus, PV has units of work or energy.

The usual definition of *one atmosphere* is the pressure exerted by a column of mercury 76 cm high. Using this definition, together with the acceleration of gravity, the density of mercury, and the relation $P = \rho g h$, we find

$$1 \text{ atm} = (13.6 \times 10^3 \text{ kg/m}^3)(9.8 \text{ m/s}^2)(0.76 \text{ m})$$
$$= 1.013 \times 10^5 \text{ N/m}^2 \tag{18-4}$$

Using this conversion factor and the fact that $1 \text{ l} = 10^3 \text{ cm}^3 = 10^{-3} \text{ m}^3$, we can convert the gas constant R to mks units:

$$R = 8.314 \text{ J/mol K} \tag{18-5}$$

In a certain sense this is still a mixed set of units, inasmuch as a mole is defined as the quantity of material whose mass in *grams* is numerically equal to the molecular mass. One way out of this difficulty would be to define an "mks mole," for which the name *kilomole* has been proposed. This unit has not found common acceptance, however, and the units of Eq. (18-5) are most commonly used.

In thermodynamic calculations it is often convenient to express R in *thermal* units of energy, rather than mechanical units. Using the conversion 1 cal = 4.186 J, we obtain

$$R = 1.99 \text{ cal/mol K} \tag{18-6}$$

As its name implies, the ideal-gas law is an idealized relationship which provides a simple mathematical *model* to describe approximately the behavior of gases. At sufficiently low pressures and high temperatures, the behavior of many gases agrees *exactly* with the ideal-gas law, within the precision of experimental observations. At low temperatures or high pressure, deviations from ideal-gas behavior are observed. All gases at sufficiently high pressure and low temperature become liquids; their behavior is then quite different from that predicted by the ideal-gas law. The conditions under which deviations from ideal-gas behavior begin to appear vary from one gas to another. We shall see in this chapter and the next how a very simple model of the microscopic structure of a gas leads to the ideal-gas law and how deviations from the law arise from conditions for which this model is not adequate.

We have presented the ideal-gas law as an empirical law, i.e., one which has been discovered experimentally. There are many "whys" which have not been answered. Why should the ideal-gas law have the very simple form of Eq. (18-2)? Why should the gas constant be the same for all gases? These questions and others are answered by the kinetic theory of gases, which we discuss next.

18-2 MICROSCOPIC MODEL OF AN IDEAL GAS

The central idea in the kinetic theory of gases is to apply the laws of mechanics to the motion of individual molecules of the gas and then to use the *average* behavior of these molecules to predict the macroscopic properties of the gas, such as pressure and temperature. We now set down several assumptions concerning the microscopic structure of a gas. These constitute the basis of a *microscopic model* which we can use to predict macroscopic results. Any prediction which agrees with experimental observation tends to justify these assumptions; any disagreement suggests a need for modification of the model.

First, we assume that a gas consists of a very large number of molecules, which may be treated as particles. If the gas is a single element or a single compound, all the particles are identical. Later we shall see how to extend the theory to mixtures of gases.

We assume that the volume of the molecules themselves is negligibly small compared to the total volume occupied by the gas. The fact that a gas may change its volume over a very wide range suggests that, on a microscopic scale, it may consist largely of empty space, the molecules themselves occupying relatively little of the total volume.

Next, we assume that the molecules are constantly in random motion. They collide with each other and with the walls of the container; as a result, the average density of molecules is the same everywhere in the container, and a molecule is as likely to be moving in any one direction as in any other.

Finally, we assume that the molecules undergo perfectly *elastic* collisions with each other and with the wall and that these collisions occur in accordance with Newton's laws. We also assume that during the motion between collisions no appreciable forces act on the molecules and that the time duration of a collision is negligible compared with the average time between collisions.

These assumptions constitute the simplest possible *kinetic-molecular* model of a gas. It may help, in visualizing this model, to think of a large number of very hard steel balls bouncing around inside a very hard steel box without losing any energy during collisions. The total volume occupied by the steel balls is very small compared to the total volume of the box. The actual molecules in the real gas are, of course, much too small to be seen with even the most powerful microscope; they are, in fact, thousands of times smaller than the wavelengths of visible light.

How should we proceed to get information out of a microscopic model such as this? The pressure a gas exerts on a wall of its container is a result of molecules colliding with the wall and exerting forces on it. We do not observe forces due to individual collisions but rather the *average* force due to many collisions. Therefore to calculate pressure we consider not the force exerted by any individual molecule but the average force resulting from a large number of collisions. Similarly, we can compute the specific heat of a gas by considering the variation of mechanical energy of the molecules with temperature. Again, we are interested not in the energy of any individual molecule but in the average energy of all the molecules.

This proposal, to obtain information about a system containing a very large number of particles by considering the *average* behavior of these particles, is a concept we have not used previously. In formulating principles of mechanics, we consider the motion of one particle, or a few particles, or a large collection of particles tied together in a very particular way, as in a rigid body. In studying heat from a macroscopic viewpoint, on the other hand, we entirely abandon the individual-particle approach and concentrate on the mass behavior of an aggregation of material. It is extremely important to understand that only by considering the average behavior of a large number of molecules can we understand the relations between the principles of mechanics, formulated in terms of individual particles, and the principles of thermodynamics, involving the macroscopic concepts of heat and temperature. *Temperature, heat, and pressure are intrinsically macroscopic concepts.* They have no meaning for an individual molecule, only for a large collection of molecules.

18-3 PRESSURE EXERTED BY AN IDEAL GAS

We now use the assumptions of the kinetic-molecular model of an ideal gas to calculate the pressure which such a gas exerts on a wall of its container. First, consider what happens when one molecule collides elastically with a wall. Suppose that the orientation of the wall is perpendicular to the x axis, as shown in Fig. 18-1. If the force exerted on the molecule during the

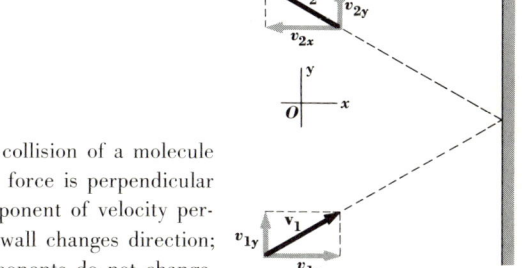

Fig. 18-1 Elastic collision of a molecule with a wall. If the force is perpendicular to the wall, the component of velocity perpendicular to the wall changes direction; the other components do not change.

collision is perpendicular to the wall, the component of velocity v_x perpendicular to the wall is reversed in direction while the components v_y and v_z are unchanged.

As a result of this collision, the x component of *momentum* of the molecule changes by an amount

$$\Delta p_x = p_{2x} - p_{1x} = -2mv_x \tag{18-7}$$

where m is the mass of one molecule. If every molecule had the same magnitude for its velocity component v_x, and if we could find the number of collisions occurring on a given area per unit time, we could calculate the time rate of change of momentum by multiplying the momentum change for each collision, given by Eq. (18-7), by the number of collisions per unit time. This would then be equal to the *force* which the area exerts on the colliding molecules; the force *on* this area would be just the negative of this.

The assumption that all the x components of velocity are equal in magnitude is of course not correct, but it is helpful to introduce it temporarily to clarify the argument. Supposing that half the molecules have x components of velocity v_x and the other half $-v_x$; how many molecules strike an area of wall A per unit time? In a time interval Δt, a particle moves in the x direction a distance $v_x \Delta t$. The particles which strike the wall during the time interval Δt are those which were at most a distance $v_x \Delta t$ from the wall at the beginning of the interval. In other words, at the beginning of Δt these particles were within a cylinder whose base area is A and whose height is $v_x \Delta t$. The volume of this cylinder is simply $A v_x \Delta t$.

If the total number of particles is N and the total volume V, the number of particles per unit volume is N/V, and the number of particles in this cylinder is

$$A v_x \Delta t \frac{N}{V} \tag{18-8}$$

Because of the randomness of the direction of velocity, half these particles are moving *toward* the wall and the other half *away from* the wall; only those moving toward the wall will strike it during the interval Δt, so the number of collisions during Δt is just one-half of Eq. (18-8). The number of collisions per unit time, obtained by dividing by Δt, is

$$\tfrac{1}{2} A v_x \frac{N}{V}$$

The total rate of change of momentum is the number of collisions per unit time multiplied by the momentum change in each collision:

$$\frac{dp_x}{dt} = \tfrac{1}{2} A v_x \frac{N}{V} (-2 m v_x) = -A m v_x^2 \frac{N}{V} = F_x \tag{18-9}$$

This expression therefore represents the average total force which the wall exerts on all the molecules which strike it. The force exerted *on* the wall is the negative of this; the pressure, which is the force per unit area, is

$$P = \frac{F_x}{A} = \frac{N m v_x^2}{V} \tag{18-10}$$

What about the assumption that v_x has the same magnitude for all molecules? This is clearly not correct. But we can separate the molecules into groups such that in each group every molecule has very nearly the same v_x. Then we calculate the rate of change of momentum and the pressure

PRESSURE EXERTED BY AN IDEAL GAS [18-3]

due to each *group* of molecules. The total pressure exerted on the wall is the sum of all the pressures due to these groups. Since the pressure due to each group depends only on the value v_x^2 for that group, the total pressure is obtained simply by using the *average* value of v_x^2 for all the molecules. We denote this average by $(v_x^2)_{av}$. Thus the total pressure is

$$P = \frac{Nm(v_x^2)_{av}}{V} \tag{18-11}$$

We can relate the average value of v_x^2 to the average of the squares of the *speeds* of the molecules. We call this latter quantity the *mean-square speed* and denote it by $(v^2)_{av}$. To establish the relation between these quantities, we first note that the speed v of a molecule is related to the components of its velocity by $v^2 = v_x^2 + v_y^2 + v_z^2$. This is true for every molecule, so it must be true for the averages for all the molecules. That is,

$$(v^2)_{av} = (v_x^2)_{av} + (v_y^2)_{av} + (v_z^2)_{av} \tag{18-12}$$

But now we recall our assumption that the direction of motion of the molecules is random; all directions of motion are equally likely. For this reason, it must be true that $(v_x^2)_{av} = (v_y^2)_{av} = (v_z^2)_{av}$. Thus,

$$(v_x^2)_{av} = \tfrac{1}{3}(v^2)_{av}$$

Using this result in Eq. (18-11), we find

$$P = \frac{1}{3}\frac{Nm(v^2)_{av}}{V} \tag{18-13}$$

Next, in an effort to make Eq. (18-13) resemble the ideal-gas equation more closely, we express the total number of molecules N in terms of the number of moles n and Avogadro's number N_0, $N = nN_0$, and multiply both sides by V. The result is

$$PV = n\tfrac{1}{3}N_0 m(v^2)_{av} \tag{18-14}$$

Comparing this with the empirical ideal-gas equation (18-2), we see that the two can be made to agree by the additional assumption that the right-hand side of Eq. (18-14) is proportional to the absolute temperature, i.e., that

$$\tfrac{1}{3}N_0 m(v^2)_{av} = RT \tag{18-15}$$

This equation is more illuminating when rewritten:

$$\tfrac{1}{2}m(v^2)_{av} = \frac{3}{2}\frac{R}{N_0}T \tag{18-16}$$

472 KINETIC THEORY OF GASES

Now $\frac{1}{2}m(v^2)_{av}$ is the average translational kinetic energy of a molecule. That is, our kinetic-theory pressure calculation agrees with the empirical ideal-gas law if and only if the translational kinetic energy is proportional to the absolute temperature. This, then, is the experimental justification for the assumption expressed in Eq. (18-15).

The constant R/N_0 appears so frequently in kinetic-theory calculations that it is usually abbreviated $R/N_0 = k$, known as *Boltzmann's constant*. Its value is

$$k = \frac{R}{N_0} = 1.38 \times 10^{-23} \text{ J/K} \tag{18-17}$$

Boltzmann's constant may be regarded as a new kind of gas constant in which the quantity of gas referred to is one molecule rather than 1 mol. With this constant, the ideal-gas equation can be expressed as

$$PV = NkT$$

where N is again the total number of molecules. Alternatively, we may say

$$P = \frac{N}{V} kT$$

which shows that the pressure P is directly proportional to the number of molecules *per unit volume* N/V and to the absolute temperature T. In terms of the constant k, the average translational kinetic energy of a molecule is simply $\frac{3}{2}kT$.

Equation (18-15) can be put in another useful form by noting that $N_0 m$ is the mass of 1 mol of molecules and is thus equal to the molecular mass: $N_0 m = M$. Thus we obtain

$$\tfrac{1}{3} M(v^2)_{av} = RT \tag{18-18}$$

An estimate of the typical speed of a molecule can be obtained from Eq. (18-18). The square root of $(v^2)_{av}$ is not an ordinary average, as would be obtained by adding all the speeds and dividing by their number, but it is the square root of the average of the squares of the speeds. This average is often called the *root-mean-square speed*, abbreviated v_{rms}. That is, if there are N molecules, and if the speed of a typical one is v_i, then, by definition,

$$v_{\text{rms}} = \sqrt{\frac{1}{N} \sum_{i=1}^{N} v_i^2} \tag{18-19}$$

From Eq. (18-18), we find that the root-mean-square molecular speed of an ideal gas is given by

$$v_{\text{rms}} = \sqrt{\frac{3RT}{M}} = \sqrt{\frac{3kT}{m}} \qquad (18\text{-}20)$$

The second form of Eq. (18-20) is obtained from the first by dividing numerator and denominator by N_0; it can also be obtained directly from Eq. (18-16).

Example
Suppose the speeds of five molecules of a gas are 1, 1, 2, 3, and 4 m/s. Find the average speed and the root-mean-square speed.

Solution
The average speed v_{av} is obtained by adding all the speeds and dividing by their number:

$$v_{\text{av}} = \frac{1+1+2+3+4}{5} = 2.20 \text{ m/s}$$

To find the root-mean-square speed v_{rms}, we square each speed, add the squares, divide by the number of molecules, and finally take the square root:

$$v_{\text{rms}} = \sqrt{\frac{1^2 + 1^2 + 2^2 + 3^2 + 4^2}{5}} = 2.49 \text{ m/s}$$

Note that v_{av} and v_{rms} are not in general equal; there is no reason for them to be equal. Later, after some discussion of the *distribution* of molecular velocities, we shall find a relationship between these two quantities when the averages are taken over a large number of molecules. The introduction of v_{rms} may seem somewhat arbitrary; its usefulness arises from its close relation to the ideal-gas law and average molecular kinetic energy.

Example
Find v_{rms} for nitrogen molecules at room temperature.

Solution
We may take room temperature to be $20°C = 293$ K. The molecular mass of nitrogen is 28 g/mol = 28×10^{-3} kg/mol. Thus, from Eq. (18-20),

$$v_{\text{rms}} = \sqrt{\frac{3(8.314 \text{ J/mol K})(293 \text{ K})}{28 \times 10^{-3} \text{ kg/mol}}} = 511 \text{ m/s}$$

This speed is about half again as great as the speed of sound in air at this temperature. One can verify that the units of this result are correct by recalling that $1 \text{ J} = 1 \text{ kg·m}^2/\text{s}^2$.

18-4 SPECIFIC HEAT OF AN IDEAL GAS

The kinetic-molecular theory of gases leads naturally to the assumption that the average kinetic energy is proportional to the absolute temperature. Thus, adding heat to a gas increases the kinetic energies of the molecules; this corroborates the view of Chap. 17 that heat is microscopic mechanical energy. We can now use this view to calculate, from the kinetic-molecular model, a prediction of the *specific heats* of gases. For the present we consider only the case in which the volume of gas is constant; this avoids any possible complication which might result from performance of mechanical work by the gas during a volume change.

According to Eqs. (18-16) and (18-17), the average translational kinetic energy is $\frac{3}{2}kT$ per molecule, or $\frac{3}{2}RT$ per mole. Since 1 mol contains a mass M equal to the molecular mass, the translational kinetic energy *per unit mass* is $3RT/2M$. To increase the temperature of a unit mass of gas by an amount ΔT, we must add an amount of energy $\frac{3}{2}(R/M)\Delta T$, and therefore we predict that the specific heat is

$$c = \frac{3}{2}\frac{R}{M} \tag{18-21}$$

This result is simpler when expressed as the *molar specific heat C*, the specific heat *per mole* rather than per unit mass. Thus the molar specific heat is expected to be

$$C = \tfrac{3}{2}R \tag{18-22}$$

With R in thermal units [Eq. (18-6)], we find $C = 2.98$ cal/mol K. This result, rather surprising in its simplicity, says that the molar specific heats of *all gases* ought to have the same value, 2.98 cal/mol K.

Table 18-1 gives some experimentally observed specific heats. It is seen from this table that the theoretical result agrees quite well with experimental values for some gases, but for others the agreement is not so good. The inert gases, whose molecules are *monatomic*, agree well with the kinetic-theory prediction. Most of the gases whose molecules are *diatomic*, on the other

Table 18-1 Specific Heats of Gases at $T = 300$ K and Constant Volume

Gas	Specific heat cal/g K	Molar specific heat, cal/mol K
He	0.755	3.02
Ne	0.153	3.1
Ar	0.075	3.0
Kr	0.036	2.9
Xe	0.0228	3.0
H_2	2.4	4.8
O_2	0.155	4.99
N_2	0.176	4.93
NO	0.166	4.97
CO	0.179	5.02
HCl	0.137	5.02
Cl_2	0.0848	6.01
CO_2	0.153	6.86
SO_2	0.117	7.5
NO_2	0.15	6.9
H_2O	0.471	6.52
$(C_2H_5)_2O$	0.416	30.8

hand, have molar specific heats of about 5 cal/mol K; those whose molecules are more complex have still larger specific heats. Evidently the actual molar specific heat depends on the *structure* of the molecules.

The fact that some molar specific heats are larger than expected must mean that there is additional energy in the system, which should be included in our calculation. One possibility is that the molecules perform *rotational* as well as *translational* motion. If so, the additional kinetic energy associated with rotational motion about the center of mass must be included in the specific-heat calculations. For example, suppose we consider a diatomic molecule not as a point mass but as a small dumbbell, with two point masses separated by a fixed distance. This model corresponds to the fact that practically all the mass of each atom is concentrated in a nucleus whose diameter is of the order of 10^{-13} cm, while in a diatomic molecule the two nuclei are separated by a much larger distance, of the order of 10^{-8} cm. This dumbbell may rotate about two mutually perpendicular axes through the center of mass, each perpendicular to the dumbbell's axis, as shown in Fig. 18-2. The moment of inertia of the dumbbell about its own axis is zero;

Fig. 18-2 Simple model of the structure of a diatomic molecule, consisting of two point masses separated by a constant distance. Two perpendicular axes of rotation are shown; the rotational motion of the molecule may have components of angular velocity in the directions of these axes.

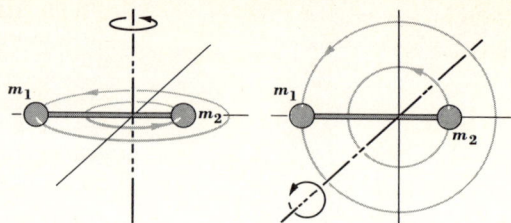

rotations about this axis are therefore irrelevant.

How much energy, on the average, is associated with this rotational motion? We have no way of knowing this in advance, but we can make a reasonable guess and see whether the results agree with observation. The average *translational* kinetic energy per molecule is $\frac{3}{2}kT$; thus an energy $\frac{1}{2}kT$ is associated, on the average, with each of the three components of velocity. We might reasonably guess that there should be an *additional* average kinetic energy $\frac{1}{2}kT$ for each of the two kinds of rotation discussed in the previous paragraph. If so, the average *total* kinetic energy (including both translational and rotational motion) is $\frac{5}{2}kT$ per molecule, or $\frac{5}{2}RT$ per mole, and the molar specific heat for a diatomic gas is

$$C = \tfrac{5}{2}R = 4.97 \text{ cal/mol K} \tag{18-23}$$

Comparing this prediction with the values in Table 18-1, we see that it agrees reasonably well with the specific heats of most *diatomic* gases; the only conspicuous exception is Cl_2. This supports our assumption that an average kinetic energy $\frac{1}{2}kT$ is associated with each relevant component of angular velocity.

The idea that each distinct kind of motion should have associated with it, on the average, a kinetic energy of $\frac{1}{2}kT$ per molecule is a very important principle called the *principle of equipartition of energy*. It is customary to call each distinct kind of motion a *degree of freedom*. A monatomic molecule has three degrees of freedom, corresponding to the three components of velocity. A diatomic molecule has five degrees of freedom, corresponding to three components of velocity and two of angular velocity. For a triatomic molecule, if the atoms are not all in a straight line, there are moments of inertia and rotational energies for all three mutually perpendicular axes. In such cases, there are six degrees of freedom, and we expect the molar specific heat to be $C = 3R = 5.98$ cal/mol K. We see in Table 18-1 that the observed specific heats of several triatomic gases are somewhat larger than this value.

Further refinements can be made in specific-heat calculations. The next step is to observe that the molecules may not be completely rigid but may have internal *vibrational* motion as well as translation and rotation. For example, a more refined model of a diatomic model would regard it not as a rigid dumbbell but as two point masses connected by a spring. Then there would be additional kinetic and potential energies associated with vibrational motion of the type discussed in Sec. 10-5. This effect accounts for the specific heat of Cl_2 shown in Table 18-1.

But if vibrational motion is important, why do not *all* diatomic gases show this effect? There is no simple answer to this question. Briefly, the reason is that the principle of equipartition of energy has limitations. Analysis of molecular motions using the principles of quantum mechanics shows that the energy associated with rotational and vibrational motion always increases in discrete steps. At sufficiently low temperatures, very few molecules have enough energy for the first step, and there is generally *less* energy associated with the vibrational motion than would be predicted by the equipartition principle. For most of the diatomic gases in Table 18-1, vibrational motion makes no contribution to the specific heat at ordinary temperatures, the notable exception in the table being Cl_2. At sufficiently high temperatures the specific heats of the other diatomic gases increase to about 6 cal/mol K.

Despite these shortcomings, the simple kinetic-molecular model is quite successful in predicting the specific heats of gases whose molecules are monatomic or diatomic. For more complex molecules, detailed considerations of the vibrational energies are necessary. In general, there may be several degrees of freedom associated with vibrational motion, and for each there is not only kinetic energy but also potential energy. As we shall see in Sec. 19-5, the average potential energy is equal to the average kinetic energy when the motion is simple harmonic. Thus in the temperature ranges where the equipartition principle holds, each vibrational degree of freedom contributes an energy kT per molecule, rather than $\frac{1}{2}kT$, and the corresponding contribution to the molar specific heat is R rather than $R/2$. In many cases quantum effects are significant, and departures from the equipartition principle must be considered carefully.

18-5 DISTRIBUTION OF MOLECULAR SPEEDS

Our calculations of the pressure and specific heat of an ideal gas depend only on average properties of the gas, especially the average of the squares of the molecular speeds. We now turn to some phenomena which require a more detailed discussion of molecular motion.

It is possible to measure speeds of *individual molecules* in a gas. A

478 KINETIC THEORY OF GASES

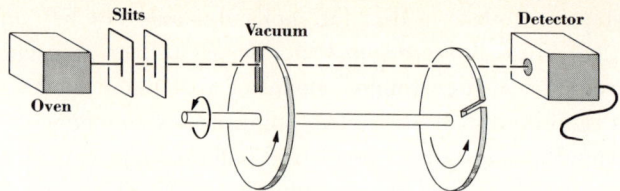

Fig. 18-3 Apparatus for producing a molecular beam and observing the distribution of molecular speeds in the beam.

simple scheme for making such measurements is shown in Fig. 18-3. In this experiment a substance is vaporized in an oven, and molecules of vapor are permitted to escape through a hole in the oven wall into an evacuated region. A series of slits is constructed to block all molecules except those in one certain direction, so as to form a well-defined beam; this beam is aimed at a series of rotating disks as shown. The entire apparatus is enclosed in a container, which is evacuated so that the molecules can move freely without colliding with air molecules. When the disks are rotating, a molecule passing through the slit in the first disk will arrive at the second at just the right time to pass through its slit only if it has a certain speed. Other molecules with different speeds will strike the second disk. This apparatus can therefore be used as a speed selector which blocks all molecules except those in a narrow range of speeds. One can then measure *how many* molecules have speeds in various ranges.

Such experiments show that the molecules do have many different speeds. By varying the angular velocity of the disk, one can observe the number of molecules falling into various speed intervals. The result of such a series of measurements is called the *speed distribution* of the molecules. It is convenient to represent such a distribution by specifying the fraction of all the molecules observed having velocities in each interval. For example, if the intervals are 0 to 100 m/s, 100 to 200, 200 to 300, and so on, the results of an experiment with 1,000 molecules might be as shown in Table 18-2 and Fig. 18-4. The height of each bar represents the fraction of all the molecules observed which have velocities in the corresponding intervals.

If smaller velocity intervals are used, say 10 m/s, the bars of the histogram are thinner and closer together, as in Fig. 18-5. Finally, in the limit of infinitesimally small intervals, we obtain a smooth curve in Fig. 18-6.

The distribution of molecular speeds in the beam is not necessarily the same as that in the oven, since the faster molecules are more likely to escape from the oven than the slower ones. It is possible to derive a simple relation between the two distributions, the details of which need not concern us here.

479 DISTRIBUTION OF MOLECULAR SPEEDS [18-5]

Table 18-2 Distribution of Molecular Speeds in a Molecular-Beam Experiment for 1,000 Molecules

Speed interval, m/s	Number in interval	Fraction of total
0–100	5	0.005
100–200	34	0.034
200–300	92	0.092
300–400	138	0.138
400–500	161	0.161
500–600	167	0.167
600–700	144	0.144
700–800	104	0.104
800–900	69	0.069
900–1,000	46	0.046
1,000–1,100	23	0.023
1,100–1,200	11	0.011
1,200–1,300	5	0.005
1,300–1,400	1	0.001

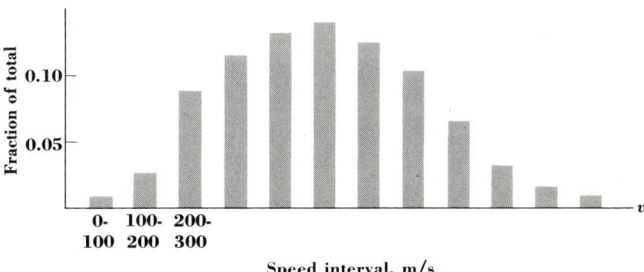

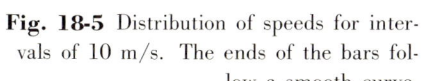

Fig. 18-4 Data of Table 18-2. The height of each bar represents the fraction of all the molecules studied which fell in each corresponding speed interval.

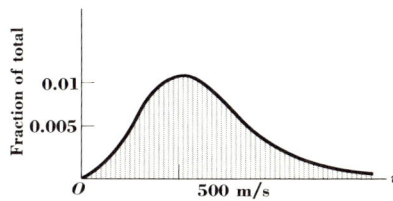

Fig. 18-5 Distribution of speeds for intervals of 10 m/s. The ends of the bars follow a smooth curve.

480 KINETIC THEORY OF GASES

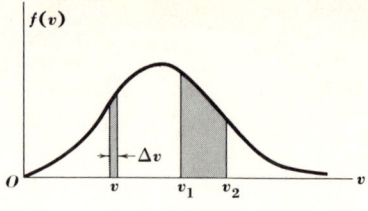

Fig. 18-6 Continuous speed-distribution curve. This curve can be obtained in principle by taking smaller and smaller speed intervals in diagrams such as Figs. 18-4 and 18-5. The fraction of all molecules having speeds between v_1 and v_2 is represented graphically by the area under the curve between these limits.

In the following discussion we assume that this correction has been made and that we are dealing with the distribution of molecular speeds for the gas in the oven.

The interpretation of the speed-distribution curve of Fig. 18-6 deserves more discussion. We may think of this curve as represented by a function $f(v)$ which gives the height of the curve for any value of v. Then we consider a small interval of width Δv between a certain speed v and a neighboring value $v + \Delta v$. In the limit as $\Delta v \to 0$, the *fraction* of all the molecules whose speeds lie in this interval is given by $f(v)\,\Delta v$. That is, if we observe N molecules in all, and if we denote by ΔN the number whose speeds are in the range Δv, then

$$\frac{\Delta N}{N} = f(v)\,\Delta v \tag{18-24}$$

Graphically, this is represented by the area of width Δv and height $f(v)$ in Fig. 18-6.

By adding up the areas of a series of these small strips, we see that the fraction of molecules having speeds between any two values v_1 and v_2 is given by

$$\frac{N_{v_1 v_2}}{N} = \int_{v_1}^{v_2} f(v)\,dv \tag{18-25}$$

which represents the area under the curve between the limits v_1 and v_2, as shown in the figure. *Every* molecule must have a speed somewhere between zero and ∞, and so the fraction of molecules having speeds which fall in this range must be unity. That is,

$$\int_0^\infty f(v)\,dv = 1 \tag{18-26}$$

The total area under the speed-distribution curve must be unity.

The speed-distribution function may also be interpreted in terms of *probability*. If we choose a molecule at random from the gas, the *probability*

Table 18-3 Calculation of v_{av} for Data of Table 18-2

Speed interval, m/s	Center v_i	Number in interval N_i	Speed × number $N_i v_i$	Fraction of total N_i/N	Speed × fraction $N_i v_i/N$
0–100	50	5	250	0.005	0.250
100–200	150	34	5,100	0.034	5.10
200–300	250	92	23,000	0.092	23.00
300–400	350	138	48,300	0.138	48.30
400–500	450	161	72,450	0.161	72.45
500–600	550	167	91,850	0.167	91.85
600–700	650	144	93,600	0.144	93.60
700–800	750	104	78,000	0.104	78.00
800–900	850	69	58,650	0.069	58.65
900–1,000	950	46	43,700	0.046	43.70
1,000–1,100	1,050	23	24,150	0.023	24.15
1,100–1,200	1,150	11	12,650	0.011	12.65
1,200–1,300	1,250	5	6,250	0.005	6.25
1,300–1,400	1,350	1	1,350	0.001	1.35
		$N = \Sigma N_i = 1,000$	$\Sigma N_i v_i = 559,300$		$v_{av} = 559.30$ m/s

$$v_{av} = \frac{1}{N}\Sigma N_i v_i = 559.300 \text{ m/s}$$

that its speed lies in a small interval Δv is given by $f(v)\,\Delta v$. The *most probable speed* is that for which the curve has its maximum value. For a speed distribution similar to that shown in Fig. 18-6, very small speeds and very large speeds are not very probable, while between these extremes there is a region of more probable speeds.

If we know the distribution of molecular speeds, we can compute the average speed, the root-mean-square speed, and similar quantities. To illustrate, we return to the example in Table 18-2. We can make a rough estimate of the average speed of the molecules described by this distribution by assuming that all the molecules in a given interval have a velocity corresponding to the center of this interval. Clearly this is not exactly correct, but we can make an approximation of the average speed on this basis. Thus we assume that 5 molecules have a velocity of 50 m/s, 34 a velocity of 150 m/s, and so forth. To compute the average, we multiply each velocity by the number of molecules having that velocity, add all these products, and divide by the total number of molecules. This calculation is arranged in tabular form in Table 18-3.

The calculation can also be performed using the fraction of the total for each interval. This corresponds to dividing by the total number N at the beginning instead of at the end. Each fraction of the total is multiplied by the corresponding velocity, and the results are added. This calculation is also shown in Table 18-3. The result is of course the same as that obtained by the other method, but this procedure has the advantage that the *total number* of molecules does not enter the calculation at all. To summarize this method: We multiply the fraction in each speed interval by the corresponding speed and add the results.

What is the corresponding operation for the continuous distribution? The fraction of molecules in the interval Δv is approximately $f(v)\,\Delta v$, and the corresponding speed is v. To compute the average, we add the products $vf(v)\,\Delta v$ for all the velocity intervals and take the limit as $v \to 0$. The result is simply the integral of $vf(v)$:

$$v_{av} = \int_0^\infty vf(v)\,dv \tag{18-27}$$

We use exactly the same reasoning to find $(v^2)_{av}$, which is given by

$$(v^2)_{av} = \int_0^\infty v^2 f(v)\,dv \tag{18-28}$$

To obtain v_{rms}, we simply take the square root of this expression.

In the next section we use these relations with a particular speed-distribution function appropriate for an ideal gas.

18-6 MAXWELL-BOLTZMANN DISTRIBUTION

During the mid-nineteenth century, Maxwell and Boltzmann derived, from theoretical considerations, a function describing the distribution of molecular speeds in an ideal gas. Their program was essentially as follows: For a gas in which the number of molecules, volume, and total energy are specified, the energy can be divided among the molecules in many different ways; i.e., there are many possible combinations of energies of individual molecules which add up to the given total energy. It is plausible to assume that all these different microscopic distributions of energy are *equally likely*. This assumption is sufficient to determine the general form of the distribution function.

The details of the derivation of this function need not concern us; the result is

$$f(v) = av^2 e^{-v^2/b^2} \tag{18-29}$$

where a and b are constants. This function has been found to agree with experimentally observed distributions obtained from molecular-beam and other experiments. For our purposes this is sufficient justification for the use of the function and sufficient support for the validity of the original assumption of equal likelihood of energy distributions.

As a first step toward understanding Eq. (18-29), we observe that it has the general shape of Fig. 18-7. For very small v, the function approaches

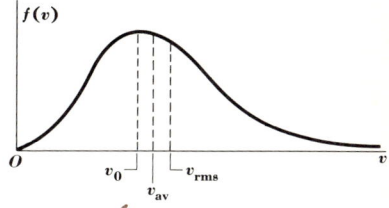

Fig. 18-7 Maxwell-Boltzmann distribution. The relative positions of the most probable speed v_0, the average speed v_{av}, and the root-mean-square speed v_{rms} are shown. What is the maximum height of the curve?

zero because of the factor v^2. For very large v, the quantity v^2 becomes very large, but the exponential function becomes small so fast that the product of the two approaches zero. According to Eq. (18-26), the total area under the curve must be unity; since the *height* of the curve at any point is proportional to a, Eq. (18-26) is satisfied only for one particular value of a. To find that value, we insert Eq. (18-29) into Eq. (18-26):

$$\int_0^\infty av^2 e^{-v^2/b^2} = 1 \tag{18-30}$$

The evaluation of this integral and others we shall encounter is a mathematical detail which need not be discussed here. For convenience, we list in Table 18-4 several integrals of this general form. Referring to the third integral in this table, we see that Eq. (18-30) can be rewritten

$$\frac{a \sqrt{\pi} b^3}{4} = 1$$

Table 18-4 Integrals Containing the Maxwell-Boltzmann Function

$$\int_0^\infty e^{-v^2/b^2} \, dv = \frac{\sqrt{\pi}}{2} b \qquad \int_0^\infty v^3 e^{-v^2/b^2} \, dv = \tfrac{1}{2} b^4$$

$$\int_0^\infty v e^{-v^2/b^2} \, dv = \tfrac{1}{2} b^2 \qquad \int_0^\infty v^4 e^{-v^2/b^2} \, dv = \frac{3\sqrt{\pi}}{8} b^5$$

$$\int_0^\infty v^2 e^{-v^2/b^2} \, dv = \frac{\sqrt{\pi}}{4} b^3$$

Thus Eq. (18-30) is satisfied only if $a = 4/(\sqrt{\pi}b^3)$. Once b is known, a is determined. Thus the Maxwell-Boltzmann distribution function can be written

$$f(v) = \frac{4v^2}{\sqrt{\pi}b^3} e^{-v^2/b^2} \qquad (18\text{-}31)$$

Let us next find the most probable value of v [the value for which $f(v)$ has its maximum value], which we denote by v_0. To find this value, we differentiate the function with respect to v and set the derivative equal to zero. The details of this calculation are left as a problem; the result is simply that the maximum occurs when $v_0 = b$. Thus the constant b has a very simple physical significance; it is the *most probable speed*.

We now compute the average and root-mean-square speeds. Using the Maxwell-Boltzmann function in Eq. (18-27) gives

$$v_{\text{av}} = \frac{4}{\sqrt{\pi}b^3} \int_0^\infty v^3 e^{-v^2/b^2}\, dv$$

Referring to Table 18-4 to evaluate the integral, we obtain

$$v_{\text{av}} = \frac{2b}{\sqrt{\pi}} = \frac{2v_0}{\sqrt{\pi}} \simeq 1.13 v_0 \qquad (18\text{-}32)$$

The average speed is just slightly larger than the most probable speed v_0. Also, from Eqs. (18-28) and (18-31),

$$(v^2)_{\text{av}} = \tfrac{3}{2} b^2 = \tfrac{3}{2} v_0^2$$

and

$$v_{\text{rms}} = \sqrt{\tfrac{3}{2}}\, b = \sqrt{\tfrac{3}{2}}\, v_0 \simeq 1.22 v_0 \qquad (18\text{-}33)$$

The relative magnitudes of these quantities are shown in Fig. 18-7.

Thus v_0, v_{av}, and v_{rms} are proportional. Furthermore, we can now relate these to the *temperature* of the gas. We know that our kinetic-molecular theory fits the observed behavior of ideal gases only if v_{rms} is related to T by Eq. (18-20). Thus, if the Maxwell-Boltzmann distribution is to describe an ideal gas, it must be true that

$$v_{\text{rms}} = \sqrt{\tfrac{3}{2}}\, b = \sqrt{\frac{3kT}{m}} = \sqrt{\frac{3RT}{M}} \qquad (18\text{-}34)$$

The constant b is thus directly related to observable properties of the gas: its temperature and molecular mass.

We can now see how the distribution of molecular speeds in a gas changes when the temperature changes. The general shape of the distribution curve remains the same, but the scales are changed. In Fig. 18-8 we show the

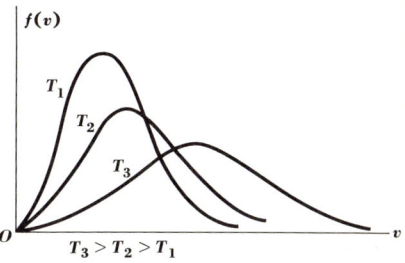

Fig. 18-8 Maxwell-Boltzmann distribution curves for various temperatures. As the temperature increases, the curve becomes flatter, and its maximum shifts to higher temperature.

Maxwell-Boltzmann distribution for three different temperatures. As temperature increases, the curve becomes flattened and extends farther in the direction of large speeds, as we should expect.

In the above discussion we have assumed that the gas is in an *equilibrium* state. This means two things. First, we assume that all parts of the gas have the same temperature, so that it is in *thermal* equilibrium. Second, we assume that all parts of the gas have the same pressure, so that it is in *mechanical* equilibrium. If these conditions are not satisfied, the Maxwell-Boltzmann distribution does not provide an accurate description of the molecular speeds, since different parts of the gas have different temperatures. The gas as a whole may then be thought of as having a mixture of several Maxwell-Boltzmann distributions corresponding to several different temperatures.

Describing a non-equilibrium state of a gas is more difficult than describing an equilibrium state, which requires only giving the temperature, pressure, and volume. For this reason, the analysis of non-equilibrium processes is more involved than that of equilibrium processes.

18-7 MOLECULAR COLLISIONS AND TRANSPORT PHENOMENA

Molecular collisions are responsible for the randomness of motion and the attainment of thermal equilibrium, but until now we have not needed any detailed information about them. We now consider very briefly some phenomena in which the details of molecular collisions are important. These processes are collectively called *transport phenomena* because they depend directly on the transport of gas molecules from one region to another through their random motions. A few examples of transport phenomena in gases are diffusion, heat conduction, viscosity, and electrical conductivity.

In any phenomenon depending on the transport of molecules, the average distance traveled by a molecule between collisions is important. This quantity is called the *mean free path*. A closely related quantity is the average number of collisions per unit time; this is related directly to the mean free path and the average molecular speed.

On the basis of a simple model, it is fairly simple to make an estimate of the average number of collisions a molecule makes per unit time. The model we present here is not entirely realistic, but it does afford general insight into how such calculations can be made. To begin, we assume that the molecules may be represented as identical rigid spheres of radius r which undergo elastic collisions. We suppose that there are N molecules in all, moving in a volume V.

How many collisions does a particular molecule make per unit time? The calculation is simplified greatly if we assume that only one molecule is in motion while the other molecules are stationary. This is not true, of course, but it is reasonable that the number of collisions a molecule makes per unit time should not depend greatly on whether the other molecules are in *random motion* in random positions or *at rest* in random positions.

When two molecules collide, the distance between their centers is $2r$. Thus if we draw a cylinder of radius $2r$ whose axis lies along the velocity of the moving molecule, any other molecule whose center is within this cylinder will collide with the moving molecule. In a time Δt a molecule whose speed is v travels a distance $v\,\Delta t$; during this interval it collides with any molecule in the cylindrical volume of radius $2r$ and length $v\,\Delta t$. The situation is illustrated in Fig. 18-9. The volume of the cylinder is $4\pi r^2 v\,\Delta t$. Since there

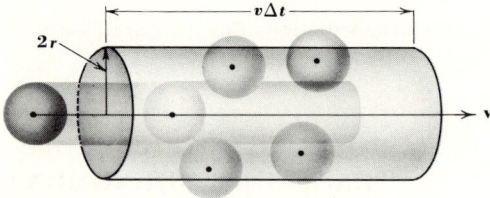

Fig. 18-9 Molecule with radius r and speed v will collide during the time interval Δt with any other molecule whose center is within the volume shown.

are N/V molecules per unit volume, the number of molecules with centers in this cylinder is $4\pi r^2 v\,\Delta t\, N/V$. Now we let Δt become very small, so that, on the average, a very small number of molecules is contained in the cylinder. In this limit, the average number of collisions per unit time is $4\pi r^2 vN/V$.

The average *time* between collisions, called the *mean free time* and denoted by τ, is the reciprocal of this expression. The mean free path, which we denote by l, is simply the product of the speed and the mean free time. That is,

$$l = v\tau = v\frac{V}{4\pi r^2 vN} = \frac{V}{4\pi r^2 N} \tag{18-35}$$

This equation has a simple interpretation. It shows that the mean free path is inversely proportional to the number of molecules per unit volume N/V. Furthermore, l is inversely proportional to πr^2, which is the area of the *silhouette* of a molecule. The molecular speed v does not appear in this result.

Equation (18-35) can be expressed in terms of macroscopic properties of the gas. We express the number of molecules N in terms of the number of moles, $N = N_0 n$, and use the ideal-gas equation in the form $V/n = RT/P$, to obtain

$$l = \frac{R}{4\pi r^2 N_0}\frac{T}{P} = \frac{k}{4\pi r^2}\frac{T}{P} \tag{18-36}$$

The numerical factor $1/4\pi$ in Eq. (18-36) is actually not correct. A more complete analysis of the number of collisions per unit time, taking into consideration the fact that all the molecules are in motion simultaneously, leads to a numerical factor $1/4\pi\sqrt{2}$. The essential features of our analysis are, however, correct.

To make direct measurements of molecular radii, we use a molecular-beam apparatus similar to that used to investigate molecular speeds. Instead of evacuating the entire apparatus, however, we permit the beam to pass through an area containing gas at a very low pressure. By varying the pressure and observing the number of gas molecules scattered out of the beam, we can make a rough determination of the molecular radius. Experiments of this sort show that the radii of many simple molecules are about 10^{-8} cm. This is of the same order of magnitude as the estimates of atomic size obtained in Chap. 16.

Example

Find the mean free path of air molecules at ordinary room temperature and atmospheric pressure, assuming a molecular radius of 10^{-10} m.

Solution

Room temperature is 300 K, and atmospheric pressure is 10^5 N/m². Using these numbers in Eq. (18-36), we find

$$l = \frac{1.4 \times 10^{-23} \text{ J/K}}{4\pi(10^{-10} \text{ m})^2} \frac{300 \text{ K}}{10^5 \text{ N/m}^2}$$
$$= 3 \times 10^{-7} \text{ m} = 3 \times 10^{-5} \text{ cm}$$

The mean free path of an air molecule at ordinary temperature is several thousand times as large as the molecular dimensions. To obtain a mean free path of 1 m, the pressure must be about 10^{-7} atm. Pressures of this magnitude are found about 100 mi above the earth's surface, in the outer extremities of our atmosphere.

As we have mentioned, several physical phenomena depend directly on the transport of gas molecules. Correspondingly, the magnitudes associated with these phenomena depend directly on the mean free path and related quantities. The simplest example is that of *diffusion*. If a quantity of one gas is introduced into one end of a container in which another gas is already present, the time required for the concentration of the new gas to increase to a given value at any other point in the container clearly depends on the average distances the molecules move between collisions. Diffusion phenomena also occur in liquids, and even in solids; in all cases, the rate of diffusion is determined by molecular speeds and by the extent to which collisions impede the progress of diffusing molecules.

Another phenomenon which depends directly on molecular collisions is thermal conductivity. When heat is conducted from one region of a gas to another, rapidly moving molecules in one region induce more rapid motion in another region. The rate at which this transfer of energy takes place depends on how far the molecules travel, on the average, between collisions. These considerations can be developed into a detailed theory of thermal conductivities of gases.

The phenomenon of viscosity, which involves forces transmitted across a fluid when different parts of the fluid are in relative motion, can be understood on a similar basis. On a microscopic scale, the force is associated with a momentum transfer from one region of fluid to another; in gases this momentum transfer is achieved by movement of molecules between the two regions. Thus a microscopic theory of viscosity can be developed. The kinetic theory of gases can be used to make quantitative calculations of various transport phenomena such as those mentioned.

18-8 LIMITATIONS OF KINETIC THEORY

In this chapter we have seen several examples of the *simplicity* of the behavior of gases. The relation between pressure, volume, temperature, and quantity of substance can be expressed by a very simple equation. The specific heats are related in a simple manner, and the distribution of molecular speeds can be described by a simple function whose form is the same for many different gases. It is therefore not surprising, at least in retrospect, that this behavior can be understood on the basis of a simple microscopic model. If the behavior is simple, the internal structure should also be simple.

The behavior of gases is not always so uncomplicated, however. At sufficiently high pressures and low temperatures, the relation between pressure, volume, and temperature may be significantly different from that predicted by the ideal-gas law. In deriving the law, we have neglected the size of the molecules and have assumed that they exert no forces on each other except during the very short time of collision. But if the total volume of the molecules is a significant fraction of the volume of the container, we may no longer neglect it. This leads to a departure from ideal-gas behavior. If molecules exert attractive forces on each other, the pressure differs from that predicted by the ideal-gas equation. If the molecules stick together sufficiently, the behavior is characteristic of a liquid or a solid, not of a gas. When the finite sizes of molecules and the attractive forces between them are included in the analysis, even very approximately, we obtain a better representation of the behavior of real gases. One procedure for doing this leads to van der Waals' equation of state, discussed in Sec. 19-2.

Our simple kinetic theory predicts, on the basis of the principle of equipartition of energy, that the specific heat of a gas is independent of its temperature. But Fig. 18-10 shows how the molar specific heat of hydrogen varies with temperature. At low temperature C has a value characteristic of a monatomic gas, indicating that the rotational kinetic energy is absent. At very high temperature C is too large for a diatomic molecule, indicating additional energy in *vibrational* motion of the molecule. This is a clear

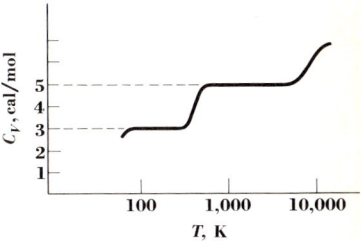

Fig. 18-10 Molar specific heat of hydrogen as a function of temperature. Note that the temperature scale is logarithmic, rather than linear.

violation of the principle of equipartition of energy. These are intrinsically quantum effects, as we mentioned in Sec. 18-4, and any kinetic-molecular model based on newtonian mechanics is powerless to deal with them.

The assumption that molecules collide elastically with the walls of the container is not completely valid. In fact, there is now good experimental evidence that the actual mechanism of collision is better described by a model in which a molecule strikes the wall, is held for a certain time, and then leaves with a speed which is *independent* of its incident speed but is governed by a speed-distribution function characteristic of the temperature *of the wall*. Still, if the gas and the container are in thermal equilibrium, then on the average the speeds before and after the collisions are equal, and the momentum transfer is the same as though each collision were elastic. Thus this more detailed description of the collision process does not invalidate the prediction about the *average* behavior of gas molecules based on the model which assumes elastic collisions.

The limitations of this simple kinetic theory of gases, then, are the limitations of the validity of the basic assumptions of the model. Clearly, we should not expect such a simple theory to predict precisely the behavior of gases under all possible circumstances. The theory can be refined to deal more accurately with real gases. But even in its simplest form, the kinetic theory of gases represents a remarkable synthesis of two fields of investigation—newtonian mechanics and heat—which historically developed along quite independent lines.

Problems

18-1 Calculate the density of nitrogen at standard temperature and pressure.

18-2 A storage tank for helium has a volume of 0.05 m^3 $= 50$ l. If the gas pressure is 100 atm at a temperature of $27°C$, how many moles of helium does the tank contain? What is the mass of helium in the tank?

18-3 Show that the density of an ideal gas is given by PM/RT, where M is the molecular mass.

18-4 An automobile tire is inflated to 24 lb/in.^{2} gauge pressure at a temperature of $20°C$. After a long drive the gauge pressure has risen to 35 lb/in.^{2}. What is the temperature? *Note:* Gauge pressure is the amount by which pressure is *greater* than normal atmospheric pressure, which is about 14.7 lb/in.^{2}.

18-5 What pressure would be necessary to compress air to a density equal to that of water, at a temperature of 0°C, if it behaves as an ideal gas? Is it likely that its behavior under these conditions is described accurately by the ideal-gas equation? Explain.

18-6 A bubble of air from a skin diver's air supply is emitted near the bottom of a lake 30 m deep, where the water temperature is 5°C. How does its size change as it rises to the surface, where the water temperature is 20°C? Which is more important, the temperature change or the pressure change? Explain.

18-7 The compression ratio (ratio of maximum to minimum volume of a cylinder) of a certain gasoline engine is 10. If air enters the cylinder at an initial temperature of 27°C, and if its temperature rises to 327°C during compression, find the final gauge pressure in pounds per square inch. (See the note in Prob. 18-4.)

18-8 Show that equal volumes of different gases under the same conditions of temperature and pressure contain the same number of molecules.

18-9 Derive a conversion factor from the liter-atmosphere to the joule (both of which are units of energy).

18-10 What is meant by "temperature of a molecule"?

18-11 In the derivation of the ideal-gas equation in Sec. 18-3, the effect of gravity on the motion of the molecules was neglected. How can this omission be justified in detail?

18-12 For a mixture of gases, show that the total pressure is equal to the sum of the pressures which the separate gases would exert if each were in the container alone. These are called the *partial pressures* of the gases comprising the mixture.

18-13 If the speeds of 10 molecules in a gas are 1, 1, 2, 2, 2, 3, 3, 4, 6, and 10 m/s, find the average and root-mean-square speeds.

18-14 Experiment shows that the dimensions of an oxygen molecule are of the order of 2×10^{-10} m. Make a rough estimate of the pressure at which the gas would exhibit noticeable deviations from ideal-gas behavior, as a result of the finite size of the molecules, at ordinary temperatures ($T = 300$ K).

18-15 For nitrogen at standard temperature and pressure, calculate the average number of molecules that collide with a 1-cm^2 area of the container wall in 1 s.

18-16 A machine is built which fires hard-steel ball bearings at a steel plate. Each ball has a mass of 0.01 kg and is fired perpendicular to the plate with a speed of 50 m/s. If five balls are fired per second, and if the collisions are elastic, find the average force exerted on the plate.

18-17 A number of hard particles move between two parallel flat rigid walls a distance d apart, as shown in Fig. P18-17. All collisions are perfectly elastic; all particles have speed v and always move perpendicular to the walls. There are N particles in all, each having mass m.

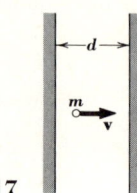

Fig. P18-17

    a What average force is exerted on each wall? What average pressure?
    b If the kinetic energy of each particle is equal to $kT/2$, calculate the specific heat of the system.
    c How does pressure vary with temperature for this system?

18-18 Consider a gas consisting of N molecules confined in a cubical box with sides of length l whose edges are parallel to the x, y, and z axes. Suppose that one-third of the molecules move parallel to each coordinate axis and none in any other direction. Furthermore, suppose that as many are moving in the $+x$ direction at any instant as in the $-x$ direction, that collisions with the wall are perfectly elastic, and that all molecules have the same speed v.
    a Calculate the pressure on any wall.
    b If the kinetic energy of a molecule is $\frac{3}{2}kT$, derive a relationship between P, V, and T for this peculiar gas.

18-19 Find the root-mean-square speeds of:
    a Oxygen molecules at 20°C
    b Helium atoms at 20°C
    c Helium atoms at 1 K

18-20 Smoke particles in the air typically have masses of the order of 10^{-17} to 10^{-15} kg. The motion of these particles resulting from collisions with air molecules, readily observed with a microscope, is called *brownian motion*. Find the root-mean-square speed of brownian motion for a particle of mass 10^{-16} kg in air at 300 K. Is the speed the same or different if the smoke particle is in hydrogen gas at the same temperature? Explain.

18-21 An object at the top of the earth's atmosphere can leave the earth if it is given an upward velocity of 11×10^3 m/s, sometimes called the *escape velocity*. What temperature would be required for most of the hydrogen molecules in the upper atmosphere to escape? What would be the necessary temperature for oxygen molecules?

18-22 On the basis of the results of Prob. 18-21, do you think it is likely that the upper atmosphere of earth, where the temperature is thought to be about 1000 K, contains appreciable quantities of hydrogen? Explain.

18-23 Compute the specific heat (per unit mass) of hydrogen gas. How does this compare with the specific heats of familiar solid and liquid materials?

18-24 A certain mixture of diatomic gases contains n_1 mol of a gas with molecular mass M_1 and n_2 mol of a gas with molecular mass M_2. Compute the molar specific heat and the specific heat per unit mass for the mixture.

18-25 At standard temperature and pressure, sulfur dioxide, SO_2, has a molar specific heat of about 7.50 cal/K. How can this value be understood on the basis of kinetic theory?

18-26 How can one understand qualitatively the large molar specific heat of diethyl ether, $(C_2H_5)_2O$, given in Table 18-1?

18-27 Suppose that the two nuclei in an oxygen molecule are separated by a distance of 2×10^{-10} m.
  a Compute the moment of inertia of the molecule about an axis perpendicular to the line joining the nuclei at its midpoint.
  b If the rotational kinetic energy is that predicted by the equipartition theorem, what is the root-mean-square angular velocity of rotation of oxygen molecules in a gas whose temperature is 300 K?

18-28 The sun contains a large quantity of hydrogen. The temperature at its surface is about 6000 K. What is the root-mean-square speed of hydrogen molecules near the surface?

18-29 The best vacuum (lowest pressure) which can be attained in the laboratory is of the order of 10^{-10} mm Hg, or about 10^{-13} atm. At this pressure and room temperature (about 300 K), how many molecules are present in a volume of 1 cm^3?

18-30 Show that in the Maxwell-Boltzmann distribution function [Eq. (18-31)] the most probable speed is equal to the constant b.

18-31 Using the result of Prob. 18-30, together with Eq. (18-34), show that a molecule having the most probable speed at a given temperature has a translational kinetic energy equal to kT.

18-32 Derive expressions for the average speed v_{av} and the most probable speed v_0 of molecules in an ideal gas in terms of the molecular mass and temperature of the gas.

18-33 Find the approximate fraction of the molecules in a gas which have speeds between v_0 and $1.1v_0$. Does this fraction depend on temperature? Explain.

18-34 A certain fictitious gas has the following distribution of molecular speeds:

$f(v) = Av$ if v is between zero and v_m, and $f(v) = 0$ if v is larger than v_m, where a and v_m are constants.

a How must the constants A and v_m be related in order for Eq. (18-26) to be satisfied? Express A in terms of v_m.

b Calculate v_{av} and v_{rms} in terms of v_m. Also find the most probable speed v_0. Are these quantities related in the same way as in the Maxwell-Boltzmann distributions? Should they be?

c Is Eq. (18-20) correct for this distribution? Explain.

d What fraction of the molecules have speeds greater than $\frac{1}{2}v_m$?

18-35 Approximately what fraction of the molecules in an ideal gas have translational kinetic energy less than $0.1kT$? *Hint:* In the Maxwell-Boltzmann function, when v is much smaller than b, the exponential factor is very close to unity and may be omitted.

18-36 Show that the Maxwell-Boltzmann distribution can be expressed as

$$f(v) = \left(\frac{2}{\pi}\right)^{1/2} \left(\frac{m}{KT}\right)^{3/2} v^2 \exp\left(-\frac{1}{2}\frac{mv^2}{kT}\right)$$

18-37 Are relativistic corrections to the masses of gas molecules significant at ordinary temperatures? At approximately what temperature do they become significant? Does this temperature depend on the molecular mass?

18-38 Two opposite walls of a box containing a gas are at different temperatures. Describe qualitatively, from a kinetic-molecular viewpoint, the process of heat conduction between the walls.

18-39 The gas that leaks into a vacuum system almost always contains a larger fraction of hydrogen and helium than that found in the outside atmosphere. How can this observation be understood on the basis of kinetic theory?

18-40 The cylindrical vacuum chamber in a certain synchrocyclotron has a diameter of about 3.5 m. If the pressure inside the chamber is of the order of 10^{-6} mm Hg, estimate the order of magnitude of the mean free path of the residual gas molecules. Is this the same as the mean free path of the protons that are accelerated in this machine? Explain.

18-41 At standard temperature and pressure, the mean free path of nitrogen molecules is about 0.8×10^{-7} m. Estimate the diameter of a nitrogen molecule.

18-42 In a voltage-regulator tube filled with argon it is desirable for the mean free path to be of the same order of magnitude as the spacing between electrodes. If this distance is 0.2 cm, approximately what should be the pressure of gas in the tube at room temperature?

18-43 The coefficient of viscosity of a fluid is ordinarily defined in terms of the forces exerted on two parallel plates with the fluid between them when they are in relative motion parallel to their planes. Discuss qualitatively what quantities characterizing the state and molecular properties of a gas should be important in determining its coefficient of viscosity.

18-44 Using the data of Prob. 18-41, find the approximate frequency with which a molecule collides with other molecules. How does this compare with the total number of collisions in 1 mol of gas in 1 s?

Properties of Matter II | 19

We have seen that some of the mechanical and thermal properties of an ideal gas can be understood on the basis of a simple model of its microscopic structure. We now discuss further several mechanical and thermal properties of various states of matter, again relating macroscopic behavior to microscopic structure. In most cases this is much more difficult to do *in detail* than for the ideal gas, but we can gain considerable *qualitative* insight into behavior through the microscopic viewpoint.

19-1 STATE OF A THERMODYNAMIC SYSTEM

When stating and discussing a physical principle or phenomenon, we often focus our attention on some particular aggregation of matter, which we call a *system*. In mechanics it is often a collection of particles or a rigid body; in thermal phenomena it may be a certain quantity of matter in a specified situation. In general, the system under consideration interacts with other matter outside the system, which we call the *surroundings*. Often this interaction is idealized in one way or another to construct a simplified model of a particular problem.

In recent chapters we have been dealing with *thermodynamic systems*, although this term has not been used explicitly. A thermodynamic system is one which may interact with its surroundings in at least two distinct ways. One of these is transfer of *heat* into or out of the system. The other or others are associated with other means of transfer of energy. In the examples to follow, this will often be performance of mechanical work by or on the system. It may also be through electromagnetic interactions, such as magnetization and demagnetization, charging and discharging a battery, and so on.

A quantity of a gas confined in a cylinder with a movable piston is a simple example of a thermodynamic system. Heat can be conducted into or out of the gas; the gas may expand and do mechanical work on its surroundings, or it may be compressed and have work done on it. Another simple system is a quantity of magnetic material such as iron. It may have heat added or taken from it, and it may gain or lose energy by being magnetized or demagnetized by a magnet. In these examples, the material itself, such as the gas or the magnetic solid, is the thermodynamic system. The piston and cylinder, the external magnet, or other apparatus is part of the surroundings of the system.

Clearly, the *state* of a thermodynamic system may change as a result of these interactions. In order to analyze these changes, we must be able to describe a state. How is this to be done? We have already seen two different approaches to the problem. In Chap. 17 we found it useful to describe a system in terms of a few readily measurable quantities. In the case of a gas these were pressure, volume, temperature, and number of moles. In other cases, other physical dimensions were used, but they were always quantities which referred to bulk or large-scale behavior. No assumptions were made concerning the microscopic structure of the system. Such a description characterizes the macroscopic state of a system; the quantities which describe the state are called its *macroscopic coordinates* or *thermodynamic coordinates*.

In Chap. 18, we took a quite different view. Looking at the *microscopic* structure of the system, in terms of atoms and molecules, we described the state of the system directly in terms of this microscopic structure. For example, a complete description of the state of a gas consists in specifying the position and velocity of every molecule, and perhaps the rotational and vibrational motions as well. Such a detailed description defines the *microscopic* state of the system.

A macroscopic description is thus given by the values of a few thermodynamic coordinates which can readily be measured, without consideration of the microscopic structure of the system. A microscopic description depends on the use of a model of the microscopic structure; a great number of quantities are needed for the description, and they are ordinarily not readily measurable quantities. Although these two points of view are quite different, we have already seen in a few simple instances that they are by no means incompatible; the kinetic-molecular theory provides a simple example of the direct relation between microscopic and macroscopic states.

Not every macroscopic state can be described as simply as may have been implied. In speaking of temperature, we have tacitly assumed that all parts of a system are at the same temperature, which is true only if the system is in thermal equilibrium. The pressure in a gas (neglecting gravity) is the same throughout the gas only if it is in mechanical equilibrium. If the system is *not* in equilibrium, then different parts are at different temperatures and pressures, and there is no such thing as a single temperature or pressure for the whole system. Clearly, such a state of affairs cannot be described as simply as an equilibrium state.

A system which is in thermal and mechanical equilibrium and whose macroscopic state can be described by a few thermodynamic coordinates which pertain to the system as a whole and do not change with time is said to be in a state of *thermodynamic* equilibrium. In almost all the discussion of Chaps. 17 and 18, except the sections on heat transfer and transport phenomena, we have tacitly assumed thermodynamic equilibrium. A nonequilibrium state cannot be described in terms of thermodynamic coordinates referring to the system as a whole.

A non-equilibrium state is not necessarily a hopeless case; it is often possible to divide the system into many smaller systems and consider each of these as an equilibrium state. But any such procedure is more involved than dealing exclusively with states of thermodynamic equilibrium. Whenever possible in the following sections, we shall deal with equilibrium states.

In working with physical quantities related to the state of a thermodynamic system, it is often convenient to distinguish between *intensive* and *extensive* quantities. An intensive quantity is one that is independent of the amount of matter present, while an extensive quantity is one directly proportional to the amount of matter. Examples of intensive quantities are temperature, density, pressure, and specific heat. Extensive quantities include volume, mass, quantity of heat, and number of moles. Another way to express the distinction between these two kinds of quantities is to imagine the system under consideration as divided into two equal parts. Any intensive quantity has the same value for each part as for the whole; an extensive quantity has *half* the value for each part that it has for the whole. This terminology is quite useful in formulating the principles of thermodynamics.

19-2 REAL GASES AND EQUATIONS OF STATE

As has just been pointed out, the macroscopic state of a gas in thermodynamic equilibrium is described by the thermodynamic coordinates P, V, and T, assuming that the quantity of gas is known. We have also found that, for a given quantity of gas, these three coordinates cannot all be varied independently; knowledge of any two is sufficient to determine the third. For an ideal gas, the relationship among the thermodynamic coordinates is the familiar ideal-gas equation,

$$PV = nRT \tag{19-1}$$

Any such equation relating the thermodynamic coordinates is called an *equation of state* for the system. Thus Eq. (19-1) is the equation of state for an ideal gas. If a system behaves as an ideal gas, the thermodynamic coordinates for every possible equilibrium state of the system must obey this equation.

The behavior of gases is not always as simple as that predicted by the ideal-gas equation. Figure 19-1 shows the experimentally observed behavior

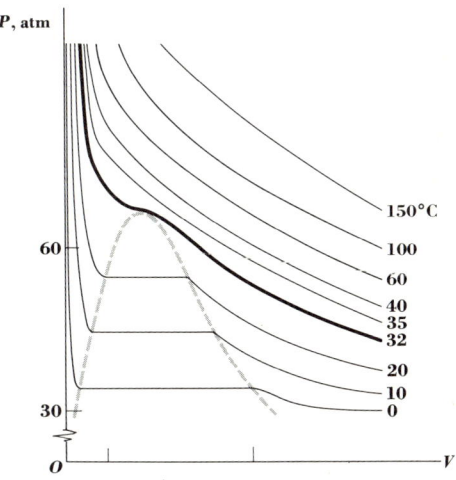

Fig. 19-1 *P-V* isotherms for carbon dioxide, showing the variation of pressure with volume for various constant temperatures. At higher temperatures, the behavior is quite similar to that of an ideal gas: $PV = $ constant. At lower temperatures, the deviations from ideal-gas behavior become more pronounced. At a temperature of about $T = 31°C$, the isotherms develop flat portions. Observation shows that under these conditions the gas is liquefying. At lower temperatures, the gas liquefies at lower pressures. The area enclosed by the broken line represents the region in which liquid and vapor are in phase equilibrium.

of a real gas, CO_2, under various conditions. Each curve shows the relation between pressure and volume for a certain constant temperature. Because these are constant-temperature curves, they are called *isotherms*; because they are drawn on a graph showing pressure versus volume, they are called *P-V isotherms*. Comparing the observed behavior of carbon dioxide with that predicted by the ideal-gas law, we note that at the higher temperatures and (especially) lower pressures, P varies approximately as $1/V$, as the ideal-gas equation predicts. At lower temperatures and higher pressures, the deviation from ideal-gas behavior becomes more conspicuous. Finally, at sufficiently low temperatures, the *P-V* isotherms develop horizontal portions, not predicted at all by Eq. (19-1). Observation shows that in this range the gas is *liquefying*; here the total volume may change greatly at constant temperature without any pressure change at all.

How can we understand these deviations from ideal-gas behavior? We recall that in developing the ideal-gas equation from a microscopic model in Chap. 18, we neglected the *volume* of the molecules and the attractive *forces* between molecules. Perhaps at sufficiently small total volume and large pressure, the volume of the molecules is no longer negligible. Also, at sufficiently low temperature, perhaps the intermolecular forces are important enough to change the velocities of the molecules significantly. If so, we must try to modify the ideal-gas equation to include these effects if we are to expect our model to exhibit close resemblance to physical reality.

For simplicity, we consider 1 mol of gas. We denote the volume of 1 mol, called the *molar volume*, by v to distinguish it from the total volume V of any general quantity of gas. The molar volume is an intensive quantity. For 1 mol, the ideal-gas equation of state takes the form $Pv = RT$. We now suppose that the molecules themselves occupy a volume b. Then each molecule is free to move not in the total volume v but only in the volume unoccupied by other molecules, which is $v - b$. Thus it is reasonable to correct for the finite dimensions of the molecules by replacing v in the ideal-gas equation with $v - b$.

The effect of the molecules exerting attractive or cohesive forces on each other is the same as though the pressure were *greater* than the actual pressure. Molecules near a wall of the container are not only *pushed* in by the wall; they are also *pulled* in by attractive forces of adjacent molecules. The net inward force on any molecule near the wall is proportional to the *density* of molecules in its neighborhood and hence inversely proportional to the molar volume. Furthermore, the total *number* of molecules close to the wall experiencing this inward force is also proportional to the density; the correction to the pressure should thus be proportional to the *square* of the density or to $1/v^2$. Therefore, we take the cohesive forces of the molecules into account by replacing P in Eq. (19-1) with the *effective pressure* $P + a/v^2$, where a is a constant.

With these two modifications the ideal-gas equation becomes

$$\left(P + \frac{a}{v^2}\right)(v - b) = RT \tag{19-2}$$

The constants a and b depend on the properties of the molecules of the particular gas we are studying and so in general are different for different gases. They are, however, *intensive* quantities, which do not depend on the quantity of gas present. In principle, if we know enough about the detailed behavior of the molecules, we should be able to predict the values of a and b theoretically. In practice, this is a complicated and difficult problem. It is usually more practical to make careful measurements of the behavior of a gas and obtain experimental values of a and b from these measurements. Such values are given in Table 19-1 for a few gases.

Table 19-1 Van der Waals' Constants for Gases

Gas	Formula	a, l^2-atm/mol^2	b, l/mol
Argon	A	1.34	0.032
Carbon dioxide	CO_2	3.59	0.043
Helium	He	0.034	0.024
Hydrogen	H_2	0.244	0.027
Nitrogen	N_2	1.39	0.039
n-Octane	C_8H_{18}	37.3	0.237
Oxygen	O_2	1.36	0.032
n-Pentane	C_5H_{12}	19.0	0.146
Water	H_2O	5.46	0.030

Equation (19-2) is known as *van der Waals' equation of state,* after its inventor. We have used rather crude arguments to develop it. Nevertheless this equation represents an improvement over the ideal-gas equation in that it results plausibly from a model which includes two effects omitted from the ideal-gas model.

When we deal not with 1 mol but with n mol of the substance, we introduce the total volume $V = nv$ to obtain

$$\left(P + \frac{an^2}{V^2}\right)(V - nb) = nRT \tag{19-3}$$

The constants a and b are the same as in Eq. (19-2).

Table 19-1 shows some interesting features of these constants. We note that a, which depends on the intermolecular forces, is much smaller for helium, an inert gas, than for any other gas. The constant b, representing the molar volume of the molecules, is considerably larger for large organic molecules than for simple molecules, but it depends surprisingly little on molecular mass for the simple molecules.

Although van der Waals' equation of state is a more accurate representation of the behavior of gases than the ideal-gas equation, there are conditions for which it too is inadequate. Many other equations of state, usually empirical in nature, have been devised to represent the behavior of matter more precisely.

The behavior of many solids and liquids can be represented approximately by an equation of state of the form

$$V = V_0(1 + \beta T - \kappa P) \tag{19-4}$$

where V_0 is a reference volume and β and κ are constants characteristic of the material. The constant β is called the *coefficient of volume expansion*, and κ is called the *compressibility* of the material. It characterizes the fractional change of volume per unit change of pressure.

In principle it should be possible to *derive* an equation of state for any substance in any state from a microscopic model, just as we derived the ideal-gas equation. In practice this is often an extremely difficult problem, and so equations of state which are at least partly empirical in nature are often used.

Since the equation of state of a material is a relation between the pressure, volume, and temperature for a fixed quantity of the material, it can be represented by means of a *surface* in a three-dimensional coordinate system with coordinates P, V, and T. Such a surface is shown in Fig. 19-2, again for the substance CO_2. To illustrate the information which can be obtained from such a surface, we can draw several lines on it, each corresponding to a constant temperature, and then look at these lines along a direction parallel to the T axis, as illustrated in Fig. 19-2. From this direction,

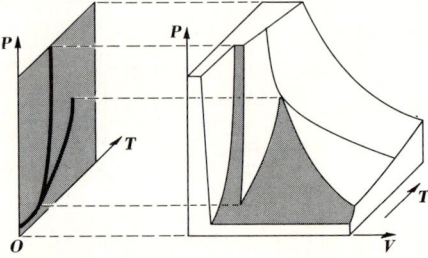

Fig. 19-2 *P-V-T* surface for CO_2. This surface may be regarded as a three-dimensional graph of the equation of state of the substance.

the lines appear exactly as in Fig. 19-1. This surface, which represents the equation of state of a substance, is called a *thermodynamic surface*.

Figure 19-2 shows that the material exhibits distinctly different kinds of behavior in different regions of pressure, volume, and temperature. The different regions correspond to different states, or *phases*, of matter—solid, liquid, and gas. We are now ready to discuss these in more detail.

19-3 PHASES OF MATTER

Many materials exist in several forms having quite different mechanical behavior but the same chemical composition. Water, for example, exists as ice, liquid water, and steam. Sulfur, in addition to having liquid and vapor states, has several quite different solid states, one being a crystalline state, another a rubbery plastic state, and a third an amorphous solid state without definite crystal structure. No known substance exhibits more than one gaseous phase; very few have more than one liquid phase, the most familiar of these few being helium, which has two. The existence of several distinct *solid* phases of a material with different crystal structures is quite common. The study of the states of matter and their relation to its microscopic structure is an interesting and useful pursuit.

The various states in which a material can exist are known as *phases*. The phase in which a particular material exists depends on the conditions, such as pressure and temperature, and, in some cases, on the past history of the material. For example, liquid water may be converted to water vapor either by heating it to its normal boiling temperature at atmospheric pressure or by reducing the pressure until the boiling point drops to the temperature of the water. When a material is transformed from one phase to another, we say that it has undergone a *phase transition* or a *change of phase*.

A phase change is often accompanied by the emission or absorption of heat, even though no temperature change is involved. To convert 1 g of ice at 0°C to 1 g of liquid water at 0°C, that is, to melt it without changing its temperature, 80 cal of heat must be added. To convert 1 g of liquid water at 100°C to the same amount of steam at the same temperature, 540 cal must be added. In neither case does the added thermal energy raise the temperature of the material; instead it changes the phase. A quantity of heat associated with a phase change is called a *latent heat*. The latent heat of *fusion* of water at normal pressure is 80 cal/g, and the latent heat of *vaporization* of water at 100°C and 1 atm pressure is 540 cal/g. Latent heats, like specific heats, are intensive quantities.

It is easy to understand in a qualitative way the microscopic basis of latent heats. In the liquid state of a material, for example, the molecules are held together by mutual attractive forces. In the gaseous state the molecules are, on the average, far enough apart for the intermolecular forces to be relatively unimportant. In order for the molecules to be pulled apart in the transition from liquid to gas, *work* must be done on them. Thus it is necessary to add energy from outside the system. This energy does not appear as additional kinetic energy but as potential energy associated with the interaction forces between the molecules; for this reason, the added energy does not increase the temperature of the substance. The latent heat of vaporization can be used to estimate the magnitude of the potential energies associated with intermolecular forces.

Under some conditions two phases exist side by side in an equilibrium state. For example, suppose we boil some water in a flask until all the air is driven out and the space above the liquid is filled with steam (water vapor).

We then stopper the flask, which now contains water in two different forms: liquid water, and above it, water vapor. This situation is known as *phase equilibrium*. A state of phase equilibrium is necessarily a state of thermodynamic equilibrium.

It is illuminating to examine the gas-liquid phase equilibrium from a microscopic viewpoint. Molecules are continually moving from one phase to another. In each phase, the molecules have a certain distribution of velocities. A few molecules at the surface in the liquid phase have enough velocity to break away from the attractive forces of the other molecules. Those molecules escape and enter the gaseous state. Simultaneously, molecules from the gaseous state are continually colliding with the surface of the liquid and returning to the liquid phase. The number of molecules which go from liquid to gas per unit time depends on the temperature, since the distribution of molecular velocities changes with temperature. The number going from gas to liquid depends on the pressure and temperature of the gas.

An equilibrium condition is reached when the two processes proceed at precisely the same rate so that, on the average, the number of molecules in either phase is not changing. Because the rates of the two processes depend on pressure and temperature, there is for any given temperature a definite pressure at which phase equilibrium is established. This pressure is called the *vapor pressure* of the substance at that temperature. Thus macroscopic equilibrium results from the competition of two microscopic processes whose effects exactly cancel each other at equilibrium.

Equilibrium between the liquid and solid phases of matter may be understood qualitatively in the same way as a liquid-gas equilibrium. Individual molecules at the interface between liquid and solid go from the solid phase to the liquid phase, while others go from the liquid phase to the solid phase. When these two processes proceed at equal rates, the two phases are in equilibrium. This is analogous to chemical equilibrium, i.e., a situation in which a reaction and its inverse proceed at exactly the same rate.

In general, equilibrium of any two phases can exist only under particular conditions of temperature and pressure. When these conditions are not satisfied, the two inverse processes do not proceed at the same rate; the amount of matter in one phase then grows, and that in the other decreases, until either the conditions for phase equilibrium are attained or all the matter is in a single phase.

It is convenient to draw a diagram showing the phases which exist for any given values of P and T. Such a *phase diagram* is shown in Fig. 19-3 for CO_2. Conditions of temperature and pressure are represented by points on this chart, and by plotting particular values of P and T we can determine whether the material is in the solid, liquid, or gas phase.

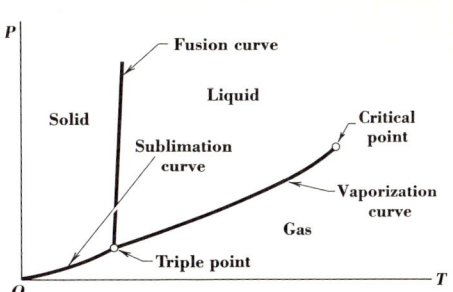

Fig. 19-3 *P-T* phase diagram for CO_2, showing fusion, vaporization, and sublimation curves and triple and critical points. Areas in this diagram represent conditions under which each phase is stable. At points on the various boundary lines, two phases can coexist in equilibrium. All three phases can be in equilibrium at only one point, the triple point.

Figure 19-3 is worthy of careful study. First, at extremely low temperatures, the material is in the solid phase, no matter what P is. At sufficiently *high* temperatures, the material is in the gaseous phase. The lines separating the various regions on the diagram represent places where two phases can be in equilibrium. Every point on the line between liquid and gas regions, for example, represents a set of conditions under which liquid and vapor can be in phase equilibrium. This curve is called the *vaporization curve* for the material. Similarly, the solid-liquid equilibrium line is the *fusion curve*. At sufficiently low pressures the solid and gas phases can be in equilibrium; the boundary representing this equilibrium is known as the *sublimation* curve.

Two points are of particular interest. First, the fusion, vaporization, and sublimation curves meet at a common point, the *triple point*, at which *all three phases* of the material are in equilibrium. The equilibrium of all three phases can occur only at this one point.

Second, the vaporization curve has a definite end point, the *critical point*, at which the distinction between liquid and gas disappears. This may seem strange; under familiar conditions, when a material goes from the liquid to the gas phase, its physical properties change quite noticeably. The density decreases considerably, the compressibility increases greatly, and the optical properties change. But as we go to higher pressures and temperatures on the vaporization curve, the density of the gaseous phase increases, becoming closer to that of the liquid. The *differences* between the physical properties of the liquid and those of the vapor at the same temperature and pressure become smaller and smaller. Finally, at the end point on the vaporization curve, the physical properties of the two phases become identical. Therefore, at any point above this, there is no phase change but only a smooth transition from one type of behavior to another.

The fact that it is possible to make transitions from liquid to vapor and back without any definite phase transition may seem at first to be at variance with common sense. We are accustomed to thinking of liquid and vapor behaviors as representing two distinctly different forms of matter. We think of a gas, for example, as quite compressible and a liquid as relatively incompressible. But at sufficiently high pressures, the compressibilities are quite similar, as are other physical properties. Thus it is reasonable that under certain conditions of temperature and pressure the distinction should disappear completely.

For most familiar gases the pressure P_c at the critical point (called the *critical pressure*) is the order of 30 to 200 atm, so critical-point behavior is not observed in everyday life. For water, $P_c = 218$ atm and $T_c = 374°C$; for CO_2, $P_c = 73$ atm and $T_c = 31.1°C$. For hydrogen and helium, the critical pressures are only a few atmospheres, but the critical temperatures are very low. In all these cases specialized laboratory equipment is needed to attain the high pressures or low temperatures at which critical-point phenomena occur.

The information contained in the *P-T* phase diagram is also contained in the thermodynamic surface for the material. If we look at this thermodynamic surface in a direction parallel to the *V* axis, the parts of the surface representing regions of phase equilibrium appear to be lines; these are the lines on the phase diagram. It is worthwhile to study the relationship between Figs. 19-2 and 19-3 carefully.

19-4 ELASTICITY

We have spoken in general terms about the *elastic* properties of matter, changes in the size or shape of matter when it is subjected to outside forces. We now try to describe in more precise language some simple kinds of elastic behavior.

Consider first the *stretching* of material under a tensile force. We can tie one end of a wire to the ceiling, attach weights to the other end, and observe how much the wire stretches for various amounts of force. The results of such an experiment can be shown graphically as in Fig. 19-4, in which elongation Δl is plotted as a function of the force F applied to each end. In this example it is found that the elongation is *proportional* to the force

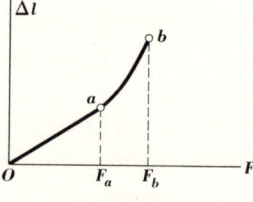

Fig. 19-4 Results of an experiment in which various forces are applied to a phosphor-bronze wire, and the corresponding elongations measured. Up to point *a*, the elongation is proportional to the force. This point represents the proportional limit. At point *b*, the wire breaks; this point represents the ultimate strength of the wire.

up to a certain point, labeled *a* on the diagram. In this region the graph is a straight line. When the force is increased beyond this point, the elongation is no longer proportional. At point *b*, the wire breaks, and the experiment is automatically concluded. The point *a* at which the elongation ceases to be proportional to the force is called the *proportional limit;* point *b* corresponds to the *ultimate strength* of the material.

In some cases it is possible to induce more exotic modes of behavior. Instead of increasing F continuously from zero to the value which breaks the wire, we might add some weight, remove it, and then add it again, to see whether the wire returns to its original length when the load is removed. Figure 19-5 shows the result of such an experiment made with a piece of

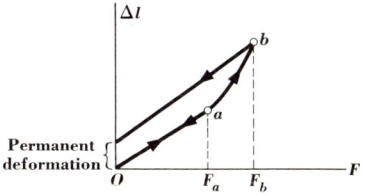

Fig. 19-5 Inelastic behavior observed in an experiment in which a soft-iron wire is stretched. The proportional limit is reached at point *a*. When a force corresponding to point *b* is applied and then removed, the wire does not return to its original length but acquires a permanent deformation.

soft-iron wire. We see that when a force corresponding to point *a* on this diagram is applied and then removed, the elongation returns to zero. If we apply a larger force corresponding to point *b* and then remove it, the elongation does not return to zero; a permanent deformation has occurred. In general, if the elongation returns to zero when the force is removed, we say that the deformation has been *elastic;* it is *reversible,* and it disappears upon removal of the force. If, however, a permanent elongation is produced, we speak of an *inelastic* deformation. The maximum force which can be applied without inelastic (irreversible) deformation is called the *elastic limit* of the material.

Often materials are employed in situations in which the force is always well below the elastic limit or the ultimate strength. It is thus useful to describe in general the behavior of materials in the region where the elongation Δl is proportional to the applied force F. In this region,

$$F = k\,\Delta l$$

where k is analogous to a spring constant and depends on the length and cross section of the wire and the material of which it is made.

A more general approach is to describe the elastic properties with a constant which depends only on the *material,* not on the dimensions of the sample. To do this, we first recognize that the force required to produce a given stretch in a sample of the material is directly proportional to the cross-section area A of the material. If one sample has twice the cross-section area of another, twice as much force is required for the same elongation; a suitable measure of the effort exerted to stretch the material is the *force per unit cross-section area F/A*, called the *tensile stress*. Tensile stress has units of force per unit area, the same units as pressure.

Furthermore, the amount the wire stretches is proportional to its original length. If two wires of equal cross-section area, one twice as long as the other, are subjected to the same tensile force, the longer wire stretches twice as much as the shorter wire. Therefore, a suitable measure of the deformation of the material, independent of the dimensions of the specimen, is the ratio of the elongation Δl to the initial length l. This ratio $\Delta l/l$, called the *tensile strain* of the material, is *dimensionless*.

When conditions are such that the elongation of the wire is proportional to the force, the tensile strain is proportional to the tensile stress, provided the cross-section area does not change significantly. Under these conditions, the ratio of tensile stress to tensile strain is a constant. Any such ratio of stress to strain is called an *elastic modulus*. In the particular case of tensile stress and strain, it is called *Young's modulus* and usually denoted by E_Y. That is,

$$E_Y = \frac{F/A}{\Delta l/l} \tag{19-5}$$

Table 19-2 shows approximate values of Young's modulus for a few familiar materials. Also shown in this table are the stresses corresponding to the elastic limits and ultimate strengths of these materials. E_Y is expressed in units of stress; it is the stress that would be required to produce $\Delta l/l = 1$, that is, an elongation equal to the original length, if the proportionality of stress and strain held this far. Actually, almost all materials break long before this, so it is better to think of $10^{-4}E_Y$ as being the stress needed to produce a strain of 10^{-4}, or $\Delta l = 10^{-4}l$. Most materials can be strained this far without exceeding their proportional limits.

The proportionality of stress and strain in elastic deformation is known as *Hooke's law*, after its discoverer, Robert Hooke, a contemporary of Newton. Like many other descriptions of the behavior of matter, it is an idealization which many materials obey approximately when stresses and strains are sufficiently small. In the present example, the cross-section area always decreases somewhat with increasing strain; this is one of several factors limiting the validity of Hooke's law.

Table 19-2 Elastic Properties of Materials

Material	E_Y, N/m²	Elastic limit, N/m²	Ultimate strength, N/m²	E_B, N/m²
Aluminum	7×10^{10}	2.0×10^8	2.2×10^8	7×10^{10}
Brass	9×10^{10}	3.9×10^8	4.7×10^8	6×10^{10}
Glass	5×10^{10}	8×10^8	10×10^8	3×10^{10}
Iron	18×10^{10}	1.5×10^8	3.0×10^8	20×10^{10}
Phosphor bronze	10×10^{10}	4.2×10^8	5.6×10^8	7×10^{10}
Steel	20×10^{10}	9.0×10^8	11.0×10^8	11×10^{10}

Example

A phosphor-bronze wire 10 m long and 1.0 mm² in cross-section area is subjected to a tensile force of 10 N. What is its elongation? How much force would be required to break the wire?

Solution

Solving Eq. (19-5) for Δl, we find

$$\Delta l = \frac{Fl}{AE_Y}$$

Using the value of E_Y given in Table 19-2 and making appropriate unit conversions, we have

$$\Delta l = \frac{(10 \text{ m})(10 \text{ N})}{(1.0 \times 10^{-6} \text{ m}^2)(10 \times 10^{10} \text{ N/m}^2)}$$

$$= 1.0 \times 10^{-3} \text{ m} = 1.0 \text{ mm}$$

The breaking stress, according to Table 19-2, is 5.6×10^8 N/m². Thus

$$5.6 \times 10^8 \text{ N/m}^2 = \frac{F}{A}$$

For an area of 10^{-6} m², we find that the ultimate strength, at which the wire will break, is reached by a force

$$F = (5.6 \times 10^8 \text{ N/m}^2)(10^{-6} \text{ m}^2) = 560 \text{ N}$$

Another elastic deformation which can be described very simply is the change in *volume* of a material when it is subjected to uniform pressure on all sides. Again we find that, within certain limits, the change in volume

510 PROPERTIES OF MATTER II

ΔV is proportional to the change in pressure ΔP. For this situation, we define stress as the change in pressure applied to the material and strain as the negative of the fractional volume change, $-\Delta V/V$. We use a minus sign because an increase in pressure corresponds to a decrease in volume ($\Delta V < 0$), and the volume strain so defined is a positive quantity. The ratio of stress to strain in this case is another elastic modulus, called the *bulk modulus*, denoted by E_B. That is,

$$E_B = \frac{\Delta P}{-\Delta V/V} = -V\frac{\Delta P}{\Delta V} \tag{19-6}$$

For some materials, especially gases, E_B depends on the original pressure, and so it is more precise to define E_B by taking the limit of Eq. (19-6) as $\Delta V \to 0$. That is,

$$E_B = -V\frac{dP}{dV} \tag{19-7}$$

The larger the value of E_B, the more pressure must be exerted to produce a given volume change. Substances with large E_B are relatively incompressible, just as materials with large E_Y are relatively unstretchable. For this reason, the *reciprocal* of the bulk modulus is called the *compressibility* of the material, denoted by κ. That is,

$$\kappa = \frac{1}{E_B} = -\frac{1}{V}\frac{dV}{dP} \tag{19-8}$$

Example
Find the bulk modulus and the compressibility of an ideal gas at constant temperature.

Solution
For an ideal gas,

$$P = \frac{nRT}{V}$$

Therefore, if T is constant,

$$\frac{dP}{dV} = -\frac{nRT}{V^2}$$

Inserting this expression in Eq. (19-7), we find

$$E_B = -V \frac{-nRT}{V^2} = \frac{nRT}{V} = P \qquad (19\text{-}9)$$

The bulk modulus for an ideal gas is equal to its pressure. This is a reasonable result; the larger the pressure, the larger the *increase* in pressure must be to produce a given fractional volume change. At high pressures a gas is less compressible than at low pressures. The compressibility is given by $\kappa = 1/P$. To emphasize the fact that we are considering elastic deformations which take place under conditions of constant temperature, we call these quantities the *isothermal bulk modulus* and the *isothermal compressibility*.

Other kinds of elastic deformations can be discussed. A discussion of deformations associated with shearing or twisting forces leads to a quantity called the *shear modulus*. Further, all the possible kinds of deformations can be discussed together, using the language of tensor analysis. These topics are not necessary for an understanding of the basic nature of the phenomena, however, and they will not be discussed here.

Additional insight into the elastic properties of matter can be gained by consideration of a model of the microscopic structure of matter. The simplest such model treats the atoms in a crystal lattice as point masses and the interatomic forces which hold them in position as springs with spring constants k representative of the variation of interatomic force with distance. Such a model is shown in Fig. 19-6 for a simple cubic lattice.

In this model, when a pair of atoms is at their equilibrium separation

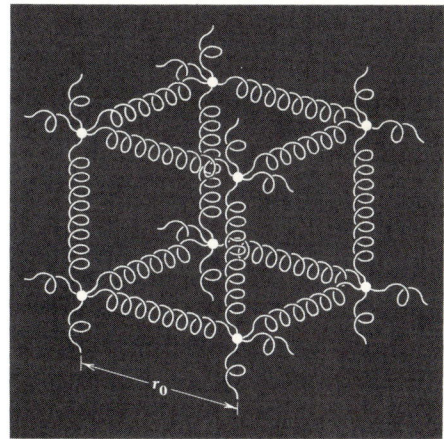

Fig. 19-6 Model describing interatomic forces in a simple cubic crystal lattice. Each spring exerts a force proportional to its displacement from its natural length r_0.

r_0, there is no force, but when they are displaced to a new separation r, the force tending to restore them to the equilibrium position is given by

$$F = -k(r - r_0) = -k\,\Delta r \tag{19-10}$$

We now derive a relation between the interatomic force constant k and the elastic moduli which characterize the macroscopic elastic behavior. As an example, we consider Young's modulus E_Y for a simple cubic lattice.

For simplicity we consider a wire in which one of the three spring directions is parallel to the length of the wire. In this case there are a number of long strings of connecting springs running the length of the wire. There is one such string for each area r_0^2 of cross section, so the number *per unit area* of cross section is $1/r_0^2$. When each spring is stretched an amount Δr, the force per spring is $k\,\Delta r$ (in magnitude); the force per unit area is the number of springs per unit area multiplied by the force per spring. That is,

$$\frac{F}{A} = \frac{1}{r_0^2} k\,\Delta r \tag{19-11}$$

The fractional change in length of the entire wire is the same as the fractional change in length of each spring; the strain of the material is thus given by

$$\frac{\Delta l}{l} = \frac{\Delta r}{r_0} \tag{19-12}$$

Using the definition of Young's modulus given by Eq. (19-5), together with Eqs. (19-11) and (19-12), we find

$$E_Y = \frac{(1/r_0^2)k\,\Delta r}{\Delta r/r_0} = \frac{k}{r_0} \tag{19-13}$$

This relation between Young's modulus and the molecular force constant is very useful in studying molecular forces, and it is remarkable that such information can be obtained so easily, by a simple measurement of an elastic modulus. For more complicated crystal structures, the calculations are more complicated, but relations similar to Eq. (19-13) can be derived for these also.

The model used to compute Young's modulus can be used to compute the bulk modulus of a material whose structure is described by this simple model. This calculation is left as a problem.

19-5 SPECIFIC HEATS OF SOLIDS

The model shown in Fig. 19-6 can be used to calculate the specific heat of a solid element whose structure it represents. The central idea in this calculation is that the atoms are not necessarily stationary at their equilibrium positions under the action of the springs but may vibrate around these positions. Each atom therefore has kinetic energy and potential energy associated with its motion. A consideration of how the *total* energy changes with temperature then leads to a calculation of the specific heat.

First, if the principle of equipartition of energy is valid in this situation, the average *kinetic* energy per atom is $\frac{3}{2}kT$, just as for a monatomic ideal gas. We do not consider rotations of the atoms in the crystal, for the same reason that we do not include the rotation of monatomic molecules in a gas. The *potential energies* of the atoms must also be considered; to include these, we compute the average potential energy per atom and compare it with the average kinetic energy given by the equipartition principle.

Consider one atom undergoing simple harmonic motion under the forces of interaction with other atoms. When this atom is displaced a small amount from equilibrium, it is acted on by a force proportional to the displacement, which tends to restore it to the equilibrium position. Although this force is the vector sum of several forces exerted by other atoms, we can represent it by means of a single effective spring constant k'. If the mass of the atom is m, then the displacement x from equilibrium at any time t can be described by

$$x = A \sin \sqrt{\frac{k'}{m}}\, t \qquad (19\text{-}14)$$

as discussed in Chap. 10. For simplicity we have assumed that $x = 0$ when $t = 0$.

The potential energy of a stretched spring is $\frac{1}{2}k'x^2$, with x given by Eq. (19-14), so the variation of V with time is given by

$$V(t) = \tfrac{1}{2} k' A^2 \sin^2 \sqrt{\frac{k'}{m}}\, t \qquad (19\text{-}15)$$

We can also use Eq. (19-14) to obtain the velocity and kinetic energy as functions of time. We find

$$v(t) = \frac{dx}{dt} = A \sqrt{\frac{k'}{m}} \cos \sqrt{\frac{k'}{m}}\, t \qquad (19\text{-}16)$$

and

$$E(t) = \tfrac{1}{2}mv^2 = \tfrac{1}{2}k'A^2 \cos^2 \sqrt{\frac{k'}{m}}\, t \tag{19-17}$$

Both $V(t)$ and $E(t)$ contain the constant factor $\tfrac{1}{2}k'A^2$ multiplied by functions of time. The two functions of time are very similar; Fig. 19-7 shows

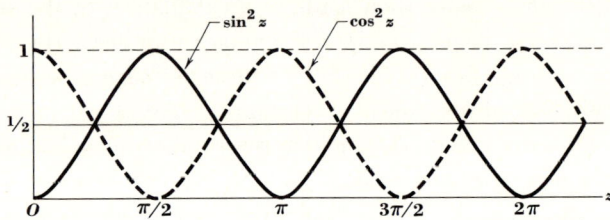

Fig. 19-7 Functions $\sin^2 z$ and $\cos^2 z$. The two graphs have the same shape but start at different parts of the cycle at $z = 0$. The average value, taken over a complete cycle, is the same for the two functions and is equal to $\tfrac{1}{2}$.

graphs of the function $\sin^2 z$ and $\cos^2 z$, where $z = (k'/m)^{1/2}t$. The two graphs have exactly the same shape, but one is displaced along the z axis relative to the other. In a complete cycle, corresponding to an interval 2π in the variable z, the *averages* of the two functions must be equal. This result can also be obtained by computing each average directly. For example,

$$(\cos^2 z)_{\text{av}} = \frac{1}{2\pi} \int_0^{2\pi} \cos^2 z \; dz = \tfrac{1}{2} \tag{19-18}$$

Similarly, $(\sin^2 z)_{\text{av}} = \tfrac{1}{2}$.

The essential result is that for this motion the average potential energy is *equal* to the average kinetic energy. Thus the average total energy per atom consists of $\tfrac{3}{2}kT$ kinetic energy and $\tfrac{3}{2}kT$ potential energy, for a total of $3kT$ per atom. The total energy per mole is then $3RT$, and the molar specific heat is

$$C = 3R = 5.97 \text{ cal/mol} \tag{19-19}$$

A remarkable feature of this calculation is that it does not depend on the atomic mass m or the force constant k'; the result thus should be applicable to a wide variety of materials. This result may be compared with Table 19-3, which gives the molar specific heats of several solid elements. Considering the simplicity of our theoretical calculation, the agreement is surprisingly good.

Table 19-3 Specific Heats of Solid Elements, $T = 20°C$

Element	Atomic mass, g/mol	Specific heat, cal/g C°	Molar specific heat, cal/mol C°
Aluminum	27.0	0.214	5.8
Copper	63.5	0.092	5.8
Iodine	126.9	0.0523	6.6
Iron	55.8	0.107	5.8
Lead	207.2	0.0306	6.3
Silver	107.9	0.558	6.6
Sulfur (monoclinic)	32.0	0.181	6.8
Uranium	238.2	0.0281	6.7

The fact that the molar specific heats of many solids are about 6 cal/mol C° is called the *rule of Dulong and Petit*. It was discovered over 100 years ago as a generalization from experimental observations; as we have just seen, it can be *derived* from a kinetic-molecular model of a solid. This is a striking example of how theoretical developments can give enhanced significance to well-known empirical rules.

As with gases, the specific heats of solids do not always agree exactly with the prediction of Eq. (19-19). At very low temperatures, specific heats nearly always become smaller. As with gases, this behavior requires quantum theory for its complete understanding. In fact, one of the problems which led to the original development of quantum mechanics was the calculation of the temperature dependence of specific heats of materials.

In some cases, especially at high temperatures, specific heats are *larger* than the prediction of our simple analysis. This increase is associated with motion of electrons relative to their nuclei, giving additional degrees of freedom. Especially in metals, such effects contribute substantially to specific heats and must be included in more detailed calculations. Our present knowledge of quantum mechanics enables us to make quite precise predictions of the specific heats of many solids over a wide range of temperatures.

19-6 THERMAL EXPANSION

The phenomenon of thermal expansion has already been mentioned in Sec. 17-3 in connection with thermometers; we now discuss it more quantitatively. It is a familiar fact that the dimensions of solid bodies change when their temperatures change. It has been found experimentally that the change ΔL in any linear dimension L of a body which results from a *small* temperature

Table 19-4 Coefficients of Expansion

Substance	α, C°$^{-1}$	Substance	β, C°$^{-1}$
Aluminum	24×10^{-6}	Bromine	1.13×10^{-3}
Brass	19×10^{-6}	Ethanol	1.12×10^{-3}
Calcite:		Mercury	0.18×10^{-3}
Parallel to axis	25×10^{-6}	Water	0.21×10^{-3}
Perpendicular to axis	-5.6×10^{-6}		
Copper	17×10^{-6}		
Glass (flameproof)	3×10^{-6}		
Ice	51×10^{-6}		
Invar	0.9×10^{-6}		
Lead	29×10^{-6}		
Oak:			
Along fiber	5×10^{-6}		
Across fiber	54×10^{-6}		
Quartz (fused)	0.4×10^{-6}		
Steel	12×10^{-6}		

change ΔT is approximately proportional to ΔT. It is also proportional to L; if two bars are made of the same material but one is twice as long as the other, the longer one expands twice as much for the same temperature change. These relations can be expressed as

$$\Delta L = \alpha L \, \Delta T \qquad (19\text{-}20)$$

where α is a proportionality factor, called the *coefficient of linear expansion*, which is different for different materials but which does not depend on the dimensions of the object. Approximate values of α for several common materials are given in Table 19-4.

Example
A steel railroad rail is laid in the spring when the temperature is 68°F. Its length at that time is exactly 90 ft. What is its length on a very hot summer day when the temperature is 104°F?

Solution
The temperature change is 36 F°, which is equivalent to $\frac{5}{9} \times 36°$ or 20 C°. Therefore $\Delta T = 20$ C°, and

$$\Delta L = (12 \times 10^{-6} \text{ C}°^{-1})(90 \text{ ft})(20 \text{ C}°) = 0.023 \text{ ft} = 0.28 \text{ in.}$$

Any units of length may be used for L and ΔL provided the *same* units are used for both; the units of α must be consistent with those of the temperature scale used. In addition, α varies somewhat with temperature, so that the proportionality expressed by Eq. (19-20) is strictly correct only in the limit when $\Delta T \to 0$. Thus a more precise statement, obtained from Eq. (19-20) by taking this limit, is

$$\frac{dL}{dT} = \alpha L \tag{19-21}$$

in which α is approximately constant for small temperature changes but in general depends on T.

Every linear dimension of an object changes with temperature in the manner shown by Eq. (19-20) or (19-21). For example, a hole in a piece of material grows with temperature in the same proportion as all other dimensions. The coefficient of linear expansion is almost always a positive quantity, but there are some interesting exceptions. A few substances under particular conditions actually shrink when heated, corresponding to negative values of α.

Sometimes changes in the *volume*, rather than the linear dimensions of an object, are of interest. It is easy to calculate volume changes, starting with Eq. (19-21), for some simple shapes. Suppose we have a cube which at a certain temperature has a length of side L. Its volume at this temperature is $V = L^3$. To find the variation of V with T, we use the chain rule for derivatives:

$$\frac{dV}{dT} = \frac{dV}{dL}\frac{dL}{dT}$$

Now $dV/dL = 3L^2$ and $dL/dt = \alpha L$, according to Eq. (19-21). Thus

$$\frac{dV}{dT} = 3L^2 \alpha L = 3\alpha V \tag{19-22}$$

Thus for small changes ΔV and ΔT we have, approximately,

$$\Delta V = 3\alpha V \Delta T$$

The quantity 3α has a role completely analogous to that of α in Eq. (19-20); it is called the *coefficient of volume expansion*.

More generally, the coefficient of volume expansion, denoted by β, is defined by

$$\frac{dV}{dT} = \beta V \tag{19-23}$$

The relationship $\beta = 3\alpha$ holds if the solid expands uniformly in all directions. There are materials for which this is not the case; Table 19-4 contains an example. For these, α depends on direction, and the relation $\beta = 3\alpha$ is not valid.

For small changes in temperature ΔT, the volume change ΔV is thus given approximately by

$$\Delta V = \beta V \Delta T \tag{19-24}$$

The precision of this relationship, as for linear expansion, depends on how much β varies with temperature.

Equations (19-23) and (19-24) also hold for volume changes in *liquids* produced by temperature changes. As with solids, we find that β is usually a positive quantity but that there are a few cases in which β is negative. For example, as liquid water is heated from 0 to 4°C, its volume actually decreases slightly; it then begins to increase again above 4°C.

Examination of the role of the interatomic forces gives additional insight into thermal expansion and the microscopic basis for this phenomenon. This matter is discussed briefly in the next section.

19-7 INTERATOMIC FORCES

In the preceding sections we have described interatomic forces in solids in terms of a simple model in which the force is proportional to the displacement from some equilibrium separation of atoms, as in Eq. (19-10). Actual interatomic-force behavior is never this simple, of course. A more realistic model might resemble Fig. 19-8. At very large r the force is zero. At $r > r_0$ the force is negative (attractive), at $r = r_0$ it is zero, and for $r < r_0$ it becomes

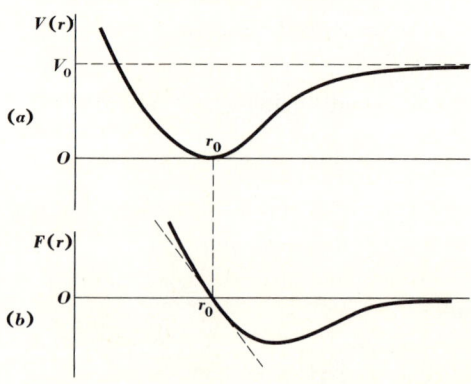

Fig. 19-8 (*a*) Typical potential-energy function corresponding to the force of interaction between two atoms. (*b*) Corresponding force. The potential energy has been defined to be zero at the equilibrium point, $r = r_0$, at which V has a minimum. The force at this point is zero. For values of r close to r_0, the force F is approximately proportional to the distance from r_0, that is, $r - r_0$. This interatomic force can be used as a basis for understanding elasticity, specific heats, and thermal expansion of solids.

positive (repulsive). A small segment of the curve near $r = r_0$ is approximately a straight line with slope $dF/dr = -k$; this is the linear approximation discussed previously.

Corresponding to this force is a potential-energy function $V(r)$, related to $F(r)$ by

$$F(r) = -\frac{d}{dr}V(r) \tag{19-25}$$

as discussed in Sec. 9-7. The potential-energy function corresponding to the force shown in Fig. 19-8b is shown in Fig. 19-8a. We have arbitrarily chosen the equilibrium point $r = r_0$ as the zero point for potential energy. The value of V at large r, labeled V_0 in the figure, represents the energy needed to pull two atoms indefinitely far apart from their equilibrium separation. Near $r = r_0$ the shape of $V(r)$ is approximately parabolic, and the effective force constant k is given by

$$k = \frac{d^2V}{dr^2} \tag{19-26}$$

just as in Sec. 10-5.

Thus the value of E_Y for a material depends on the shape of $V(r)$ near r_0. If the *potential well*, as this minimum in $V(r)$ is called, is narrow and curves sharply near its minimum, k is large, E_Y is also large, and the material is relatively unstretchable. If the potential well is broad and flat, the reverse is true.

The "depth" of the potential well V_0 is closely related to melting and vaporization. A material with a large V_0 must be heated to a relatively high temperature for substantial numbers of atoms to acquire enough energy to pass into the liquid or vapor phase, while materials with smaller V_0 have lower melting and boiling temperatures.

Thermal expansion can also be understood qualitatively on the basis of the shape of the $V(r)$ curve near r_0. As indicated in the discussion of specific heats in Sec. 19-5, the atoms in the crystal lattice vibrate about their equilibrium positions with amplitudes that increase with increasing temperature. If this motion is really simple harmonic, an atom is displaced to one side of the equilibrium position as much of the time as to the other side, and the *average* displacement is zero. In this case the average separation between atoms does not change with amplitude of oscillation and is therefore independent of temperature.

But now suppose the potential energy $V(r)$ is not symmetric about $r = r_0$ but rises more steeply for $r < r_0$ than for $r > r_0$, as with the curve in Fig.

19-8a. In this case the motion is also asymmetric; the maximum excursion to separations greater than r_0 is larger than that to separations less than r_0. Thus the *average* separation becomes greater than r_0. As the amplitude of oscillation becomes larger and larger, corresponding to higher and higher temperatures, the average position shifts more and more, and the material expands.

Thus the essential point in understanding thermal expansion is the increasingly large departures from simple harmonic motion which occur with increasing temperature. As in the discussion of Young's modulus, we expect a material with a narrow, deep potential well to have relatively small thermal-expansion coefficients, and conversely.

To conclude this section we show in Fig. 19-9 sketches of potential-

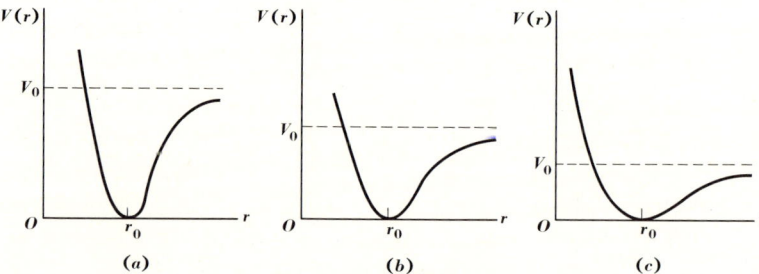

Fig. 19-9 Potential-energy curves for interatomic force in (a) a ceramic material, (b) a metal, and (c) a polymer.

energy curves for three materials—a ceramic, a metal, and a polymer such as polyethylene. The narrow deep well for the ceramic accounts for its large Young's modulus (small stretchability), high melting temperature, and small thermal-expansion coefficient. The shallow, broad well for the polymer accounts for its small E_Y (making it easily stretched), low melting temperature, and large thermal-expansion coefficient. Together, these properties provide an interesting example of the enhanced understanding of the behavior of materials that can be gained from microscopic models.

Problems

19-1 The equation of state of a system whose thermodynamic coordinates are P, V, and T is a relationship among them such that if any two are known, the third can be determined. Are there any cases in which the volume of a thermodynamic system is *not* uniquely determined by its pressure and temperature? Explain.

19-2 The equation of state of a thermodynamic system can be represented graphically by plotting curves of pressure versus *density* (rather than versus volume) for each of several constant temperatures. The resulting curves are called *pressure-density isotherms*. Sketch several pressure-density isotherms for an ideal gas. Indicate the highest and lowest temperatures.

19-3 Plot a series of P-V isotherms for a material whose equation of state is Eq. (19-4).

19-4 A magnetic substance may be regarded as a thermodynamic system whose coordinates are magnetic field H, magnetization M, and temperature. Magnetic field is an intensive quantity, while magnetization is extensive. For many paramagnetic substances these quantities are related by the equation $M = mCh/T$, where m is the mass of material and C is a constant characteristic of the material. Sketch a series of H-M isotherms for such a system, making H the vertical axis. Indicate the direction of increasing T.

19-5 A block of aluminum is heated from 0 to $10°C$ in a device which keeps its volume constant. If the equation of state is Eq. (19-4), and if the pressure at $0°C$ is 1 atm, what is the pressure at $10°C$?

19-6 How is the constant b in van der Waals' equation related to the volume of a molecule?

19-7 For oxygen at standard temperature and pressure, which of the two "corrections" in van der Waals' equation is more important?

19-8 Consider 0.10 kg of water at an initial temperature of $-20°C$. Heat is added to the water at a constant rate of 50 cal/s. Plot a graph showing temperature as a function of time from the beginning until all the water vaporizes.

19-9 At $100°C$ and atmospheric pressure, the latent heat of vaporization of water is about 540 cal/g. Assuming this energy is used to pull molecules apart, estimate the *binding energy* for a pair of molecules. Express this energy in joules, using the conversion 1 cal = 4.186 J. Is it correct to assume that *all* the latent heat is used to separate molecules? Explain.

19-10 On the basis of a kinetic-molecular model, explain why the boiling temperature of a liquid always increases with pressure.

19-11 Foods are sometimes preserved by *freeze-drying*, a process in which the food is first quick-frozen and then the water is extracted. Is the water removed more quickly if the food is placed in vacuum or if it is surrounded by very dry air containing no water vapor? Explain.

19-12 A brass block whose mass is 10 kg is suspended from the ceiling by a steel wire 10 m long and 0.5 mm² in cross-section area.
   a Find the stress, the strain, and the elongation of the wire.
   b Regarding the wire as a spring, find its force constant.
   c The block is pulled 1 mm below its equilibrium position and released. Describe the resulting motion in detail.

19-13 A steel rod 2 mm square and 5 m long is used to suspend an abstract sculpture from the ceiling. The sculpture has a mass of 50 kg. Find the stress, the strain, and the elongation of the rod.

19-14 If the steel rod in Prob. 19-13 is used instead to suspend the sculptor from the ceiling, what is the maximum mass of sculptor which could be supported?

19-15 A certain steel cable $\frac{1}{8}$ in. in diameter has a Young's modulus of $E_Y = 30 \times 10^6$ lb/in.² and a breaking stress of 0.20×10^6 lb/in.². Two parallel 50-ft lengths of this cable, with rungs of aluminum tubing fastened between them, are used to make a lightweight cable ladder for cave exploration.
   a How much does the ladder stretch under the weight of a 200-lb caver?
   b How much weight will the ladder support without breaking? What is its elongation just before it breaks if the proportional limit is very close to the ultimate strength?

19-16 A certain nylon rope used for rock-climbing has a diameter of $\frac{7}{16}$ in. and an effective cross-section area of about 0.15 in.². Its length is 120 ft. Under the weight of a 150-lb climber, it stretches approximately 3 ft. Find the stress and strain under these conditions, and compute Young's modulus. Compare with values of E_Y for steel cables.

19-17 Steel railroad rails are sometimes welded into long continuous lengths. If a rail was welded on a hot summer day when the temperature was 30°C, what is the tensile stress in midwinter, when the temperature is −10°C? Compare this with the breaking stress.

19-18 A high-pressure vessel contains 1,000 cm³ of water at atmospheric pressure. How much pressure is required to decrease the volume by 1.0 cm³? For water, $E_B = 2.1 \times 10^9$ N/m².

19-19 For a substance whose equation of state is $P(V - nb) = nRT$, calculate the temperature coefficient of volume expansion, the isothermal compressibility, and the isothermal bulk modulus.

19-20 A bottle contains 5.0×10^{-4} m³ of ethanol. The bottle, with its top open, is placed in a pressure chamber, and it is observed that when the pressure is increased to 10 atm, the volume decreases by 5.0×10^{-7} m³. Find the stress, the strain, and the bulk modulus for ethanol.

523 PROBLEMS

19-21 A block of copper in the shape of a cube 0.10 m on a side is subjected to a pressure of 5.0×10^6 N/m². The bulk modulus of copper is 14×10^{10} N/m².
    a What is this pressure in atmospheres?
    b Find the change in volume of the block resulting from this pressure.

19-22 The elastic moduli all have the units of *pressure*. Find the bulk modulus at atmospheric pressure in atmospheres of:
    a Water
    b Air

19-23 Although there are exceptions, liquids typically have bulk moduli smaller than those of solids by a factor of 10 to 100. How can this difference be understood on the basis of microscopic structure?

19-24 For a system containing liquid and vapor phases of the same substance in phase equilibrium, the isothermal bulk modulus may be zero. Explain.

19-25 By means of x-ray diffraction experiments it has been established that iron has a cubic crystal structure with lattice spacing 2.86×10^{-10} m. Using this and the observed Young's modulus for iron, compute the microscopic spring constant, assuming a simple cubic lattice. Actually, the lattice is ordinarily body-centered cubic. Discuss qualitatively whether this would change the calculation significantly.

19-26 For a material with a simple cubic crystal structure which can be described by the model shown in Fig. 19-6, compute the bulk modulus in terms of the microscopic spring constant and the lattice spacing. How are the bulk modulus and Young's modulus related for this simple model?

19-27 How does the microscopic spring constant obtained in Prob. 19-25 compare in order of magnitude with the molecular spring constants used to describe interatomic forces in Sec. 10-5?

19-28 In a material whose crystal lattice is simple cubic, with lattice spacing 2.0×10^{-10} m and microscopic spring constant 20 N/m, suppose that an atom undergoes simple harmonic motion in a direction parallel to one of the crystal axes. If the total energy of the atom is $3kT$, find the amplitude of oscillation at:
    a 300 K
    b 1 K

19-29 For the situation of Prob. 19-28, at what temperature does the amplitude become equal to one-half the lattice spacing? Is it likely that the material is still in the solid state at this temperature?

19-30 Suppose microscopic atomic forces are not directly proportional to displacement from equilibrium, as indicated by Eq. (19-10), but can be described more precisely by $F = -kx - cx^3$, where x is the displacement from equilibrium and k and c are constants.
  a Plot a graph of F versus x.
  b Is the motion which occurs under this force simple harmonic motion?
  c Qualitatively, should the average kinetic energy for this motion be greater or less than the average potential energy? Explain.

19-31 In the situation of Prob. 19-30, how (qualitatively) does the specific heat differ from what it would be if the constant c were zero?

19-32 Following the discussion of Sec. 19-6, suppose that the behavior of the microscopic springs can be represented by a spring constant k_1 for displacement of an atom toward one of its neighbors but a smaller constant k_2 for displacements farther away than equilibrium.
  a Show that the maximum displacements A_1 and A_2 on the two sides are related by $k_1 A_1^2 = k_2 A_2^2$.
  b Compute the *average* position over one complete cycle of the motion.
  c If the total energy is equal to $3kT$, compute the average position (displacement from equilibrium) as a function of T. From this, compute the coefficient of linear expansion of the material.

19-33 It is desired to construct a thermometer from a glass bulb, a section of tubing of 0.1 mm inside diameter, and mercury. What should the volume of the bulb be if the range 0 to $100°C$ on the scale is to cover a distance of 10 cm?

19-34 A railroad rail 10 m long is laid when the temperature is $0°C$. How much space should be left for expansion if the maximum temperature anticipated is $40°C$?

19-35 An aluminum piston is fitted in a steel cylinder of diameter 4 in. At $20°C$ the space between piston and cylinder is 0.005 in. all around. How hot must the piston and cylinder become for the clearance to be zero?

19-36 A surveyor's 100-ft steel tape is correct at a temperature of $68°F$. If a distance is measured as 65.43 ft on a hot summer day when the temperature is $95°F$, what is the true distance?

19-37 A bar of aluminum-bronze alloy which is 0.500 m in length at $20°C$ was found to increase in length by 0.85 mm when heated to $120°C$. Find the coefficients of linear and volume expansion of the material.

19-38 Although there are exceptions, liquids typically have much larger coefficients of volume expansion than solids. Discuss this observation on the basis of the microscopic structure of the materials.

19-39 A glass bottle of nominal capacity 250 cm^3 is filled brim full of water at $20°C$. If the bottle and contents are heated to $50°C$, how much water runs over?

PERSPECTIVE V

The central theme in the past three chapters has been understanding the relationships between various aspects of the macroscopic behavior of matter and its microscopic structure. When the structure can be represented by a simple microscopic model, the relationships can be made quite precise and quantitative; this is the case for the ideal gas. For nonideal gases, solids, and liquids, the relations become successively less quantitative and more descriptive, but always the microscopic view provides additional understanding of macroscopic behavior.

We began with temperature and heat, both intrinsically macroscopic concepts. In discussing measurement of temperature, quantity of heat, and heat transfer, a microscopic viewpoint is not necessary; in fact, these were discussed quantitatively with considerable success many years before their microscopic basis was understood. The macroscopic approach has definite limitations, however, and more complete understanding of the properties of matter requires a combination of the two views.

In one simple case, the ideal gas, the structure of matter can be represented by a very simple microscopic model. To relate this model to observed macroscopic properties, we introduced a new concept, that of computing the macroscopic behavior from the average behavior of a large number of particles. This procedure provided the bridge between microscopic and macroscopic quantities in the derivation of the equation of state and specific heats of gases, both of which agree with observed behavior reasonably well. The same procedure provided an understanding of the specific heats of solids. The elastic properties of matter and thermal expansion, as we have pointed out, are directly related to the variation of intermolecular forces with distance. Lacking precise information about these, *we have to be content with a macroscopic formulation and a qualitative understanding of their microscopic basis.*

In the next two chapters we consider in detail two quantities which, like temperature and heat, are macroscopic quantities describing the average behavior of a system but which have their roots in microscopic structure. One of these, internal energy, provides the basis for a precise relation between macroscopic mechanical energy and the energy associated with the microscopic motion and position of molecules. The other, entropy, characterizes the degree of disorder or randomness of a system and provides a quantitative expression of the tendency of all natural processes to proceed in a definite direction.

Heat and Work | 20

In this chapter we consider in some detail the relation between mechanical energy and heat. We begin with physical phenomena in which macroscopic mechanical energy disappears and heat appears. These lead naturally to the concept of internal energy of a thermodynamic system and to the expression of its relation to other forms of energy in the first law of thermodynamics. We then consider applications of this law to various processes involving changes of state of a thermodynamic system, including the operation of an internal-combustion engine.

20-1 CONVERSION OF ENERGY

In the discussion of mechanical energy in Chap. 9 it was found that in some mechanical systems the forces can be associated with potential energy in such a way that the total mechanical energy of the system, including both kinetic and potential energies, is constant. We called these *conservative* forces and systems to emphasize the fact that mechanical energy is conserved. In such a system, when kinetic energy disappears, an equal amount of potential energy appears, and vice versa.

But in other dynamical systems, such as those involving dynamic friction or inelastic collisions, there are forces which cannot be described in terms of potential energy. If kinetic energy disappears as a result of work done against friction, this energy is lost; it never reappears as mechanical energy. Such systems are called *nonconservative* or *dissipative systems*, indicating that in such systems the total mechanical energy decreases.

In a dynamic situation involving friction, *heat* is always produced. When the brakes of a moving car are applied, the brake shoes and drums become hot. When the oil in the torque converter of an automatic transmission is churned by the rotating members, it becomes hot; in fact, provision must be made for cooling this oil. Some people are even able to start a fire by rubbing two sticks together! In all these situations, and in many others, mechanical energy disappears as heat is produced.

These observations are not nearly so astounding nowadays as they were in the early nineteenth century, when such pioneers as Count Rumford and James Joule first began to investigate the relation between mechanical energy and heat. We now have a vast and convincing body of evidence that heat is microscopic mechanical energy; this fact was used to calculate the specific heat of a gas from a kinetic-molecular model in Chap. 18. For complex systems, however, it is not always possible to calculate in detail the relation between mechanical energy and heat, and so it is very useful to analyze this relation from a *macroscopic* viewpoint.

Joule made many careful observations of situations in which mechanical energy was dissipated by means of inelastic collisions, friction, stirring a fluid, and so forth. These experiments demonstrated conclusively the *proportionality* of the mechanical energy lost and the thermal energy, or heat, produced. To be more precise, Joule found that dissipating a certain quantity of mechanical energy in a substance produces the same change in the state of the substance as adding a proportional quantity of heat. Specifically,

$$4.186 \text{ J} = 1 \text{ cal}$$

Joule himself did not attain this degree of precision; the value given has been obtained by the painstaking work of many later experimenters.

Perhaps the most famous of Joule's experiments is one in which a paddle wheel is turned in a container of water, with the expenditure of a measured amount of mechanical work. The amount of water and its change in temperature are also measured; thus the relation between work and heat produced can be determined. For example, suppose we have 1,000 g of water. By measuring the torque on the paddle wheel and the number of revolutions, we determine that 16,800 J of work is done on the water. At the end of the process, the water temperature has risen 4 C°. To produce this temperature change by adding heat directly would require 4000 cal. We thus conclude that 4000 cal is equivalent to 16,800 J, or 1 cal = 4.2 J, a figure reasonably close to the accepted value. In any actual experiment, of course, allowances must be made for the heat capacity of the container and heat loss to the surroundings.

20-2 INTERNAL ENERGY

As observed in Chap. 9 the concept of kinetic energy is not very useful by itself because there are relatively few situations in which the kinetic energy of a system is constant. In many other cases, however, a potential energy may be introduced, defined so that the *sum* of kinetic and potential energies is constant. In nuclear reactions ordinary mechanical energy is not conserved, but the total mass of the system also changes; if we associate a definite amount of energy with the *mass* of the system, then the total energy, including kinetic, potential, and mass energy, is conserved.

The central theme here is to generalize the concept of energy by defining new kinds of energy to try to make the principle of conservation of energy valid in a wider and wider variety of situations. Thus it is natural to try to define a new form of energy corresponding to heat, which we call *internal energy*. As already pointed out, internal energy is associated with microscopic mechanical energy. This is not a suitable definition because it is not an *operational* definition; microscopic mechanical energy cannot be measured directly. Thus it is best to formulate first a macroscopic definition of internal energy which provides a means of measuring this quantity or, at least, changes in it.

As an aid to defining internal energy precisely, we make use of the concept of a thermodynamic system, introduced in Sec. 19-1. We consider a system which can exchange energy with its surroundings by two different means: by addition or removal of heat and by the performance of work. In general, both these mechanisms can be two-way processes. We can add heat to the system or take heat out of it; correspondingly, work can be done on the system, or the system itself can do work on its surroundings. In most of the systems to be considered here, work means ordinary mechanical work. There are, however, thermodynamic systems in which energy can be transferred by other means. For example, energy can be added to or subtracted from a magnetic material by magnetizing or demagnetizing it, as well as by adding or subtracting heat. For this system, *work* refers to electromagnetic energy transmission and is thus a more general concept than simple mechanical work. For every thermodynamic system, however, one of the means of energy transfer must be *heat*.

Now we consider a change in the state of the thermodynamic system during which a quantity of heat Q is added to the system, which simultaneously does an amount of work W on its surroundings. We regard Q and W as algebraic quantities which may be either positive or negative. We define Q to be positive when heat is added to the system and negative when heat is removed. Similarly, W is positive when the system does work on its surroundings, as when a gas expands against a piston, and negative when work is done on the system, as when a gas is compressed. As a result of the heat and work exchanges, the internal energy of the system, denoted by U, changes by an amount ΔU. We now *define* ΔU so that the *total* energy of the system and its surroundings is conserved.

If no work is done and energy is transferred only by means of heat, the change in internal energy must be equal to the amount of heat added; in this case $\Delta U = Q$. On the other hand, if no heat is added and energy is exchanged only by the system doing work W on its surroundings, its internal energy must *decrease* by a corresponding amount. In this case, then, $\Delta U = -W$. In general, if both heat and work are exchanged with the surroundings, the change in internal energy of the system is given by

$$\Delta U = Q - W \tag{20-1}$$

This equation forms the basis for the definition of internal energy of a thermodynamic system, for it shows how to calculate the change in internal energy in any process in which the state of the system changes.

The sign conventions used for Q and W above may seem a bit strange, inasmuch as positive values of Q correspond to energy added to the system while positive W means energy leaves the system. These conventions are arbitrary, of course, but they are established by common usage, and we shall adhere to them. Some texts, however, do define Q or W with the opposite sign, so caution is needed.

Another way of expressing Eq. (20-1) is $Q = \Delta U + W$, which may be interpreted as saying that when a quantity of heat Q is added to the system, some of it is used to increase the internal energy by an amount ΔU and the remainder leaves the system again as work W done by the system. Either mechanical units of energy (such as joules) or thermal units (such as calories) may be used in Eq. (20-1), so long as all quantities are expressed in the *same* units.

Equation (20-1) provides a definition of change of internal energy. It does not specify the actual value of U for any state, but only the *change* in U for any change of state. This is not a serious deficiency. In defining potential energy of a mechanical system, we are at liberty to specify an arbitrary value for the potential energy at a particular position; we then calculate changes in the potential energy for different positions. Similarly, we can specify an arbitrary value for the internal energy of a thermodynamic system in a given state; we then compute the change ΔU which occurs when the system goes to another state. In some simple cases, such as an ideal gas, we may be able to calculate the internal energy directly from the microscopic structure of the material, but in general this is not necessary.

So far, it may appear that internal energy is a fictitious quantity invented so that energy will be conserved in processes involving both heat and work. No new principles have been introduced, only a new *definition*. Internal energy does, however, have a real fundamental significance, which lies in the experimentally established fact that the change of internal energy of any thermodynamic system for any change of state depends only on the initial and final states and not on the details of the processes which led from one state to the other. That is, the internal energy of a thermodynamic system depends only on its state, and not on the processes it has undergone before reaching this state.

This statement may sound deceptively simple and obvious. We have already remarked that internal energy is associated with microscopic mechanical energy of the system; this clearly depends only on the *state* of the system, and not on its past history. But we have demonstrated in detail the connection between internal energy and microscopic mechanical energy only in one very simple case, the ideal gas. We have no right to assume for a more general thermodynamic system that the change in internal energy of a system, as defined by Eq. (20-1), depends only on the initial and final states until we have established experimentally with a wide variety of systems that this is in fact the case. Thus the discovery that the internal energy of any thermodynamic system depends only on its state is a fundamental one and by no means trivial or obvious.

As an example, suppose an ideal gas under initial conditions of pressure, volume, and temperature P, V, and T, respectively, is to be taken to a state with the same volume V but twice the initial pressure $2P$. Because of the ideal-gas equation of state, the final temperature is $2T$. The simplest process to achieve this change of state is simply to add heat, holding the volume constant, until the pressure has risen to $2P$ and the temperature to $2T$. No mechanical work is involved because there is no volume change; thus the internal energy change is equal to the heat added, which we can measure.

Another possibility is to heat the gas while permitting it to expand just enough to keep the pressure constant at P while the temperature rises to $2T$ and then compress it back to the original volume, adding or subtracting whatever heat is needed to keep the temperature at $2T$. The final pressure is again $2P$. We can measure the net heat input Q to the system and the net work W performed by it during this sequence and compute ΔU from Eq. (20-1). Experiment shows that ΔU is the same as for the constant-volume process and indeed for *all possible* processes leading from the initial to the final state.

It follows that if the system is taken from its initial to final state by one process and then back to the initial state by another process, the *net* change in internal energy of the system must be zero. It is impossible to produce a change in the internal energy of the system without changing its state. *Each possible state has a definite internal energy*.

This observation is of the utmost importance. If this were not the case, we could take a system from one state to another by adding a certain amount of internal energy and then take it back to its original state by subtracting a *greater* amount of internal energy. The system would then be back where it started, but we would have made a profit in energy by removing more than we put in. By repeating the process, we could produce as much energy as we please; we could build a perpetual-motion machine! The reason this is not possible is that internal energy depends only on the state; when the system is restored to its original state, the net internal energy change must be zero.

The principle that the internal energy of a system depends only on its *state* is known as the *first law of thermodynamics*. Although Eq. (20-1) leads to a definition of internal energy and is not a statement of a new physical principle, the fact that the internal energy depends only on the state of the system *is* a new principle. For a few simple systems, it can be derived from a microscopic model; such a derivation is given for an ideal gas in Sec. 20-5. In general, though, the principle must be regarded as an experimental fact, established by many observations.

Two additional remarks are in order. First, it should be clear that internal energy is an *extensive* quantity, proportional to the amount of material present in the system. Second, because the internal energy depends only on the state of the system, it can be regarded as a function of the thermodynamic coordinates of the system. For a gas with thermodynamic coordinates P, V, and T, we emphasize this functional relationship by writing $U(P,V,T)$. Of course, these coordinates are not all independent but are related by the equation of state. Thus we may regard U as a function of *any two* of the three thermodynamic coordinates. In some problems it is convenient to go one step further and regard U itself as a thermodynamic coordinate. For example, for a gas we may regard U and V as independent thermodynamic coordinates; when values are specified for these quantities, the state of the system is determined, and T and P can be found. Many other combinations are possible.

20-3 CHANGES OF STATE OF AN IDEAL GAS

To illustrate the first law of thermodynamics, we consider several kinds of processes for a particularly simple thermodynamic system, a gas in a cylinder with a piston (Fig. 20-1).

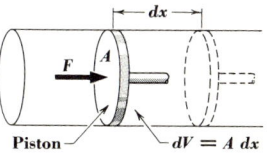

Fig. 20-1 Gas confined in a cylinder pushes with pressure P on a flat piston whose area is A. The total force F exerted on the piston is $F = PA$. When the piston is displaced outward an infinitesimal distance dx, the work done by the gas is $dW = P\,dV$.

We consider first the *work* done by the gas during a change of volume. Let the gas pressure be P and the area of the piston A. When the gas expands, pushing the piston a small distance dx, how much work does the gas do?

If the gas exerts a total force F on the piston, the work dW done during this expansion is

$$dW = F\,dx \tag{20-2}$$

since the force and displacement are in the same direction. The total force is the product of the pressure and the area: $F = PA$. Furthermore, the displacement dx is related to the volume change dV of the gas by the equation $dV = A\,dx$.

Using these relations in Eq. (20-2), we find

$$dW = PA\frac{dV}{A} = P\,dV \tag{20-3}$$

Although we have obtained Eq. (20-3) by considering a particularly simple situation, it is not hard to see that similar calculations could be made for *any* infinitesimal displacement of the boundary surface of the gas, with the same result. So the work dW done by the gas in *any* small volume change dV is given by $dW = P\,dV$.

For a finite volume change from V_1 to V_2, the total work done by the system can be found by integrating Eq. (20-3):

$$W = \int dW = \int_{V_1}^{V_2} P\,dV \tag{20-4}$$

It is often useful to represent thermodynamic processes graphically; one useful procedure is to plot pressure as a function of volume, the so-called

Fig. 20-2 Variation of pressure with volume during expansion of a gas from volume V_1 to V_2. The work done by the gas during a small change in volume dV is represented by the shaded area; the total work done during the expansion is the total area under the curve bounded by the vertical lines at V_1 and V_2.

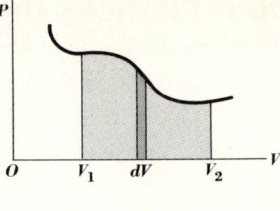

P-V diagram. An example is shown in Fig. 20-2; P is usually placed on the vertical axis. On such a diagram Eqs. (20-3) and (20-4) have a simple graphical significance. $dW = P\, dV$ is the area of the small shaded strip, and the total work W done during the expansion from V_1 to V_2, given by Eq. (20-4), is the total area bounded by the curve and the two vertical lines corresponding to V_1 and V_2.

If the pressure remains constant during the expansion, as in Fig. 20-3,

Fig. 20-3 When a gas expands in such a way that the pressure is constant, P does not vary with V. In this case, the total work done during an expansion from V_1 to V_2 is represented by the rectangular area shown: $W = P(V_2 - V_1)$.

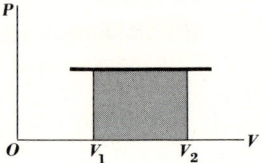

the process is called *isobaric*. In that case P may be taken outside the integral in Eq. (20-4):

$$W = P \int_{V_1}^{V_2} dV = P(V_2 - V_1) \tag{20-5}$$

When P is not constant, we must know how it varies as a function of V in order to evaluate the integral. For example, suppose the system is an ideal gas and, instead of keeping P constant, we keep the *temperature* constant. The pressure is related to the volume by $P = nRT/V$. The work done by an ideal gas in an *isothermal* expansion from volume V_1 to V_2 at constant temperature T is then given by

$$W = \int_{V_1}^{V_2} P\, dV = nRT \int_{V_1}^{V_2} \frac{dV}{V} = nRT \ln \frac{V_2}{V_1} \tag{20-6}$$

This process is shown graphically in Fig. 20-4. To keep the temperature constant, it is necessary to add heat during the expansion; otherwise, the

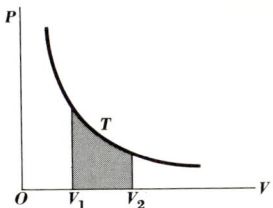

Fig. 20-4. When an ideal gas expands under conditions of constant temperature, pressure is inversely proportional to volume. The work done by the gas during this expansion is represented graphically by the shaded area. The amount of work done is given by Eq. (20-6).

temperature would drop. The cooling of gases on expansion is a familiar phenomenon; an example is air escaping from an air hose at a filling station.

One can measure the amount of heat required to keep the temperature constant during an isothermal expansion. For example, one might heat the gas with an electric heater and measure the electric energy input to the heater. Careful measurements show that the amount of heat required is exactly equal to the amount of work done by the gas during its expansion. Thus, according to Eq. (20-1), the internal energy of the gas *does not change* in such a process.

Other similar experiments with gases whose behavior approximates that of an ideal gas have shown that in all situations in which the temperature is constant, the internal energy is also constant. We conclude that the *internal energy of an ideal gas depends only on its temperature* and not on the pressure or volume. This result is also predicted by the kinetic-molecular model, but it can be obtained experimentally without considering the microscopic structure of the gas at all.

The fact that the internal energy of an ideal gas depends only on its temperature has a number of interesting consequences, one of which is exhibited by the situation shown in Fig. 20-5. A thermally insulated box

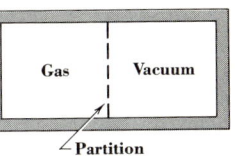

Fig. 20-5 Thermally insulated box with removable center partition. Initially, a quantity of gas is confined in the left side, while the right side is completely evacuated. When the partition is removed, the gas expands to fill the entire box. Because the walls of the box do not move, the gas does no work on its surroundings during this expansion. It is thermally insulated, so there is no transfer of heat with the surroundings. If the gas is ideal, the temperature does not change during this process. For most real gases, the temperature decreases; this is called the Joule-Thomson effect.

is divided into two equal volumes by a removable partition. A quantity of an ideal gas is initially confined in one side, while the other side is evacuated. Then the partition is removed, so that the gas can expand to fill the whole volume. We now ask how the temperature of the gas changes during this process.

Because of the construction of the box, the gas does no work on its surroundings during the expansion. (An expansion which occurs without performance of work is called a *free expansion*.) Also, if the box is thermally insulated, no heat is transferred to or from the system. Thus $W = Q = 0$, and according to Eq. (20-1), the internal energy does not change. For an ideal gas U depends only on T; since U is constant, T must also be constant. *The temperature of an ideal gas does not change during a free expansion.*

As already pointed out, the ideal gas is a model which describes the behavior of many real gases reasonably well; the statement that U depends only on T is part of this model. Not all thermodynamic systems exhibit this simple behavior. In general, U may depend on pressure as well as temperature. For such a system, U is still constant during a free expansion, for the reason given above, but T may change. For most gases under ordinary conditions, T decreases during a free expansion, although there are some exceptions. The cooling which occurs on expansion is called the *Joule-Thomson effect* after its discoverers; it is used in one process for liquefying gases by cooling.

The definition of ΔU given by Eq. (20-1) and the general formulas for computing work, Eqs. (20-2) to (20-4), are valid for any thermodynamic system whose coordinates are (P, V, T) and not just for an ideal gas. Equation (20-6) is, of course, more specific; it is valid only for an ideal gas.

20-4 SPECIFIC HEATS OF AN IDEAL GAS

In Chap. 18 we used the kinetic-molecular model to calculate the specific heat of an ideal gas. We assumed that the volume of the gas is constant, so all the heat added to the gas remains as internal energy and none leaves the system as work done on the surroundings. To emphasize that this specific heat refers to a constant-volume (isovolumic or *isochoric*) process, we use the symbol C_V for the molar specific heat at *constant volume*.

Often it is convenient to heat a substance under conditions of constant *pressure* rather than constant *volume*. The specific heat is not the same under these conditions; as heat is added at constant pressure, the system expands, doing work against its surroundings. Therefore *additional* heat must be added to compensate for this loss of energy. We define a new specific heat C_P,

SPECIFIC HEATS OF AN IDEAL GAS [20-4]

which refers to heating a substance at *constant pressure* (isobarically). Ordinarily we expect C_P to be larger than C_V.

It is easy to calculate the relation between these two specific heats for an ideal gas, using the first law of thermodynamics. We consider an infinitesimal change of state in which heat dQ is added, the internal energy of the system changes by dU, work dW is done by the system, and its temperature rises by dT. For a process which takes place at constant volume, $dW = 0$ and $dQ = dU$. Thus, for an isovolumic process,

$$dQ = dU = nC_V \, dT \tag{20-7}$$

Now we make the important observation that for an ideal gas the change in internal energy is given by

$$dU = nC_V \, dT \tag{20-8}$$

even if V is not constant, because the internal energy of an ideal gas depends only on its temperature. When V changes, dW is not zero, and dQ is not equal to dU, but it is still true that $dU = nC_V \, dT$.

When P, rather than V, is held constant, the amount of heat which must be added is

$$dQ = dU + dW \tag{20-9}$$

The change in internal energy is still given by Eq. (20-8). The work is $dW = P \, dV$; to express it in terms of dT, we differentiate the ideal-gas equation, using the fact that P is constant:

$$\begin{aligned} PV &= nRT \\ P \, dV &= nR \, dT = dW \end{aligned} \tag{20-10}$$

Combining Eqs. (20-8) and (20-10) with Eq. (20-9), we find

$$\begin{aligned} dQ &= nC_V \, dT + nR \, dT \\ &= n(C_V + R) \, dT \end{aligned} \tag{20-11}$$

This equation shows that the quantity $C_V + R$ is the molar specific heat for the process, which we have called the specific heat at constant pressure C_P. Thus we have found that

$$C_P = C_V + R \tag{20-12}$$

This is a remarkably simple relation. It is particularly pleasing because the results of the kinetic-molecular calculations of specific heats in Chap. 18 are simple multiples of the gas constant R. For an ideal monatomic gas,

$C_V = 3R/2$ and $C_P = 5R/2$. For an ideal diatomic gas, $C_V = 5R/2$ and $C_P = 7R/2$.

The relation between the two specific heats is given by their *ratio*, usually denoted by γ:

$$\gamma = \frac{C_P}{C_V} \tag{20-13}$$

For an ideal monatomic gas, $\gamma = 5/3$; for an ideal diatomic gas, $\gamma = 7/5$. If γ is known, both C_V and C_P for an ideal gas can be determined by use of Eq. (20-12).

Example
If $1/2$ mol of oxygen is heated from 0 to $100°C$, how much heat must be added (*a*) when the volume is held constant and (*b*) when the pressure is held constant if the initial pressure is 1.00×10^5 N/m² (about 1 atm)? In part *b*, how much work does the gas do?

Solution
Assuming that oxygen behaves as an ideal diatomic gas, we have $C_V = 5R/2$ and $C_P = 7R/2$. For part *a*, an isovolumic process,

$$Q = nC_V \Delta T$$
$$= 1/2(5/2)(8.314 \text{ J/K})(100 \text{ K})$$
$$= 1{,}040 \text{ J} = 249 \text{ cal}$$

For part *b*, an isobaric process,

$$Q = nC_P \Delta T$$
$$= 1/2(7/2)(8.314 \text{ J/K})(100 \text{ K})$$
$$= 1{,}455 \text{ J} = 348 \text{ cal}$$

The work done by the gas in part *b* with P constant, is $W = P(V_2 - V_1)$. Using the ideal-gas equation,

$$W = nRT_2 - nRT_1$$
$$= 1/2(8.314 \text{ J/K})(100 \text{ K})$$
$$= 415 \text{ J} = 99 \text{ cal}$$

The *difference* between the two amounts of heat is equal to work done in the isobaric process. This must be so because the change in the internal energy of the system is the same in both cases.

The two specific heats of an ideal gas are particularly simple, inasmuch as they are both constants independent of the pressure and temperature of the gas. Few actual thermodynamic systems are so simple in this respect. We have already discussed in Chap. 17 situations in which specific heats depend on temperature; intermolecular forces cause the specific heat of a nonideal gas to depend on pressure as well, since the molecules have both potential and kinetic energy.

20-5 ADIABATIC PROCESS FOR AN IDEAL GAS

As another application of the first law of thermodynamics, we consider an expansion or compression of an ideal gas during which it is thermally insulated from its surroundings so that no heat is conducted in or out. A process which occurs without transfer of heat to or from the system is called an *adiabatic* process. When an ideal gas expands adiabatically, it loses internal energy because it does work on its surroundings. Therefore its temperature must *decrease*. In an adiabatic compression, the temperature rises; this effect is responsible for the high temperature in a diesel engine which causes the fuel to ignite spontaneously when injected into the cylinders.

We now calculate the relation between volume and temperature for an adiabatic expansion or compression of an ideal gas. For an adiabatic process for any substance, $dQ = 0$; thus Eq. (20-1) requires that $dU = -dW$. For an ideal gas this becomes

$$nC_V dT = -P\,dV = -\frac{nRT}{V}dV \tag{20-14}$$

Dividing by n and rearranging,

$$\frac{dT}{T} + \frac{R}{C_V}\frac{dV}{V} = 0 \tag{20-15}$$

The coefficient R/C_V can be expressed in terms of γ as follows. From Eq. (20-12), $R = C_P - C_V$. Therefore,

$$\frac{R}{C_V} = \frac{C_P - C_V}{C_V} = \frac{C_P}{C_V} - 1 = \gamma - 1 \tag{20-16}$$

Equation (20-15) then becomes

$$\frac{dT}{T} + (\gamma - 1)\frac{dV}{V} = 0 \tag{20-17}$$

This is a differential equation relating small changes dT and dV in temperature and volume. To find a general relation between T and V, we integrate:

$$\ln T + (\gamma - 1)\ln V = \text{constant} \tag{20-18}$$

Using the properties of logarithms, we can rewrite this as

$$TV^{\gamma-1} = \text{constant} \tag{20-19}$$

That is, the quantity $TV^{\gamma-1}$ has the same value for all stages of an adiabatic process. If we start with initial conditions T_1 and V_1 and end at T_2 and V_2, then

$$T_1 V_1^{\gamma-1} = T_2 V_2^{\gamma-1} \tag{20-20}$$

Example
Air in a cylinder of a diesel engine, with an initial temperature of 300 K, is compressed to one-sixteenth its original volume. Find the final temperature.

Solution
From Eq. (20-20), we have

$$T_2 = T_1 \left(\frac{V_1}{V_2}\right)^{\gamma-1}$$

Air is mostly a mixture of diatomic gases, so we may take $\gamma = 7/5 = 1.4$. Therefore,

$$T_2 = (300)(16)^{0.4} = (300)(3.03) = 909 \text{ K}$$

Fractional powers of numbers like those appearing above can be evaluated easily with a log table or a log-log duplex slide rule.

A relation between P and V for an adiabatic process can be derived from Eq. (20-19) by using the ideal-gas equation to eliminate T. The derivation of this relation is left as an exercise; the result is

$$PV^{\gamma} = \text{constant}$$

or

$$P_1 V_1^\gamma = P_2 V_2^\gamma \tag{20-21}$$

Equation (20-21) shows that the pressure drops more rapidly with increasing volume in an adiabatic expansion than in an isothermal expansion, since for an adiabatic expansion P is proportional to $1/V^\gamma$ rather than $1/V$. Adiabatic and isothermal expansions of an ideal gas are shown graphically in Fig. 20-6.

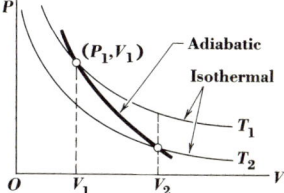

Fig. 20-6 *P-V diagram showing adiabatic and isothermal expansions starting from initial conditions (P_1, V_1). In the isothermal case, the temperature is constant at T_1; during the adiabatic expansion, the temperature drops to T_2. The isotherm for temperature T_2 is also shown.*

20-6 MICROSCOPIC VIEW OF AN ADIABATIC PROCESS

According to the macroscopic definition of internal energy, the change in internal energy in an adiabatic process is equal to the negative of the work which the system does during the process. For an ideal gas, we can also calculate from a *microscopic* viewpoint the work done during an adiabatic process and the change in the kinetic energies of the molecules. We find that the total change in kinetic energy is the negative of the work done by the gas. This establishes directly that the internal energy of an ideal gas is identical with the total kinetic energy of the molecules.

In making this calculation, we use directly the formulations of Sec. 18-3, where the kinetic-molecular model was used to find the pressure exerted by an ideal gas. In this case, however, instead of considering collisions of molecules with an area A of wall, we place the gas in a cylinder with a piston whose area is A and consider the collisions of molecules with the surface of this piston. For an adiabatic expansion, we suppose that the piston is moving with a speed u. For reasons which will soon become clear, we assume that u is very much smaller than the average molecular speed.

Following the procedure of Sec. 18-3, we consider all molecules having an x component of velocity v_x. When such a molecule collides elastically with a stationary wall, it rebounds with a velocity whose x component is $-v_x$. But if the wall is moving, the x velocity after impact is no longer $-v_x$. In an elastic collision the *relative* velocity between the two objects has the same magnitude before and after the collision. When the piston moves with speed u, as shown in Fig. 20-7, the x component of relative velocity before

Fig. 20-7 When a molecule whose x component of velocity is v_x collides with a piston moving with speed u as shown, the x component of velocity of the molecule relative to the piston is $v_x - u$. The velocity relative to the piston after an elastic collision is $-(v_x - u)$; the velocity of the molecule relative to the container is $-(v_x - u) + u$.

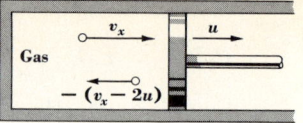

collision is $v_x - u$; after the impact it is $-(v_x - u)$, so the velocity of the molecule relative to the *container* is

$$-(v_x - u) + u = v_x + 2u = -(v_x - 2u) \tag{20-22}$$

That is, the *speed* of the molecule is reduced by $2u$.

Correspondingly there is a loss of kinetic energy. Before the collision with the piston, the kinetic energy is $\frac{1}{2}mv_x^2$. Afterward, it is

$$\tfrac{1}{2}m(v_x - 2u)^2 = \tfrac{1}{2}mv_x^2 - 2mv_x u + 2mu^2$$

That is, the kinetic energy *changes* by an amount

$$\Delta E = -2mv_x u + 2mu^2 = -2mv_x u \left(1 - \frac{u}{v_x}\right) \tag{20-23}$$

When the speed of the piston is much smaller than the speeds of the molecules, u/v_x is much smaller than unity and we can discard the second term in parentheses in Eq. (20-23).

Thus each molecule loses an amount of kinetic energy $2mv_x u$ when it collides with the moving piston. The gas loses kinetic energy by this mechanism at a *rate* equal to the loss of kinetic energy per molecule multiplied by the number of collisions with the piston per unit time. According to Eq. (18-8) and the equation which follows it, the number of collisions per unit time is $\frac{1}{2}Av_x N/V$. Thus the rate of change of kinetic energy is

$$\frac{dE}{dt} = -2mv_x u \frac{\frac{1}{2}Av_x N}{V} = -\frac{Amv_x^2 N}{V} u \tag{20-24}$$

Comparing the last form of Eq. (20-24) with Eq. (18-9), which gives the total force exerted on area A, we see that the quantity $Amv_x^2 N/V$ is the total force, therefore

$$\frac{dE}{dt} = -Fu \tag{20-25}$$

But Fu is exactly equal to the rate at which work is done on the piston by the gas. Thus we conclude that the rate at which the gas loses energy is equal to the rate at which it does work, and the total loss of internal energy is equal to the total work done.

If we were content with a macroscopic view of the behavior of ideal gases, this calculation would be unnecessary. We would simply define internal energy by means of Eq. (20-1) and establish experimentally that the internal energy depends only on the state of the system. But, in addition, the development just given shows conclusively that the internal energy of an ideal gas is identical with the kinetic energy of molecular motion.

20-7 THERMODYNAMIC PROPERTIES OF MATTER

Although in this chapter the first law of thermodynamics has been discussed mostly with reference to changes of state of ideal gases, it can be applied as well to other thermodynamic systems. The great power of thermodynamics, in fact, lies in its applicability to substances and systems whose behavior is much more complicated than that of ideal gases. The equation of state for such a system may be very complex, and in some cases it is necessary to represent the equation graphically, as in Fig. 19-1, or by means of tables.

When the system under consideration is not an ideal gas, it is not in general true that the internal energy depends only on the temperature. For many systems U is also a function of pressure. The relation $dU = nC_V dT$, which holds for *all* processes for an ideal gas (even if the volume is not constant), is true for such nonideal systems *only when the volume is constant*. Otherwise, the change in internal energy in a given process may also depend on the pressure change. In practical problems, such as the design of steam turbines, gasoline engines, and jet engines, it is often necessary to measure the changes in internal energy with temperature and pressure for a given substance and tabulate the results. These tables can then be used to find the internal energy for any given state for further calculations with the first law of thermodynamics.

Similarly, as remarked in Sec. 20-4, the specific heats of the material may vary with T and P, and this may necessitate tabulating the values for various conditions for use in calculations. These complications, however, do not invalidate the basic principle under discussion, the existence of an internal energy which depends only on the thermodynamic coordinates of the system.

20-8 QUASI-STATIC PROCESSES

It was pointed out in Sec. 19-1 that a macroscopic state of a system can be described by its thermodynamic coordinates only when that state is one of thermodynamic equilibrium. The equation of state of a system has meaning only for thermodynamic equilibrium states. In a non-equilibrium state, when different parts of the system are at different temperatures or different pressures, the equation of state and the concepts of pressure and temperature for the system as a whole have no meaning.

But what happens when the system makes a transition from one state to another? We have computed the work done by an expanding ideal gas in a cylinder; during this expansion, the system *cannot* be in an equilibrium state, for otherwise it would not be able to accelerate the piston to set it into motion. When heat is conducted into a system, there must be a departure from thermal equilibrium, since conduction of heat depends on a difference in temperature between two points. Thus a transition from one state to another *must* involve non-equilibrium states. Yet in calculating the work done by an expanding gas, we use the equation of state of the gas, assuming implicitly that it is always in an equilibrium state. Is there any way to reconcile this apparent inconsistency?

There is a way, and it involves a very important though somewhat subtle point. We have assumed, without really saying so, that the system is always *very close* to an equilibrium state. If there is only an *infinitesimal* difference between the temperatures at two points of a system and at most an infinitesimal difference in pressure from one point to another, then only an infinitesimal error will be made by assuming that the system is always in an equilibrium state. Furthermore, by conducting heat sufficiently slowly or by doing work sufficiently slowly, we can make a transition from one state to another in such a way that at every stage in the process the system approximates a possible equilibrium state as closely as we like.

A process during which a system is never more than infinitesimally far away from an equilibrium state is referred to as a *quasi-static* or *quasi-equilibrium* process. Such processes represent idealizations which can never be exactly achieved in nature but which can be approximated very closely under some circumstances. Thus it is often useful to use a quasi-static process as a model to represent a real physical process. For example, in the next section a series of quasi-static processes is used to represent what takes place in an internal-combustion engine. In this case, the model is only a crude approximation of reality, but it is still sufficiently accurate to provide useful information.

What factors determine whether or not a process may be considered quasi-static? There is no simple answer to this question. When the speed of a moving piston is small compared with the speeds of most molecules, there is likely to be only a small departure from mechanical equilibrium;

this condition was used in Sec. 20-6 in the analysis of an adiabatic process for an ideal gas. The approach to thermal equilibrium is directly related to thermal conductivity, which varies widely from one material to another.

A quasi-static process is *reversible* in the sense that a process which takes place under quasi-static conditions can always be made to proceed in the opposite direction by an infinitesimal change in the conditions. For example, a quasi-static expansion can be reversed by an infinitesimal increase in pressure; a quasi-static flow of heat can be reversed by changing a temperature infinitesimally. The concept of a reversible process is one of the central ideas in the formulation of the second law of thermodynamics, a basic principle governing the directionality of physical processes. This subject is discussed in Chap. 21.

It is easy to imagine changes of state which involve more spectacular departures from equilibrium than the quasi-static processes just discussed. Consider a gas in a cylinder with a piston which is jerked partway out of the cylinder with a speed greater than the speed of most of the molecules. Some molecules are "left behind," and the gas does not do as much work during this expansion as it would if conditions of mechanical equilibrium were maintained. If such a process is carried out adiabatically, the change in internal energy is not the same as though it had been carried out in a quasi-equilibrium manner.

A non-equilibrium process cannot be represented on a P-V diagram, since the intermediate states are not equilibrium states. Sometimes, however, it is desirable to indicate on the diagram that a non-equilibrium process has taken place. This can be shown as in Fig. 20-8. In this figure the thin solid line represents a quasi-static process going from (P_1, V_1) to (P_2, V_2), and the

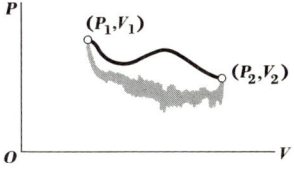

Fig. 20-8 Quasi-static process leading from initial conditions (P_1, V_1) to final conditions (P_2, V_2) is represented on a P-V diagram by the line shown. In such a process, the system is never more than an infinitesimal distance from an equilibrium state. A process which involves non-equilibrium states may be represented symbolically by a broad shaded line, as shown. The work done by the system and the heat transferred to the surroundings are in general not the same as for an equilibrium process with the same initial and final states; the change in internal energy, however, is the same in both cases.

546 HEAT AND WORK

broad shaded line indicates a non-equilibrium process between the same initial and final equilibrium states. In general, the work done by the system and the heat transferred to its surroundings are different in the two cases. If the initial and final states are the same, however, the change in internal energy must be the same in both cases, whether a non-equilibrium process has occurred or not. The first law of thermodynamics is valid for non-equilibrium processes.

An important feature of the first law is that it facilitates calculations of the relation between initial and final states of the system without detailed knowledge of the process which led from one to the other. This advantage may be compared with the usefulness of the principle of conservation of mechanical energy in mechanics, with which it is often possible to relate two different states of motion without calculating a complete description of the motion.

20-9 A PRACTICAL PROBLEM

To conclude this chapter, we consider a practical problem which involves several principles we have discussed, a calculation of the efficiency of an internal-combustion engine. What is meant by *efficiency?* Clearly, the efficiency of an engine has something to do with the effectiveness with which it uses its fuel. When gasoline burns in an engine, heat is produced. The engine transforms some of this heat into mechanical work; the more work it gets from a given amount of fuel, the more *efficient* it is said to be. More precisely, the burning of a certain quantity of gasoline results in the production of a quantity of heat Q, and as a result the engine produces an amount of mechanical work W; the efficiency η of the engine is defined as

$$\eta = \frac{W}{Q} \tag{20-26}$$

It is assumed that Q and W are measured in the same units, either mechanical or thermal units of energy.

In any situation where heat is transformed partly into mechanical energy, the nature of the process is often emphasized by use of the term *thermodynamic efficiency*. The following discussion gives a calculation of the thermodynamic efficiency of an engine based on a very simple model.

Briefly, the succession of processes which takes place in a four-cycle gasoline engine is as follows: First, air which has been mixed with gasoline vapor in the carburetor enters the cylinder at approximately atmospheric pressure and temperature (intake stroke). Second, it is compressed to a small fraction of its original volume (compression stroke). Third, the mixture is ignited, causing a rapid rise in temperature (combustion), and permitted to expand, pushing the piston down in the cylinder (power stroke). Finally, the exhaust valve opens, and the gas expands further through the exhaust system (exhaust stroke).

We now construct a very simple model to represent this succession of processes. We assume that the substance in the cylinder is a constant quantity of an ideal gas and that all processes are quasi-static. The maximum volume of the cylinder is V, and its volume when the piston is all the way in is V/r, where r is the *compression ratio* of the engine; typically r has a value between 6 and 11.

The sequence of processes is shown on a P-V diagram in Fig. 20-9.

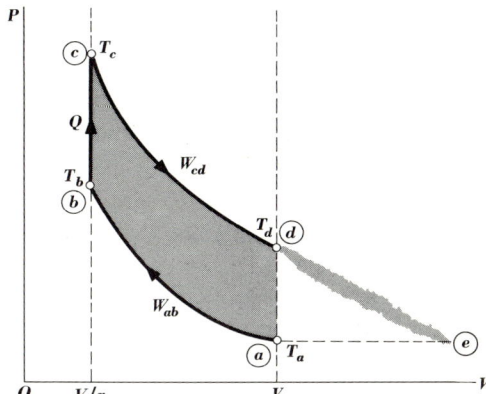

Fig. 20-9 P-V diagram for the sequence of processes occurring in a gasoline engine. In the idealized model discussed, all processes are assumed to be quasi-static except the exhaust stroke $d \to e$, during which no useful work is done.

Beginning at the point labeled a, the gas is brought into the cylinder (intake) at atmospheric pressure and temperature (P_a, T_a). Then the following processes occur:

$a \to b$ (compression): The gas is compressed adiabatically from volume V to V/r, during which its temperature rises to T_b.

$b \to c$ (combustion): The mixture is ignited, resulting in the addition of a quantity of heat Q. We assume that this process occurs instantaneously, so that the temperature rises abruptly from T_b to T_c; the pressure also rises.

$c \to d$ (power): The gas expands quasi-statically and adiabatically from volume V/r back to the original volume V, during which the temperature drops from T_c to T_d.

$d \to e$ (exhaust): The gas expands to some unknown final temperature and volume in the exhaust system. No *useful* work occurs during this expansion.

In order to calculate the efficiency, we must compute the net work which the gas does during this sequence of processes. Clearly, no work is done during process $b \to c$; no *useful* work is done during process $d \to e$, since this is work done on the outside atmosphere rather than on the piston of the engine. So the total useful work is $W = W_{ab} + W_{cd}$. Since processes $a \to b$ and $c \to d$ are adiabatic, the work done in each is simply the negative of the change of internal energy. Because the gas is ideal, the change of internal energy is determined readily from the temperature change by Eq. (20-8). Thus we find

$$W = -nC_V(T_b - T_a) - nC_V(T_d - T_c) \tag{20-27}$$

We note that W_{ab} is negative, while W_{cd} is positive and larger in magnitude than W_{ab}.

The heat added during process $b \to c$ can be related simply to the change in temperature, since this process takes place at constant volume:

$$Q = nC_V(T_c - T_b) \tag{20-28}$$

Using the last two equations with Eq. (20-26), we obtain

$$\eta = \frac{W}{Q} = \frac{T_a - T_b + T_c - T_d}{T_c - T_b} = 1 - \frac{T_d - T_a}{T_c - T_b} \tag{20-29}$$

This is a reasonably simple expression for the efficiency, but we can obtain an even simpler one by expressing η in terms of the volumes. This can be done by applying Eq. (20-20) to each of the two adiabatic processes. We find

$$T_a V^{\gamma-1} = T_b \left(\frac{V}{r}\right)^{\gamma-1} \qquad T_c \left(\frac{V}{r}\right)^{\gamma-1} = T_d V^{\gamma-1} \tag{20-30}$$

$$T_b = T_a r^{\gamma-1} \qquad T_c = T_d r^{\gamma-1} \tag{20-31}$$

To use this result to simplify Eq. (20-29) we need only subtract the first of Eqs. (20-31) from the second:

$$T_c - T_b = r^{\gamma-1}(T_d - T_a) \tag{20-32}$$

Using this in Eq. (20-29), we finally find

$$\eta = 1 - \frac{1}{r^{\gamma-1}} \tag{20-33}$$

This is a remarkably simple result. To the extent that our model is an adequate description of the processes occurring in a gasoline engine, the efficiency depends only upon the compression ratio. Equation (20-33) shows that η is always less than unity. This must be so, of course, for we can never get out more energy than was put in. But the larger r is, the closer η is to unity. Thus, for maximum efficiency, it is desirable to have the compression ratio as large as possible. Practical difficulties such as mechanical strength and fuel-combustion properties impose limitations on r.

This example illustrates the use of some of the principles developed in this chapter. It also introduces a subject which must be discussed in more detail, the problem of efficiency of conversion of heat into work. The gasoline engine *wastes* some of its heat in the exhaust cycle. Is it possible, by clever design, to build an engine which converts *all* its heat into useful work?

No one has ever succeeded in making such an engine, and we now believe that it is impossible to do so. The principles we have discussed so far do not forbid such an achievement, but perhaps there is some new principle which limits the possible efficiencies of engines and similar devices. There is indeed such a principle; it is the *second law of thermodynamics*, discussed in Chap. 21.

Problems

20-1 A lamp bulb operating on direct current consumes electric energy at a rate given by $P = EI$, where P is the power in watts, E is the potential difference across the bulb in volts, and I is the current in amperes. If $V = 110$ V and $I = 0.9$ A, find the power. If this bulb is immersed in 5 kg of water contained in a copper bucket of mass 2 kg, how much does the temperature rise in 3 min?

20-2 A certain mountaineer ascended from Lupine Meadows (elevation 6,800 ft) to the summit of Grand Teton (13,766 ft) and returned. His weight is 170 lb.
   a If the climber's body were 100 percent efficient in converting food to mechanical energy (which of course it is not), how much food energy would he need to reach the summit? (One food-value calorie is actually a kilocalorie, equal to 10^3 cal.)
   b If the climber is thermally insulated, how much does his temperature rise during the descent?

20-3 A ball of putty of mass m and initial speed v collides with another ball of putty of the same mass but initially at rest. The two stick together. If the specific heat of putty is c, how much does the temperature rise as a result of the collision?

20-4 Niagara Falls is about 50 m high. If the potential energy lost by water going over the falls is converted completely into heat, how much does its temperature rise? What effects might increase or decrease this temperature rise?

20-5 The two compartments of a thermally insulated container are separated by a partition; one contains gas, the other vacuum. The partition is suddenly broken, and the gas fills the container. Prove that the total internal energy of the gas does not change during the process. Is this statement still valid if the gas is not an ideal gas? Explain.

20-6 In the situation of Prob. 20-5, is the conclusion altered if initially the two parts contain quantities of the same gas at different temperatures and pressures? Explain.

20-7 A heat engine is any device for converting heat into mechanical work. A certain engine took 100 cal from a source of heat during expansion of a substance in the engine. It then gave off 50 cal of heat while the substance was compressed. At the end of the process the substance was in the same thermodynamic state as at the beginning. How much work did the engine do during this process? Express the result in mechanical units.

20-8 A factory worker with a little knowledge of thermodynamics proposes to heat the coffee in his thermos bottle by shaking it. In this process, does the temperature actually rise? Is work done on the coffee? Is heat added to it? Does its internal energy increase?

20-9 A certain refrigerator freezes ice cubes at the rate of 3.6 kg/h, starting with water at 20°C. Its electric-power input is 100 W. At what rate must it expel heat to the air from its condenser?

20-10 Is the work done by a gas during an infinitesimal expansion dV given by $dW = P\,dV$ even if the gas is not ideal? Does this relation hold for a liquid or a solid? Explain.

20-11 At 100°C and 1 atm, water has a latent heat of vaporization of 540 cal/g. The density of water vapor under these conditions is 0.597 kg/m^3. What fraction of the latent heat goes into work done by the vapor, and what fraction goes into increased internal energy?

20-12 At a pressure of 1 atm, the solid and vapor phases of carbon dioxide are in equilibrium at -78.5°C. At this temperature the density of the solid (Dry Ice) is 1.53 g/cm^3, and the latent heat of sublimation is 137.9 cal/g. If the vapor behaves as an ideal gas, what fraction of the latent heat is used as work done by the material and what fraction to increase its internal energy? Is it likely that the ideal-gas equation really describes adequately the behavior of the vapor under these conditions? Explain.

20-13 How much work is done by n mol of a gas described by the van der Waals' equation of state [Eq. (19-3)] during an isothermal expansion from volume V_1 to V_2?

20-14 For a magnetic material whose thermodynamic coordinates are H, M, and T, as discussed in Prob. 19-4, the definition of internal energy in differential form is $dU = dQ + H\,dM$. That is, the work done by the substance is $dW = -H\,dM$. If the equation of state is $M = mCH/T$, how much work does the substance do during an isothermal magnetization from M_1 to M_2? (This work is not mechanical work in the usual sense. It may be electric energy added to the circuit which powers the electromagnet associated with the magnetic field H.)

20-15 For a gas, is C_P ever smaller than C_V? Explain.

20-16 Show that the molar specific heats of an ideal gas can be expressed in terms of γ and R as follows:

$$C_V = \frac{R}{\gamma - 1} \qquad C_P = \frac{\gamma R}{\gamma - 1}$$

Are these expressions correct for a nonideal gas? Explain.

20-17 A cylinder with a piston contains 0.32 kg of oxygen at a pressure of 3 atm and a temperature of 27°C. It undergoes the following series of processes: (1) It is heated at constant volume until the pressure doubles. (2) It expands isothermally until the pressure drops to the original value. (3) It is cooled isobarically back to the original temperature.
  a Show this series of processes on a P-V diagram. Label all relevant values of P, V, and T.
  b Make a table showing the heat added to the gas, the work done by it, and the change in its internal energy for each of the steps.
  c Calculate the total heat added, the total work done, and the total change in internal energy for the whole sequence.

20-18 An ideal gas undergoes the following series of processes, starting at (P_0, V_0, T_0): (1) It is heated isovolumically to pressure $2P_0$. (2) It is heated isobarically to volume $2V_0$. (3) It is cooled isovolumically to pressure P_0. (4) It is cooled isobarically to volume V_0.
  a Represent the sequence on a P-V diagram, labeling each end point with the appropriate values of the thermodynamic coordinates.
  b Make a table showing the work done by the gas, the heat added to it, and the change in its internal energy during each step.
  c Find the net work done by the gas during the whole cycle, the net heat absorbed, and the net change in internal energy.

552 HEAT AND WORK

20-19 A cylinder fitted with a piston contains 2.8 g of nitrogen at a pressure of 1.0×10^5 N/m^2 and a temperature of 27°C.

    a   If the gas is heated isobarically to 327°C, find the heat added, the work done by the gas, and the change in its internal energy.

    b   If, instead, the gas is heated at constant volume to 327°C and then permitted to expand isothermally to the final volume of part *a*, find Q, W, and ΔU for each of the two steps and the *total* Q, W, and ΔU.

20-20 For the magnetic material discussed in Prob. 20-14, one can define two specific heats analogous to C_V and C_P for a gas. These may be called C_M and C_H, corresponding to processes in which M and H, respectively, are held constant. For such a system, which of these specific heats is larger? Explain.

20-21 For the magnetic material discussed in Probs. 19-4, 20-14, and 20-20, suppose that C_M is a constant which is independent of H and T and that the internal energy U depends only on T. Derive an expression for C_H, and show that it is *not* a constant.

20-22 Suppose it is desired to change the state of an ideal gas from (P, V, T) to $(\frac{1}{2}P, 2V, T)$. This can be accomplished either by (1) an isothermal expansion or (2) isovolumic cooling to $\frac{1}{2}P$, followed by isobaric heating and expansion to $2V$. Show directly that neither Q nor W is the same for (1) and (2) but that ΔU is the same in both cases.

20-23 Show that in an adiabatic process for an ideal gas the quantity PV^γ is constant.

20-24 Derive a relation between P and T in an adiabatic process for an ideal gas.

20-25 Derive an equation relating M and T for the adiabatic magnetization of a magnetic material if U depends only on T, C_M is constant, and the equation of state is $M = mcH/T$. Show that an adiabatic magnetization is always accompanied by a rise in temperature.

20-26 A cylinder contains 0.1 g of helium at 27°C and 1 atm. The gas is compressed adiabatically until the volume is one-eighth the original volume. Find the final temperature and pressure.

20-27 A gas expands to twice its original volume. Is the work done by the gas greater if the process is isothermal or if it is adiabatic? Explain.

20-28 The bulk modulus of an ideal gas is measured under two different conditions: (1) constant temperature and (2) adiabatic compression. Show that the *isothermal* bulk modulus is equal simply to P but that the *adiabatic* bulk modulus is γP.

20-29 An ideal gas expands adiabatically from V_1 to V_2, starting at temperature T_1. Find the work done by the gas in terms of T_1, the volume ratio V_1/V_2, and constants.

20-30 An ideal gas expands from volume V_1 to V_2. The process may be isobaric, isothermal, or adiabatic. Sketch each of these on a P-V diagram. For which is Q the greatest? The least? For which are W greatest and least? For which are ΔU greatest and least?

20-31 Repeat the discussion of Sec. 20-6 for motion of the piston corresponding to *compression* of the gas. Show that the work done on the gas equals the increase in its internal energy.

20-32 A certain spring has the property that at a constant temperature the force is proportional to the elongation; the spring constant depends on temperature and is, in fact, directly proportional to the absolute temperature. Thus the force F required to stretch the spring a distance x at temperature T is $F = kTx$, where k is a constant. The specific heat of the spring was measured under various conditions of T and x, keeping x constant for any individual measurement, and was found to be a constant independent of T and x.
  a Formulate the first law of thermodynamics for this system.
  b Derive an expression for the specific heat which would be measured if F, rather than x, were held constant.
  c Suppose the spring is stretched adiabatically. Derive an expression relating temperature and elongation for such a process.

20-33 One mole of an ideal monatomic gas is confined in a cylinder with a piston at initial conditions (P_0, V_0, T_0). The piston is suddenly pulled out just rapidly enough for the pressure to drop to $P_0/3$ and remain at this value until a new equilibrium state is reached. Assuming the system is thermally insulated, find the final equilibrium volume and temperature. Is this a quasi-static process? Explain.

20-34 Is it reasonable to regard the expansion of hot gases in the cylinders of an automobile engine as a quasi-static process? One criterion for the maintenance of mechanical equilibrium, as pointed out in the text, is that the motion of the piston must be much slower than the average molecular speeds. An estimate of maximum piston speeds can be obtained for an engine with 4-in. stroke and maximum speed of 5,000 r/min by assuming that the piston undergoes simple harmonic motion with amplitude 2 in. and frequency 5,000 min^{-1}. Compare this maximum piston speed with molecular speeds for air at the temperature encountered in the engine (several hundred degrees Celsius).

Second Law of Thermodynamics 21

Many natural processes proceed more readily in one direction than in the opposite direction. Investigation of various irreversible processes leads to a verbal formulation of the second law of thermodynamics, governing the direction of natural processes. Irreversibility is also associated with an increase in disorder. The concept of entropy is introduced as a quantitative measure of the disorder of a system; this is then used for a mathematical formulation of the second law of thermodynamics. It is shown that the second law can be used to define a temperature scale which is not dependent on the properties of any particular material. Finally, a microscopic meaning of the concept of entropy, in terms of the statistical probability of the state, is discussed briefly.

21-1 DIRECTIONALITY OF NATURAL PROCESSES

Some processes involving a change of state of a thermodynamic system, with accompanying transfer of heat and work, can be carried out as readily in one direction as another. An example is the quasi-static adiabatic expansion of a gas, discussed in Chap. 20. In the situation shown in Fig. 21-1, the piston confining the gas in the cylinder is held in equilibrium by the weight of the mass placed on it. By removing this mass, a very small amount at a time, we can permit the gas to expand so slowly that at any instant its state differs only infinitesimally from an equilibrium state. If the mass is now replaced, again very gradually, the whole process is reversed, and the final state is the same as the initial one, provided that friction between the piston and the cylinder is negligible. Such a process is called a *reversible* process.

Fig. 21-1 Reversible adiabatic expansion. The cylinder is thermally insulated, so no heat flows in or out of the gas. If the weights are removed in very small increments, the gas remains nearly in a state of thermal and mechanical equilibrium, and the process is quasi-static. Such a process is also reversible; if the weights are replaced, again in small increments, the system returns to its original state, going through a sequence of states which are very nearly equilibrium states.

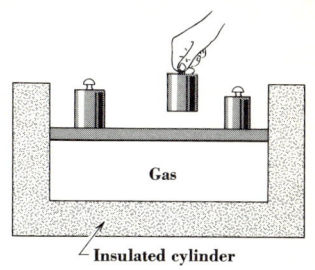

It is easy to think of processes which are not so readily reversible. Heat flows spontaneously from a hot body to a colder body in contact with it, and there is no easy way to reverse this flow. When two bottles containing different gases are connected, the gases mix spontaneously, but there is no simple way to make them separate again. It is easy to convert a given amount of mechanical energy entirely into heat by means of friction, but considerably more elaborate measures are necessary to convert heat into mechanical energy, and no one knows how to do this with 100 percent efficiency. Each of these processes seems to have a preferred direction. We call these *irreversible* processes.

None of the principles we have studied thus far tells us *why* such processes should be irreversible. If heat were sometimes to flow from a cold body to a hot body, it would not violate the first law of thermodynamics, but we know that this does not happen spontaneously. In order to make it happen, we must add work from the outside, as in an electric refrigerator. There must be some new principle involved which has to do with the *preferred direction* of natural processes. The main objective of this chapter is to discover and understand this principle and some of its simple applications.

Another characteristic of irreversible processes in nature is the tendency toward increased microscopic *disorder*. By disorder we mean randomness, or chaos, or molecular disorganization. Any partial sorting or arranging of molecules according to position, velocity, or kind of molecule increases the organization or *order* of the system and decreases its disorder.

It may seem at first glance that disorder has nothing to do with irreversibility, but in fact the two are very directly related. In all the examples of irreversible processes mentioned above, the tendency is toward a state having greater microscopic disorder. Conversion of macroscopic mechanical energy to heat, for example, involves a transformation from a well-organized motion, in which a body moves as a whole with macroscopic kinetic energy, to a situation where the kinetic energy is associated with random molecular motions. This latter motion is clearly less orderly than the former.

The mixing of the two gases is another example of a tendency toward increasing disorder. In the initial situation the molecules of the two gases are sorted into containers, but at the end of the process they are mixed, i.e., in a less sorted and more disordered condition. In the flow of heat from a hot body to a cold body, we begin with a state in which the "hot" molecules have, on the average, higher speeds than the molecules of the cold body. Thus they are partially sorted according to speed. When the bodies reach thermal equilibrium, the order has disappeared; the final state is more disordered than the initial one.

Thus we see emerging a relationship between the tendency of nature to proceed to a state of greater *disorder* (familiar to anyone who tries to keep his desk neat) and the *irreversibility* of some physical processes. In the next several sections, we investigate various attempts to make processes go in the "unnatural" directions to see with what effectiveness this can be done, and in so doing we discover the fundamental limitations which lead to the second law of thermodynamics.

21-2 CARNOT CYCLE

It has been mentioned that it is easy to convert work completely into heat but much more difficult (or perhaps impossible) to convert heat completely into work. Any device which takes in heat and produces work is called a *heat engine*. We now want to investigate the *thermodynamic efficiency* of heat engines which take heat from a source at a given temperature and convert some of it into work.

Consider an ideal gas at absolute temperature T expanding isothermally from volume V_1 to V_2. The work done by n mol of gas during this expansion, as calculated in Sec. 20-3, is

$$W = nRT \ln \frac{V_2}{V_1} \tag{21-1}$$

Because the internal energy of an ideal gas does not change in an isothermal process, the amount of heat which has to be added is exactly equal to the work which the gas does. In this case, a given amount of heat is converted *entirely* into mechanical work.

The isothermal expansion of a gas is not, however, a suitable process upon which to base any kind of general discussion of heat engines because after the absorption of heat and performance of work, the system is in a different state from that in which it started. This process cannot be continued indefinitely; we eventually run out of compressed gas and space to put it after it expands. In order to discuss this problem in a meaningful way, we must consider a *cyclic* process, a series of processes such that the final state of the thermodynamic system is the same as the initial state. In a cyclic process there may be a number of steps during which heat is absorbed or given off and work is done, but at the end the system must be in the same state as at the beginning. For similar reasons, we shall consider only heat engines which work with a constant quantity of matter.

There are of course infinitely many different cyclic processes. In designing an idealized model of a heat engine, we pose the following problem: A heat engine takes heat Q_1 from a source, or reservoir, of heat at a constant temperature T_1, performs some work W, and perhaps rejects some heat Q_2 to a heat reservoir at a lower constant temperature T_2. What is the maximum efficiency such an engine can have?

We first note that the conversion of heat into work is an "unnatural" process, in the sense that nature seems to prefer irreversible conversion of work into heat. Thus we should avoid all irreversible processes which go in the "natural" direction and use only reversible, quasi-static processes. This restriction is sufficient to determine the nature of the cycle almost uniquely. To avoid irreversible heat flow, heat transfer must occur only when the working substance in the engine is at the same temperature as one of the reservoirs. At all other times, when the temperature is different from T_1 or T_2, there must be no heat transfer. That is, each process must be either *isothermal* at T_1 or T_2 or else *adiabatic*.

These are the characteristics of the *Carnot cycle*, named for its inventor, Sadi Carnot (1796–1832), a French engineer who was one of the early pioneers in the field of thermodynamics. At the beginning of the cyclic process the working substance is at temperature T_1. We put the system in thermal contact with the heat reservoir at T_1 and permit it to expand *isothermally* and quasi-statically at T_1. Otherwise, if the temperature of the system were permitted to vary, irreversible heat flow would occur. Thus the first step is for the system to absorb a quantity of heat Q_1 at constant temperature T_1, during which it expands from volume V_a to V_b. This step is represented by the line joining points a and b in Fig. 21-2.

Next, we remove the system from contact with the source of heat and permit it to undergo a further expansion to volume V_c. This expansion, like the first, is carried out quasi-statically and reversibly. There is no transfer of heat, and therefore it is an *adiabatic* process, represented by the line joining points b and c in Fig. 21-2. During this adiabatic expansion the temperature drops to a lower value T_2.

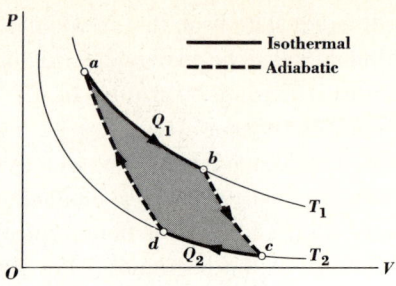

Fig. 21-2 Carnot cycle for an ideal gas on a P-V diagram. Processes $a \rightarrow b$ and $c \rightarrow d$ are reversible isothermal processes at temperatures T_1 and T_2, respectively, during which quantities of heat Q_1 and Q_2 are added to the gas; Q_2 is negative. Processes $b \rightarrow c$ and $d \rightarrow a$ are reversible and adiabatic. The net work done by the gas during the entire cycle is represented by area $abcda$.

We are now at a crucial point. Eventually the system has to be brought back to point a if we are to have a cycle. But it clearly will not do simply to retrace the path from point c to b and then to a. If we were to do that, there would be no net heat absorbed and no net work done by the system. We have to find a way back to point a which involves *less* work and *less* heat than the first two steps, so that a net amount of heat is absorbed and a net amount of work is done during the cycle. Further, in order to obey the ground rules we have laid down, this must involve only isothermal or adiabatic processes. Therefore the only possibility is to put the system into thermal contact with another reservoir at constant temperature T_2 and compress it *isothermally* and quasi-statically to volume V_d. During this isothermal compression at temperature T_2, the system loses heat. By convention, Q is always the quantity of heat *added to* the system, so we call this quantity of heat Q_2 and recognize that it is negative. Finally, a quasi-static *adiabatic* compression brings the system back to its original state. These processes constitute the Carnot cycle of Fig. 21-2.

We are now ready to calculate the *efficiency* of the Carnot cycle. The essential question is: How much of the heat Q_1 added to the system is converted into useful work? The thermodynamic efficiency η is defined, just as in Sec. 20-9, as

$$\eta = \frac{W}{Q_1} \tag{21-2}$$

where W is the *net* work done during the cycle. Since there is no net change in internal energy in the course of one complete cycle, the first law of thermodynamics, applied to the whole cycle, requires that $Q_1 + Q_2 - W = 0$. Thus the efficiency becomes

$$\eta = \frac{Q_1 + Q_2}{Q_1} = 1 + \frac{Q_2}{Q_1} \tag{21-3}$$

We note that when W is positive, Q_2 is smaller than Q_1 in magnitude; thus Q_2/Q_1 is always between zero and -1, and η is always between zero and unity, as of course it must be.

The efficiency η can also be expressed in terms of the temperatures of the two reservoirs. To develop this relationship, we now apply the first law of thermodynamics to individual parts of the cycle. Because the internal energy of an ideal gas does not change during an isothermal process, it must be true that the heat Q_1 added at temperature T_1 equals the work W_{ab} done during the isothermal expansion $a \to b$. That is,

$$Q_1 = W_{ab} = nRT_1 \ln \frac{V_b}{V_a} \tag{21-4}$$

Similarly,

$$Q_2 = W_{cd} = nRT_2 \ln \frac{V_d}{V_c} \tag{21-5}$$

Each expression in Eq. (21-5) is negative. Substituting for Q_1 and Q_2 in Eq. (21-3), we find

$$\eta = 1 + \frac{nRT_2 \ln (V_d/V_c)}{nRT_1 \ln (V_b/V_a)} = 1 - \frac{T_2 \ln (V_c/V_d)}{T_1 \ln (V_b/V_a)} \tag{21-6}$$

In the last expression we have divided out the common factors and used the fact that $\ln(1/a) = -\ln a$.

Finally, because both *adiabatic* processes ($b \to c$ and $d \to a$) occur between the same two temperatures, the volume ratios must be equal. That is,

$$\frac{V_c}{V_b} = \frac{V_d}{V_a} \quad \text{or} \quad \frac{V_c}{V_d} = \frac{V_b}{V_a} \tag{21-7}$$

as can easily be verified in detail with Eq. (20-20). Using this relation in Eq. (21-6), we see that the ratio of the two logarithms is just unity, and we finally obtain

$$\eta = 1 - \frac{T_2}{T_1} \tag{21-8}$$

This simple result shows that for this idealized cyclic process, the efficiency depends only on the ratio of the absolute temperatures of the two reservoirs, as measured on the ideal-gas scale discussed in Sec. 17-4. When the difference in temperature is very large, the ratio T_2/T_1 is very small, and the efficiency is close to unity. When the two temperatures are nearly equal, the efficiency is almost zero. The efficiency can be exactly unity only if $T_2 = 0$. Intuitively, it seems unlikely that one could maintain, at absolute zero, a reservoir to which indefinitely large amounts of heat could be added without changing the temperature. Furthermore, it can be shown, with the aid of principles to be developed later in this chapter, that it is never possible to attain absolute zero. Thus the efficiency of such an engine is always less than unity, but it *approaches* unity as the temperature of the colder heat reservoir approaches absolute zero.

Another useful relationship is obtained by comparing Eq. (21-8) with Eq. (21-3). From these we see that

$$\frac{Q_2}{Q_1} = -\frac{T_2}{T_1} \tag{21-9}$$

That is, the ratio of heat rejected to heat absorbed is just the negative of the ratio of the two corresponding temperatures. The minus sign arises because Q_2 is negative.

Although we have discussed the Carnot cycle with reference to an ideal gas, *any* material may be used as the working substance. As an example, consider a substance whose behavior is described by the P-V isotherms in Fig. 21-3. As remarked in Sec. 19-2, this pattern is typical of a material

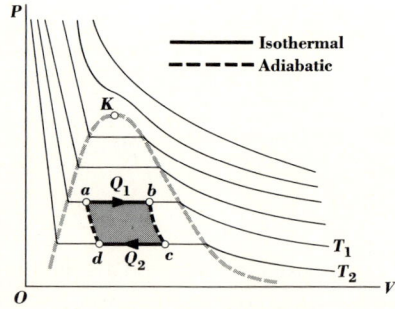

Fig. 21-3 P-V isotherms for CO_2 in the region in which liquefaction occurs. In the area bounded by the broken line, liquid and vapor are in equilibrium. To the right is the vapor phase; to the left, liquid. Above the critical point, labeled K, the distinction between liquid and vapor disappears. A possible Carnot cycle operating entirely in the region of liquid-vapor equilibrium is shown. As in Fig. 21-2, the heavy solid lines are reversible isothermal processes, the broken lines reversible adiabatic processes.

which undergoes a gas-liquid phase transition. The steep curves for small V are typical of liquid behavior; the flatter portions for large V are typical of gas behavior. The horizontal regions represent states of liquid-vapor equilibrium. For such states the volume can be increased without change of pressure or temperature by adding heat and vaporizing some of the liquid. The region of liquid-vapor equilibrium is bounded by the dashed line.

A Carnot cycle using this system in the liquid-vapor equilibrium region is shown by the heavy line in Fig. 21-3. Process $a \to b$ is an isothermal vaporization, also isobaric in this case, during which heat Q_1 (latent heat of vaporization) is absorbed and some liquid vaporizes. Process $b \to c$ is an adiabatic expansion during which the liquid-vapor mixture drops to temperature T_2. Process $c \to d$ is an isothermal isobaric partial condensation at temperature T_2, during which the system liberates a quantity of heat corresponding to the latent heat of vaporization of the condensing vapor under these conditions. Finally, the process $d \to a$ is an adiabatic compression back to the original state.

Calculation of the thermodynamic efficiency of this Carnot cycle is not so simple as in the case of the ideal gas, because the equation of state is not known. If the family of curves in Fig. 21-3 and the latent heat of vaporization at various temperatures are known, however, the efficiency can be computed numerically. Remarkably, it turns out that the efficiency agrees with Eq. (21-8). This result, in fact, turns out to be correct for every Carnot cycle, no matter what substance is used as the thermodynamic system. That is, *every Carnot cycle operating between two given temperatures has the same efficiency*. We have not yet seen any fundamental reason for this remarkable fact, but there is a reason. It is the second law of thermodynamics, to be discussed later in this chapter.

Furthermore, our speculation that in constructing an engine with no irreversible processes we have found the most efficient possible engine will turn out to be correct. The second law of thermodynamics can also be used to show that *no engine operating between two given temperatures can be more efficient than a Carnot engine operating between the same temperatures*. Before discussing these matters further, however, we discuss another aspect of the Carnot cycle.

21-3 REFRIGERATORS

We have just discussed an idealized scheme for converting heat partially into mechanical work, reversing the natural process of converting work into heat. We now consider the problem of taking heat out of a cold material and adding it to a hotter material, reversing the natural tendency of heat to flow from hotter to colder objects.

Just as any general device for converting heat into mechanical work is referred to as an *engine*, any device for removing heat from a cold place and adding it to a hotter place is called a *refrigerator*. An ordinary household refrigerator does this; it removes heat from its interior and expels it into the surrounding air, warming the air. In order to accomplish this, a refrigerator must have additional energy, such as electric energy, supplied to it. The heat expelled to the air is the *sum* of the energy from the power source and that removed from the cold material inside the refrigerator.

The engine employing the Carnot cycle, discussed in the previous section, can easily be adapted for use as a refrigerator. Each step in the process is reversible; therefore it is possible to reverse the entire cycle. This process is illustrated diagrammatically in Fig. 21-4, which is the same as Fig. 21-2

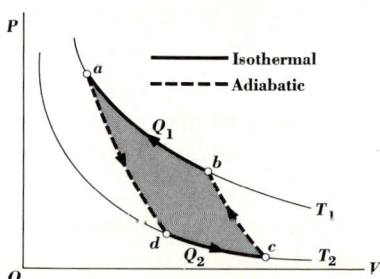

Fig. 21-4 Carnot refrigerator cycle. This is identical with Fig. 21-2 except that the directions of all the processes are reversed. As a result, a net quantity of work is done on the gas, and it gives off a net quantity of heat.

except that each step runs in the opposite direction. We permit a substance to undergo an isothermal expansion ($d \rightarrow c$) at a cold temperature, in contact with the interior of the refrigerator. This is followed by an adiabatic compression ($c \rightarrow b$) during which the temperature rises. We then put the working substance in contact with the higher-temperature surroundings, to which it gives heat during an isothermal compression ($b \rightarrow a$). Finally, an adiabatic expansion ($a \rightarrow d$) leads back to the original state.

Because each process in this cycle is the reverse of that in the Carnot heat engine, all quantities of work and heat are the negatives of those in the original process. Thus a *positive* quantity of heat Q_2 is added to the system at the cold temperature T_2, and a *negative* quantity Q_1 is added at the higher temperature T_1, corresponding to the fact that heat is *given off* by the system at T_1. During the cycle a net amount of work is done on the system. Again, $W = Q_1 + Q_2$, but now W and Q_1 are negative, while Q_2 is positive.

Just as we defined the efficiency of an engine as a quantitative measure of its effectiveness in converting heat into work, we can characterize the effectiveness of a refrigerator in taking heat from a cold place to a hot place by the ratio of the heat removed from the cold place to the amount of work expended to remove it. This ratio is called the *performance coefficient* of the refrigerator; it is sometimes abbreviated K. Symbolically,

$$K = -\frac{Q_2}{W} = -\frac{Q_2}{Q_1 + Q_2} \tag{21-10}$$

The minus sign is introduced to make K a positive quantity, since Q_2 is positive but W is negative.

Equation (21-9), relating the two quantities of heat to the two temperatures, is still valid; using this with Eq. (21-10), we obtain

$$K = \frac{T_2}{T_1 - T_2} \tag{21-11}$$

When the two temperatures are nearly the same, K is large; not much work is required if the temperature difference is small, but K never becomes infinitely large unless $T_1 = T_2$. In other words, it is never possible for this device to take heat from a cold place and add it to a hot place with *no* addition of work.

21-4 SECOND LAW OF THERMODYNAMICS

We return now to the problem of heat engines. The cyclic engines discussed above have the common feature that not all the heat goes into mechanical work; some of it is always thrown away at a lower temperature. This is a common characteristic of all heat engines. Gasoline and diesel internal-combustion engines waste heat in the hot exhaust gases. A steam engine and a steam turbine both expel steam at a temperature lower than the input temperature, after it has expanded and done work against the piston or turbine blades. A jet engine wastes heat in the hot exhaust gases. None of these engines is able to operate in a cyclic process by taking heat from a source at a given temperature and converting *all* of it into mechanical work; some heat is always wasted.

Since no one has ever been able to construct a cyclic heat engine which converts heat into work with 100 percent efficiency, we are led to suspect that there is some general principle which forbids it. Generalizing from the above observations of actual heat engines, we make the following statement: *It is impossible to construct a heat engine which operates in a cyclic process and which has no other effect than to take heat from a reservoir at one temperature and do an equivalent amount of mechanical work without discarding any heat at a lower temperature.* This generalization, first stated by Kelvin, in 1851, is one statement of the *second law of thermodynamics*.

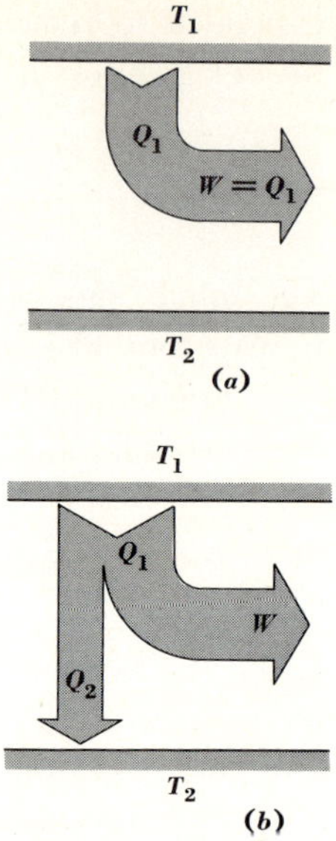

Fig. 21-5 (*a*) Heat engine with 100 percent efficiency would take heat Q_1 from a reservoir at temperature T_1 and convert it completely to an equivalent amount of work W. The second law of thermodynamics states that it is impossible in principle to construct such an engine. (*b*) Real engine absorbs heat Q_1 from a reservoir at temperature T_1, rejects a smaller quantity of heat Q_2 to a reservoir at a lower temperature T_2, and does an amount of work $W = Q_1 - Q_2$. The efficiency is $\eta = W/Q_1$.

Compared with many physical laws, this generalization has a rather queer form. It is not a quantitative relationship between physical quantities but rather a statement that a certain operation is impossible. This may seem to make it a less useful and powerful law than the quantitative statements to which we are accustomed, but it is actually very powerful. Furthermore, we can make a more quantitative statement which embodies the content of this generalization, with the aid of the concept of *entropy*, to be introduced in the next section.

A corresponding generalization can be made concerning refrigerators. It is a matter of experience that one cannot make a device which takes heat from a cold place and delivers the same quantity of heat to a hot place without requiring any expenditure of work. All refrigerators which have been built, regardless of their principles of operation, require that work be put into the system, so that the amount of heat expelled to the hot place is greater than that taken from the cold place. *It is impossible to make a device which operates in a cyclic process and has no other effect than the removal of a quantity of heat from a body at a certain temperature and delivery of this heat to another place at a higher temperature.* This statement was first made by Clausius in 1850.

This is another way of saying that heat spontaneously flows from a hot place to a cold place and that the only way to reverse this flow is by the expenditure of work from an outside agency. There is no such thing as a "workless" refrigerator.

Fig. 21-6 (*a*) Workless refrigerator would take heat from a cold reservoir at temperature T_2 and deliver it all to a reservoir at a higher temperature T_1 without the addition of work. The second law of thermodynamics states that it is impossible in principle to construct such a refrigerator. (*b*) Real refrigerator absorbs a quantity of heat Q_2 from a cold reservoir at temperature T_2, has work W done on it, and delivers a larger quantity of heat Q_1 to a reservoir at higher temperature T_1. The performance coefficient of the refrigerator is $K = -Q_2/W$.

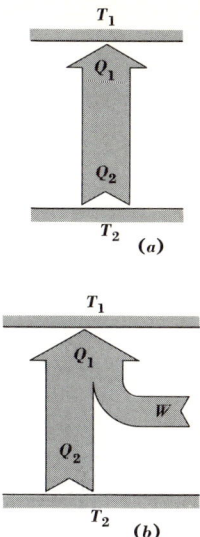

Although the statements concerning refrigerators and heat engines may seem to be quite unrelated, a little thought shows that they are very closely related indeed. If a device could be made which violates the refrigerator principle, it could be used in combination with a heat engine to make a machine which violates the second law of thermodynamics as stated with regard to heat engines. For one could simply take the heat wasted by a heat engine, use a workless refrigerator to transport this heat back to the hot

Fig. 21-7 If a workless refrigerator could be constructed in violation of the Clausius statement of the second law, it could be used in conjunction with a real heat engine to make a composite machine constituting a heat engine with an overall efficiency of 100 percent, in violation of the Kelvin statement of the second law.

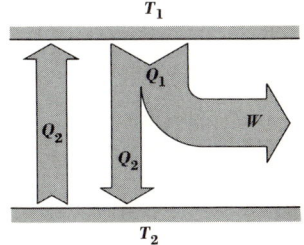

reservoir, and add it to the engine again. Thus the heat engine and the workless refrigerator together would constitute a system which violates the second law of thermodynamics.

Further analysis shows that this relationship is *reciprocal*; if one could make a heat engine which was 100 percent efficient, it would also be possible to use this engine to make a workless refrigerator by using it to drive an ordinary refrigerator. Thus the impossibility of making a perfect heat engine and the impossibility of making a workless refrigerator are completely *equivalent*, and both statements are called the *second law of thermodynamics*.

Still other equivalent statements can be made. Two of these are the following: All reversible engines operating between two given temperatures have the same efficiency. Correspondingly, for two given reservoir temperatures, *no engine whatever* can have an efficiency greater than that of the ideal Carnot engine, which is a reversible engine. These statements can be derived from the second law of thermodynamics as stated for a heat engine or a refrigerator; their derivations are left as problems.

Inventors who have no knowledge of physics frequently invent on paper (but never successfully build) devices for generating power based on violations of the laws of thermodynamics. Any device which delivers more energy than it receives constitutes a violation of the *first* law of thermodynamics; this is referred to as a perpetual-motion machine of the *first* kind. An example of a perpetual-motion machine of the *second* kind is the oft-proposed scheme to drive ships across the ocean by extracting heat from the ocean, cooling it infinitesimally, and converting all this heat into mechanical work. This is a clear violation of the second law, and it should not be surprising that no one has ever been able to make it work. This is not to be confused, however, with the proposal to extract energy from the oceans by making use of the temperature gradients that exist between the surface and great depths. These schemes do *not* violate the second law, and they may well be developed into a practical energy source.

Considering the Carnot engine, we note that if the temperature T_2 of the cold reservoir is very small, the efficiency is very close to unity. If we could have a reservoir at *absolute zero*, that is, $T_2 = 0$, then we could make a heat engine with an efficiency of unity. However, analysis of the Carnot refrigerator shows that its performance coefficient becomes vanishingly small as the temperature of the reservoir from which heat is taken approaches absolute zero. Therefore, it is impossible to construct a reservoir at $T_2 = 0$. This is not to say, of course, that it is not possible to come *very close* to absolute zero; temperatures of less than 0.001 K have been achieved.

21-5 ENTROPY

The tendency of a system to proceed toward a state of greater disorder or randomness, discussed in Sec. 21-1, is related to the two "impossibility" statements of the second law of thermodynamics in Sec. 21-4. For example, a heat engine which would violate the second law of thermodynamics would take heat from a reservoir, thereby decreasing its disorder by decreasing the random thermal motion. There would be no corresponding increase in disorder in any part of the system, and thus the engine would achieve a net *decrease* in the disorder of the system and its surroundings. On the basis of the second law, we may therefore speculate that it is impossible to produce an overall decrease in the disorder of a system plus its surroundings. To give this statement meaning, we need a suitable quantitative measure of disorder. The concept of *entropy* provides such a measure.

To introduce entropy, we consider the change in the disorder of a system which results from a quasi-static addition or removal of heat. Adding heat to a system increases the disorder by increasing the random molecular motion. But the effectiveness of a given quantity of heat in increasing disorder depends also on the initial *temperature* of the system. Adding heat to a very hot system increases its disorder relatively little, since the molecules are already in rapid motion. But the addition of the same quantity of heat to a very cold system with little molecular motion produces, relatively speaking, a much greater increase in molecular motion and thus in disorder. So it is reasonable to define the change in disorder of the system when a quantity of heat Q is added to the system at temperature T as Q/T. Introducing the symbol S to represent the measure of disorder of the system, we say that the *change* in disorder ΔS resulting from the reversible addition of a quantity of heat Q at temperature T is

$$\Delta S = \frac{Q}{T} \qquad (21\text{-}12)$$

The quantity S is called the *entropy* of the system.

Equation (21-12) defines entropy operationally, just as internal energy was defined operationally in Chap. 20. As with internal energy, this equation provides a means of calculating the *change* in S resulting from the quasi-static addition of heat to a system. Equation (21-12) has meaning only for quasi-static processes, since *temperature* has no meaning for a non-equilibrium system as a whole. Of course, changes in disorder can also occur in non-equilibrium, irreversible processes; we shall see later how to compute entropy changes in such processes.

We now discuss several calculations of entropy changes, in order to exhibit the usefulness of this quantity. We consider first the entropy changes in a Carnot heat engine. The engine absorbs a (positive) quantity of heat Q_1 at temperature T_1 and a (negative) quantity Q_2 at temperature T_2. The total change in entropy of the engine during one cycle is, according to Eq. (21-12),

$$\Delta S = \frac{Q_1}{T_1} + \frac{Q_2}{T_2} \qquad (21\text{-}13)$$

But according to Eq. (21-9), this is zero. Thus, during a complete cycle, the net entropy change *of the engine* is zero.

What about the entropy change of the surroundings? The heat reservoir at T_1 *loses* an amount of heat Q_1 and therefore experiences a *decrease* in entropy, i.e., a change $\Delta S = -Q_1/T_1$. But the cold reservoir at T_2 undergoes an *increase* of entropy $\Delta S = -Q_2/T_2$ because it gains heat $-Q_2$, where Q_2 is negative. Thus the net entropy change of the *surroundings* is also zero. This should not be terribly surprising. We recall that the Carnot engine contains only reversible processes. Perhaps in any reversible process the total entropy of system plus surroundings does not change; perhaps the total entropy increases only in *irreversible* processes.

Another simple example is the calculation of entropy change for the reversible isothermal expansion of a gas. If the gas is ideal, the heat Q added during an isothermal expansion is $Q = nRT \ln(V_2/V_1)$, and the corresponding entropy change is

$$\Delta S = \frac{Q}{T} = nR \ln \frac{V_2}{V_1} \qquad (21\text{-}14)$$

That is, when a gas expands isothermally and reversibly, its entropy *increases*. The disorder of the system increases correspondingly. The temperature is constant, so the average molecular speeds do not change. But when the gas expands, the molecules are allowed to move in a larger volume, so there is an increase in randomness of position.

The definition of ΔS given by Eq. (21-12) can easily be generalized to situations in which the temperature changes during the quasi-static addition of heat. We have only to consider an increment dS of entropy resulting from a small reversible addition of heat dQ in which T is very nearly constant. Then

$$dS = \frac{dQ}{T} \qquad (21\text{-}15)$$

and the total entropy change for a given process is

$$\Delta S = \int_{S_1}^{S_2} dS = \int_{T_1}^{T_2} \frac{dQ}{T} \qquad (21\text{-}16)$$

As an example, if we heat n mol of a material with a constant molar specific heat C from temperature T_1 to T_2, then $dQ = nC\,dT$, so the entropy change of the material is

$$\Delta S = \int_{T_1}^{T_2} \frac{nC\,dT}{T} = nC \ln \frac{T_2}{T_1} \qquad (21\text{-}17)$$

if the heat is added quasi-statically. This example, like the previous one, shows that entropy is an *extensive* thermodynamic quantity, proportional to the quantity of matter in the system.

Next, consider a process which is clearly irreversible, the flow of heat from a hot body to a colder body. Suppose a body at temperature T_1 is connected by a thin metallic rod to a body at a lower temperature T_2; heat flows from the hot body to the cold. Suppose that this heat flow is slow enough for each body to remain *internally* very nearly in thermal equilibrium. The rod, of course, is *not* in thermal equilibrium, since one end is at T_1 and the other at T_2. Now suppose a quantity of heat Q flows from the hot body. So far as this body is concerned, the flow of heat is reversible, since there are no finite temperature gradients within the body. The net change in its entropy is $\Delta S_1 = -Q/T_1$, a *decrease* in entropy. Similarly, the cold body undergoes an *increase* in entropy $\Delta S_2 = Q/T_2$. The net change in entropy for the whole system is

$$\Delta S = \Delta S_1 + \Delta S_2 = Q\left(\frac{1}{T_2} - \frac{1}{T_1}\right) \qquad (21\text{-}18)$$

Since T_2 is less than T_1, this entropy change is a positive quantity. Thus, in irreversible flow of heat from a hot body to a colder one, the total entropy increases. As pointed out in Sec. 21-1, this also corresponds to an increase in the disorder of the system.

Entropy has been introduced as a measure of the disorder of a system, and it is therefore reasonable that S should be a quantity which, like internal energy, depends only on the *state* of a system, not on the process by which it arrived at that state. Correspondingly, the *change* in entropy of a system in a transition from one equilibrium state to another should depend only on the initial and final states, not on the details of the process which led from one to the other.

These statements can be checked experimentally for any system by examining several different processes leading from given initial to final states and computing the entropy change for each. All available evidence indicates that the entropy change of a system is in fact the same for all possible processes between given initial and final states. In addition, it is possible to prove that if the second law of thermodynamics is valid, this *must* be the case. The proof will not be given here, but it adds support to the second law and to entropy as a function only of the state of a system. Thus it is established that the entropy of a system depends only on its state, and not on the process by which it reached that state.

We recall that internal energy, while it is a quite different thermodynamic quantity, also has this important property. Any thermodynamic quantity which depends only on the state of the system is called a *function of state* or a *state function*. Other state functions can be devised from combinations of these.

The fact that the entropy of a system depends only on its state makes it possible to use Eq. (21-12) or (21-16) to calculate entropy changes for *irreversible* processes. To calculate ΔS for an irreversible process between

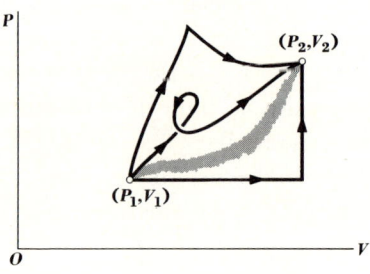

Fig. 21-8 *P-V* diagram showing different processes between given initial (P_1, V_1) and final (P_2, V_2) states for a thermodynamic system. The work W done by the system and the heat Q absorbed depend on the initial and final states and on the path leading from one to the other. Thus, in general, Q and W are different for the various processes shown. The internal-energy and entropy changes ΔU and ΔS do not depend on the path but only on the initial and final states.

given initial and final equilibrium states, we need only devise a *reversible* process or a series of reversible processes between the same two states and calculate the entropy change for the reversible process. The entropy change for the irreversible process must be the same as for the reversible one between the same states.

As an example, consider the free expansion of an ideal gas, discussed in Sec. 20-3. A box is divided by a removable partition into two compartments of equal size. One compartment contains a quantity of an ideal gas; the other is evacuated. Now we suddenly remove the partition, keeping the system insulated so that no heat can flow in or out. What is the entropy change during this process?

First we determine the final state. Because the gas does no work on its surroundings, and because no heat is permitted to enter, the internal energy does not change. Therefore the temperature is also constant, since it is an ideal gas. Thus the volume doubles, the pressure decreases by half, and the temperature remains constant. The process clearly is irreversible, but the same final state could be achieved by a reversible isothermal expansion. During such an expansion, as we have seen several times, an amount of heat

$$Q = nRT \ln \frac{V_2}{V_1} = nRT \ln 2$$

must be added to the system. The entropy change during the process is

$$\Delta S = \frac{Q}{T} = nR \ln 2 \qquad (21\text{-}19)$$

The final state is the same as that of the irreversible process when the partition was removed from the box, so the entropy change must be the same for the two processes. Thus the entropy change for the irreversible free expansion is $\Delta S = nR \ln 2$.

In all the above examples, we observe that in a situation involving only *reversible* processes the total entropy of the system plus its surroundings is constant. Whenever an *irreversible* process is involved, leading to an increase in the disorder in the system and its surroundings, the entropy change is always a positive quantity. This and other observations support the view that entropy is a suitable measure of the disorder of a system.

21-6 ENTROPY AND THE SECOND LAW

It is now possible to formulate the second law of thermodynamics in a precise quantitative manner, using the concept of entropy. To do this, we observe that any device that violates either statement of the second law in Sec. 21-4 achieves a net *decrease* in the total entropy of system plus surroundings. For the refrigerator this is immediately obvious, since a workless refrigerator would produce precisely the effect opposite that of the irreversible heat flow described in Sec. 21-5. A heat engine with 100 percent efficiency would decrease the entropy of the heat source without any corresponding increase of entropy elsewhere.

Conversely, if any process could be found in which the total entropy change of the system plus surroundings is negative, it could be used to violate one or the other of the statements of the second law of thermodynamics in Sec. 21-4. Therefore, a statement of the second law which is equivalent in all respects to those previously given is the following: *In any process involving a thermodynamic system interacting with its surroundings, the total entropy change of the system and surroundings can never be negative. If only reversible processes occur, the entropy change is zero; otherwise it is positive.*

It is essential to keep in mind that the statement refers to the net entropy change in the system *and surroundings*. It would not be correct in general to say that the total entropy change of the system itself must always be zero or positive. We have, in fact, seen a number of processes in which the entropy change of the system can be negative. But the *total* entropy change, including the surroundings as well as the thermodynamic system itself, can never be negative.

An immediate corollary is that when the system is isolated from its surroundings, the entropy change of the system must be zero or positive. *The entropy of an isolated system can never decrease.*

This tendency toward an increase in entropy and thus in disorder agrees with everyday observations concerning the decomposition of organic matter. Dry leaves and corpses turn to dust; a book which contains a highly organized sequence of symbols turns into a shapeless mass of pulp when left out in the rain. Some processes in living plants and animals may seem at first glance to violate the second law of thermodynamics, since they start with relatively unstructured materials and create highly structured forms. This is an increase in the order or organization of the system, corresponding to a decrease in entropy. But these are not *isolated* systems. A green plant, for example, needs sunlight to live. Its entropy can decrease only if somewhere else there is a corresponding increase in entropy. In this case there is a positive entropy change associated with irreversible heat transfer through radiation from the (very hot) sun to the (much cooler) earth. Thus such life processes are consistent with the second law of thermodynamics.

21-7 THERMODYNAMIC TEMPERATURE SCALE

In introducing the concept of temperature in Chap. 17, we defined temperature scales based on changes in the physical properties of materials with temperature. We also indicated the desirability of defining a temperature scale which *does not* depend on the specific behavior of a particular material. We are now in a position to construct such a scale. To do this we simply operate a Carnot heat engine between two reservoirs at different temperatures. Because all reversible engines operating between two given temperatures have the same efficiency, the particular material used in the Carnot engine is of no consequence. The two quantities of heat Q_1 and Q_2 transferred from the respective reservoirs to the engine are, as we have seen, proportional to the absolute temperatures. That is,

$$\frac{Q_1}{Q_2} = -\frac{T_1}{T_2} \tag{21-20}$$

This equation can be used to *define* the temperature ratio T_1/T_2. One may arbitrarily assign a numerical value to one temperature, say T_1; then the numerical value of any other temperature T_2 can be found by operating a Carnot engine between the two temperatures, measuring the two quantities of heat Q_1 and Q_2, and using Eq. (21-20).

An alternative procedure, and the one ordinarily followed, is to choose arbitrarily a value for the *difference* between two temperatures. For example, suppose that T_1 is the boiling temperature of water at atmospheric pressure and T_2 its freezing point. For the thermodynamic temperature scale to be consistent with the usual Celsius scale, it must be true that $T_1 = T_2 + 100°$. For a Carnot engine operated between these two temperatures,

$$\frac{Q_1}{Q_2} = -\frac{T_2 + 100°}{T_2}$$

Solving this for T_2, we find

$$T_2 = -\frac{Q_2(100°)}{Q_1 + Q_2}$$

which, with the measured values of Q_1 and Q_2, defines the freezing temperature of water on the thermodynamic temperature scale. Experiment shows that $T_2 = 273.2°$, in agreement with the value obtained from the scale based on the gas thermometer.

The thermodynamic temperature scale was first proposed in 1848 by William Thomson, a British physicist (later Lord Kelvin); it is named the Kelvin temperature scale in his honor. Although it agrees very closely with the gas-thermometer scale, it does not depend on the properties of any particular material. Thus this temperature scale alone really deserves the name *absolute temperature scale*.

It should be pointed out that it is not actually necessary to construct an engine to use this scheme for defining a temperature scale. All that is needed is a sufficiently detailed empirical description of the properties of the material being considered, for example, P-V-T relationship and specific heats as functions of P, V, and T, and the calculations appropriate for the Carnot cycle can be carried out on paper.

21-8 ENTROPY AND PROBABILITY

Thus far we have not attempted to give a precise *microscopic* definition of entropy. It is true that we have interpreted entropy in terms of the degree of microscopic disorder of the system; this, in fact, was the idea which initially led to the introduction of entropy. But the actual *definition* of entropy we have been using is entirely a macroscopic one which has nothing to do directly with the motions of individual molecules. We now discuss very briefly how entropy can be defined microscopically, in terms of the microscopic state of the system as defined by the positions and velocities of its molecules.

The microscopic definition of entropy is based upon the concept of probabilities of thermodynamic states. To illustrate, we consider molecules of an ideal gas in a container with fixed volume. Following the kinetic-molecular model of an ideal gas, we regard the molecules as moving at random in the container. We assume that the position of each molecule is random; at a given time, it is as likely to be at any point in the volume as at any other. In addition, the direction of the velocity is random; any direction is as likely as any other. The speeds are distributed according to some known distribution function; not all speeds are equally likely, but each speed interval has a definite probability. For an ideal gas, this probability distribution is the Maxwell-Boltzmann distribution.

Suppose we now construct an imaginary wall to divide the container into two equal volumes. On the average, half the molecules are in each side, but there are times when *more* than half are in one side and *less* than half in the other. Similarly, there may be very rare occasions when *all* the molecules are in one side and none at all are in the other. This is exceedingly unlikely, but it is not, strictly speaking, impossible. We can say only that at any instant the *probability* that all the molecules are in one side of the container is very much smaller than the probability that they are nearly equally distributed.

In this example there is a connection between the *probability* of the state and the degree of *disorder*. The state in which all the molecules are in one half of the box has more order and less entropy than the state in which they have the same temperature and speed distribution but occupy the entire box, because in the former state the molecules move with random motion in a smaller space. But the states in which the molecules are spread nearly uniformly over the entire box are more likely to occur than those in which all the molecules are in one side. Thus the more *disordered* states are the more *probable* states, and the states with the most order are the least probable. Another way of looking at this situation is to note that there are many microscopic distributions which give a fairly even distribution of molecules between two sides of the box but a much smaller number of arrangements which lead to concentration of all the molecules in one side of the box.

Perhaps a more familiar example of the same thing is the disorder of papers on the top of a desk. If papers are placed on the desk at random, there is a small chance that they accidentally end up in neat stacks but a much larger chance that they will be mixed up and disordered. Of all possible arrangements, there are many more disordered ones than ordered ones, relatively speaking. If one throws paint randomly at the canvas, the probability that the result will be the "Mona Lisa" is rather small. Of all the possible distributions of paint spatters, very few have anything to do with art.

Because entropy is directly related to disorder, there must be a direct relationship between the entropy of a state and its probability of occurrence. The greater the probability, the greater the disorder, and therefore the greater the entropy. It is possible to show, in fact, that if W is the probability for a given state and S its entropy, the two are related very simply by

$$S = k \ln W \tag{21-21}$$

where k is the same Boltzmann constant which appeared in the kinetic theory of gases in Chap. 18. Establishing this relationship in general is in the province of statistical mechanics; we shall not attempt a detailed derivation here, but it is illuminating to see how it works out in the particular example we have been discussing.

Consider an ideal gas confined in a volume V. Assuming that the molecules move independently and randomly, we can calculate the probability at any given instant that *all* the molecules are in a smaller volume V_1 within V. First, the probability of one molecule's appearing in the volume V_1 at any specific time is equal to the ratio of this volume to the *total* volume: V_1/V. What is the probability that *all* the molecules are in this volume at any given time? If there are N molecules, we must have a coincidence of N different independent events (the presence of individual molecules in V_1), each of which has probability V_1/V. The probability that all of a collection of independent events happen at once is the *product* of the probabilities of the separate events. Thus the probability that all the molecules are in volume V_1 is $(V_1/V)^N$. Denoting this probability by W_1, we have

$$W_1 = \left(\frac{V_1}{V}\right)^N \tag{21-22}$$

Similarly, the probability W_2 that at a given time all the molecules are in a volume V_2 within V is

$$W_2 = \left(\frac{V_2}{V}\right)^N \tag{21-23}$$

To establish the connection between these probabilities and entropy, we recall that when an ideal gas expands from volume V_1 to V_2, the entropy change is

$$\Delta S = nR \ln \frac{V_2}{V_1}$$

We can express this in terms of molecular quantities, using the relations $n = N/N_0$ and $R = N_0 k$, as in Chap. 18:

$$\Delta S = Nk \ln \frac{V_2}{V_1} \tag{21-24}$$

Returning to Eqs. (21-22) and (21-23), we see that the logarithm of the volume ratio can be obtained simply by dividing one of these equations by the other and taking logs, as follows:

$$\frac{W_2}{W_1} = \frac{(V_2/V)^N}{(V_1/V)^N} = \left(\frac{V_2}{V_1}\right)^N$$

$$\ln \frac{W_2}{W_1} = \ln \left(\frac{V_2}{V_1}\right)^N = N \ln \frac{V_2}{V_1} \tag{21-25}$$

Comparing this result with Eq. (21-24), we find

$$\Delta S = k \ln \frac{W_2}{W_1} = k \ln W_2 - k \ln W_1 \tag{21-26}$$

This suggests defining the entropy of each state as $S = k \ln W$, in agreement with Eq. (21-21). Thus, at least in this simple example, we have established directly the relationship between the entropy of a state and its statistical probability.

We now arrive at a very interesting and perhaps somewhat disturbing conclusion. From a microscopic viewpoint, the progress of isolated systems to states of higher entropy is not a result which is absolutely certain to occur but only one which is overwhelmingly *probable*. The second law of thermodynamics now emerges as a *statistical* law; we cannot say with absolute certainty that the entropy of an isolated macroscopic system can *never* decrease but only that it is extremely *unlikely* to decrease. Like Gilbert and Sullivan, we must change "never" in the second law to "hardly ever." There are, in fact, situations in which observable and measurable fluctuations from states of thermodynamic equilibrium do occur spontaneously. It should be clear that ordinary macroscopic thermodynamics is powerless to deal with such situations, and for them the more detailed methods of statistical mechanics are necessary.

Problems

21-1 Discuss the following phenomena with respect to their reversibility or irreversibility: burning the page of a book, melting an ice cube in a glass of warm water, discharging a charged capacitor through a resistor, gas rushing into a vacuum, mixing alcohol and water, breaking a glass bottle, spilling a paragraph of hand-set lead type. Would the reversal of any of these phenomena lead to a violation of the *first* law of thermodynamics? Explain.

21-2 A block is given an initial upward velocity on a rough (not frictionless) inclined plane. It attains a certain maximum height and then slides back to the starting point. Explain why this is an irreversible process.

21-3 Does the liquid or solid phase of a material in equilibrium under given conditions have greater disorder? Explain.

21-4 An ideal gas undergoes a cycle represented on a *P-V* diagram as in Fig. P21-4.
  a Find the work done by the gas and the heat added to it during each part of the cycle.
  b If heat which leaves the gas during any part of the cycle is considered lost, calculate the thermodynamic efficiency of this process.

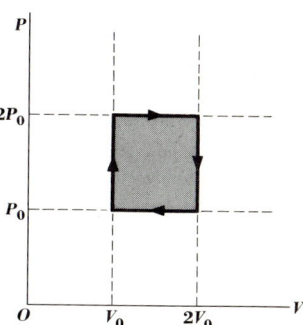

Fig. P21-4

21-5 A heat engine contains 5 mol of a gas having $\gamma = 1.5$. It undergoes a Carnot cycle, operating between two reservoirs whose temperatures are 400 and 300 K. The initial pressure (point *a* in Fig. 21-2) is 10 atm, and during the initial isothermal expansion the pressure drops to 5 atm.
  a Make a table showing the heat absorbed, work done, and change in internal energy during each step. Be sure all quantities have correct algebraic signs.
  b Find the total heat absorbed and the total work done during the cycle.
  c Calculate the thermodynamic efficiency from the above results. Compare with the prediction of Eq. (21-8).

21-6 A heat engine operates in an ideal Carnot cycle between the temperatures of 327 and 127°C. If it absorbs 3000 cal from the hot reservoir, how much heat does it give off to the cooler reservoir? How much mechanical work does it do (in joules)?

21-7 A certain Carnot engine operates between 200 and 300 K.
 a What is the thermodynamic efficiency?
 b If it is desired to improve the thermodynamic efficiency, is it better to increase the temperature of the hot reservoir by 50 K or to decrease the temperature of the cold reservoir by 50 K? Explain.

21-8 Consider a two-stage turbine. Steam enters the first turbine at 327°C. It is exhausted at 227°C and enters the second turbine at this temperature. Finally, it is exhausted to the atmosphere at a temperature of 127°C. Calculate the Carnot efficiency of this process.

21-9 A Carnot engine is operating between temperatures T_1 and T_2. An inventor proposes to improve the efficiency of the situation as follows: An engine is run between T_1 and an intermediate temperature T_a, where $T_2 < T_a < T_1$. All heat expelled at T_a is given to another engine, which then operates between T_a and T_2. Both engines are considered to be ideal Carnot engines. The inventor's brother contends that the efficiency cannot be improved in this manner. Is he correct? Explain. Compute the overall efficiency of the inventor's scheme.

21-10 A substance undergoes a reversible cyclic process represented by a closed path on a P-V diagram. Prove that the area enclosed by this path represents the net work done *by* the substance during the cycle if the path is traversed in a clockwise sense and the net work done *on* the substance if counterclockwise (assuming P is the vertical axis, V the horizontal). Is this true if the substance is not an ideal gas? Explain.

21-11 Describe how a Carnot cycle might be constructed using a magnetic material of the type discussed in several of the problems of Chap. 20. Represent such a cycle graphically on an H-M diagram, with H the vertical axis. In what sense should the cycle be traversed if the system is to operate as a heat engine?

21-12 Describe a Carnot cycle using, as a thermodynamic system, the spring discussed in Prob. 20-32. Show the cycle graphically on an F-x diagram, with F the vertical axis.

21-13 Give several reasons why the efficiency of a practical heat engine might be less than that of an ideal Carnot engine.

21-14 An ideal Carnot refrigerator freezes ice cubes at the rate of 5 g/s, starting with water at the freezing point. Heat is given off to the room at 30°C.
 a What electric power input is required?
 b At what rate is heat expelled to the room?

21-15 Estimate the performance coefficient of a common household refrigerator. Explain how you obtained this estimate.

21-16 To cool her kitchen on a hot summer day a housewife leaves the refrigerator door open. What actually happens to the temperature of the room? Explain.

PROBLEMS

21-17 Derive an equation relating the efficiency η of a Carnot heat engine and the performance coefficient K when the engine is run backward as a refrigerator.

21-18 If a heat engine with 100 percent efficiency could be built, in violation of the "engine" statement of the second law of thermodynamics, show in detail how it could be used in conjunction with another machine to make a workless refrigerator, in violation of the "refrigerator" statement of the second law.

21-19 In some localities it is economical to heat homes in winter by use of a heat pump, which extracts heat from some source, such as a large body of water, and delivers it to the interior of the house. Is such a scheme compatible with the first law of thermodynamics? With the second law? Explain.

21-20 How does the performance coefficient for a practical refrigerator differ from that for an ideal Carnot refrigerator? What are some possible reasons for this difference?

21-21 Heat is added quasi-statically to 100 g of ice at 0°C until it is all melted. What is the change in entropy of the ice? Does this correspond to an increase or decrease in disorder? Explain.

21-22 Find the total change in entropy of 2.0 kg of water heated from ice at $-20°C$ to steam at 100°C if the specific heats of water and ice are independent of temperature.

21-23 In Prob. 21-21, if the heat comes from a large body whose temperature is 20°C, what is the entropy change of the source? The *total* entropy change of water and source?

21-24 An ideal gas expands adiabatically and reversibly from initial volume V_1 to final volume V_2. Considering the gas as a thermodynamic system, find the change in entropy of (*a*) the system and (*b*) the surroundings during this process.

21-25 One end of a copper rod is in contact with a heat reservoir at 127°C; the other end is in contact with a reservoir at 27°C. During a certain time interval, 1200 cal of heat flows from the hot end to the cooler end.
   a Find the entropy change of each reservoir and the total entropy change.
   b Does the entropy of the copper rod change? Explain.

21-26 A paddlewheel driven by a 0.5-hp motor is used to churn a barrel of water containing 200 kg of water. The initial water temperature is 27°C.
   a Find the approximate entropy change of the water during the first minute. Also find the total entropy change of system plus surroundings during this interval.
   b Is the entropy change during the tenth minute the same as that during the first? Explain.

21-27 During a free expansion of an ideal gas there may be no heat transfer into the system, and yet its entropy increases. Explain how this is possible.

21-28 Suppose that the entropy of n mol of an ideal gas is S_0 in a specified state (P_0, V_0). Derive an expression for the entropy S of the gas in any other state in terms of the thermodynamic coordinates (P, V) of that state.

21-29 As discussed in the text, the entropy of a system is determined, at least for states of thermodynamic equilibrium, by the thermodynamic coordinates of the state. Could S be used as a thermodynamic coordinate? For an ideal gas, for example, can the state of the system be specified by giving values of S and T? Explain.

21-30 For a Carnot cycle, sketch a graph showing the relationship between the entropy of the system and its temperature. Make T the vertical axis, S the horizontal. Label the points corresponding to points a through d in Fig. 21-2. For a complete cycle, this graph must always be a closed path. Why? What significance does the *area* enclosed by the path have?

21-31 Show how the efficiency of a cyclic process can be obtained graphically from a T-S diagram for the process, such as that discussed in Prob. 21-30.

21-32 Prove that the T-S diagram for a Carnot cycle always has the same shape, no matter what the working substance (ideal gas, liquid-vapor equilibrium, magnetic material, stretched spring, etc.)

21-33 An ideal Carnot engine is operated between the normal freezing and boiling points of water, and its efficiency is observed to be $\eta = 0.270$. If the difference between these temperatures is defined to be $100°$, find the actual temperatures on the thermodynamic temperature scale, as determined from the Carnot efficiency.

21-34 A chemical firm has just developed a new compound called *glop*. To publicize it, the company proposes to make thermometers calibrated by means of a thermodynamic temperature scale based on the melting and boiling temperatures of glop at atmospheric pressure. It was found that a certain Carnot engine takes in 1000 cal at the boiling temperature and expels 800 cal at the melting temperature.
   a Find the ratio of the two temperatures.
   b If it is arbitrarily decided to make the two temperatures differ by $10°$, find the melting and boiling temperatures on the absolute-thermodynamic glop scale.
   c A chemist observes with a Celsius thermometer that the temperatures are 127 and $227°$, respectively. Derive a general equation relating the Celsius and glop temperature scales.

21-35 Consider six molecules in random motion in a box. The box is divided by a partition into two equal volumes.
   a In how many different ways can the six molecules be divided between the two sides?

b How many of the divisions in part *a* correspond to all six molecules' being in the left side? Five molecules? Four? Three? Two? One? None?

c Find the probability that at any instant of time all the molecules are in the left side. Repeat for the other numbers. What number is most probable?

21-36 A container is divided by a heat-conducting wall into two compartments having equal volumes. The outside of the container is thermally insulated. Each compartment contains 0.1 mol of an ideal diatomic gas at temperature $T = 300$ K. Suppose that a quantity of heat Q is conducted through the wall, and as a result the temperature of one side rises by $\Delta T = 0.001$ while that of the other side drops by the same amount. The probability of occurrence of this state is different from that for the state in which the two temperatures are equal. The *ratio* of the probabilities can be expressed in terms of the entropy difference between the two states, using Eq. (21-21) or (21-26).

a How much heat Q must be transferred to produce the specified temperature changes?

b What entropy change of the entire system results from the heat transfer of part *a*? An approximate value can be obtained by considering that one side gains heat Q at an average temperature $T + \Delta T/2$ while the other side loses Q at an average temperature $T - \Delta T/2$. Of course, an exact calculation can be performed without this approximation.

c What is the ratio of the probabilities of occurrence of the two states?

d Why is the method suggested in part *b* for calculating the entropy change only an approximation?

e Does the temperature fluctuation become more or less probable if the amount of gas in each side is decreased, say to 10^{-12} mol? Explain.

Mechanical Waves | 22

The general concept of a wave and methods for describing a wave in mathematical language are introduced. The speed of transverse waves in an elastic medium is shown to be related to the mechanical properties of the medium. The transmission of energy and momentum by a wave is discussed with two particular examples, and then sinusoidal waves, in which each particle of the medium undergoes simple harmonic motion, are introduced. The possibility of the simultaneous existence of two waves in the same medium is discussed; this leads to the principle of superposition and an analysis of standing waves and normal modes of vibration of a mechanical system. Finally, several generalizations of the wave concept and of the methods of analysis of wave motion are discussed briefly.

22-1 THE NATURE OF WAVES

A wave is any disturbance from an equilibrium state which moves, or *propagates*, with time from one region of space to another. Examples of wave motion are numerous, both in everyday experience and in the various physical sciences, and are found in all branches of physics. The classical and time-worn example of wave motion is that of waves on the surface of a still pond produced by dropping a stone into the water. The disturbance created by the stone spreads out horizontally in all directions along the surface; observation reveals, however, that individual particles of water do not move horizontally with the wave crests and troughs but have mostly up-and-down displacements around their original positions. Thus *matter* is not transported outward from the origin of the disturbance but instead the *disturbance* is propagated outward.

An even more commonplace example is sound. Any source of sound produces variations in the pressure of air; these variations are propagated from one region of space to another, but the air molecules do not themselves move to the distant points. A third example of wave motion is exhibited when one ties one end of a rope or string to a fixed object, pulls it taut, and then shakes the free end. The result is a "wriggle" on the string which moves along its length. In this case, the displacements of individual particles in the string are perpendicular, or *transverse*, to its length, while the disturbance travels parallel to the length.

The foregoing are examples of mechanical waves, and in each case some aggregation of matter serves as a *medium* for the transmission of the wave. In the first case it was water, in the second air, in the third a stretched string. Each of these systems had a certain equilibrium state, and in each the wave involved a departure from the equilibrium state at various points in the medium. There are also waves in which no material medium is involved. One important example is an electromagnetic wave. Here the disturbance from equilibrium is the temporary presence of electric and magnetic fields in a certain region. The principles of electrodynamics show that electric and magnetic fields which change with time induce electric and magnetic fields in neighboring regions, and so an electromagnetic disturbance is propagated through space. Although the physical nature of electromagnetic waves is quite different from that of mechanical waves, they share a number of common features and can be described in the same general language.

An example of wave motion further removed from ordinary experience is the wavelike nature of fundamental particles. Here again, there is no medium for transmission of a wave, but one may think of a particle as a disturbance from an equilibrium (vacuum) state which propagates from one region of space to another as the particle moves. The concept of particles as having a fundamental wave nature is one of the cornerstones of present-day quantum mechanics.

In each of the above examples, *energy* must be supplied to the system to produce a wave. Thus a wave can transport energy from one region to another; correspondingly, momentum and angular momentum can also be carried. The transmission of energy by electromagnetic waves is familiar, and the destructive power of ocean surf is a convincing demonstration of the energy transported by water waves. In Sec. 22-5 we consider in detail the mechanism of energy transport on a string.

As frequently happens in the formulation of physical principles, the most familiar phenomena are not always the simplest ones to analyze in detail. It is often instructive and useful to select a particularly simple situation to exhibit basic ideas, uncluttered by undue mathematical complication. We select as our prototype of wave motion the propagation of waves on a stretched string or rope. This is one of the simplest illustrations of wave motion, and in the next several sections we discuss it in considerable detail.

We tie one end of a long string to a fixed object and then, keeping the tension constant, impart some motion to the other end, as shown in Fig. 22-1. Suppose, in particular, that with the rope initially at rest, the free

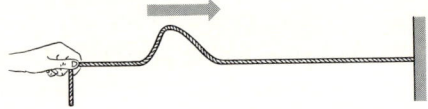

Fig. 22-1 Wave pulse can be produced on a stretched string by giving one end of the string a transverse motion.

end is moved quickly upward and then back to its original position. The resulting motion, called a *wave pulse,* can conveniently be studied by means of a multiple-flash photograph, which permits observation of the shape of the rope at a number of successive times. Figure 22-2 shows a sketch made from such a photograph.

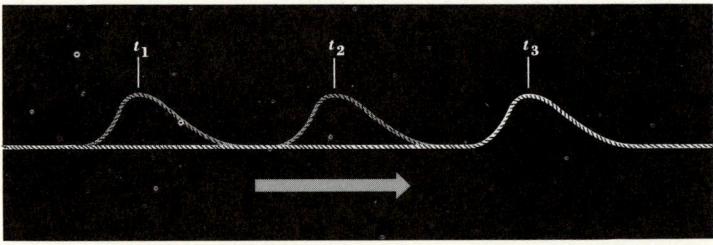

Fig. 22-2 Single wave pulse produced as in Fig. 22-1, shown at several successive times as it moves along the length of the string. As the disturbance moves, its shape does not change. Measurements show that the disturbance propagates with a definite speed along the string.

Several features of this figure are striking. First, the disturbance maintains its shape as it travels along the string. Second, by measuring the position of the crest of the pulse at several times, we can establish that the pulse moves along the string with a definite *speed*. Experiments with pulses of various sizes and shapes show that the wave speed is independent of the shape of the pulse and that the shape is maintained during propagation. The speed of propagation of waves depends on the tension in the string, as well as on the distribution of mass along its length. This dependence is investigated in detail in Sec. 22-3. Further observation shows that the displacement of a particle in the string from its equilibrium position is always perpendicular to the string; the particles do not move parallel to the string. Any wave for which the displacement is always perpendicular to the direction of propagation is called a *transverse* wave.

Even in the simple cases in which the wave propagates in one definite direction, as on a stretched string, the behavior is not always as simple as that which has been described. There are situations in which a wave pulse produced at one end of a string changes in shape as it travels along the string; this phenomenon is called *dispersion*. In this chapter we confine our attention to wave motion in which propagation occurs in one fixed direction and in which dispersion does not occur. In particular, the major portion of our discussion is focused on the stretched string, as a prototype of mechanical wave motion.

22-2 MATHEMATICAL DESCRIPTION OF WAVES

Before proceeding with an analysis of the dynamical basis of wave motion, it is essential to develop efficient ways of *describing* wave motion. In describing the motion of a point, we establish a system to represent the *position* of the point at any instant of time; i.e., we assign coordinates. When the point moves, the change in each coordinate is some function of time; if the functions of time are known, the motion is completely described.

To describe wave motion is a somewhat more subtle problem. For a transverse wave on a string, for example, it is necessary, in order to describe the state of the system at any time, to specify the position of *every point* of the string, not just one or a few points as in the case of simpler mechanical systems. The configuration at any one instant of time can be described by means of a function. In Fig. 22-3 the x axis coincides with the equilibrium position of the string. We neglect any sag in the string caused by gravity,

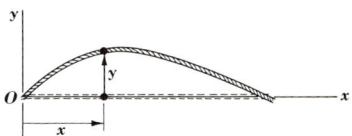

Fig. 22-3 Coordinate system for a mathematical description of transverse waves on a string. The x axis is the equilibrium position of the string; the displacement of a point on the string with equilibrium position x is denoted as y. The equilibrium positions of the string and of the particular point whose displacement is indicated are shown as dotted lines.

586 MECHANICAL WAVES

and we assume that it is stretched very nearly in a straight line. During any wave motion, various points are displaced perpendicular to the x axis. Denoting the equilibrium position of any point by x and its displacement by y, we can describe the position of the entire string at a particular time t_1 with a function $y = f_1(x)$. Similarly, the position of the rope at a different time t_2 is given by $y = f_2(x)$, where $f_2(x)$ is in general a function different from $f_1(x)$.

This procedure looks extremely clumsy, since we need a different function for each instant of time. But all these functions are closely related if the wave maintains the same shape as it travels. The curve which represents the function $f_2(x)$ has exactly the same shape as that which represents the function $f_1(x)$, but it is displaced a distance along the x axis equal to the distance the wave travels in the time interval $t_2 - t_1$. If the wave travels with speed c, the displacement is $c(t_2 - t_1)$.

To specify precisely how the functions for various times are related, we consider a pulse which at time $t = 0$ is at the position shown in Fig. 22-4.

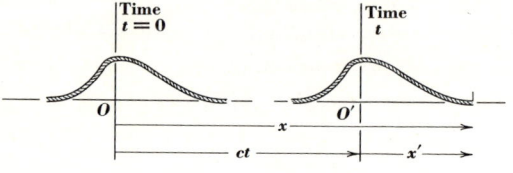

Fig. 22-4 In time T a wave pulse moves a distance ct. A single function can be used to describe a wave pulse at two times by using two coordinate systems displaced relative to each other by a distance ct, as shown. The two horizontal coordinates are related by $x = x' + ct$.

At a later time t, the pulse has moved a distance ct to the second position shown. We now introduce a new axis system whose origin is O', displaced to the right a distance ct with respect to the first. If the function $y = f(x)$ describes the pulse at time $t = 0$, then the *same* function describes the pulse at some later time, with respect to the displaced coordinate system O'. Thus at time t the pulse is described relative to O' by the function $f(x')$.

From the figure we see that the two coordinate systems are related by the equation $x' = x - ct$. Thus a wave which is described relative to O at time $t = 0$ by $f(x)$ is described in this *same* coordinate system at a later time t by the function $f(x - ct)$, which simply means the original function with x replaced everywhere by $x - ct$.

Example

Suppose a transverse wave pulse on a rope is described at time $t = 0$ by the equation

$$y = \frac{1}{1 + x^2}$$

where x and y are both measured in meters. If the wave speed is 2 m/s, write an expression which describes the shape of the wave at $t = 1$ s, $t = 2$ s, and for any general value t.

Solution
In general, as just pointed out, the shape of the rope at any time t is given by

$$y = \frac{1}{1 + (x - ct)^2}$$

In the first case, $ct = (2 \text{ m/s})(1 \text{ s}) = 2$ m, and the expression at time $t = 1$ s is

$$y = \frac{1}{1 + (x - 2)^2}$$

Similarly, at time $t = 2$ s, it is

$$y = \frac{1}{1 + (x - 4)^2}$$

In these expressions, the waveshape has moved 2 and 4 m, respectively, from its position at time $t = 0$. The configuration of the rope at these three times is shown in Fig. 22-5.

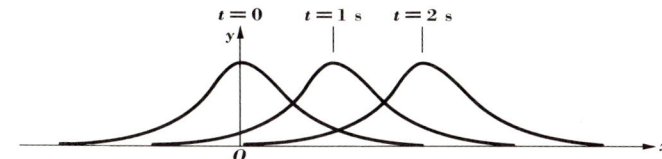

Fig. 22-5 Motion of a transverse wave pulse. The pulse is shown at three successive times.

In general, *every function* which contains x and t only in the particular combination $x - ct$, where c is a constant, represents a wave which propagates in the $+x$ direction without changing its shape. If this is not completely clear from the foregoing discussion, it may be helpful to consider how rapidly an observer would have to move in order to keep up with the peak of a wave pulse. To keep abreast of a certain point on a wave, an observer must move fast enough for the function $f(x - ct)$ describing the displacement to be constant at the moving point he is observing. In other words, for him $x - ct$ must be constant. But this can happen only if the point he is observing moves according to the equation $x = ct + \text{constant}$. Thus the waveshape moves with a speed c in the $+x$ direction.

A few additional examples of functions having this characteristic are

$$(x - ct)^2 \qquad \sin k(x - ct) \qquad \log k[(x - ct)^2 + a]$$

Each of these, as pointed out, contains x and t only in the particular combination $x - ct$. On the other hand, the functions

$$x^2 - c^2 t^2 \qquad xct$$

contain x and t in combinations other than $x - ct$.

By precisely the same reasoning it can be established that any function $y = f(x,t)$ containing x and t only in the combination $x + ct$ represents a wave disturbance which travels in the *negative x* direction. Such a disturbance might be created, for example, by keeping the left-hand end of our rope fixed and giving the right-hand end a displacement of the type discussed previously.

Various kinds of information can be obtained from the function describing a wave, which is usually called simply a *wave function*. For example, the velocity of transverse motion of a particle of the rope at any position x is simply the rate of change of y with respect to t at this point. Denoting the transverse speed of the particle at the position x and time t by $v(x,t)$, we see that

$$v(x,t) = \frac{\partial y}{\partial t} \tag{22-1}$$

The notation $\partial y/\partial t$, we recall, is read "partial derivative of y with respect to t" and is used as a reminder that y is a function of two variables, x and t; the partial derivative with respect to t is obtained by regarding all other variables as constants while taking the derivative. Physically, this corresponds to observing a particular point (one value of x) on the string. Similarly, the *slope* of the string at any point is $\partial y/\partial x$. One can also find the acceleration of any point, the curvature, and so on.

The derivatives of y with respect to x and t, representing the slope and velocity at a given point, respectively, are closely related; it is worthwhile to observe the nature of this relationship. Figure 22-6 shows a small section of string at two times separated by an interval Δt. In this interval, the wave

Fig. 22-6 Wave pulse shown at two times separated by an interval Δt. The displacement $c\,\Delta t$ of the wave and the displacement Δy of a point on the string are shown. From the small triangle it can be seen that $\partial y/\partial x = -\Delta y/c\,\Delta t$. In the limit as $\Delta t \to 0$, this leads to Eq. (22-2).

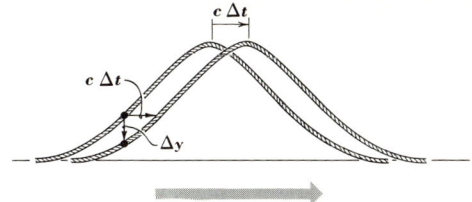

moves a distance $c\,\Delta t$ to the right. The point shown moves downward a distance Δy. The slope of the section of string is given by $\partial y/\partial x$; the slope can also be obtained from the small triangle and is $-\Delta y/c\,\Delta t$. The minus sign is needed because the slope is positive while Δy is negative in this case. Combining these two expressions and taking the limit as $\Delta t \to 0$, we find

$$\frac{\partial y}{\partial x} = -\frac{1}{c}\frac{\partial y}{\partial t} \tag{22-2}$$

This same result can be obtained more formally as follows: For the case of a wave traveling in the $+x$ direction, let $u = x - ct$. Then the two derivatives can be written, using the chain rule, as

$$\frac{\partial y}{\partial x} = \frac{dy}{du}\frac{\partial u}{\partial x} = \frac{dy}{du} \quad \text{and} \quad \frac{\partial y}{\partial t} = \frac{dy}{du}\frac{\partial u}{\partial t} = -c\frac{dy}{du} \tag{22-3}$$

since $\partial u/\partial x = 1$ and $\partial u/\partial t = -c$. Thus, again

$$\frac{\partial y}{\partial x} = -\frac{1}{c}\frac{\partial y}{\partial t}$$

Similarly, it is easy to show that for a wave traveling in the $-x$ direction, the two derivatives are related by

$$\frac{\partial y}{\partial x} = \frac{1}{c}\frac{\partial y}{\partial t} \tag{22-4}$$

To go one step further, we can derive a relation between the *second* derivatives of y with respect to x and t. For a wave in the $+x$ direction, Eq. (22-2) states that the derivative of y with respect to x is equal to $-1/c$ times its derivative with respect to t. Thus it is also true that the derivative of either side of Eq. (22-2) with respect to x is $-1/c$ times the derivative of either side (since it is an equation) with respect to t. In particular,

$$\frac{\partial}{\partial x}\left(\frac{\partial y}{\partial x}\right) = -\frac{1}{c}\frac{\partial}{\partial t}\left(-\frac{1}{c}\frac{\partial y}{\partial t}\right) \quad \text{or} \quad \frac{\partial^2 y}{\partial x^2} = \frac{1}{c^2}\frac{\partial^2 y}{\partial t^2} \tag{22-5}$$

Similarly, one can start with Eq. (22-4) and repeat the argument to show that Eq. (22-5) is also true (without change of sign) for a wave in the $-x$ direction. Thus this equation is correct for both kinds of waves and is more general than Eq. (22-2) or (22-4).

Equation (22-5) is, in fact, a very important and useful result. It is a differential equation, and its status is similar to that of the differential equation for the harmonic oscillator, Eq. (10-2). It does not tell us directly what motion occurs in any particular situation, but it provides an equation whose *solutions* represent possible motions. Thus a study of the functions which satisfy Eq. (22-5) is in effect a study of the possible wave motions which can occur on a stretched string. A few functions representing wave motions of special importance are discussed in the following sections.

The entire body of mathematical language introduced in this section can also be applied directly to the description of wave motion in which the displacements of individual particles of the mechanical medium are *parallel* to the direction of motion of the wave. For example, when one end of a steel bar is struck a sharp blow with a hammer, the compression wave travels along the length of the bar. As the wave pulse passes a given point, molecules at this point are displaced from their equilibrium positions; the progress of the wave can be described by the displacement from equilibrium at each point as a function of time. Thus, in Fig. 22-7, it may happen that at a certain

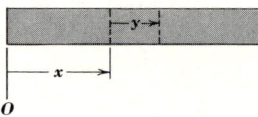

Fig. 22-7 Coordinate system for a mathematical description of longitudinal waves in an elastic substance. The coordinate x describes the equilibrium position of a certain point in the material, and y represents the instantaneous displacement of this point from its equilibrium position. By convention, y is positive when the displacement is in the direction of increasing x.

time t a molecule whose equilibrium position is given by the coordinate x is displaced a distance y from this equilibrium position in a direction parallel to the direction of progress of the wave. At a somewhat later time, molecules further along the length of the rod will be displaced a corresponding amount from their equilibrium positions. Thus the wave function $f(x,t)$ again represents the displacement from equilibrium at each point as a function of time, the only difference being that in this case the displacement is parallel to the direction of propagation of the wave rather than perpendicular to it. All the equations developed so far, up to and including Eq. (22-5), are valid for this kind of motion as well as for the transverse displacements of points on a string.

22-3 DYNAMICS OF WAVES ON A STRING

Thus far, our discussion of wave motion has mostly been phenomenological. The existence of waves on a stretched string propagating with a definite speed and without change of shape is an observed fact. It is now appropriate to try to understand such waves on the basis of the principles of mechanics, which must govern the motion of particles in the string.

For this purpose, we consider the motion of a section of string which is participating in a wave motion. Figure 22-8 shows a section whose length

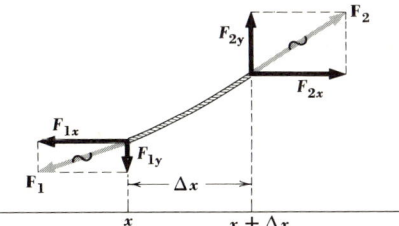

Fig. 22-8 Free-body diagram for a section of string whose length in its equilibrium position is Δx. The force at each end of the string is tangent to the string at the point of application; each force is represented in terms of its x and y components.

in the equilibrium position would be Δx. In the figure this section is displaced from the equilibrium position (represented by the x axis) and stretched, so its actual length is somewhat larger than this. Assuming that the string is uniform, we denote its mass per unit length in the equilibrium position by μ; this quantity is called the *linear mass density*. Thus the mass of this section is $m = \mu \, \Delta x$.

We now proceed to apply Newton's second law to this section of string. Neglecting gravity, we see that the forces are those shown in Fig. 22-8. The force applied to each end is in the direction of the string at the corresponding point; these forces can be resolved into their x and y components. Under the assumption that the displacements are always transverse, there is no component of acceleration, or hence of force, along the x axis. Thus it always must be true that $F_{1x} = F_{2x}$; these symbols refer to the *magnitudes* of the corresponding components, disregarding sign. Furthermore, these quantities must be equal to the tension T in the string when it is not displaced from equilibrium. That is,

$$F_{1x} = F_{2x} = T \tag{22-6}$$

The transverse acceleration of this section of string is produced by the transverse component of force, $F_{2y} - F_{1y}$. To obtain the magnitudes of these forces, we note that the ratio F_{1y}/F_{1x} is equal to the slope of the string at the point x, and correspondingly for the other end. Thus we obtain the relations

$$\frac{F_{1y}}{F_{1x}} = \left(\frac{\partial y}{\partial x}\right)_x \qquad \frac{F_{2y}}{F_{2x}} = \left(\frac{\partial y}{\partial x}\right)_{x+\Delta x} \tag{22-7}$$

where the notation indicates that the derivatives are evaluated at the points x and $x + \Delta x$, respectively. Combining Eqs. (22-6) and (22-7) gives

$$F_{2y} - F_{1y} = T\left[\left(\frac{\partial y}{\partial x}\right)_{x+\Delta x} - \left(\frac{\partial y}{\partial x}\right)_x\right] \tag{22-8}$$

Applying $F = ma$, we now equate the net transverse force given by Eq. (22-8) to the product of the mass of the section and its acceleration:

$$F_{2y} - F_{1y} = \mu \Delta x \frac{\partial^2 y}{\partial t^2} \tag{22-9}$$

The question may arise whether the acceleration $\partial^2 y/\partial t^2$ should be evaluated at the point x or the point $x + \Delta x$. Since the next step is to let $\Delta x \to 0$, this is not really an important question.

Combining Eqs. (22-8) and (22-9), we have

$$\frac{(\partial y/\partial x)_{x+\Delta x} - (\partial y/\partial x)_x}{\Delta x} = \frac{\mu}{T}\frac{\partial^2 y}{\partial t^2} \tag{22-10}$$

We see that in the limit as $\Delta x \to 0$, the left side of this equation simply becomes the derivative with respect to x of $\partial y/\partial x$, which is $\partial^2 y/\partial x^2$; in this limit we therefore have

$$\frac{\partial^2 y}{\partial x^2} = \frac{\mu}{T}\frac{\partial^2 y}{\partial t^2} \tag{22-11}$$

Any wave function $f(x,t)$ which describes a physically possible wave on the string must satisfy Eq. (22-11).

This equation has precisely the same form as Eq. (22-5), which was derived from purely kinematic considerations. The fact that we have been able to obtain this equation from Newton's laws of motion demonstrates that waves of the type under discussion are in fact compatible with the principles of mechanics. Furthermore, comparing Eqs. (22-11) and (22-5), we see that the wave speed c is related to the mechanical properties μ and T in that $1/c^2 = \mu/T$, or

$$c = \sqrt{\frac{T}{\mu}} \tag{22-12}$$

Increasing the tension in the string increases the wave speed, but increasing the linear mass density decreases the speed.

22-4 SINUSOIDAL WAVES

Thus far, the discussion of wave motion in a string has centered around the idea of a single wave pulse, localized at any given time in a certain region on the string, in which successive particles in the string undergo temporary displacements from equilibrium and then return. Some of the most important and interesting kinds of wave motion, however, are those in which each particle undergoes a motion which is *periodic* in the sense of Chap. 10. We consider first a situation in which the string is so long that during the time of observation the wave motion never reaches the far end of the string, so that effects due to the fixed point need not be considered. Suppose now that the near end is given a periodic motion with period T and corresponding frequency $f = 1/T$.

Since the motion of the end of the string is periodic, the wave pattern on the string at any instant is also repetitive. During one period T of the motion, the wave travels along the string a distance cT. Let us denote this distance by $\lambda = cT$. At the end of the time interval T, the end of the string begins to repeat the motion of the first interval. By the end of the second period, the wave has traveled a total distance 2λ, but since the motion of the end point during the second period is the same as that during the first period, the wave pattern on the string consists of two separate patterns, each of length λ, identical in shape. If the end of the string is at $x = 0$, then the shape of the string between $x = 0$ and $x = \lambda$ is exactly the same as that between $x = \lambda$ and $x = 2\lambda$. An example of a periodic motion of the end of the string and the resulting repetitive waveform is given in Fig. 22-9, which shows several states in the progress of the wave along the string. This figure also illustrates the relationship $\lambda = cT$.

The distance λ is usually called the *wavelength*; it is equal to the distance between any pair of corresponding points on two successive repetitions of the waveform. The relationship

$$\lambda = cT = \frac{c}{f} \tag{22-13}$$

is valid for any periodic wave motion.

An important example of periodic wave motion occurs when the end

594 MECHANICAL WAVES

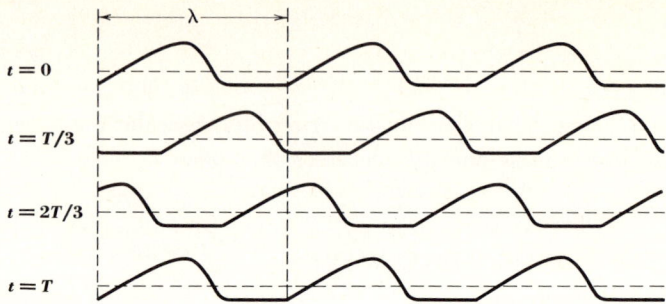

Fig. 22-9 Periodic wave produced by periodic motion of the end of a string, beginning at time $t = 0$. In a time T, the wave moves a distance λ.

of the string is given a *sinusoidal* motion. It was noted in Chap. 10 that in some respects sinusoidal motion is the simplest periodic motion of a point, and so it is natural to consider the wave motion it induces. Let the end of the string be at $x = 0$, and suppose this point is given a transverse displacement $A \sin 2\pi ft$, a simple harmonic motion with amplitude A and frequency f. Then, at the point $x = 0$, the wave function is

$$y(0,t) = A \sin 2\pi ft \tag{22-14}$$

To find the complete wave function, we recall from Sec. 22-2 that the function for a wave propagating in the $+x$ direction must contain x and t only in the particular combination $x - ct$, which is equivalent to stating that the coefficient of t in the wave function must be $-c$ times the coefficient of x. Here the coefficient of t is $2\pi f$, so the coefficient of x must be $-2\pi f/c$, and the general wave function is

$$y(x,t) = A \sin \left(2\pi ft - \frac{2\pi fx}{c} \right)$$
$$= A \sin 2\pi f \left(t - \frac{x}{c} \right) \tag{22-15}$$

As a check, we substitute Eq. (22-15) into Eq. (22-5), which must be satisfied by any possible wave function. It is in fact satisfied; the calculation is left as a problem.

This procedure for obtaining the general wave function corresponding to sinusoidal motion of the end of the string is perfectly straightforward and legitimate, but it may lack something in intuitive appeal. The significance of the wave described in Eq. (22-15) may be appreciated more fully by noting first that every particle in the string undergoes simple harmonic motion; for a fixed value of x, corresponding to a particular point on the string, the displacement is a sinusoidal function of time. The motions of various points, of course, differ in *phase;* they are out of step with respect to each other, but nevertheless each point moves with simple harmonic motion. Second, at any instant the shape of the string is described by a sinusoidal function of x. That is, for any fixed value t, y is a sinusoidal function of x.

An alternative form of Eq. (22-15), obtained with the help of Eq. (22-13), is

$$y(x,t) = A \sin 2\pi \left(\frac{t}{T} - \frac{x}{\lambda} \right) \qquad (22\text{-}16)$$

In this form, the wave function shows immediately that when t changes by an amount equal to T, the argument of the sine function changes by 2π, corresponding to a complete cycle of motion of any point. Similarly, when x changes by an amount equal to λ, the argument of the function again changes by 2π, corresponding to a complete repetition of the waveform.

This discussion can be adapted to the case of a wave propagating in the $-x$ direction. Such a wave cannot, of course, be produced by shaking the end of the string at $x = 0$, since there is no string to the left of this point. But we may easily imagine a wave motion somewhere in the middle of the string, arising from sinusoidal motion of a point to the right of the region we are observing. Since any wave propagating in the $-x$ direction must contain x and t only in the combination $x + ct$, a possible wave function for a sinusoidal wave of amplitude A, frequency f, and speed c is

$$y(x,t) = A \sin 2\pi f \left(t + \frac{x}{c} \right) \qquad (22\text{-}17)$$

which can also be written

$$y(x,t) = A \sin 2\pi \left(\frac{t}{T} + \frac{x}{\lambda} \right) \qquad (22\text{-}18)$$

One further generalization of wave functions for sinusoidal waves is sometimes useful. It was observed in Chap. 10 that a simple harmonic motion described by the equation

$$y = A \sin(2\pi ft + \phi) \qquad (22\text{-}19)$$

has the same general character as one described by

$$y = A \sin 2\pi ft \qquad (22\text{-}20)$$

The two motions have the same frequency, period, and amplitude; they differ only in that they are out of step with respect to each other. When ϕ, called the *phase angle*, is equal to π or $180°$, the two motions are precise opposites of each other; otherwise they are out of step by some fraction of a period. From the foregoing discussion, it should be clear that if the free end of a string is given a motion corresponding to Eq. (22-19), the resulting wave is described by the wave function

$$y(x,t) = A \sin\left[2\pi f\left(t - \frac{x}{c}\right) + \phi\right] \tag{22-21}$$

which is a wave differing from that described by Eq. (22-15) only in that at every point along the string, at any instant, the two waves are out of step by the same fraction of a period. The quantity ϕ is called the phase angle of the wave, as previously.

In summary, when a point at the end of the string is given a periodic motion, every point on the string undergoes the same periodic motion at a later time determined by its distance from the end of the string. The resulting form is also periodic in nature, repeating itself in a distance λ called the wavelength. The wavelength is related to the wave speed and the period by $\lambda = cT$. In particular, when the periodic disturbance is sinusoidal, every point on the string undergoes simple harmonic motion with the same amplitude and period, and the shape of the string at any instant of time is described by a sine wave. Period and frequency have precisely the meanings they have for simple harmonic motion; the additional quantities needed to describe wave motion on a string are the speed of the wave c and the wavelength λ.

22-5 ENERGY IN WAVE MOTION

It was remarked in Sec. 22-1 that energy is always associated with wave motion and that a wave can transport energy from one region of space to another. We are now in a position to discuss in detail the energy associated with the wave motion of a stretched string and the mechanism by which it can convey energy.

It is clear that there must be kinetic energy associated with wave motion on a string. In particular, consider a section of string whose unstretched length is Δx and whose mass is $\mu \Delta x$. The transverse speed of this element at any instant is $\partial y/\partial t$, so its kinetic energy is

$$E = \tfrac{1}{2}\mu \, \Delta x \left(\frac{\partial y}{\partial t}\right)^2 \tag{22-22}$$

Alternatively, we say that the string possesses a kinetic energy *per unit* length of $\frac{1}{2}\mu(\partial y/\partial t)^2$. This quantity is a *linear energy density;* it describes the kinetic energy per unit length of string, just as the quantity μ describes the mass per unit length.

The fact that a string undergoing wave motion has potential energy is not quite so obvious. Briefly, the reason for the existence of potential energy is that a section of string which is displaced in such a way that it is not parallel to the equilibrium position must be stretched, and to stretch it requires work. Consider again a section of string of length Δx. Figure 22-10 shows

Fig. 22-10 (*a*) Section of string of length Δx in equilibrium position. (*b*) Same section, displaced (with a slope $\partial y/\partial x$ with respect to the equilibrium position) by moving the right end a distance Δy. In this final position, the vertical component of force on the right end is F_y; the average force exerted during the displacement is $\frac{1}{2}F_y$, since the force increases uniformly from zero to F_y during the displacement Δy. The horizontal component of force T does not vary.

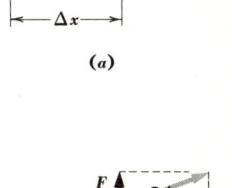

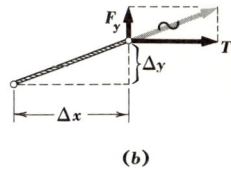

such a section in two positions, one parallel to the axis, in which it is not stretched relative to its equilibrium length, and another in which it is stretched. In Fig. 22-10*b*, the two ends of the section have transverse displacements which differ by an amount Δy. If we assume the section to be short, so that its shape can be represented by a straight line without appreciable error, then $\Delta y = \Delta x(\partial y/\partial x)$.

It has been shown previously that the transverse force at a point on the string has a magnitude given by $F_y = T(\partial y/\partial x)$. When the section Δx is moved from the position of Fig. 22-10*a* to that of Fig. 22-10*b* by moving the right end, the *average* transverse force applied to the right end is

$$(F_y)_{av} = \tfrac{1}{2} T \frac{\partial y}{\partial x}$$

since the force varies uniformly from zero to a maximum of $T(\partial y/\partial x)$ as the right end is displaced. The work required for this displacement is

$$W = (F_y)_{av} \Delta y = \tfrac{1}{2} T \Delta x \left(\frac{\partial y}{\partial x}\right)^2 \tag{22-23}$$

This is the potential energy of this section of string if potential energy is defined as zero at the equilibrium position. The potential energy *per unit length* is $\frac{1}{2}T(\partial y/\partial x)^2$.

The *total* energy per unit length, including both kinetic and potential energies, is

$$\tfrac{1}{2}\mu\left(\frac{\partial y}{\partial t}\right)^2 + \tfrac{1}{2}T\left(\frac{\partial y}{\partial x}\right)^2 \tag{22-24}$$

To find the total mechanical energy H possessed by a section of the string between points x_1 and x_2 at any instant, we need only integrate Eq. (22-24) between these limits, using derivatives of the wave function describing the particular wave under consideration.

A related problem is the mechanism by which energy is transferred along the string from one region to another. Consider a point P on the string as in Fig. 22-11a; the energy given to the part to the right of P is simply the

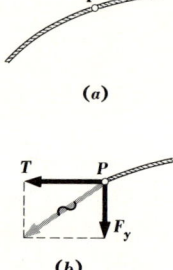

Fig. 22-11 (*a*) Energy is transported past a point P on the string; work is done on the string to the right of P as a result of the transverse force at P and its displacement. (*b*) Free-body diagram showing the force exerted on the string at P. Only the component F_y does work, since there is no displacement in the direction of T.

work done on this part by the string to the left of P. The force on the right side of the string at P is shown in Fig. 22-11b. The horizontal component of this force does no work, because the displacement is always transverse; the vertical component is

$$F_y = T\frac{\partial y}{\partial x} \tag{22-25}$$

The rate at which this force does work on the right side is simply the product of F_y and the vertical *velocity*. Denoting the power, or rate at which work is done, on the right side as P, we obtain

$$P = F_y v = -T\frac{\partial y}{\partial x}\frac{\partial y}{\partial t} \tag{22-26}$$

For example, if the wave is a sinusoidal one propagating in the $+x$ direction, as described by Eq. (22-15), then

$$\frac{\partial y}{\partial x} = -\frac{2\pi f}{c} A \cos 2\pi f\left(t - \frac{x}{c}\right)$$

$$\frac{\partial y}{\partial t} = 2\pi f A \cos 2\pi f\left(t - \frac{x}{c}\right)$$

$$P = 4\pi^2 f^2 A^2 \frac{T}{c} \cos^2 2\pi f\left(t - \frac{x}{c}\right) \tag{22-27}$$

For this wave P is always positive, indicating that energy is always transferred from left to right, never the reverse. The rate of transfer of energy at any given point varies from a minimum of zero to a maximum of $4\pi^2 f^2 A^2 T/c$. The *average* rate of transfer of energy is this expression multiplied by the average value of $\cos^2 2\pi f(t - x/c)$. By the method described in Sec. 19-5, it can be shown that this average value is $\frac{1}{2}$. Hence the rate of transfer of energy is proportional to the square of the amplitude of the wave and the square of the frequency, and it also depends on the mechanical properties of the system.

This whole calculation can be repeated for a sinusoidal wave propagating in the $-x$ direction, as described by Eq. (22-17). In this case, the result is the same except that the sign is changed. This indicates, of course, that energy is transferred in the negative x direction, as expected.

Another fundamental question concerned with energy in wave motion is the *speed* with which energy is transferred. First, if the system is such that waves propagate without changing their shape, i.e., without dispersion, it can be shown that the speed of transfer of energy is equal to the wave speed c. In wave propagation with dispersion, the usual technique is to discuss the behavior of the system for sinusoidal waves of various frequencies; the wave speed is different for different frequencies. When a complex wave is represented as a combination of sinusoidal waves, the speed of propagation of energy is not in general the same as the speed of any one of the individual waves if dispersion is present.

It is possible to define a new quantity, the *group velocity*, to characterize the speed of propagation of energy in such a situation. If the wave is a combination of waves with different frequencies close to a certain central value f, with correspondingly close values of λ and c, it can be shown that the group velocity v_g is given by

$$v_g = c - \lambda \frac{dc}{d\lambda} \tag{22-28}$$

where all quantities are evaluated using the central f, c, and λ. This is the speed with which energy is transferred.

22-6 LONGITUDINAL WAVES

Several aspects of wave motion in a mechanical system have been discussed, using a stretched string for illustration. Another important kind of wave motion occurs in an elastic solid or fluid when a particular region is given a displacement *parallel* to the direction of propagation of the wave. The result is called a *longitudinal* wave; such a wave is as easy to analyze as a transverse wave on a string, and many of the results for the string can be taken over directly or with only minor modifications to longitudinal waves.

Consider a long steel rod; when one end is given a sudden displacement, as might be achieved by striking it with a hammer, the immediate effect is a compression of the material in the region of the end. This compression induces stresses on the neighboring regions, leading to compression in them. The result is a compression wave which travels along the length of the rod. Figure 22-12 shows an example of such a wave. In this figure, the very heavily

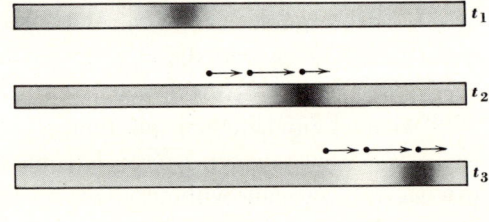

Fig. 22-12 Longitudinal wave pulse in an elastic rod, produced by displacing one end of the rod and then restoring it to its original position. The wave pulse is shown at several successive times; the shading represents the change in density in various regions of the rod. The instantaneous displacements of various points from equilibrium are shown by small arrows above the rod. The relationship between the density and the displacements should be studied carefully.

shaded regions have greater than average density. The actual particles of material, of course, do not travel down the length of the rod; what travels is the pattern of compression and associated stress. The displacements of individual particles which lead to this compression and stress follow the direction of propagation of the wave; hence it is natural to call such a wave motion a *longitudinal* wave.

Another simple example is shown in Fig. 22-13. A sliding piston at one end of a pipe containing a gas is moved back and forth. The result is a longitudinal wave which propagates in the direction of the pipe. In many

Fig. 22-13 Longitudinal wave in a pipe filled with gas, resulting from motion of a piston at one end of the pipe, as shown. Displacements of gas molecules are in the direction of the pipe; the variation in density with position at one instant is shown, and the corresponding displacements are indicated by small arrows.

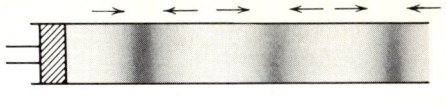

cases of longitudinal wave motion, such as the above examples, the pattern of the wave disturbance propagates without changing its shape and with a definite speed, just like transverse waves on a string. As before, we call these *nondispersive* waves.

A longitudinal wave can be described with a wave function, using the scheme discussed at the end of Sec. 22-2. To proceed with a detailed analysis of longitudinal waves, we apply Newton's laws of motion to a small element of the medium having a length in the equilibrium state of Δx. The forces on this element are longitudinal forces on its ends, associated with the elastic deformation of the medium. Thus the first natural step is to relate the force or stress at any point x to the wave function $y(x,t)$ describing the displacement of this point from its equilibrium position.

Considering an element of length Δx whose left end has equilibrium position x and whose right end has equilibrium position $x + \Delta x$, we see that when $y(x,t)$ and $y(x + \Delta x, t)$ are equal, the element is neither stretched nor compressed and there is no force on either end. But if the two displacements are different, the length is no longer Δx. In particular, the *change* in length is given by $y(x + \Delta x, t) - y(x,t)$, and the fractional change in length, or *strain*, is

$$\frac{y(x + \Delta x, t) - y(x,t)}{\Delta x}$$

Taking the limit as $\Delta x \to 0$, we see that the strain at any point is simply $\partial y/\partial x$.

Now, for simplicity, we assume that the strains are always sufficiently small for Hooke's law to be obeyed. That is, the strain is proportional to the stress (in this case force per unit area) applied to the ends of the element. The proportionality constant, as discussed in Sec. 19-4, is the elastic modulus for the material, which we denote simply by E. Thus, when Hooke's law is obeyed,

$$E = \frac{F/A}{\partial y/\partial x} \qquad \text{or} \qquad F = AE\frac{\partial y}{\partial x} \tag{22-29}$$

where A is the cross-section area of the elastic substance. The force transmitted across any cross-section area of a material is proportional to the elastic modulus, the area, and the rate at which the displacement from equilibrium changes from one point to another in the medium.

We now use Eq. (22-29), together with Newton's second law, to derive a differential equation governing the motion. Considering again an element of length Δx, we see that the net force applied to this element is

$$F = AE\left[\left(\frac{\partial y}{\partial x}\right)_{x+\Delta x} - \left(\frac{\partial y}{\partial x}\right)_{x}\right] \tag{22-30}$$

where the notation indicates, as usual, that the derivatives are to be evaluated at the points $x + \Delta x$ and x, respectively. This is now to be equated to the mass times the acceleration of this element. The mass is expressed conveniently in terms of the mass density ρ, or mass per unit volume. The volume of the element is $A\,\Delta x$, so its mass is $A\rho\,\Delta x$. The acceleration, of course, is $\partial^2 y/\partial t^2$, so the equation of motion for the element is

$$AE\left[\left(\frac{\partial y}{\partial x}\right)_{x+\Delta x} - \left(\frac{\partial y}{\partial x}\right)_{x}\right] = A\rho\,\Delta x\frac{\partial^2 y}{\partial t^2} \tag{22-31}$$

in which we recall that both the force and acceleration act along the length of the elastic medium, in the direction of propagation. The cross-section area A divides out of this equation. We now divide both sides by Δx and note, as before, that in the limit as $\Delta x \to 0$, the left-hand side is simply E multiplied by the derivative with respect to x of $\partial y/\partial x$, which is the second derivative. That is,

$$\frac{\partial^2 y}{\partial x^2} = \frac{\rho}{E}\frac{\partial^2 y}{\partial t^2} \tag{22-32}$$

This is a pleasing and important result. The differential equation (22-32) has precisely the same form as that obtained for the string; in addition, it has the same form as the equation obtained in Sec. 22-2 as a general equation which any wave function must satisfy. This proves that it is indeed possible and consistent with Newton's laws for a longitudinal wave to propagate in an elastic medium. Furthermore, comparing Eq. (22-32) with Eq. (22-5), we immediately see that the wave speed is given by

$$c = \sqrt{\frac{E}{\rho}} \tag{22-33}$$

For example, if the elastic medium is a steel rod of uniform cross section, in which case the appropriate elastic modulus is Young's modulus, $E_Y = 20 \times 10^{10}$ N/m², $\rho = 7 \times 10^3$ kg/m³, and the wave speed is

$$c = \sqrt{\frac{20 \times 10^{10} \text{ N/m}^2}{7 \times 10^3 \text{ kg/m}^3}} = 5.3 \times 10^3 \text{ m/s}$$

some 15 times as great as the speed of sound in air at ordinary temperatures.

An application of Eq. (22-33) of more immediate practical interest occurs when the elastic medium is a long column of an ideal gas confined in a pipe and the wave is *sound*. Even at equilibrium, there are stresses everywhere within this elastic medium, since forces must be applied from outside by the walls of the pipe to contain the gas. But in the equilibrium state the stress is the same at both ends of any element of gas. Thus the only modification necessary in the formulation is to reinterpret Eq. (22-29) as the *change* in force produced by the non-equilibrium situation. Then Eq. (22-30) still gives the net force on the element of length Δx, and the rest of the formulation continues exactly as before. This time the appropriate elastic modulus is the *bulk* modulus for the material, which is in fact the only elastic modulus having any meaning for fluids.

It was shown in Sec. 19-4 that for volume changes at constant temperature the bulk modulus for an ideal gas is simply $E_B = P$ [Eq. (19-9)]. Furthermore, the density ρ of an ideal gas can be expressed simply, in terms of the thermodynamic coordinates, by replacing n in the ideal-gas equation of state with m/M, where m represents the total mass of gas and M is the molecular mass. Making this substitution and rearranging, we find

$$\frac{m}{V} = \rho = \frac{PM}{RT}$$

Substituting this and the relation $E_B = P$ into Eq. (22-33), we find

$$c = \sqrt{\frac{RT}{M}} \tag{22-34}$$

For example, the average molecular mass of air is about 29 g/mol = 29×10^{-3} kg/mol, so the predicted speed in air at ordinary temperature, say $T = 300$ K, is

$$c = \sqrt{\frac{(8.314 \text{ J/mol K})(300 \text{ K})}{29 \times 10^{-3} \text{ kg/mol}}} = 293 \text{ m/s}$$

As a matter of fact, the measured value of the speed of sound in air at this temperature is 348 m/s, larger than the predicted value by 20 percent. What went wrong?

The resolution of this apparent disagreement forms an interesting chapter in the development of the theoretical understanding of wave phenomena. In applying Eq. (22-33), we used the bulk modulus of an ideal gas under conditions of constant temperature, certainly a reasonable assumption for an ordinary laboratory measurement. But suppose that in a wave motion the expansions and compressions take place so rapidly that there is no time for thermal equilibrium to be maintained during the motion. If this is the case, an alternative assumption which may be more appropriate is that the expansions and compressions take place *adiabatically* rather than isothermally. In that case, PV is not constant, as for an isothermal process, but instead PV^γ is constant, as discussed in Sec. 20-5; then, for an adiabatic process for an ideal gas,

$$\frac{dP}{dV} = -\gamma \frac{\text{constant}}{V^{\gamma+1}} = -\gamma \frac{P}{V}$$

Use of this expression in Eq. (19-7), which defines the bulk modulus in general, yields

$$E_B = \gamma P$$

which we call the *adiabatic bulk modulus*. This result shows that the bulk modulus measured under adiabatic conditions is larger than that measured isothermally by a factor equal to the ratio of the specific heats of the gas, $\gamma = C_P/C_V$.

The corresponding modification in the expression for the wave speed yields

$$c = \sqrt{\frac{\gamma RT}{M}} \qquad (22\text{-}35)$$

Using $M = 29 \times 10^{-3}$ kg/mol, $T = 300$ K, and $\gamma = 7/5$, the value appropriate for diatomic gases, we obtain

$$c = \sqrt{\frac{7/5(8.314 \text{ J/mol K})(300 \text{ K})}{29 \times 10^{-3} \text{ kg/mol}}} = 347 \text{ m/s}$$

which agrees well with the experimental value quoted above.

This agreement substantiates the assumption that the expansions and compressions are adiabatic; this could not have been predicted in advance from the principles of macroscopic thermodynamics. To understand in detail why the processes are more nearly adiabatic than isothermal requires an examination of the microscopic process of heat transfer which leads to thermal equilibrium, a subject too complex to be discussed in further detail here.

The entire discussion of sinusoidal waves on a string and energy relations for waves on a string can be applied directly to longitudinal waves in an elastic substance. The details of these formulations will not be given here, but some of them are incorporated in the problems.

22-7 SUPERPOSITION OF WAVES

In most of the wave phenomena discussed thus far, the wave motion originates at one end of the medium and propagates in one direction. What happens if two wave motions produced in different regions of the elastic medium simultaneously travel *toward* each other? On the stretched string, for example, what kind of motion results when both ends of the string are moved so as to produce wave pulses? Clearly, as long as the two pulses are separated, this situation is no more complicated than the other. But what happens when they reach the same part of the string and overlap?

Figure 22-14 shows a series of sketches made from a multiple-flash

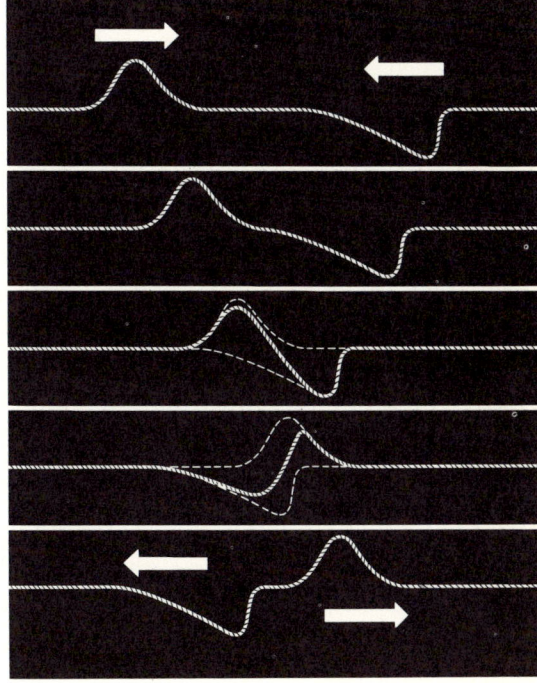

Fig. 22-14 Two pulses with different shapes moving toward each other are produced in two separate sections of a string. The position of the string is shown at several successive times. The positions of the individual pulses at the time they overlap are shown as broken lines. During these times, the actual displacement of the string, shown as a solid line, is the sum of displacements corresponding to the separate pulses. After the two pulses pass, they proceed with their original shapes.

photograph of a string in which two differently shaped pulses are produced at the two ends, move toward each other, and finally pass each other. As in previous cases, each pulse moves without changing its shape. After they pass, each continues to move with its original shape, as though the other were not present. More striking, though, is what happens when the two pulses overlap. During the time of overlap, careful measurements show that the *total displacement* of any point in the string is just the sum of the two displacements corresponding to the separate pulses. That is, each pulse moves exactly as though the other pulse were not there at all; the displacement of any point in the string at any instant can be obtained by finding the displacement of the point corresponding to each of the separate pulses at that instant and adding these two separate displacements.

Similar experiments could be carried out, although perhaps not in such a straightforward manner, for other wave situations, such as longitudinal waves in an elastic solid or a gas. The results can again be understood on the assumption that each pulse moves independently and that when two pulses occupy the same part of the elastic medium, they do not interfere with each other's motions. In general, when two wave motions are present simultaneously in an elastic medium, the displacement of a point at one instant is equal to the *sum* of the displacements of that point corresponding to the two separate waves at that instant. This important and useful generalization is called the *principle of superposition*. Although introduced here with respect to mechanical waves, it has applications to other types of waves as well. For example, the fact that two beams of light may cross and continue unchanged is an example of the principle of superposition for electromagnetic waves.

Although the principle of superposition has been introduced as a generalization from experimental observations, it is not really a new independent principle but can be derived from laws of wave motion already discussed. Suppose $f_1(x,t)$ is a wave function describing one possible wave motion of an elastic medium; it must satisfy the general differential equation (22-5). Similarly, let $f_2(x,t)$ represent another possible wave function, also a solution of Eq. (22-5). If each function separately is a solution of the differential equation, their *sum* also satisfies this equation. This statement is almost obvious from the form of the equation. To prove it formally, we let $F(x,t) = f_1(x,t) + f_2(x,t)$, take the derivatives indicated in the differential equation, and show that $F(x,t)$ satisfies it if the individual functions f_1 and f_2 satisfy it. Thus the principle of superposition is a consequence of the *form* of the differential equation (22-5).

Shaking both ends of a string at once is not by any means the only way of producing two simultaneous wave disturbances. Consider the situation of Fig. 22-15. A wave pulse originates somewhere to the right of the picture and travels to the left toward the fixed end of the string, which is tied to a wall. The various parts of Fig. 22-15 show what happens to the pulse.

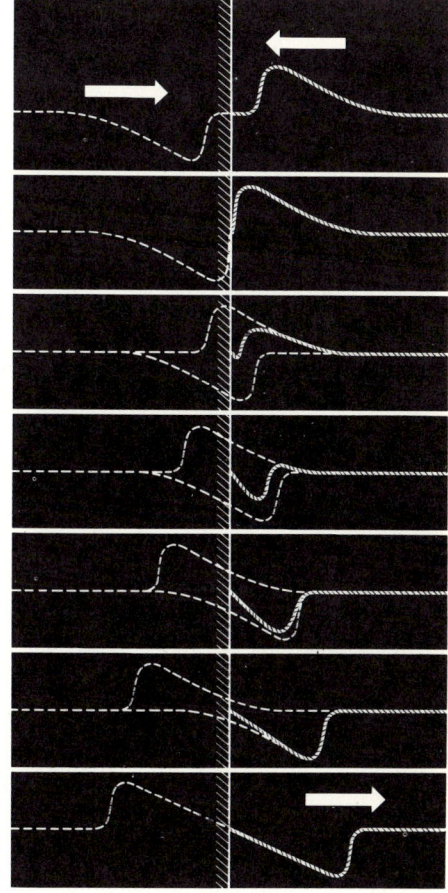

Fig. 22-15 As a wave pulse arrives at a fixed point on the string, the reaction force at the point generates a reflected pulse having the same shape as the original but inverted and moving in the opposite direction. The initial and reflected pulses are shown separately as broken lines; the actual displacement of the string, which is the sum of these, is shown as a solid line.

As it arrives at the wall, the motion appears somewhat complicated, but the final result is an inverted pulse moving away from the wall and having the same shape as the original. What has happened is that the wave has been *reflected* at the fixed end. The phenomenon of reflection of waves is probably more familiar in other contexts. The reflection of a sound wave by a rigid surface produces echoes; reflection of an electromagnetic wave produces the image in a mirror. The pulse phenomenon shown in Fig. 22-15 is a simple example of reflection of a mechanical wave on a string.

The mechanism of reflection of a wave on a string at a fixed point can be understood in more detail by considering the forces at this point. As the pulse arrives at the end of the string, it exerts an upward force on the fixed point, but the wall to which the string is attached supplies an equal and opposite force so that the point does not move. This reaction force then generates a wave pulse having the same shape as that which arrived but inverted and progressing in the opposite direction. The reflected pulse produced in this manner must be such that the displacements corresponding to the incoming and reflected pulses always add to zero at the fixed point. This can be true at all times only if the two pulses have the same shape.

A reflection also occurs if the end of the string is mounted in such a way that there is no vertical force at the end (Fig. 22-16); the ring maintains

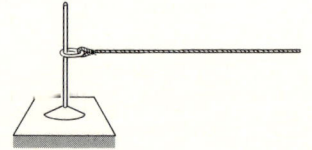

Fig. 22-16 Mechanism for anchoring the end of a string so that the tension is maintained without a transverse force at the end. When a wave pulse arrives at this point, the reflected pulse generated has the same shape as the incoming pulse and progresses in the opposite direction without inversion.

the tension of the string but does not exert a vertical force because it slides without friction. The absence of vertical force means that the end of the string moves farther than it would with more string attached. Roughly speaking, the inertia of this section of the string causes it to "overshoot" compared to the motion of the original wave pulse; this generates another wave traveling in the opposite direction. Again the wave pulse has the same shape as the incident pulse, but this time it is not inverted. In general, a wave pulse is reflected with inversion at a fixed end of a string and without inversion at a "free" end.

The phenomenon of reflection can readily be described in terms of wave functions for the various waves. Suppose $y = f(t + x/c)$ represents a wave function for a pulse traveling to the left, toward the fixed end of the string at $x = 0$. The function $-f(t - x/c)$ represents a wave traveling in the opposite direction, inverted (because of the negative sign) and having the same shape (because the function is the same). The total displacement at any point at any instant is given by

$$f\left(t + \frac{x}{c}\right) - f\left(t - \frac{x}{c}\right) \tag{22-36}$$

At the point $x = 0$ this reduces to $f(t) - f(t)$, and this is obviously always zero, as must be the case if this end of the string is stationary. Similarly, the total disturbance corresponding to reflection of a wave from a free end, where there is no inversion, can be expressed in the form $f(t + x/c) + f(t - x/c)$ if again the end at which the reflection takes place is at $x = 0$.

22-8 STANDING WAVES

The motion which results from reflection of sinusoidal waves is of particular interest. Suppose a sinusoidal wave moving toward the left and described by the wave function $y_1(x,t) = A \sin 2\pi f(t + x/c)$ is reflected by a fixed end at $x = 0$. As just pointed out, the reflected wave is described by the wave function $y_2(x,t) = -A \sin 2\pi f(t - x/c)$, a wave having the same amplitude, frequency, and speed but inverted and progressing in the opposite direction.

According to the principle of superposition, the total displacement y of a point of the string at any given instant is the sum of these two functions, $y = y_1 + y_2$. This sum can be simplified considerably by using the trigonometric identity for the sine of the sum of two angles:

$$\sin(a \pm b) = \sin a \cos b \pm \cos a \sin b$$

Expanding each wave function with this identity and adding, we obtain

$$y_1 = A \sin 2\pi ft \cos \frac{2\pi fx}{c} + A \cos 2\pi ft \sin \frac{2\pi fx}{c}$$

$$y_2 = -A \sin 2\pi ft \cos \frac{2\pi fx}{c} + A \cos 2\pi ft \sin \frac{2\pi fx}{c}$$

$$y = y_1 + y_2 = 2A \cos 2\pi ft \sin \frac{2\pi fx}{c}$$

$$= 2A \cos 2\pi ft \sin \frac{2\pi x}{\lambda} \qquad (22\text{-}37)$$

What is the meaning of this expression? Clearly, it is not in the standard form for a wave propagating in either the $+x$ or $-x$ direction. Each point undergoes simple harmonic motion with frequency f, but the amplitude of the motion is different for different values of x. For values of x such that $\sin(2\pi x/\lambda) = \pm 1$, the amplitude of the motion is $2A$, while for values of x such that $\sin(2\pi x/\lambda) = 0$, the amplitude is zero; these points never move at all! Thus Eq. 22-37 does not represent a progressive wave but one with

610 MECHANICAL WAVES

a stationary pattern which becomes larger and smaller with time. Such a wave is called a *standing wave*.

The stationary points on the string are called *nodes*. They are located at values of x such that $\sin(2\pi x/\lambda) = 0$. The sine is zero when the argument is an integral multiple of π, so the positions of the nodes are determined by the equation

$$\frac{2\pi x}{\lambda} = n\pi \quad \text{or} \quad x = n\frac{\lambda}{2} \quad n = 0, 1, 2, 3, \ldots \quad (22\text{-}38)$$

Successive nodes are spaced one-half wavelength apart. In an exactly similar manner, it can be shown that the points of maximum amplitude, called *antinodes*, are spaced midway between the nodes. Figure 22-17 shows a standing wave resulting from the reflection at a fixed point of a sinusoidal wave. The string is shown at several different positions, and spacings of the nodes and antinodes are indicated.

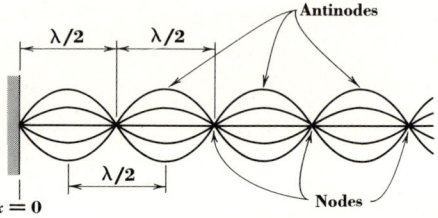

Fig. 22-17 Standing wave resulting from reflection of a sinusoidal wave at a fixed point. The position of the string is shown at several times. The amplitude of the pattern becomes larger and smaller with time, but it does not move along the length of the string; hence the term *standing wave*. The nodes, or points at which no motion occurs, and the antinodes, or points of maximum displacement, are shown.

This entire discussion can be repeated for reflection at a free end of a string. In this case, the reflection takes place without inversion, and it can be shown that if the incoming wave function is the same y_1 discussed above, the equation for the resulting standing wave is

$$y(x,t) = 2A \sin 2\pi ft \cos \frac{2\pi x}{\lambda} \quad (22\text{-}39)$$

which represents a standing wave of the type shown in Fig. 22-18; in this case the end of the string is an antinode rather than a node. At a free end there is no transverse force, so according to Eq. (22-25), the slope at this point must always be zero. This can be verified from Eq. (22-39) and also from Fig. 22-18. Derivation of Eq. (22-39) is left as a problem.

Everything said about reflections and standing waves in a stretched string

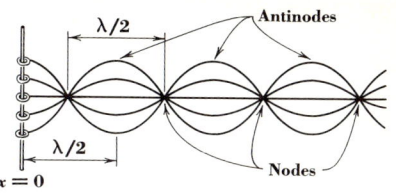

Fig. 22-18 Standing waves produced by reflection of a sinusoidal wave at a free end. In this case, the end is an antinode rather than a node because the wave is reflected without inversion. Positions of nodes and antinodes have the same relative spacing as in Fig. 22-17.

can also be applied directly to longitudinal vibrations in an elastic substance. Analogous to the stationary end of a string is a stationary end of a solid rod clamped rigidly to a support or an immovable barrier at the end of a pipe in which a longitudinal wave travels. In these cases, the requirement that the fixed end be a node leads again to inversion of the reflected wave with respect to the incoming wave, in the sense that displacements in the reflected wave are in the direction opposite those of the incoming wave.

Proceeding further, we now ask what kind of wave motion is possible on a stretched string if *both* ends are held fixed and some initial motion is imparted to the string. In this case, reflections take place at both ends, and these may produce standing waves. Clearly, however, a sinusoidal standing wave can exist only if *both* ends of a string correspond to nodes of the standing wave, and this happens only when the total length of string, say L, is an integral number of half wavelengths. That is, when both ends are stationary, a sinusoidal standing wave can exist only if

$$L = n\frac{\lambda}{2} \qquad n = 1, 2, 3, \ldots \tag{22-40}$$

The corresponding frequencies of possible standing waves can be found by using the relation $c = \lambda f$ in Eq. (22-40); the result is

$$f = n\frac{c}{2L} \qquad n = 1, 2, 3, \ldots \tag{22-41}$$

This interesting result shows that when both ends of a string of length L are held fixed, a sinusoidal standing wave can exist only if the frequency is a whole-number multiple of the quantity $c/2L$. The frequencies given by Eq. (22-41) therefore constitute a set of characteristic frequencies with which the string can vibrate. Motion with any one of these natural frequencies is called a *normal mode* of vibration of the system, and the corresponding frequency is a normal-mode frequency. Standing waves corresponding to several normal modes are shown in Fig. 22-19.

According to the principle of superposition, such a system can vibrate

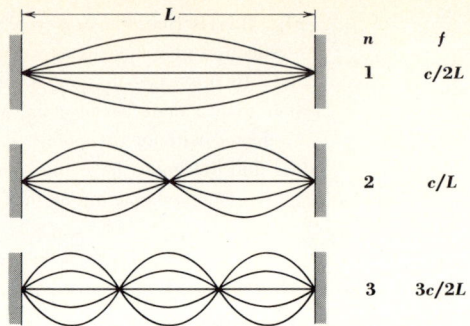

Fig. 22-19 Normal modes of vibration for a string of length L fixed at both ends. The frequency and the value of n in Eq. (22-41) for each normal mode are shown. It is seen that for each mode, n is equal to the number of antinodes and is less by 1 than the number of nodes.

in several normal modes simultaneously. Which normal modes occur in any particular situation depends on the conditions under which the wave motion is originally produced. The most familiar examples are the sounds produced by stringed instruments when the strings are initially displaced from equilibrium by plucking, bowing, or striking with hammers. After a string is set into motion, its vibrations are such that they can always be described as a combination of the normal modes discussed above. Hence any musical sound produced by a repetitive motion of this type can always be described as a superposition of sinusoidal motions with various frequencies which are integral multiples of some lowest, or *fundamental*, frequency. The various frequencies are called *harmonics* of the fundamental frequency. Musicians sometimes call them *overtones*.

Sinusoidal standing waves can be produced in many other simple situations. In the case of longitudinal waves in a pipe containing a gas, the reflection may be produced by having one or both ends open, corresponding to antinodes at the corresponding ends. Organ pipes, whistles, and all other wind instruments, including the human voice, make use of standing waves in some kind of enclosure. Several problems at the end of this chapter deal with the analysis of frequencies of standing waves in various systems.

22-9 EXTENSIONS AND GENERALIZATIONS

We conclude this chapter with several miscellaneous remarks which indicate directions in which the theory of mechanical waves can be extended. The first of these concerns the subject of partial reflections. Reflections in a stretched string have been discussed for two particular cases, in which an end is either stationary or free in the transverse direction. A third possibility is that the string may be tied to another string having different physical properties. In this case the tension is the same in both strings, but the linear mass densities are in general different, so the wave speed is not the same in the two strings.

In Fig. 22-20, suppose a wave pulse originates to the left of the figure,

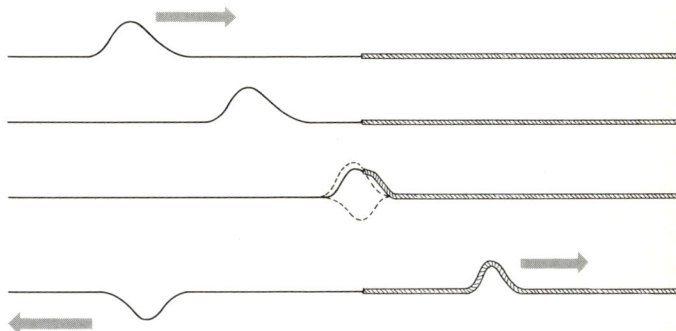

Fig. 22-20 Partial reflection of a wave pulse at a boundary point between two strings having different linear mass densities. In this case, the initial pulse is on the light string, and the reflected pulse traveling back on the light string is inverted with respect to the original pulse. The pulse transmitted to the heavy string is not inverted. The figures show that the wave speed is larger in the light string than in the heavy one, in agreement with Eq. (22-12).

on the light string. Experiment shows that the progress of the pulse is as illustrated. As the pulse reaches the boundary between the strings, a reflected pulse is formed whose amplitude is smaller than that of the incoming pulse; in addition, a pulse is transmitted to the heavier string. The wave speed is smaller in the heavier string, so the transmitted pulse is foreshortened. What has occurred, then, is a *partial reflection* in which the incoming wave is partly transmitted into the new medium and partly reflected. In this case, the reflected wave is inverted; this is reasonable, since in the limit when the heavy string becomes infinitely heavy, the result must reduce to that in which the end of the light string is held stationary.

Similarly, when a wave originates on the heavy part of the string, it is partly transmitted without inversion and partly reflected without inversion from the boundary between the strings. In the limit when one string is much lighter than the other, this approaches the situation in which the heavy string has a completely free end.

In certain cases, it is possible to compute the amplitudes of the transmitted and reflected waves if the amplitude of the incoming wave is known. For example, if the boundary point is at $x = 0$ and the incoming wave is a sinusoidal wave described in the region in which $x < 0$ by the wave function

$$y_i = A \sin 2\pi f\left(t - \frac{x}{c_1}\right) \qquad (22\text{-}42)$$

then the transmitted wave (in the region $x > 0$) is

$$y_t = B \sin 2\pi f\left(t - \frac{x}{c_2}\right) \qquad (22\text{-}43)$$

and the reflected wave (in the region $x < 0$) is

$$y_r = C \sin 2\pi f\left(t + \frac{x}{c_1}\right) \qquad (22\text{-}44)$$

where we indicate explicitly that the wave speeds c_1 and c_2 are different in the two sections of string. The total wave function on the left side is $y_i + y_r$. The constants B and C depend on the amplitude A of the incoming wave and can be determined by observing that the two functions $y_i + y_r$ and y_t must agree at the point $x = 0$, since otherwise the two sections would not join. Furthermore, the *slopes* must be equal at this point; otherwise the point would have an infinitely large acceleration. The remainder of the determination of B and C is left as an exercise. It turns out that they are directly proportional to A and depend on the wave speeds:

$$\frac{B}{A} = \frac{2c_2}{c_1 + c_2} \qquad \frac{C}{A} = \frac{c_2 - c_1}{c_1 + c_2} \qquad (22\text{-}45)$$

These amplitude ratios are called the *transmission coefficient* and *reflection coefficient*, respectively, terms used in describing many kinds of wave phenomena, especially in optics.

Another phenomenon common to optics and transverse mechanical waves is that of *polarization*. Although in the discussion of transverse waves on a string it has been assumed that the displacements are in a single plane, they could equally well be in a plane perpendicular to this one. *Any* transverse displacement of a point on a string can be described as the vector sum of two displacements in two mutually perpendicular directions. If all displacements lie in one plane, the wave is said to be *plane polarized;* all the transverse waves discussed so far fall in this category. But other varieties of motion can be synthesized by combining two plane-polarized waves with a phase difference; then each particle of the string, instead of moving to and fro in a straight line, describes a circular or elliptical path. If a wave is formed from two plane-polarized waves of equal frequency and amplitude at right angles to each other and differing in phase by 90°, each particle of the string moves in a circular path whose radius is equal to the amplitude of either of the plane-polarized component waves. This wave presents the appearance of a rotating helix and is called a *circularly polarized* wave. Such a wave transports not only energy but also *angular momentum* from one end of the string to the other.

Polarization is a phenomenon common to all transverse waves. The most important examples of polarization occur in electromagnetic radiation, which is a transverse wave in the sense that the electric and magnetic fields representing the disturbance from equilibrium are perpendicular to the direction of propagation of the wave. Although the polarization of electromagnetic waves is of considerably more practical importance than that of mechanical waves, the basic ideas are shared by both. For longitudinal waves, on the other hand, there is no such thing as polarization.

The entire discussion of waves in this chapter has been concerned with waves which travel with a definite speed and without changing their shape. The fact that waves possessing these properties can exist hinges on two kinds of basic assumptions. First, we assume that the speed of a wave does not depend on the particular shape of the wave disturbance. In a wave pulse, the speed is independent of the amplitude and duration; in a sinusoidal wave, it is independent of the amplitude and frequency. As detailed analysis of the mechanism of wave propagation has shown, these characteristics are related to elastic properties of the medium, specifically to behavior which can be described by Hooke's law. If this law is not obeyed precisely, the resulting differential equation does not have the same form as Eq. (22-5) and may be more complicated. In this case, the wave speed may depend on the details of the waveshape, and in general *dispersion* occurs.

Another general assumption we have made is that there exists no mechanism for loss of mechanical energy through friction during the propagation of a wave. In the case of the stretched string, there may be forces associated with air resistance and internal friction, but such forces were not included in the development of the differential equation (22-11). Similar idealizations were made, without being stated explicitly, for longitudinal waves in elastic substances; for example, viscous friction was not included.

If losses of mechanical energy due to friction are not negligible, a wave cannot propagate with a constant shape. A wave pulse suffering mechanical-energy losses due to friction and propagating from left to right on a string might appear at successive times as shown in Fig. 22-21. The precise manner in which the amplitude of the pulse decreases depends on the details of the friction. If, as often happens, the friction is associated with viscosity, the frictional force at any point may be proportional to the speed of that point.

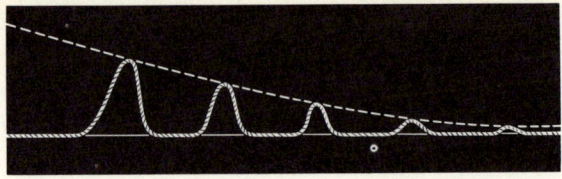

Fig. 22-21 Wave pulse which suffers attenuation. If there were no attenuation, the crest of the wave would move along a straight line; with attenuation, it follows the broken line. When attenuation results from frictional forces proportional to the velocity at each point, the broken line is an exponential function.

In this case it can be shown that the height of the pulse is a decreasing exponential function of position. That is, as the pulse travels, its peak describes not a straight line but an exponential curve, as shown in the figure. Any wave which decreases in amplitude as it is propagated is said to be *attenuated*.

We can write a wave equation to describe a sinusoidal wave attenuated by frictional forces proportional to velocity. The amplitude, instead of being a constant, is a decreasing exponential function of x. Thus a possible wave function describing a wave propagating in the positive x direction with attenuation resulting from friction is

$$y(x,t) = Ae^{-x/R} \sin 2\pi f\left(t - \frac{x}{c}\right) \tag{22-46}$$

in which R is a constant, with dimensions of length, which characterizes how rapidly the amplitude of the wave decreases with x. It is easy to show that in a distance equal to R the amplitude decreases from its initial value to $1/e$ of this initial value.

The phenomena of dispersion and attenuation are important in the propagation of electromagnetic waves. In many kinds of work involving pulses of very short duration, including radar, detectors which count elementary particles, and communication systems, dispersion and attenuation pose real problems. Still, even in these areas, there are many situations in which the behavior of waves can be idealized by neglecting these quantities; the discussion of wave propagation then becomes similar to that given in this chapter for mechanical waves. Here, as in all other branches of physics, a simple, idealized model may yield directly useful results in situations in which the model is adequate; in other situations, the model must be refined to take into account effects originally neglected. But even then the simple model provides a great deal of insight into the general behavior of the system.

Problems

22-1 Suppose it is desired to produce on a string a wave pulse having the form shown in Fig. P22-1.
   a If the wave speed is 50 m/s, describe in detail the motion which should be given to the end of the string to produce this pulse.
   b How is the answer to part *a* changed if the wave speed is 10 m/s?

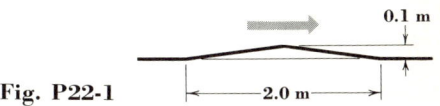

Fig. P22-1

22-2 One end of a stretched string is moved transversely with a uniform speed of 1.0 m/s until it is displaced 0.1 m; it is then returned to the original position with the same uniform speed. The resulting wave pulse is found to travel with a speed of 5.0 m/s.
   a Sketch the shape of the string at $t = 0, 0.1, 0.2, 0.3,$ and 0.4 s.
   b What is the *length* of the wave pulse on the string?
   c Draw a graph showing transverse velocity as a function of position on the string for $t = 0.4$ s.
   d What force must be applied to the end of the string?

22-3 A rope 40 m long has a total mass of 2.0 kg. If it is stretched horizontally to a tension of 500 N, what is the speed of transverse waves on the rope?

22-4 A rope 10 m long is hung from the ceiling. The mass of the rope is 0.5 kg, and an object of mass 0.5 kg is tied to the bottom end.
   a How does wave speed vary with position on the rope?
   b If a wave pulse of length 0.1 m is produced at the bottom end, what is its length when it reaches the top end?

22-5 A certain wave motion on a string is represented by the function $y = A \sin k(x - ct)$, where A, k, and c are constants.
   a Derive an expression for the transverse velocity of a point on the string as a function of its equilibrium position x and time t.
   b Derive an expression for the slope of the string at point x and time t.

22-6 Verify that the wave function of Prob. 22-5 satisfies Eq. (22-5) by computing the required derivatives and substituting.

22-7 Prove that for any transverse wave on a string the slope at any point x is equal to the ratio of the instantaneous transverse speed of the point to the wave speed.

22-8 In transverse wave motion on a string, any section of string which instantaneously forms a straight line, i.e., has zero curvature, must be moving at that instant with constant speed. Explain physically why this is so.

22-9 It is impossible in practice to produce a waveform on a string which has "corners," i.e., points at which the slope of the string changes discontinuously. Explain physically why this is so.

22-10 Show that the function $f(x,t) = x^2 + 4axt + fa^2t^2$ satisfies Eq. (22-5) if one assumes a certain relationship between the constant a and the wave speed c. What is this relationship?

22-11 An exuberant freshman on a vacation is standing in a canoe on a smooth lake. He rocks the canoe back and forth, making five complete cycles in 5 s. He observes that the resulting wave travels a distance equal to the length of the canoe (about 5 m) in 2 s. What is the wavelength?

22-12 One end of a stretched rope is given a periodic transverse motion with a frequency of 10 s^{-1}. The rope is 50 m long, has total mass 0.5 kg, and is stretched with a tension of 400 N.
 a Find the wavelength of the resulting waves.
 b If the tension is doubled, how must the frequency be changed in order to maintain the same wavelength?

22-13 In vacuum, electromagnetic waves have a speed of about 3×10^8 m/s.
 a Find the wavelength of radiation emitted by radio station KDKA, whose frequency is $1{,}020 \text{ kHz} = 1{,}020 \times 10^3 \text{ s}^{-1}$.
 b Find the frequency of visible light in the orange portion of the spectrum, with wavelength about 600×10^{-9} m.

22-14 Show that Eq. (22-15) can be written in the following alternative forms:
$$y(x,t) = A \sin 2\pi \left(\frac{t}{T} - \frac{x}{\lambda} \right) = A \sin \omega \left(t - \frac{x}{c} \right)$$

22-15 Write a wave function for a sinusoidal wave propagating in the $+x$ direction with amplitude 0.1 m, wavelength 2.0 m, and frequency 5 s^{-1} if the point $x = 0$ has zero displacement at time $t = 0$. What is the wave speed?

22-16 Solve Prob. 22-15 if the point $x = 0$ has its maximum positive displacement at time $t = 0$.

22-17 The wave described in Prob. 22-15 is produced by giving a sinusoidal motion to the end of the rope at $x = 0$. Derive an expression for the force which must be applied to this point as a function of time. The expression may also contain the linear mass density and tension.

22-18 The equation of a sinusoidal wave on a string is $y = 0.05 \sin 2\pi(0.1x + 2.0t)$, where the distances are measured in meters and t is in seconds. Draw a diagram showing the position of the string at $t = 0$ and at $t = \frac{1}{8}$ s. From this diagram find the wavelength, speed, and amplitude of the wave.

22-19 For the wave function of Prob. 22-18, find the maximum transverse speed and the maximum slope of the string.

22-20 Show that the wave functions given by Eqs. (22-15) and (22-17) satisfy Eq. (22-5).

22-21 Compute the total energy of the wave pulse of Prob. 22-1 if $T = 500$ N and $\mu = 0.2$ kg/m.

22-22 Show that for a sinusoidal wave on a string, the average kinetic energy per unit length is equal to the average potential energy per unit length.

22-23 In Prob. 22-1, find the rate of transfer of energy past a given point during the time the pulse is passing the point. Also compute the time required for the entire pulse to pass this point, and show that the product of these two quantities is equal to the total energy obtained in Prob. 22-21.

22-24 For a sinusoidal wave on a string, show that the instantaneous rate of energy transfer past a given point is greatest when the point has zero displacement and least when it has maximum displacement.

22-25 A longitudinal wave in a steel rod is described by the function $y = 2 \times 10^{-4} \sin 2\pi(5{,}000t - \frac{1}{2}x)$, where the distances are in meters and t is in seconds. Find the period, frequency, wavelength, and amplitude of the wave.

22-26 In Prob. 22-25 find the maximum stress in the rod and the maximum velocity of a particle in the rod.

22-27 For a sinusoidal longitudinal wave in an elastic medium, show that the points of maximum displacement are points of minimum stress, and conversely. This should be done both analytically, using Eq. (22-29), and descriptively, by considering relative displacements of neighboring points.

22-28 Assuming adiabatic conditions, compute the speed of sound at 300 K in:
  a Oxygen
  b Hydrogen
  c Helium

22-29 For a sinusoidal longitudinal wave in an elastic solid, derive a relationship between the maximum stress in the material and the amplitude, frequency, and speed of the wave. For a 10-kHz (10^4 s^{-1}) wave in steel, what is the largest amplitude which will not exceed the ultimate strength of the material?

22-30 Derive an expression analogous to Eq. (22-24) for the total energy per unit length of a longitudinal wave in a solid rod of uniform cross section.

22-31 Under what conditions is a longitudinal wave pulse in an elastic solid reflected from an end with inversion? Without inversion?

22-32 Verify the fact that when a sinusoidal wave on a string is reflected from a free end, the resulting standing wave is described by Eq. (22-39).

22-33 Find the frequencies of standing waves with wave speed c which can occur on a string of length L if one end is fixed but the other is anchored as in Fig. 22-16. In particular, show that the possible frequencies are odd-integer multiples (1, 3, 5, . . .) of a lowest, or fundamental, frequency. Sketch the standing-wave patterns of the three lowest frequency modes.

22-34 A steel rod of length L is clamped to a stationary support at its center and has both ends free. Find the normal-mode frequencies, and sketch the corresponding standing-wave patterns for the first three modes.

22-35 In Prob. 22-34, find the fundamental frequency of a steel rod 1 m in length.

22-36 Find the normal-mode frequencies of an organ pipe of length L which is open at both ends. What length of pipe is needed if the fundamental frequency is to be $16\ \text{s}^{-1}$? (This is approximately the frequency of the note one octave below the lowest C on a piano; it is the lowest note found on most large organs.)

22-37 Answer the questions of Prob. 22-36 if the pipe is open at one end but closed at the other, as is often the case with large wood pipes.

22-38 If a pipe organ is tuned at a temperature of $20°C$, by what fraction does the pitch change if the air is supplied to the pipes from a blower in the basement, where the temperature is $10°C$? Is this a perceptible change?

22-39 If a guitar string is tuned to a certain note, how much must the tension be changed if the pitch is to be raised one octave, which corresponds to doubling the frequency?

22-40 Complete the calculation of the transmission and reflection coefficients for a sinusoidal wave at a boundary between two strings having different linear mass densities, as discussed in Sec. 22-8. Assume that to the left of the boundary point ($x = 0$) the wave is represented by the wave function $y_i + y_r$, where y_i is the incoming wave [Eq. (22-42)] and y_r is the reflected wave [Eq. (22-44)], and assume that for points to the right of the boundary point the wave function is Eq. (22-43). The conditions stating that the wave functions, as well as their slopes, must agree at the point $x = 0$ lead to two simultaneous equations for the constants A, B, and C. If the incoming amplitude A is known, these equations can be solved for B and C.

22-41 Repeat the calculation of Prob. 22-40 for partial reflection of a sinusoidal longitudinal wave at a joint between two rods of equal diameters but different materials. Again, the two wave functions must agree at the point $x = 0$, and the stress, obtained from Eq. (22-29), must be the same on both sides of the joint. Again, these two conditions lead to two simultaneous equations, from which B and C can be found in terms of A.

EPILOGUE

In the last three chapters we have discussed a number of very general and powerful principles. The first law of thermodynamics provides a generalization of the principle of conservation of energy to include internal energy of matter and a method for dealing with this energy even without detailed knowledge of the structure of matter. The second law of thermodynamics gives, in very general terms, a principle governing the direction in which a thermodynamic process tends to proceed.

In the formulation of both these principles we relied heavily on observations of a wide variety of physical phenomena. Both laws are new generalizations from observation; neither is derived from principles previously presented. It is true that these laws can be derived for systems which can be represented by simple microscopic models, but their validity for more general kinds of systems must be regarded as an experimental fact. That the principles of thermodynamics can be applied to such a wide variety of physical systems makes them among the most powerful and useful of all the laws of physics.

Finally, we have examined some of the most important characteristics of wave phenomena. Although the discussion has been confined mainly to mechanical waves, wave phenomena not only occur in nearly all branches of physics but form one of the most important threads that extend through all of physics and give the discipline coherence and unity.

This brings us to a pause in our study of the fundamentals of physics. We have been concerned in this volume primarily with the fundamental principles of mechanics and elementary thermodynamics. The student will pursue the study of electricity and magnetism, optics, atomic and nuclear physics, and perhaps other fields in other books. It is hoped that the subject matter, point of view, and methods of analysis of the present volume have not only provided the necessary facts but have also developed the rigorous and critical pattern of thinking necessary for the continuation of this study.

In conclusion, we urge the student to continue to look for the unifying themes in physics and the other physical sciences. Among these are the use of idealized models to simplify complex situations, the relation of microscopic and macroscopic descriptions, and such fundamental concepts as mass, energy, momentum, force, and the associated conservation principles. In these common themes are found the great power and beauty of physical science.

APPENDIX A
PHYSICAL CONSTANTS

Fundamental Physical Constants

Name	Symbol	Value
Gravitational constant	G	6.673×10^{-11} N·m²/kg²
Speed of light	c	2.998×10^{8} m/s
Avogadro's number	N_0	6.025×10^{23} molecules/mol
Gas constant	R	8.314 J/mol K
Boltzmann constant	k	1.380×10^{-23} J/K
Mechanical equivalent of heat	J	4.186 J/cal
Mass of electron	m_e	9.108×10^{-31} kg
Mass of neutron	m_n	1.675×10^{-27} kg
Mass of proton	m_p	1.672×10^{-27} kg
Charge of electron	e	1.601×10^{-19} C
Planck's constant	h	6.625×10^{-34} J·s

Other Constants

Name	Symbol or abbreviation	Value
Standard temperature and pressure	STP	0°C and 1 atm
Standard atmospheric pressure	1 atm	1.013×10^{5} N/m²
Volume of 1 mol of gas at STP	v	2.242×10^{-2} m³
Acceleration of gravity (nominal)	g	9.80 m/s²
Density of dry air at STP	ρ	1.293 kg/m³
Density of pure water at STP	ρ	1.000×10^{3} kg/m³
Ice point of pure water at 1 atm	T_0	273.15 K

APPENDIX B
CONVERSION FACTORS

Length

	Meter	Inch	Foot	Mile
1 meter =	1	39.37	3.281	6.214×10^{-4}
1 inch =	2.540×10^{-2}	1	8.333×10^{-2}	1.578×10^{-5}
1 foot =	0.3048	12	1	1.894×10^{-4}
1 mile =	1,609	6.336×10^{4}	5,280	1

1 angstrom (Å) = 10^{-10} meter (m) = 10^{-8} centimeter (cm)
1 micrometer (μm) = 10^{-6} m
1 nanometer (nm) = 10^{-9} m

Volume

	Cubic meter	Cubic inch	Cubic foot
1 cubic meter =	1	6.102×10^{4}	35.31
1 cubic inch =	1.639×10^{-5}	1	5.787×10^{-4}
1 cubic foot =	2.832×10^{-2}	1,728	1

1 liter (l) = 10^{-3} m³ = 10^{3} cm³
1 U.S. gallon = 231 in.³

Mass

	Kilogram	Slug	Atomic mass unit
1 kilogram =	1	6.852×10^{-2}	6.024×10^{26}
1 slug =	14.59	1	8.789×10^{27}
1 atomic mass unit =	1.660×10^{-27}	1.137×10^{-28}	1

On earth, 1 kilogram (kg) has a weight of approximately 2.205 lb.

CONVERSION FACTORS

Time

	Second	Minute	Hour	Day	Year
1 second =	1	1.667×10^{-2}	2.778×10^{-4}	1.157×10^{-5}	3.169×10^{-8}
1 minute =	60	1	1.667×10^{-2}	6.944×10^{-4}	1.901×10^{-6}
1 hour =	3,600	60	1	4.167×10^{-2}	1.141×10^{-4}
1 day =	8.640×10^{4}	1,440	24	1	2.738×10^{-3}
1 year =	3.156×10^{7}	5.259×10^{3}	8.766×10^{3}	365.2	1

Force

1 newton (N) = 0.2248 lb
1 dyn = 10^{-5} N

Speed

	Meter per second	Kilometer per hour	Foot per second	Mile per hour
1 meter per second =	1	3.600	3.281	2.237
1 kilometer per hour =	0.2778	1	0.9113	0.6214
1 foot per second =	0.3048	1.097	1	0.6818
1 mile per hour =	0.4470	1.609	1.467	1

60 mi/h = 88 ft/s
1 furlong/fortnight = 1.662×10^{-4} m/s

Pressure

	Newton per square meter	Centimeter of mercury	Atmosphere	Pound per square inch
1 newton per square meter =	1	7.501×10^{-4}	9.869×10^{-6}	1.450×10^{-4}
1 centimeter of mercury =	1,333	1	1.316×10^{-2}	0.1934
1 atmosphere =	1.013×10^{5}	76	1	14.70
1 pound per square inch =	6.895×10^{3}	5.171	6.805×10^{-2}	1

Energy

	Joule	Kilowatt-hour	Calorie	Electron-volt	Foot-pound	British thermal unit
1 joule =	1	2.778×10^{-7}	0.2389	6.242×10^{18}	0.7376	9.481×10^{-4}
1 kilowatt-hour =	3.600×10^{6}	1	8.601×10^{5}	2.247×10^{25}	2.655×10^{6}	3.413
1 calorie =	4.186	1.163×10^{-6}	1	2.613×10^{19}	3.087	3.986×10^{-3}
1 electron-volt =	1.601×10^{-19}	4.450×10^{-26}	3.827×10^{-20}	1	1.182×10^{-19}	1.519×10^{-22}
1 foot-pound =	1.356	3.766×10^{-7}	0.3239	8.464×10^{18}	1	1.285×10^{-3}
1 British thermal unit =	1.055	2.930×10^{-4}	252.0	6.585×10^{21}	779.9	1

1 MeV = 10^6 eV
1 GeV = 10^9 eV

Power

	Watt	Horsepower	Foot-pound per second	British thermal unit per hour
1 watt =	1	1.341×10^{-3}	0.7376	3.413
1 horsepower =	745.7	1	550	2.545
1 foot-pound per second =	1.356	1.818×10^{-3}	1	4.628
1 British thermal unit per hour =	0.2930	3.929×10^{-4}	0.2161	1

1 watt = 1 J/s
1 kilowatt (kW) = 10^3 W

APPENDIX C
THE SOLAR SYSTEM

Body	Mass, 10^{24} kg	Diameter, 10^6 m	Period of orbit	Radius of orbit, 10^9 m
Sun	1.97×10^6	1,390		
Moon	0.0735	3.476	27.3 d	0.38
Mercury	0.328	5.140	88.0 d	58
Venus	4.82	12.62	224.7 d	108
Earth	5.98	12.76	365.3 d	149
Mars	0.634	6.86	687.0 d	228
Jupiter	1,880	143.6	11.86 yr	778
Saturn	563	120.6	29.46 yr	1,426
Uranus	86.1	53.4	84.02 yr	2,869
Neptune	99.9	49.7	164.8 yr	4,495
Pluto	0.5 (?)	0.4 (?)	247.7 yr	5,900

APPENDIX D
THE GREEK ALPHABET

Name	Capital	Lowercase	Name	Capital	Lowercase
Alpha	A	α	Nu	N	ν
Beta	B	β	Xi	Ξ	ξ
Gamma	Γ	γ	Omicron	O	o
Delta	Δ	δ, ∂	Pi	Π	π
Epsilon	E	ϵ	Rho	P	ρ
Zeta	Z	ζ	Sigma	Σ	σ, ς
Eta	H	η	Tau	T	τ
Theta	Θ	θ, ϑ	Upsilon	Υ	υ
Iota	I	ι	Phi	Φ	ϕ, φ
Kappa	K	κ	Chi	X	χ
Lambda	Λ	λ	Psi	Ψ	ψ
Mu	M	μ	Omega	Ω	ω

APPENDIX E
COMMON LOGARITHMS

10	0	1	2	3	4	5	6	7	8	9
10	0000	0043	0086	0128	0170	0212	0253	0294	0334	0374
11	0414	0453	0492	0531	0569	0607	0645	0682	0719	0755
12	0792	0828	0864	0899	0934	0969	1004	1038	1072	1106
13	1139	1173	1206	1239	1271	1303	1335	1367	1399	1430
14	1461	1492	1523	1553	1584	1614	1644	1673	1703	1732
15	1761	1790	1818	1847	1875	1903	1931	1959	1987	2014
16	2041	2068	2095	2122	2148	2175	2201	2227	2253	2279
17	2304	2330	2355	2380	2405	2430	2455	2480	2504	2529
18	2553	2577	2601	2625	2648	2672	2695	2718	2742	2765
19	2788	2810	2833	2856	2878	2900	2923	2945	2967	2989
20	3010	3032	3054	3075	3096	3118	3139	3160	3181	3201
21	3222	3243	3263	3284	3304	3324	3345	3365	3385	3404
22	3424	3444	3464	3483	3502	3522	3541	3560	3579	3598
23	3617	3636	3655	3674	3692	3711	3729	3747	3766	3784
24	3802	3820	3838	3856	3874	3892	3909	3927	3945	3962
25	3979	3997	4014	4031	4048	4065	4082	4099	4116	4133
26	4150	4166	4183	4200	4216	4232	4249	4265	4281	4298
27	4314	4330	4346	4362	4378	4393	4409	4425	4440	4456
28	4472	4487	4502	4518	4533	4548	4564	4579	4594	4609
29	4624	4639	4654	4669	4683	4698	4713	4728	4742	4757
30	4771	4786	4800	4814	4829	4843	4857	4871	4886	4900
31	4914	4928	4942	4955	4969	4983	4997	5011	5024	5038
32	5051	5065	5079	5092	5105	5119	5132	5145	5159	5172
33	5185	5198	5211	5224	5237	5250	5263	5276	5289	5302
34	5315	5328	5340	5353	5366	5378	5391	5403	5416	5428
35	5441	5453	5465	5478	5490	5502	5514	5527	5539	5551
36	5563	5575	5587	5599	5611	5623	5635	5647	5658	5670
37	5682	5694	5705	5717	5729	5740	5752	5763	5775	5786
38	5798	5809	5821	5832	5843	5855	5866	5877	5888	5899
39	5911	5922	5933	5944	5955	5966	5977	5988	5999	6010
40	6021	6031	6042	6053	6064	6075	6085	6096	6107	6117
41	6128	6138	6149	6160	6170	6180	6191	6201	6212	6222
42	6232	6243	6253	6263	6274	6284	6294	6304	6314	6325
43	6335	6345	6355	6365	6375	6385	6395	6405	6415	6425
44	6435	6444	6454	6464	6474	6484	6493	6503	6513	6522
45	6532	6542	6551	6561	6571	6580	6590	6599	6609	6618
46	6628	6637	6646	6656	6665	6675	6684	6693	6702	6712
47	6721	6730	6739	6749	6758	6767	6776	6785	6794	6803
48	6812	6821	6830	6839	6848	6857	6866	6875	6884	6893
49	6902	6911	6920	6928	6937	6946	6955	6964	6972	6981
50	6990	6998	7007	7016	7024	7033	7042	7050	7059	7067
51	7076	7084	7093	7101	7110	7118	7126	7135	7143	7152
52	7160	7168	7177	7185	7193	7202	7210	7218	7226	7235
53	7243	7251	7259	7267	7275	7284	7292	7300	7308	7316
54	7324	7332	7340	7348	7356	7364	7372	7380	7388	7396

COMMON LOGARITHMS

55	0	1	2	3	4	5	6	7	8	9
55	7404	7412	7419	7427	7435	7443	7451	7459	7466	7474
56	7482	7490	7497	7505	7513	7520	7528	7536	7543	7551
57	7559	7566	7574	7582	7589	7597	7604	7612	7619	7627
58	7634	7642	7649	7657	7664	7672	7679	7686	7694	7701
59	7709	7716	7723	7731	7738	7745	7752	7760	7767	7774
60	7782	7789	7796	7803	7810	7818	7825	7832	7839	7846
61	7853	7860	7868	7875	7882	7889	7896	7903	7910	7917
62	7924	7931	7938	7945	7952	7959	7966	7973	7980	7987
63	7993	8000	8007	8014	8021	8028	8035	8041	8048	8055
64	8062	8069	8075	8082	8089	8096	8102	8109	8116	8122
65	8129	8136	8142	8149	8156	8162	8169	8176	8182	8189
66	8195	8202	8209	8215	8222	8228	8235	8241	8248	8254
67	8261	8267	8274	8280	8287	8293	8299	8306	8312	8319
68	8325	8331	8338	8344	8351	8357	8363	8370	8376	8382
69	8388	8395	8401	8407	8414	8420	8426	8432	8439	8445
70	8451	8457	8463	8470	8476	8482	8488	8494	8500	8506
71	8513	8519	8525	8531	8537	8543	8549	8555	8561	8567
72	8573	8579	8585	8591	8597	8603	8609	8615	8621	8627
73	8633	8639	8645	8651	8657	8663	8669	8675	8681	8686
74	8692	8698	8704	8710	8716	8722	8727	8733	8739	8745
75	8751	8756	8762	8768	8774	8779	8785	8791	8797	8802
76	8808	8814	8820	8825	8831	8837	8842	8848	8854	8859
77	8865	8871	8876	8882	8887	8893	8899	8904	8910	8915
78	8921	8927	8932	8938	8943	8949	8954	8960	8965	8971
79	8976	8982	8987	8993	8998	9004	9009	9015	9020	9025
80	9031	9036	9042	9047	9053	9058	9063	9069	9074	9079
81	9085	9090	9096	9101	9106	9112	9117	9122	9128	9133
82	9138	9143	9149	9154	9159	9165	9170	9175	9180	9186
83	9191	9196	9201	9206	9212	9217	9222	9227	9232	9238
84	9243	9248	9253	9258	9263	9269	9274	9279	9284	9289
85	9294	9299	9304	9309	9315	9320	9325	9330	9335	9340
86	9345	9350	9355	9360	9365	9370	9375	9380	9385	9390
87	9395	9400	9405	9410	9415	9420	9425	9430	9435	9440
88	9445	9450	9455	9460	9465	9469	9474	9479	9484	9489
89	9494	9499	9504	9509	9513	9518	9523	9528	9533	9538
90	9542	9547	9552	9557	9562	9566	9571	9576	9581	9586
91	9590	9595	9600	9605	9609	9614	9619	9624	9628	9633
92	9638	9643	9647	9652	9657	9661	9666	9671	9675	9680
93	9685	9689	9694	9699	9703	9708	9713	9717	9722	9727
94	9731	9736	9741	9745	9750	9754	9759	9763	9768	9773
95	9777	9782	9786	9791	9795	9800	9805	9809	9814	9818
96	9823	9827	9832	9836	9841	9845	9850	9854	9859	9863
97	9868	9872	9877	9881	9886	9890	9894	9899	9903	9908
98	9912	9917	9921	9926	9930	9934	9939	9943	9948	9952
99	9956	9961	9965	9969	9974	9978	9983	9987	9991	9996

APPENDIX F
VALUES OF TRIGONOMETRIC FUNCTIONS

Angle		Sine	Cosine	Tangent	Angle		Sine	Cosine	Tangent
Degree	Radian				Degree	Radian			
0°	0.000	0.000	1.000	0.000					
1°	0.018	0.018	1.000	0.018	46°	0.803	0.719	0.695	1.036
2°	0.035	0.035	0.999	0.035	47°	0.820	0.731	0.682	1.072
3°	0.052	0.052	0.999	0.052	48°	0.838	0.743	0.669	1.111
4°	0.070	0.070	0.998	0.070	49°	0.855	0.755	0.656	1.150
5°	0.087	0.087	0.996	0.088	50°	0.873	0.766	0.643	1.192
6°	0.105	0.105	0.995	0.105	51°	0.890	0.777	0.629	1.235
7°	0.122	0.122	0.993	0.123	52°	0.908	0.788	0.616	1.280
8°	0.140	0.139	0.990	0.141	53°	0.925	0.799	0.602	1.327
9°	0.157	0.156	0.988	0.158	54°	0.942	0.809	0.588	1.376
10°	0.175	0.174	0.985	0.176	55°	0.960	0.819	0.574	1.428
11°	0.192	0.191	0.982	0.194	56°	0.977	0.829	0.559	1.483
12°	0.209	0.208	0.978	0.213	57°	0.995	0.839	0.545	1.540
13°	0.227	0.225	0.974	0.231	58°	1.012	0.848	0.530	1.600
14°	0.244	0.242	0.970	0.249	59°	1.030	0.857	0.515	1.664
15°	0.262	0.259	0.966	0.268	60°	1.047	0.866	0.500	1.732
16°	0.279	0.276	0.961	0.287	61°	1.065	0.875	0.485	1.804
17°	0.297	0.292	0.956	0.306	62°	1.082	0.883	0.470	1.881
18°	0.314	0.309	0.951	0.325	63°	1.100	0.891	0.454	1.963
19°	0.332	0.326	0.946	0.344	64°	1.117	0.899	0.438	2.050
20°	0.349	0.342	0.940	0.364	65°	1.134	0.906	0.423	2.145
21°	0.367	0.358	0.934	0.384	66°	1.152	0.914	0.407	2.246
22°	0.384	0.375	0.927	0.404	67°	1.169	0.921	0.391	2.356
23°	0.401	0.391	0.921	0.425	68°	1.187	0.927	0.375	2.475
24°	0.419	0.407	0.914	0.445	69°	1.204	0.934	0.358	2.605
25°	0.436	0.423	0.906	0.466	70°	1.222	0.940	0.342	2.747
26°	0.454	0.438	0.899	0.488	71°	1.239	0.946	0.326	2.904
27°	0.471	0.454	0.891	0.510	72°	1.257	0.951	0.309	3.078
28°	0.489	0.470	0.883	0.532	73°	1.274	0.956	0.292	3.271
29°	0.506	0.485	0.875	0.554	74°	1.292	0.961	0.276	3.487
30°	0.524	0.500	0.866	0.577	75°	1.309	0.966	0.259	3.732
31°	0.541	0.515	0.857	0.601	76°	1.326	0.970	0.242	4.011
32°	0.559	0.530	0.848	0.625	77°	1.344	0.974	0.225	4.331
33°	0.576	0.545	0.839	0.649	78°	1.361	0.978	0.208	4.705
34°	0.593	0.559	0.829	0.675	79°	1.379	0.982	0.191	5.145
35°	0.611	0.574	0.819	0.700	80°	1.396	0.985	0.174	5.671
36°	0.628	0.588	0.809	0.727	81°	1.414	0.988	0.156	6.314
37°	0.646	0.602	0.799	0.754	82°	1.431	0.990	0.139	7.115
38°	0.663	0.616	0.788	0.781	83°	1.449	0.993	0.122	8.144
39°	0.681	0.629	0.777	0.810	84°	1.466	0.995	0.105	9.514
40°	0.698	0.643	0.766	0.839	85°	1.484	0.996	0.087	11.43
41°	0.716	0.656	0.755	0.869	86°	1.501	0.998	0.070	14.30
42°	0.733	0.669	0.743	0.900	87°	1.518	0.999	0.052	19.08
43°	0.751	0.682	0.731	0.933	88°	1.536	0.999	0.035	28.64
44°	0.768	0.695	0.719	0.966	89°	1.553	1.000	0.018	57.29
45°	0.785	0.707	0.707	1.000	90°	1.571	1.000	0.000	∞

APPENDIX G
MATHEMATICAL FORMULAS

TRIGONOMETRY

Definitions

$$\sin \theta = \frac{y}{r}$$

$$\cos \theta = \frac{x}{r}$$

$$\tan \theta = \frac{y}{x}$$

$$\cot \theta = \frac{1}{\tan \theta}$$

$$\sec \theta = \frac{1}{\cos \theta}$$

$$\csc \theta = \frac{1}{\sin \theta}$$

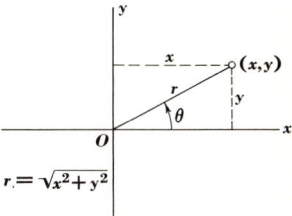

Fig. G-1

Identities

$$\sin^2 A + \cos^2 A = 1$$

$$\tan A = \frac{\sin A}{\cos A}$$

$$\sin \left(\frac{\pi}{2} \pm A \right) = \cos A$$

$$\cos \left(\frac{\pi}{2} \pm A \right) = \mp \sin A$$

$$\sin (-A) = -\sin A$$

$$\cos (-A) = \cos A$$

$$\tan (-A) = -\tan A$$

$$\sin (\pi \pm A) = \mp \sin A$$

$$\cos (\pi \pm A) = -\cos A$$

$$\sin (A \pm B) = \sin A \cos B \pm \cos A \sin B$$

$$\cos (A + B) = \cos A \cos B \mp \sin A \sin B$$

$$\tan (A \pm B) = \frac{\tan A \pm \tan B}{1 \mp \tan A \tan B}$$

$$\sin 2A = 2 \sin A \cos A$$
$$\cos 2A = \cos^2 A - \sin^2 A = 2 \cos^2 A - 1 = 1 - 2 \sin^2 A$$

Law of Sines

$$\frac{\sin A}{a} = \frac{\sin B}{b} = \frac{\sin C}{c}$$

Law of Cosines

$$a^2 = b^2 + c^2 - 2bc \cos A$$

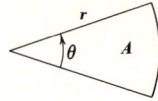

Fig. G-2

Area of Sector

$$A = \tfrac{1}{2} r^2 \theta$$

CALCULUS

Derivatives

If y is a function of x, $y = f(x)$, then the derivative of y with respect to x is defined as

$$\frac{dy}{dx} = \lim_{\Delta x \to 0} \frac{f(x + \Delta x) - f(x)}{\Delta x} = \lim_{\Delta x \to 0} \frac{\Delta y}{\Delta x}$$

Alternative notations:

$$\frac{dy}{dx} = f'(x) = D_x y = y'$$

Derivatives of elementary functions:

$$\frac{d}{dx} x^n = n x^{n-1} \qquad \frac{d}{dx} e^{ax} = a e^{ax}$$

$$\frac{d}{dx} (\sin ax) = a \cos ax \qquad \frac{d}{dx} (\ln ax) = \frac{1}{x}$$

$$\frac{d}{dx} (\cos ax) = -a \sin ax$$

Chain rule: If y is a function of u and u is a function of x, then

$$\frac{dy}{dx} = \frac{dy}{du} \frac{du}{dx}$$

633 MATHEMATICAL FORMULAS

Second derivative:

$$\frac{d^2y}{dx^2} = \frac{d}{dx}\left(\frac{dy}{dx}\right) = \frac{d}{dx}f'(x) = f''(x) = D_x^2 y$$

Integrals

If the interval from a to b is divided into N subintervals, of which a typical one is Δx_i, then the integral of a function $f(x)$ from a to b is defined as

$$\int_a^b f(x)\,dx = \lim_{\substack{\Delta x_i \to 0 \\ N \to \infty}} \sum_{i=1}^N f(x_i)\,\Delta x_i$$

If $F(x)$ is any function whose derivative is $f(x)$, $f(x) = dF(x)/dx$, $F(x)$ is called an *antiderivative* of $f(x)$, or an *indefinite integral* of $f(x)$.

$$\int_a^b f(x)\,dx = F(b) - F(a) = \Big[F(x)\Big]_a^b$$

If $f(x) = dF(x)/dx$, then $F(x) = \int f(x)\,dx$ [indefinite integral of $f(x)$].

Miscellaneous indefinite integrals:

$$\int \frac{dx}{\sqrt{a^2 - x^2}} = \sin^{-1}\frac{x}{a}$$

$$\int x \exp ax^2\,dx = \frac{1}{2a} \exp ax^2$$

$$\int x^3 \exp ax^2\,dx = \left(\frac{x^2}{2a} - \frac{1}{2a^2}\right) \exp ax^2$$

Series Expansions
Binomial theorem:

$$(a + b)^n = a^n + na^{n-1}b + \frac{n(n-1)}{2!}a^{n-2}b^2$$
$$+ \frac{n(n-1)(n-2)}{3!}a^{n-3}b^3 + \cdots$$

$$\frac{1}{1+x} = 1 - x + x^2 - x^3 + x^4 \cdots \qquad -1 < x < 1$$

$$\sin x = x - \frac{x^3}{3!} + \frac{x^5}{5!} \cdots$$

$$\cos x = 1 - \frac{x^2}{2!} + \frac{x^4}{4!} \cdots$$

$$e^x = 1 + x + \frac{x^2}{2!} + \frac{x^3}{3!} + \cdots$$

Vector Algebra

If $\mathbf{A} = A_x\mathbf{i} + A_y\mathbf{j} + A_z\mathbf{k}$ and $\mathbf{B} = B_x\mathbf{i} + B_y\mathbf{j} + B_z\mathbf{k}$, then

$$\mathbf{A} + \mathbf{B} = (A_x + B_x)\mathbf{i} + (A_y + B_y)\mathbf{j} + (A_z + B_z)\mathbf{k}$$

$$\mathbf{A} \cdot \mathbf{B} = A_xB_x + A_yB_y + A_zB_z$$

$$\mathbf{A} \times \mathbf{B} = (A_yB_z - A_zB_y)\mathbf{i} + (A_zB_x - A_xB_z)\mathbf{j} + (A_xB_y - A_yB_x)\mathbf{k}$$

Mathematical Constants

$$\pi = 3.142$$
$$e = 2.718$$
$$\sqrt{2} = 1.414$$
$$\sqrt{3} = 1.732$$
$$1 \text{ rad} = 57.3°$$

Answers to Odd-Numbered Problems

Chapter 1

1-1 $\dfrac{\text{length}}{\text{time}^2}$

$\dfrac{1}{\text{time}}$

$\dfrac{\text{mass}}{\text{length}^3}$

1-3 Inconsistent

1-5 a: $\dfrac{\text{length}}{\text{time}^2}$

b: length
c: time

1-7 (a) No; T should increase with l.

(b) $T = 2\pi \left(\dfrac{l}{g}\right)^{1/2}$

1-9 $T_F = \tfrac{9}{5} T_C + 32$

1-11 1.50×10^{-10} kg-m²/s²

1-13 10; yes; no

1-15 78

1-17 No

Chapter 2

2-1 (a) 5 cm

(b) (5 cm, 53°8′)

2-3 (a) $(-5\sqrt{2}\text{ m},\ -5\sqrt{2}\text{ m})$

(b) (5 m, 0)

(c) $(-3\text{ ft},\ 3\sqrt{3}\text{ ft})$

2-5 (b) (4 in., 0)

$(-2\text{ in.},\ 2\sqrt{3}\text{ in.})$

$(2\text{ in.},\ 2\sqrt{3}\text{ in.})$

(4 in., 60°)

2-7 $(-2, -7, 5);\ 78^{1/2}$

2-9 (a) $(-6\text{ in.})\mathbf{j}$

(b) $(3\text{ in.})\,(\mathbf{i} - \mathbf{j})$

(c) 0

2-11 $a(\pm \mathbf{i} \pm \mathbf{j} \pm \mathbf{k})$

three; $\cos^{-1} \tfrac{1}{3},\ \cos^{-1} \tfrac{1}{3},\ 180°$

2-13 $\pm \dfrac{(\mathbf{i} - \mathbf{j} + \mathbf{k})}{\sqrt{3}}$

2-15 $\mathbf{i} \times \mathbf{i} = \mathbf{j} \times \mathbf{j} = \mathbf{k} \times \mathbf{k} = 0$

$\mathbf{i} \times \mathbf{j} = -\mathbf{j} \times \mathbf{i} = \mathbf{k}$

$\mathbf{j} \times \mathbf{k} = -\mathbf{k} \times \mathbf{j} = \mathbf{i}$

$\mathbf{k} \times \mathbf{i} = -\mathbf{i} \times \mathbf{k} = \mathbf{j}$

2-17 1

2-21 No; triple vector product is not associative.

Chapter 3

3-1 (a) -5 cm
  (b) -15 cm
  (c) 5 cm
  (d) -5 cm

3-3 (a) 14 cm/s
  (b) 6 cm/s, 24 cm/s
  (c) 18 cm/s^2
  (d) 12 cm/s^2, 24 cm/s^2

3-5 No; it becomes negligible as $\Delta t \to 0$

3-7 $x = 2t^3 - 3t^4 + 3$

3-11 11 ft/s^2; 352 ft; 15.5 s; no

3-13 48 ft/s

3-15 (a) 20 s
  (b) 1760 ft
  (c) 120 mi/h

3-17 (a) $\mathbf{r} = (3t + 5)\mathbf{i} + (\frac{1}{2}t^2 + 3t - 4)\mathbf{j}$
  (c) $v_x = 3$, $v_y = t + 3$, at $t = 4$ s, $v = 7.6$ at $66.8°$
  (d) $a_x = 0$, $a_y = 3$ (constant)

3-19 (a) Counterclockwise
  (b) $v_x = -\omega R \sin \omega t$, $v_y = \omega R \cos \omega t$
  (d) $a_x = -\omega^2 R \cos \omega t$, $a_y = -\omega^2 R \sin \omega t$
  (e) $\mathbf{r} = R(\mathbf{i} \cos \omega t + \mathbf{j} \sin \omega t)$, $\mathbf{a} = -\omega^2 R(\mathbf{i} \cos \omega t + \mathbf{j} \sin \omega t)$

3-21 At ends, $\mathbf{v} = \mathbf{0}$, $\mathbf{a}$ tangent to arc
  At center, $\mathbf{v}$ horizontal, $\mathbf{a}$ vertically upward

3-23 (a) 120 ft; 20 s
  (b) $36.9°$ upstream; 25 s

3-25 (a) Speed relative to the air
  (b) N5.75°E
  (c) 497 mi/h; 36 s late

Chapter 4

4-1 1,210 lb

4-3 1,125 lb; friction of road against rear tires

4-5 (a) 3 m/s^2, $+x$ direction
  (b) 4 m/s^2, $+y$ direction
  (c) 5 m/s^2, $53.1°$ counterclockwise from $+x$ direction
  (d) (3 m/s^2, 4 m/s^2)

4-7 (a) $3ma$ between 1 and 2
    $2ma$ between 2 and 3
    ma between 3 and 4
  (b) $4ma$ forward on wheels; $3ma$ backward at coupler

4-9 1.39×10^3 m/s^2

4-11 (a) $-m\omega^2 A \sin \omega t$

4-13 (a) 4.47 s
  (b) 10 m

4-15 No; $\Sigma \mathbf{F} = m(\mathbf{a} + \mathbf{A})$.

4-17 (a) $\mathbf{v} = 2t^2 \mathbf{i} + 2t^3 \mathbf{j}$
    $\mathbf{r} = \frac{2}{3}t^3 \mathbf{i} + \frac{1}{2}t^4 \mathbf{j}$
  (c) $\mathbf{r} = (\frac{2}{3}t^3 + 2)\mathbf{i} + (\frac{1}{2}t^4 + 3)\mathbf{j}$
  (d) $\pm \frac{16}{3}$ m

4-19 (a) Yes; accelerations are directed toward each other.
  (b) Distance from initial position inversely proportional to mass

4-21 1.5 m/s^2

Chapter 5

5-1 Yes; no

5-3 For any θ, $a = g \tan \theta$; no, because g changes; no, because there g is not involved.

5-5 2.45 m/s^2; 50 kg

5-7 40 cm/s

5-13 74.5 ft/s; 2.32 s

5-15 1070 ft

5-17 (a) 47.7°
  (b) 13.8 cm

5-23 Monkey and bananas accelerate upward at same rate.

5-25 $a_1 = \dfrac{2m_2 - m_1}{m_1 + 4m_2} g$

$a_2 = \dfrac{2m_1 - 4m_2}{m_1 + 4m_2} g$

$T = \dfrac{3m_1 m_2}{m_1 + 4m_2} g$

Same

5-27 (a) $\mu = \tan \theta$
(b) Kinetic

5-29 (a) $\dfrac{mg}{\cos \theta - \mu \sin \theta}$
(b) $\theta = \cot^{-1} \mu$

5-31 (a) $m_2 > \mu m_1$
(b) $a = \dfrac{m_2 - \mu m_1}{m_1 + m_2} g$,

$T = \dfrac{(\mu + 1) m_1 m_2 g}{m_1 + m_2}$

(c) $t = (2h/a)^{1/2}$

5-33 (a) 2.78 m/s^2
(b) 8.38 m/s^2

5-35 (a) 785 N
(b) 37.8 m

Chapter 6

6-1 62.8 ft/s; 188 s^{-1}
6-3 No; it changes in direction
6-5 $v = (lg \sin \theta \tan \theta)^{1/2}$
6-7 (a) $7.27 \times 10^{-5} \text{ s}^{-1}$
(b) 465 m/s
(c) 3.38×10^{-2} m/s
6-9 5.07×10^3 s
6-11 3.12 m/s; 3.12 s^{-1}
6-13 $\theta = \tan^{-1} \dfrac{v^2}{gR}$
6-15 $T_1 = \tfrac{1}{2} m \omega^2 a + mg$
$T_2 = \tfrac{1}{2} m \omega^2 a - mg$
6-17 (b) 14.0 m/s
6-19 Circular path of radius mv/B, constant speed

6-23 0.129
6-25 (a) $a_r = -2.5 \text{ m/s}^2$ (max),
$a_\theta = 0.25 \text{ m/s}^2$
(b) 2.51 m/s, $5.72°$ from radial direction
(c) 0.256

Chapter 7

7-1 4.25×10^4 kg-m/s
7-3 11.4 m/s
7-5 20 m/s
7-7 0, **v**
7-9 Equal speeds $v/\sqrt{2}$, at angles $45°$ to original direction
7-11 $R/\sqrt{2}$ away from center; at $45°$ to original direction
7-13 Initial velocity horizontal, 424 m/s
7-15 (a) $\dfrac{Nmv_0}{M + Nm}$

(b) $mv_0 \left(\dfrac{1}{M + m} + \dfrac{1}{M + 2m} \right.$
$\left. + \cdots + \dfrac{1}{M + Nm} \right)$

7-19 10.8×10^3 N
7-21 $\dfrac{Mg}{u}$
7-23 (a) $m = M + M_f(1 - t/T)$
(b) $v = u \ln(1 - \gamma t/T)$
$x = -u(t - T/\gamma) \ln(1 - \gamma t/T) + \text{constant, where}$
$\gamma = M_f/(M + M_f)$
7-25 (a) 0.5×10^{-4}
(b) 15%
(c) Factor of 2.30
7-27 1.08×10^{-18} m/s
7-29 About 0.02 m/s
7-31 (a) 0.566 kg-m/s, $\perp$ to cushion
(b) 0.566 kg-m/s, $\perp$ to cushion
(c) Opposite to (b)

Chapter 8
8-1 (a) 5.19×10^{-11} N
(b) 5.19×10^{-11} N
8-3 (a) 2.67×10^{-8} N
(b) 1.33×10^{-11} m/s^2
8-5 4.70, 4.55, 5.50, 3.75 g/cm^3
8-7 $g = \frac{4}{3}\pi G\rho R$
8-9 3.42×10^8 m
8-11 -0.31×10^{-6}, or $-0.31 \times 10^{-4}\%$
8-13 $\dfrac{GM}{y\sqrt{y^2 + a^2}}$
8-15 7.90×10^3 m/s; no
8-17 142×10^6 mi
8-19 9.35×10^3 s
8-21 $\sqrt{\dfrac{3\pi}{\rho G}}$
8-23 $F = \dfrac{Gm_1 m_2}{R^3}$
$K = \dfrac{2\pi}{(GM)^{1/2}}$
8-25 $T^2 = \dfrac{4\pi^2}{GM} R^{N+1}$
8-27 $mv^2 = kx(x + l)$
8-29 $\left(\dfrac{2B}{A}\right)^{1/6}$, $\dfrac{-A^2}{4B}\left(\dfrac{B}{A}\right)^{1/6}$
8-31 5.93×10^6 m/s
8-35 (a) No motion
(b) Linear motion, constant velocity
(c) Circular motion, constant speed
(d) Helical motion, constant speed
8-37 9.58×10^7 m/s
8-39 $\mathbf{E} + \mathbf{v} \times \mathbf{B} = 0$
$v = \dfrac{E}{B}$

Chapter 9
9-1 86.6 J
9-3 Equal
9-5 No; force always perpendicular to velocity
9-9 937 lb
9-11 800 ft-lb/s = 1.45 hp
9-13 107 ft/s
9-17 120 ft
9-19 $h = \dfrac{5R}{2}$
9-21 $2mg\dfrac{\sin\theta}{k} + 2d$
9-23 $E = \dfrac{(qBr)^2}{2m}$
$E_{\max} = \dfrac{(qBr)^2}{2m}$
$\omega = \dfrac{qB}{m}$
9-25 $h/2$
9-27 4.00×10^5 J; 2.28×10^5 J
9-29 (a) $e^2 h_0$
(b) $e^{2n} h_0$
(c) $\sqrt{2h_0/g}$; $2e\sqrt{2h_0/g}$
(d) $2e^n\sqrt{2h_0/g}$;
$\left(\dfrac{2}{1-e} - 1\right)\sqrt{\dfrac{2h_0}{g}}$
because $1 + e + e^2 + \cdots = \dfrac{1}{1-e}$
9-31 $-(7A/60)(A/B)^{5/6}$
9-33 (a) $-2x^2 + x^3 - 2x^4$
(b) 448
(c) 448
9-35 (a) $F_x = -4kx(x^2 + y^2)$; $F_y = -4ky(x^2 + y^2)$
(b) $4k(x^2 + y^2)^{3/2}$, $\tan^{-1}\dfrac{y}{x}$
(c) $F_r = -4kr^3$; $F_\theta = 0$
9-39 40 kg
9-41 2000 m

Chapter 10
10-1   4 s; $\frac{1}{4}$ s^{-1}; 1.57 s^{-1}
10-3   (a) 0.1 m; 1.59 s^{-1}; 0.628 s
      (b) $x = (0.1 \text{ m}) \sin 10 \text{ s}^{-1} t$
10-7   0.02 m
10-11 $T = \pi \left(\dfrac{m}{k_1}\right)^{1/2} + \pi \left(\dfrac{m}{k_2}\right)^{1/2}$
10-13   Period decreases with increasing amplitude.
10-15   0.419 s; 0.033 m; 2.39 Hz
10-17   Increases
10-19 $\omega = \sqrt{2ag}$; no
10-23   (b) $\dfrac{b}{a}$
       (c) $\dfrac{a^4}{b^3}$
       (d) $2\pi \left(\dfrac{b^3 \mu}{a^4}\right)^{1/2}$
10-25   (a) 20
       (b) No; 0.002 m from "equilibrium" position

Chapter 11
11-1 2.3×10^6 m from center; 4.1×10^6 m below surface
11-3   1/1,838 the distance from proton to electron
11-5 0.484×10^{-10} m from O atom
11-7   ($\frac{1}{2}$ cm, $\frac{1}{2}$ cm, $\frac{1}{2}$ cm)
11-9   (a) 20 m in front of rear car
       (b) 50,000 kg·m/s
       (c) $16\frac{2}{3}$ m/s
       (d) 50,000 kg·m/s
11-11   (a) $d/4$ from ball of mass 3m; $\frac{3}{4}\mathbf{v}$
        (b) $3m\mathbf{v}$
11-13   (a) When ball is free
        (b) Same in both cases
11-17 $\dfrac{4m_1 m_2 E}{(m_1 + m_2)^2}$
11-19   (a) $\mathbf{r}_1 = \mathbf{R} - \dfrac{m_1 \mathbf{r}}{m_1 + m_2}$
        $\mathbf{r}_2 = \mathbf{R} + \dfrac{m_2 \mathbf{r}}{m_1 + m_2}$
        (b) $(m_1 + m_2)\dfrac{d\mathbf{R}}{dt}$
11-21 $(1 + \dfrac{m}{M})E$
11-23   (a) 1.0 m/s; 0.1 J
        (b) 2.0×10^{-3} kg·m^2; 0.1 J
11-25 $\frac{1}{2}M(a^2 + b^2)$
11-29 $\frac{1}{3}M(a^2 + b^2)$
11-33   (a) $\sqrt{\dfrac{4m_1 gh}{2m_1 + m_2}}$; $\dfrac{1}{R}\sqrt{\dfrac{4m_1 gh}{2m_1 + m_2}}$
        (b) 1.67 m/s; 16.7 s^{-1}
11-35 $\sqrt{\dfrac{2(m_1 - m_2)gh}{m_1 + m_2 + \frac{1}{2}M}}$

Chapter 12
12-1   Six
12-3   (a) 3.96×10^7 mi^2/h
       (b) 14,100 mi/h
12-5 $\dfrac{2\pi ab}{(a - \sqrt{a^2 - b^2})v_0}$
12-7 $\dfrac{k}{mv_0^2} + \sqrt{\left(\dfrac{k}{mv_0^2}\right)^2 + b^2}$
12-9   (a) No; the force is not central.
       (b) Yes; the string does no work.
12-11   (a) $\dfrac{I_1 \omega_0}{I_1 + I_2}$
        (b) $E_f = \dfrac{E_0 I_1}{I_1 + I_2}$
        Mechanical energy is lost in friction.
12-13   Same; $v_\theta \to 0$ as $r \to \infty$
12-15   (a) $v_0/2$
        (b) $v_0/6R$
        (c) $7mv_0^2/24$; no; inelastic collision

12-17 (a) $\dfrac{6mv_0}{(4M+3m)d}$

(b) $\dfrac{6mv_0}{\sqrt{2}(4M+3m)d}$

12-19 $\dfrac{m\pi}{m+\frac{1}{2}M}$; 3.5°; no

12-21 Yes

12-25 (a) 0.001 m/s^2; 0.05/s^2

(b) 100 s

(c) 5.0/s

12-27 $\dfrac{mg}{m+\mu}$ $\dfrac{m\mu g}{m+\mu}$

where $\mu = \dfrac{M}{2}\dfrac{R_2{}^2}{R_1{}^2}$

12-29 (a) No; otherwise there would be zero torque on pulley.

(b) $\dfrac{m_1}{m_1+m_2+I/R^2}g$

$\dfrac{m_1 m_2 + m_1 I/R^2}{m_1+m_2+I/R^2}g$

$\dfrac{m_1 m_2}{m_1+m_2+I/R^2}g$

(c) No

(d) $\dfrac{m_1-\mu m_2}{m_1+m_2+I/R^2}g$

$\dfrac{m_1 m_2(1+\mu) + m_1 I/R^2}{m_1+m_2+I/R^2}g$

$\dfrac{m_1 m_2(1+\mu) + \mu m_2 I/R^2}{m_1+m_2+I/R^2}g$

12-31 $\dfrac{2Fdn_1}{n_2 I}$; no; $\dfrac{2Fd}{I}$

Chapter 13

13-5 $(3gl)^{1/2}$; $(2gl)^{1/2}$

13-7 $8g/5$; drops away from bird

13-9 (b) $\theta = \cos^{-1}\left(\dfrac{a}{b}\right)$; no

(c) Yo-yo slides without rolling.

13-11 (a) $\dfrac{\frac{2}{3}F}{m}$

(b) $\dfrac{\frac{2}{3}F}{mg-F}$

(c) Yes; $F > mg$

13-13 Sphere first; cylindrical shell last

13-15 $\dfrac{g\sin\theta}{1+a^2/2b^2}$

13-19 $mg + \dfrac{mv^2 R}{2d^2}$

$\dfrac{mv^2 R}{2d^2}$ down $\dfrac{mv^2}{d}$ inward

13-21 Assembly rotates about axis at end of fork, with angular velocity $2\omega R^2/d^2$.

13-23 $L' = 8.4\times 10^{-6}|\mathbf{R}\times\mathbf{P}|$

13-25 (a) $\dfrac{(3g/2l)^{1/2}}{2\pi}$

(b) $(3/2)^{1/2}$ faster

(c) 1.49 m

13-29 $mg \pm ml^2 \sin\theta \cos\theta \dfrac{\omega^2}{d}$

13-31 (a) $MR^2\omega^2 \sin\theta \cos\theta/8d$

(b) 200 N if $d=0.1$ m

13-33 5.78 tons; (5.78 tons, 5 tons)

13-35 57.7 lb; 200 lb up at base, 57.7 lb horizontal at top

13-37 353 lb; 250 lb up, 250 lb outward

Chapter 14

14-1 (a) 4.3×10^{-8} s

(b) 10.4 m

14-5 0.942×10^{-7} s

14-7 86.6 m

14-11 $\sqrt{(\Delta x)^2 - c^2(\Delta t)^2}$

Chapter 15
15-1 No
15-3 2.60×10^8 m/s; 63% too large
15-9 (a) 5.72×10^{11} J
(b) 6.30×10^3 kg
15-11 (a) $\omega = qB/m$
(b) $\omega = qBc^2/E = (qB/m)(1 - v^2/c^2)^{1/2}$
(c) 20×10^6 r/s

Chapter 16
16-1 About 4×10^4 N
16-3 0.98×10^5 N/m^2; 0.97 atm
16-5 4.9×10^4 N
16-7 180 m^3; 3.50 m; 3.41 m
16-9 8/9; yes; icebergs are fresh water.
16-11 0.309 m/s; 2.42×10^{-3} m^3/s
16-13 $x = 2\sqrt{h(h_{max} - h)}$
16-15 5.15 m/s; 0.162 m^3/s
16-17 6.75 atm; 4.0 m/s; 0.0314 m^3/s
16-21 2 N/m^3
16-23 0.1 m/s
16-25 1500; laminar

Chapter 17
17-1 2.68×10^{-26} kg
17-3 1.2×10^{-13} cm
17-5 Nearly independent of mass
17-7 0.0012; 0.9988
17-9 2.31×10^{-10} m
17-11 $-459.8°$F
17-13 (a) 1.94×10^{-3}/C°; 100 Ω
(b) 119.4 Ω
17-15 9480 cal
17-17 0.4 kg
17-19 0.107 cal/g C°
17-21 (a) cal/kg (C°)^4
(b) $\frac{1}{4}MA(T_2^4 - T_1^4)$
17-23 0.058 g/s
17-25 102.2°C
17-27 200 cal/s
17-29 84 cal/s
17-31 $q = 4\pi k \dfrac{T_2 - T_1}{1/R_2 - 1/R_1}$
$T(r) = T_1 + (T_2 - T_1)\dfrac{1/r - 1/R_1}{1/R_2 - 1/R_1}$
17-33 $T(x) = T_1 + (T_2 - T_1)\dfrac{\ln(1 + 3x/L)}{\ln 4}$
17-35 6.08×10^5 cal/s
17-37 1.45×10^{-7} m/s
17-39 No; rate of heat flow varies from point to point. Increasing; decreasing.

Chapter 18
18-1 1.25 kg/m^3
18-5 8.10×10^8 N/m^2; no
18-7 279 lb/in.^{2}
18-9 1 l-atm = 101.3 J
18-11 Potential energy due to gravity is very much smaller than kinetic energy.
18-13 3.40 m/s; 4.29 m/s
18-15 3.55×10^{23}/s
18-17 (a) $\dfrac{Nmv^2}{d}$ $\dfrac{Nmv^2}{Ad}$
(b) $\dfrac{k}{2m}$
(c) $P = \dfrac{NkT}{Ad}$
18-19 (a) 477 m/s
(b) 1,350 m/s
(c) 79.0 m/s
18-21 10^4 K, 10^5 K
18-23 2.4 cal/g C°; considerably larger
18-27 (a) 1.06×10^{-45} kg-m^2
(b) 1.98×10^{13}/s
18-29 2.4×10^6
18-33 0.083; no
18-35 0.022

18-37 10^{11} to 10^{13} K, depending on molecular mass

18-41 4×10^{-10} m

Chapter 19

19-1 Yes; for example, a system with two phases in equilibrium.

19-5 499 atm

19-7 "a" correction about twice as large (percent) as "b"

19-9 6.8×10^{-20} J; no, some used as work done against surroundings

19-11 In vacuum

19-13 1.22×10^8 N/m^2
 6.1×10^{-4}
 3.1 mm

19-15 (a) 0.163 in.
 (b) 4,800 lb; 3.92 in.

19-17 9.6×10^7; less than 10 percent of ultimate strength

19-19 $\dfrac{1}{T + bP/R}$

$\left(1 - \dfrac{nb}{V}\right)\dfrac{1}{P}$

$\dfrac{P}{1 - nb/V}$

19-21 (a) 49.5 atm
 (b) 3.5×10^{-8} m^3

19-25 51.5 N/m

19-27 Same order or magnitude

19-29 2420 K; no

19-31 Smaller, because $E_{av} > V_{av}$

19-33 4.60×10^{-8} m^3

19-35 228°C

19-37 1.7×10^{-5}/C°

19-39 1.51 cm^3

Chapter 20

20-1 0.82 C°

20-3 $\dfrac{v^2}{8c}$

20-5 Yes; no heat transfer or external work

20-7 209 J

20-9 124 cal/s

20-11 0.075; 0.925

20-13 $nRT \ln \dfrac{V_2 - nb}{V_1 - nb} + an^2 \left(\dfrac{1}{V_2} - \dfrac{1}{V_1}\right)$

20-15 No; the gas would have to contract when heated.

20-17 (b) (1) $Q = \Delta U = 62{,}400$ J
 $W = 0$
 (2) $Q = W = 34{,}600$ J
 $\Delta U = 0$
 (3) $Q = -87{,}400$ J
 $W = -25{,}000$ J
 $\Delta U = -62{,}400$ J
 (c) $Q = W = 9{,}600$ J
 $\Delta U = 0$

20-19 (a) $Q = 874$ J
 $W = 250$ J
 $\Delta U = 624$ J
 (b) $Q = \Delta U = 624$ J
 $W = 0$
 $Q = W = 346$ J
 $\Delta U = 0$
 $Q = 970$ J
 $W = 346$ J
 $\Delta U = 624$ J

20-21 $C_H = C_M + \dfrac{cH^2}{T^2}$

20-25 $M_2{}^2 - M_1{}^2 = 2m^2 c C_M \ln \dfrac{T_2}{T_1}$

20-27 Isothermal

20-29 $\dfrac{nRT_1}{\gamma - 1}\left[1 - \left(\dfrac{V_1}{V_2}\right)^{\gamma - 1}\right]$

20-33 $V_0 \dfrac{\gamma + 3}{\gamma + 1} \quad T_0 \dfrac{\gamma + 3}{3(\gamma + 1)}$
 No

Chapter 21

21-3 Liquid

21-5 (a) $a\text{-}b$: $W = Q = 11{,}500$ J
 $\Delta U = 0$

b-c: $W = -\Delta U = 8{,}314$ J
$Q = 0$
c-d: $W = Q = -8{,}640$ J
$\Delta U = 0$
d-a: $W = -\Delta U = -8{,}314$ J
$Q = 0$
(b) $W = Q = 2{,}860$ J
(c) 0.25

21-7 1/3
(b) Decrease T_2
21-9 Overall efficiency is $1 - T_2/T_1$ in both cases.

21-17 $\eta = \dfrac{1}{K+1}$

21-19 Yes; yes
21-21 29.3 cal/K; increase
21-23 -27.3 cal/K; $+2.0$ cal/K
21-25 (a) -3.0 cal/K; 4.0 cal/K; 1.0 cal/K
(b) No; its state does not change.
21-29 Yes; yes
21-33 270°; 370°
21-35 (a) 64
(b) 1, 6, 15, 20, 15, 6, 1
(c) 0.016, 0.094, 0.234, 0.313, 0.234, 0.094, 0.016; 3

Chapter 22

22-1 (a) 5.0 m/s upward for 1/50 s, then 5.0 m/s downward for 1/50 s
(b) 1.0 m/s upward for 1/10 s, then 1.0 m/s downward for 1/10 s

22-3 100 m/s
22-5 (a) $-kcA \cos k(x - ct)$
(b) $kA \cos k(x - ct)$
22-11 2.5 m
22-13 (a) 294 m
(b) 5×10^{14} Hz
22-15 $y = 0.1 \sin 2\pi(5t - \tfrac{1}{2}x)$ (x, y in meters, t in seconds) 10 m/s

22-17 $-\dfrac{\pi}{10} T \cos 10t$

(T in newtons, t in seconds)
22-19 0.628 m/s; 0.0314
22-21 10 J
22-23 250 J/s; 1/25 s; 10 J
22-25 1/5,000 s; 5,000 Hz; 2 m; 2×10^{-4} m

22-29 $\dfrac{2\pi A E_y f}{c} = 2\pi A f (PE_y)^{1/2}$

0.463×10^{-3} m
22-31 Stationary end; free end

22-33 $f = n\dfrac{c}{4L}$

$n = 1, 3, 5, \ldots$

22-35 2,650 Hz
22-37 5.37 m
22-39 Increase by factor of 4

Index

Aberration, stellar, 393
Absolute pressure, 418
Absolute temperature, 445–446, 572–573
Absolute value of a vector quantity, 34
Absolute zero, 446, 566
Acceleration, 56–59, 66–69
   angular, 133, 147, 324–328, 342–348
   average, 56, 66–69
   of center of mass, 287, 338
   centripetal, 135–139, 146
   in circular motion, 135–139
   components of, 67
   constant, 83–84
      motion with, 59–62
   coriolis, 151
   of gravity, 104–106
   instantaneous, 57, 67
   in polar coordinates, 147–151
   relative, 71–72
   in space, 67–69
Action and reaction, 94–97
Addition:
   of forces, 87, 89–94
   of vectors, 31–38
Adiabatic bulk modulus, 603–605
Adiabatic process, 539–541, 557–559
   microscopic view, 541–543
Air table, 82
Alphabet, Greek, 627
Amorphous solid, 438–439
Amplitude of harmonic motion, 246
Angular acceleration, 133–147
Angular frequency, 247

Angular momentum, 312–331
   conservation of, in planetary motion, 314
   of planet, 313–315
   with respect to center of mass, 337–342
   of rigid body, 323–331, 337–361
   vector nature, 350, 359–360
   vector nature of, 328–331
   in wave motion, 615
Angular velocity, 133, 145–147
   of precession, 351
   of rigid body, 294, 323–328, 340–361
   spin, 351
Antinode, 610
Apogee, 317, 333
Archimedes, 419
Areal velocity (*see* Sector velocity)
Aristotle, 6, 80
Atmosphere, 417
Atom, 436
   structure of, 318–323
Atomic mass, 440–442
Atomic mass unit, 441
Atomic weight, 441
Attenuation, 615–616
Atwood machine, 129, 310, 327–328
Average speed for gas molecules, 481–482
Average velocity, 53, 63
Avogadro's number, 442

Balancing, dynamic, 358
Ballistic trajectories, 110–114
   with air resistance, 114

645

646   INDEX

Barometer, 416–417
Bernoulli equation, 423–428
Binomial theorem, 633
Boltzmann's constant, 472
Boundary layer, 433
Boyle's law, 464
Brahe, Tycho, 185
British thermal unit, 450
Bulk modulus, 510
   adiabatic, 603–605
   isothermal, 603–605
Buoyancy, 419

Calculus formulas, 632–633
Caloric fluid, 448
Calorie, 448–450
Calorimetry, 448–451
Carnot cycle, 556–563
Cartesian coordinates, 25–29
Cavendish balance, 180
Celsius temperature, 443
Center of mass, 277–285
   acceleration of, 287–338
   coordinate system, 288–291, 337–342
   gravitational torque, 341–342
   motion of, 285–288
   relativistic, 400
Center-of-momentum system, 289
Centigrade temperature, 443
Central force, 189–192
Central-force field, 235
Centripetal acceleration, 135–139, 146
Charles' law, 464–465
Circular motion, 132–135
Coefficient of friction, 120–122
Coefficient of linear expansion, 516
Coefficient of restitution, 168, 228
Coefficient of viscosity, 429
Coefficient of volume expansion, 502, 517–518
Collision, relativistic, 397–401
Collisions, 157, 167–170
   in center-of-mass systems, 289–291
Commutative law:
   for vector addition, 32
   for vector product, 45
Components of a vector, 34–38
Compound, 436
Compressibility, 502, 510
Compression ratio, 547
Conduction, 452–456
Conductivity, thermal, 453
   table, 454

Conservation:
   of energy, 219–226, 235–237
   of mass, 85
      and energy, 406
   of momentum, relativistic, 397–401
Conservative force field, 228–235
Constant acceleration, 83–84
Constants:
   mathematical, 634
   numerical, 13–14
   physical, table, 623
   proportionality, 13
Contact force, 119–122
   components of, 119
Continuity equation, 422
Contraction of length, 388
Convection, 457
Conversion factors, 624–626
Coordinate axes, 27
Coordinate systems, 26
Coordinate transformation, galilean, 378–379
Coordinates:
   of center of mass, 280, 288–291
   of a point, 27–29
   polar, 28–29
   of rigid body, 293, 337–338
   thermodynamic, 497
Copernicus, 6
Coriolis acceleration, 151
Correspondence principle, 407
Critical damping, 266–267
Critical point, 505–506
Cross product, 42–48
Crystal lattice, 439
Crystalline solid, 438–439
Curvature of particle in magnetic field, 197
Cycle, thermodynamic, 556–563
Cyclotron, 196

Damped oscillations, 262–267
Deduction, 4
Definition, operational, 9–10
Degrees of freedom, 476
Del, 234
Derivative, 54, 63
   partial, 234
   of vector product, 147
Diatomic gas, specific heat of, 474–477
Diatomic molecule, 254–262
Diesel engine, 540
Difference, notation for, 20
Differential equation, 244, 251

INDEX

Differential principle, 155–156
Diffusion, 488
Dilation of time, 381–386
Dimensional consistency, 11–12
Directionality, 534, 554–556, 566–577
Disorder, 555, 566–577
Dispersion, 615–616
Displacement, 29–31
   addition of, 31–34
Dissipative force, 527
Dissociation energy, 260
Distribution of molecular speeds, 477–485
Distribution function for molecular speeds, 480–485
Distributive law:
   for scalar product, 40
   for vector product, 45
Dot product, 39–42
Dulong and Petit, rule of, 515
Dynamic balancing, 358
Dynamics, 79
   relativistic, 397–410
   of wave motion, 591–605

Earth:
   mass of, 183–627
   radius of, 627
   rotation of, 99, 152
   shape of, 188
Efficiency, 546–549, 556–561, 563–566
Elastic collision, 168, 227, 292
Elastic deformation, 507
Elastic limit, 507
Elastic modulus, 508–512
   table, 509
Elasticity, 506–512
   microscopic model of, 511–512
Electric charge, 192–197
Electric field, 192–194
Electric force, 193
Electron, 437
Element, 436
Endothermic reaction, 260
Energy:
   conservation of, 219–226, 235–237, 530
   equipartition principle, 476, 489–490
   in harmonic motion, 250–252
   and heat, 448, 458, 474, 527–541
   internal, 529–543
   kinetic, 209–213
      of rigid body, 294–295
      of system of particles, 291–293

Energy:
   and mass, 236
   nonmechanical, 235–237
   potential, 215–226
   relativistic, 404–407
   rest, 405–406
   in wave motion, 583, 596–600
Energy density in wave motion, 597
Engine, heat, 556–566
Engineering, nature of, 8–9
Entropy, 566–577
   and probability, 573–577
   and second law, 571–572
Epicycloid, 6
Equation of motion, relativistic, 402
Equation of state:
   of ideal gas, 464–467
   of real gas, 498–500
   of solid, 501–502
   van der Waals', 500–501
Equations in physics, 13–14
Equilibrium, 122–124
   phase, 504
   of rigid body, 361–364
   stable, 228
   thermal, 447, 485, 497
   thermodynamic, 498, 543–546
   unstable, 229
Equipartition of energy, 476, 489–490
Equivalence, principle of, 410
Escape velocity, 230–231
Ether, the, 377
Event, description of, 381
Exothermic reaction, 261
Expansion, thermal, 515–520
Extensive quantity, 498
External force, 285–286, 324, 338, 340

Fahrenheit temperature, 444
Fields:
   force, 183–185, 192–197
   scalar, 185
   vector, 185
First law of thermodynamics, 532
Fission, nuclear, 261
Flow line, 420
Flow tube, 421–428
Fluid flow, 420–423
Foot-pound, 204
Force:
   central, 189–192
   centrifugal, 141

Force:
    centripetal, 139–145
    conservative, 228–235, 527
    definition of, 86–87
    electromagnetic, 124–125
    external, 285–286, 324, 340
    gravitational, 124–125
    impulsive, 172
    interatomic, 231, 257–262, 518–520
    internal, 285–286, 324–340
    nature of, 124–126
    nuclear, 125
    superposition of, 87, 89–94
    workless, 224
Force diagram, 96, 114–117
Force field, conservative, 220–235
Force fields, 183–185, 192–197
Forced oscillations, 267–270
Formulas, mathematical, 631
Frame of reference, 26
Frames of reference, 98–100, 161
Free-body diagram, 96, 114–117
Free expansion, 535–536
Free fall, 106–108
    with air resistance, 108–110
Frequency:
    angular, 247
    of harmonic motion, 247
    of molecular vibration, 256
    of pendulum, 254
    of wave motion, 593
Friction, 119–122
    coefficient of, 120
    kinetic, 122
    static, 122
Friction force, 344, 345
Function of state, 570
Fundamental frequency, 612
Fusion:
    heat of, 450, 503
    nuclear, 261
Fusion curve, 505

Galilean transformation, 378–379
Galilei, Galileo, 14–16, 80–81
Gas:
    ideal (*see* Ideal gas)
    kinetic theory of, 464–490
Gas constant, universal, 465–467
Gasoline engine, 546–549
Gauge pressure, 418

Gay-Lussac's law, 464–465
Gedankenexperimente (thought experiments), 381
Geometry, non-euclidean, 8
Gimbal, 349
Gradient, 234
    temperature, 454–456
Gravitation, 179–183
Gravitational constant, 181
Gravitational field, 184
Gravitational torque, 341–342
Gravity, acceleration of, 104–106, 182–183
Greek alphabet, 627
Group velocity, 599–600
Gyroscope, 348–353

Harmonic, 612
Harmonic oscillator, 242–252
Heat:
    and energy, 458
    of fusion, 450, 503
    nature of, 448
    of vaporization, 450, 503
Heat engine, 556–566
Heat transfer, 451–457
Hertz, 247
Hooke's law, 508
Horsepower, 214
Hydrostatics, 414–420

Ideal gas:
    equation of state, 464–467
    kinetic theory of, 464–477
    microscopic model of, 467–469
    specific heat, 474–477
Ideal-gas constant, 465–467
Idealized models, 14–17
Identities, trigonometric, 631–632
Impact parameter, 320
Impulse, 170–173
Impulsive force, 172
Incompressible flow, 420
Induction, 4
Inelastic collision, 168, 228, 292
Inelastic deformation, 507
Inertia, 79–80
    moment of, 295–302
Inertial frame of reference, 98–100, 161, 375–378
Initial conditions, 248
Integral principle, 155

Integrals, 633
Intensive quantity, 498
Internal-combustion engine, 546–549
Internal energy, 529–543
   of ideal gas, 535–536
Internal force, 285–286, 324, 340
Invariance of physical laws, 99–100, 375–378, 397, 403
Ion, 440
Irreversible process, 555–556, 568–572
Isobaric process, 534
Isolated system, 161, 285–286
Isotherm, 499
Isothermal bulk modulus, 511, 603–605
Isothermal process, 534

Joule, 204, 210
Joule-Thomson effect, 535–536

Kelvin temperature, 446, 572–573
Kepler, J., 6
Kepler's laws, 185–189, 312–315
Kilocalorie, 450
Kilomole, 466–467
Kinematics, 31, 77
Kinetic energy, 210
   relative to center of mass, 292
   relativistic, 379–380, 404–407
   of rigid body, 294–295
   rotational, of gas molecules, 457–476
   of system of particles, 291–293
   vibrational, of gas molecules, 476
   and work, 210–213
Kinetic-molecular model of gas, 467–469
Kinetic theory of gases, limitations of, 489–490

Laminar flow, 421
Latent heat, 450, 503
   of fusion, 450, 503
   of vaporization, 450, 503
Laws of motion, Newton's, 78–97
Left-handed coordinate system, 48
Length:
   contraction of, 386–388
   relativity of, 386–388
Lever arm, 315, 316, 324, 329
Light, speed of, 377–378, 380
Line of flow, 420
Line integral, 207–209

Linear expansion, 515–516
Liquids, structure of, 439–440
Logarithms, table, 628–629
Longitudinal wave, 590, 600–605
Lorentz force law, 197
Lorentz transformation, 381, 389–394

Macroscopic view, 17–18
Magnetic field, 194–197
Magnetic force, 194–197
Magnetic material, 521–552, 578
Magnetic resonance, 367–368
Magnitude of a vector quantity, 34
Manometer, 418, 419, 426
Mass:
   center of, 277–285
   conservation of, 85–86
   definition of, 84–85
   of earth, 183, 627
   and energy, 86, 236
   reduced, 256, 308
   relativistic, 402
   total, of system, 85, 278
   variable, 162–165
   and weight, 104–106
Massless particles, 166, 407
Mathematical description of waves, 585, 591
Mathematical formulas, 631
Mathematics, nature of, 8
Maxwell-Boltzmann distribution, 482–485
Mean free path, 486–498
Mean free time, 487
Mean-square speed, 471–474
Measurement, 9–13
Mechanical equivalent of heat, 528
Mechanics of fluids, 414–433
Medium for wave propagation, 376–377, 583–584
Meter, definition of, 10
Michelson-Morley experiment, 377
Microscopic model of ideal gas, 467
Microscopic view, 17–18
Mode, normal, 611
Models, 14–17
Modulus:
   adiabatic, 603–605
   bulk, 509–511
   elastic, 508–512
   isothermal, 603–605
Molar volume, 500
Mole, 441

Molecular beam, 478
Molecular force, 231, 257–262
Molecular mass, 440–442
Molecular speed distribution, 477–485
Molecular vibrations, 254–262
Molecular weight, 441
Molecule, 436
Moment:
    of a force, 315, 316, 323–324, 329–330, 361–364
    due to gravity, 341–342
    of inertia, 295–302
        of various bodies, 301
    of momentum, 315
    with respect to center of mass, 339
Moment arm, 315, 316, 324, 329
Momentum:
    angular (see Angular momentum)
    conservation of, 97, 158–162
    electromagnetic, 174
    relativistic, 397–401
    of system of particles, 278–279
    total, 160–161
Momentum vector diagram, 162
Monatomic gas, specific heat of, 474–477
Most probable speed, 481
Motion along a line, 52–62
Multiplication:
    of vector by scalar, 33
    of vectors, 39–48

Natural state of motion, 81
Nature of forces, 124–126
Neutrino, 167
Neutron, 437
Newton's laws of motion, 78–97
Node, 610
Non-equilibrium process, 545–546
Nonturbulent flow, 421
Nonviscous flow, 421
Normal force, 115, 119–122
Normal mode, 611
Nuclear fission, 261
Nuclear force, 438
Nuclear fusion, 261
Nucleus, 318–323, 437
Nutation, 353

Observation, 8
Observer:
    motion of, 13
    role in relative velocity, 72

Operational definition, 9–10
Orbital angular momentum, 353–356
Orbits for inverse-square forces, 192
Order:
    long-range, 439
    short-range, 440
Origin of coordinate system, 26
Overdamping, 266–267
Overtone, 612

Parabolic trajectory, 110–114
Parallel-axis theorem, 299, 355–356
Paramagnetism, 521, 551, 552, 578
Partial derivative, 234
Partial reflection, 612–614
Particle, massless, 166–167, 407
Pendulum:
    Kater's, 369
    physical, 356–358
Performance coefficient, 563
Perigee, 317, 333
Period:
    of harmonic motion, 246
    of molecular vibration, 256
    of pendulum, 254
    of satellite, 187–189
    of wave motion, 593
Periodic motion, 242–270
Perpendicular-axis theorem, 300
Perpetual motion machine, 566
Phase:
    in harmonic motion, 248–249
    in wave motion, 595–596
Phase diagram, 504–506
Phase equilibrium, 504
Phase transition, 503
Phases of matter, 502–506
Photon, 166–167
Physical constants, table, 623
Physical quantity, 9–10
Physical theory, nature of, 6–9
Physics, objects and methods of, 3
Planets, motion of, 185–189, 312–318
Plasma, 440
Point, position of, 25–29
Poise, 429
Poiseuille's law, 432
Polar coordinates, 28, 132
    unit vectors in, 148
    velocity and acceleration in, 147–151
Polarization, 614–615
    circular, 615
    plane, 614

651 INDEX

Position vector, 62
    relative, 70
Potential energy, 215–219
    gravitational, 216, 230, 232–235
    and reversible work, 221
    of spring, 216–217
    and work, 217–218, 220–221
Potential well, 519
Power, 214
    in wave motion, 598
Power series, 634
Powers of ten, 19
Precession, 350–353
Pressure:
    absolute, 418
    of gas, microscopic view, 469–474
    gauge, 418
    variation with height, 415–418
Principal axes of inertia, 360
Principal moments of inertia, 360
Principle of equivalence, 410
Principle of superposition, 606
Probability:
    and molecular speeds, 480–485
    of thermodynamic states, 573–577
Products of vectors, 39–48
Projection of a vector, 39–40
Propagation of waves, 582–585
Proper length, 388
Proper time, 386
Proportion, 13
Proportional limit, 507
Proportionality constant, 13
Proton, 437
Ptolemy, Claudius, 6

Quantity of heat, 447–451
Quantum, 166
Quasi-equilibrium process, 543–546, 557
Quasi-static process, 543–546, 557

Radial force, 317
Radian, 133, 139
Radiation, 457
Range of ballistic missile, 112–113
Reaction, 94–97
Rectangular coordinates, 28
Reduced mass, 256, 308
Reflection, 607–612
    partial, 612–614
Reflection coefficient, 614

Refrigerator, 561–566
    workless, 564–566
Relative acceleration, 71–72
Relative motion, 69–72
Relative velocity, 69–72
    nonrelativistic, 379
    relativistic, 391–393
Relativity:
    general theory of, 409
    of length, 386–388
    and newtonian mechanics, 407–410
    principle of, 375–378
    of time, 381–386
Reservoir, heat, 556–563
Resonance, 269–270
    magnetic, 367–368
Rest energy, 405–406
Rest mass, 402
Restitution, coefficient of, 168, 228
Reversible process, 545, 554, 557
Reversible work, 221
Reynolds number, 432–433
Right-hand rule, 43, 48, 146
Right-handed coordinate system, 48
Rigid body, 293–305, 323–331, 337–364
Rocket, 163–165
Root-mean-square speed, 472–474
Rotation:
    of earth, 99, 152
    of rigid body, 293–305, 323–331, 337–362

Satellites, 187–189, 312–318
Scalar field, 185
Scalar product, 39–42
Scalar quantity, 25
Scattering, 197
    Rutherford, 318–323
Scattering angle, 320
Second, definition of, 10
Second law of thermodynamics, 549, 563–576
Sector velocity, 313–314
Semielastic collision, 168–228, 292
Significant figures, 19
Simple harmonic motion, 242–252
Simple pendulum, 252–254
Simplicity of physical theory, 7
Simultaneity, 381–383
Sinusoidal functions, 245–246
Sinusoidal waves, 593–596, 609–612
Slope in wave motion, 588
Solar system, 6–7
    data, 627
    idealized model of, 16

Solution of differential equation, 244–245, 251
Sound, speed of, 600–605
Specific heat, 448
    of ideal gas, 474–477
        constant pressure, 536–539
        constant volume, 536
    of solids, 513–515
    table of, 449, 515
    temperature dependence of, 450–451, 539
Specific heats of gases, table, 476
Speed, 64
    of light, 377–378, 380
    transverse, 588
    of wave propagation, 584, 593, 602–605
Spin angular momentum, 353–356
Spin angular velocity, 351
Stable equilibrium, 228
Standard temperature and pressure, 465
Standing wave, 609–612
State of thermodynamic systems, 496–498, 531–543
State functions, 570
States of matter, 438
Static equilibrium, 123
Statics:
    of fluids, 414–420
    of rigid body, 361–364
Steady state:
    in fluid flow, 421
    in heat flow, 454–456
Steiner's theorem (see Parallel-axis theorem)
Stellar aberration, 393
Strain, 508–512
Streamline, 421
Stress, 508–512
String tension, 117
Structure of matter, 436–440
Sublimation curve, 505
Subtraction of vectors, 33
Sum, Σ notation for, 20–22
Summation index, 21
Superelastic collision, 168, 228
Superposition of forces, 87, 89–94
Superposition of waves, 605–612
Synchronization of clocks, 385
System:
    isolated, 161, 285–286
    of particles, 277–293
    thermodynamic, 496–498, 529

Technology, nature of, 8–9
Temperature, 442–444, 446

Temperature gradient, 454–456
Temperature scale, 443–444
    absolute, 446–447, 573
    Celsius, 443
    Fahrenheit, 444
    Kelvin, 446
    thermodynamic, 572–573
Tensile strain, 508
Tensile stress, 508
Tension in string, 117
Test particle, 183
Theory, physical, 5
Thermal conductivity:
    of gases, 488
    table, 454
Thermal equilibrium, 447, 485, 497
Thermodynamic coordinates, 497
Thermodynamic efficiency, 546–549, 556–561, 563–566
Thermodynamic equilibrium, 498, 543–546
Thermodynamic properties, 543, 566–571
Thermodynamic surface, 502
Thermodynamic system, 496–498, 529
Thermodynamic temperature scale, 572–573
Thermodynamics:
    first law, 532
    second law, 549, 563–576
    zeroth law, 447
Thermometer, 443–447
    gas, 445–446
Thought experiment, 381
Thrust of a rocket, 165
Time:
    dilation of, 381–386
    relativity of, 381–386
Torque, 43, 47, 316–317, 323–324, 329–330, 361–364
    due to gravity, 341–342
    with respect to center of mass, 339
Torricelli, 416
Torricelli's theorem, 426–427
Trajectories, ballistic, 110–114
    with air resistance, 114
Transformation:
    galilean, 378–379
    Lorentz, 381, 389–394
Transmission coefficient, 614
Transport phenomena, 485–489
Transverse force in wave motion, 598
Transverse wave, 585, 591–593
Trigonometric functions, table, 630
Trigonometric identities, 631–632
Triple point, 505

Tube of flow, 421–427
Tunnelling, 261
Turbulent flow, 421, 432–433

Ultimate strength, 507
Underdamping, 266–267
Unit conversion factors, 624–626
Unit vectors, 35–37
Units, 9–12
   British system, 11
   consistency of, 11–12
   derived, 10
   mks system, 11
Universal gravitation, 179–183
Unstable equilibrium, 229

Vapor pressure, 504
Vaporization, heat of, 450, 503
Vaporization curve, 505
Variable mass, 162–165
Vector, 29–31
Vector addition, 31–38
Vector angular velocity, 145–147
Vector diagram, momentum, 162
Vector field, 185
Vector product, 42–48
   derivative of, 147
Vector quantity, 25, 48–49
Velocity, 52–56, 62–65
   angular, 133, 145–147
     (*See also* Angular velocity)
   average, 53–63
   of center of mass, 279
   components of, 64
   instantaneous, 54, 63
   in polar coordinates, 147–151
   relative, 69–72

Velocity:
   sector, 313–314
   in space, 62–66
   tangent to path, 63–64, 67
Velocity gradient, 429
Velocity transformation, relativistic, 391–393
Venturi tube, 425–426
Vibrations, molecular, 254–262
Viscosity, 428–432
   of gases, 488
Volume expansion, coefficient of, 502, 517–518

Watt, 214
Wave, longitudinal, 600–605
Wave equation, 589–590, 592, 602
Wave function, 587–588
   for reflection, 608–612
Wave pulse, 584, 605–608, 613–616
Wavelength, 593
Waves:
   mathematical description, 585–591
   nature of, 582–585
Weight, 104–106
Work, 40, 203–209
   dependence on path, 209, 232–235
   as integral, 205, 207
   and kinetic energy, 209–213
   and potential energy, 217–218
   as scalar product, 40, 206
   of thermodynamic system, 533–536
Workless constraint, 224
Workless force, 224
Workless refrigerator, 564–566

Young's modulus, 508

Zeroth law of thermodynamics, 447

FREQUENTLY USED MATHEMATICAL FORMULAS

$\sin^2 A + \cos^2 A = 1$

$\sin (90° \pm A) = \cos A$

$\cos (90° \pm A) = \mp \sin A$

$\sin (180° \pm A) = \mp \sin A$

$\cos (180° \pm A) = -\cos A$

$\sin (A \pm B) = \sin A \cos B \pm \cos A \sin B$

$\cos (A \pm B) = \cos A \cos B \mp \sin A \sin B$

$\sin 2A = 2 \sin A \cos A$

$\cos 2A = \cos^2 A - \sin^2 A = 2 \cos^2 A - 1 = 1 - 2 \sin^2 A$

$\sin (-A) = -\sin A$

$\cos (-A) = \cos A$

$\tan (-A) = -\tan A$

$\dfrac{d}{dx} x^n = n x^{n-1}$

$\dfrac{d}{dx} (\sin ax) = a \cos ax$

$\dfrac{d}{dx} (\cos ax) = -a \sin ax$

$\dfrac{d}{dx} e^{ax} = a e^{ax}$

$\dfrac{d}{dx} (\ln ax) = \dfrac{1}{x}$

$\pi = 3.142$

$e = 2.718$

$\sqrt{2} = 1.414$

$\sqrt{3} = 1.732$

$1 \text{ rad} = 57.3°$

$\sin 30° = \cos 60° = \tfrac{1}{2}$

$\cos 30° = \sin 60° = \dfrac{\sqrt{3}}{2} = 0.866$

$\sin 45° = \cos 45° = \dfrac{1}{\sqrt{2}} = 0.707$

$\tan 30° = \cot 60° = \dfrac{1}{\sqrt{3}} = 0.577$

$\tan 60° = \cot 30° = \sqrt{3} = 1.732$

$\tan 45° = \cot 45° = 1$

(*See Appendix G for a more complete list.*)